INDUSTRIAL WATER TREATMENT PROCESS TECHNOLOGY

INDUSTRIAL WATER TREATMENT PROCESS TECHNOLOGY

PARIMAL PAL
Professor and Formerly Head of the Department of Chemical Engineering, National Institute of Technology Durgapur, India

ELSEVIER
Butterworth-Heinemann
An imprint of Elsevier
elsevier.com

Butterworth-Heinemann is an imprint of Elsevier
The Boulevard, Langford Lane, Kidlington, Oxford OX5 1GB, United Kingdom
50 Hampshire Street, 5th Floor, Cambridge, MA 02139, United States

British Library Cataloguing-in-Publication Data
A catalogue record for this book is available from the British Library

Library of Congress Cataloging-in-Publication Data
A catalog record for this book is available from the Library of Congress

ISBN: 978-0-12-810391-3

For Information on all Butterworth-Heinemann publications
visit our website at https://www.elsevier.com/books-and-journals

Publisher: Joe Hayton
Acquisition Editor: Ken McCombs
Editorial Project Manager: Peter Jardim
Production Project Manager: Kiruthika Govindaraju
Designer: Greg Harris

Typeset by MPS Limited, Chennai, India

DEDICATION

Dedicated to the memory of my grandfather Rampada Pal of Barabelun, the great philanthropist and educationist.

CONTENTS

PREFACE

As in the past century, today water treatment continues to be one of the top 20 research areas in terms its importance and impact on human society. Considering the limited availability of fresh water for human consumption amidst the world's growing population and increasing levels of water pollution, maintaining a sustainable supply of fresh, safe water remains a challenge for both the scientific community and policy planners. Over the years, several books on water resources and water treatment have been published, the majority of which are from a civil engineering point of view with a focus on traditional design, construction, supply, and distribution. These texts have limited information on recent developments in water treatment, particularly those based on membrane-separation principles. Industry-specific treatment technologies have also been largely overlooked. Moreover, water-treatment issues are rarely covered comprehensively in currently available texts. This book fills those gaps and covers up-to-date developments in all major water-treatment methods, water-treatment issues such as temporal and spatial variations of availability of water, water pollution, and the latest available technologies for possible abatement. This text also addresses sustainable water-treatment technology and how to evolve a sustainable management scheme along with basic principles of separation—purification and their use in a successful treatment technology. Finally this resource guides readers on how to choose the appropriate technology for not only treatment with an end-of-discharge pipe approach but also preventive approaches including innovative schemes that comply with regulations and ethics. The aim of this text is to serve as a resource for planning, developing, commissioning, operating, and maintaining a sustainable, modern, efficient water-treatment plant.

This book also emphasizes membrane-based hybrid treatment technologies from separation principles for design, construction, economic viability, and hence sustainability in an attempt to increase confidence in nonconventional design and commissioning of next-generation water-treatment plants.

This book will help readers select the best scheme and set up and successfully operate an efficient and modern water-treatment plant based on geopolitical and socioeconomic conditions. Students at the undergraduate and research levels will find this book useful as will water engineers, membrane suppliers, government policymakers, public health engineers, and those responsible for providing safe potable water to people and protecting surface water from hazardous industrial discharge.

Parimal Pal

January 1, 2017

ACKNOWLEDGMENTS

At the outset, I thank Kenneth P. McCombs and Peter Jardim of Elsevier Science for kind support, practical suggestions, and guidance throughout the course of writing this book. They made it really easy and comfortable for me while still keeping me engaged in writing. The tireless and sincere efforts of Kiruthika Govindraju and other editorial staff members of Elsevier Science eventually culminated into the final shape of the book.

The roles my doctoral and postdoctoral scholars, especially Dr. Ramesh Kumar, Sankha Chakrabortty, and Jayato Nayak, played in water-related investigations based on my ideas in our Environment and Membrane Technology laboratory, the findings of which have been included in this book in many cases, were very much appreciated. Special thanks to Jayato for many illustrations. I am thankful to Dr. Jayabrata Pal and Dr. Piyush Roy for ideas on waterborne-health-related issues of water quality. Thanks to Prof. Mousumi Roy for her ideas on water resource-management issues. Special thanks to Madhubonti Pal for contributing the chapter on nanotechnology-based water treatment. Thanks are also due to all authors and their publishers for kind permission for use of quite a few illustrations.

I owe a lot to my dear and near ones including my daughter Shrabana for patience and support during the course of writing this book.

Parimal Pal

CHAPTER 1

Introduction

1.1 INTRODUCTION TO THE ISSUES OF ACCESS TO SAFE DRINKING WATER

Ensuring universal access to clean water is considered one of the biggest challenges of the 21st century. While the planet earth has sufficient water for the world's population, the water sources are not always available where they are needed or are unsafe. More than 40% of the world's population faces water shortages. Furthermore, 97.5% of all water on Earth is salt water, leaving only 2.5% as fresh water, and nearly 70% of that fresh water is locked up in glaciers. The remaining water is present as soil moisture, or is found in deep underground aquifers as groundwater. Thus virtually only 1% of the planet's fresh water is available for direct human consumption, and the world's ever-growing population has put unprecedented pressure on this limited supply. In fact, 663 million people today do not have access to safe drinking water and 2.4 billion people lack access to sanitation and toilet. By 2025, water withdrawal will increase by 50% in developing countries and by 18% in developed countries, suggesting that by that time, two-thirds of the world's population will be living in areas of high water stress. This withdrawal of water is predominantly from underground aquifers. Feeding the growing population will be a big challenge as one-third of the global food production today is in the areas of high water stress. Water scarcity affects all spheres of human life. One-third of the schools globally do not have access to safe drinking water and sanitation, and the same number of hospitals in developing countries is deprived of safe water for health and sanitation. This water problem is largely due to the existence of very limited freshwater resources that are distributed very unevenly. Increasing water demand by an ever-growing population and continued water pollution by human activities is only aggravating water scarcity across the globe. Furthermore, a child dies every 90 seconds due to a water-related disease [1–4].

Industrial Water Treatment Process Technology.
DOI: http://dx.doi.org/10.1016/B978-0-12-810391-3.00001-1

1.2 WORLDWIDE TEMPORAL AND SPATIAL VARIATION OF WATER RESOURCES

Worldwide variation in temporal and spatial distribution of water has aggravated the problem of water scarcity. The crisis has been further aggravated by lack of investment in water treatment, recovery, and reuse. People have mainly continued to tap existing natural water resources instead of recovering and reusing water through adoption of novel scientific and technological methods. This approach is used in the regions facing severe water scarcity such as the Middle East and the North African countries where oil refinery and petrochemical projects utilize very large amounts of water. But where an investment of just $1 provides a $4 economic return, it is economically and scientifically attractive and justified to go all out in implementing innovative water treatment projects promising full recovery and reuse of the precious resource of the earth [1]. The problems of temporal and spatial variation are likely to continue or even increase, according to the IPCC (Intergovernmental Panel of Climate Change), which predicts less precipitation in arid and semiarid areas of low latitudes and more precipitation in high latitude areas following climate change in the coming decades.

As per the estimates of the FAO (2000), the world's total water resource stands at 43,750 km^3/year with wide continental variation in spatial distribution. While America has the highest share at 45% of the world's total fresh water, Africa has only 9%. Europe and Asia have 15.5% and 28%, respectively, of the world's total fresh water. The availability of fresh water per inhabitant in America is 24,000 m^3/year, whereas it is only 3400 m^3/year in Asia. In Europe, availability per inhabitant is 9300 m^3/year and in Africa it is 5000 m^3/year. Country-wise variation is even more prominent. The availability of fresh water in Kuwait is only 10 m^3/inhabitant/year against 100,000 m^3/inhabitant/year in Canada, Iceland, Gabon, and Suriname. The availability of 1000 m^3/inhabitant/year is considered the minimum required for sustaining life and ensuring agricultural production in countries requiring irrigation for agriculture, whereas 500 m^3/inhabitant/year is considered as water scarcity. The 10 poorest countries in terms of water resources per inhabitant are Bahrain, Jordan, Kuwait, Libyan Arab Jamahirya, Maldives, Malta, Qatar, Saudi Arabia, United Arab Emirates, and Yemen. Because of their very geographical locations, nine giant countries on the other hand hold 60% of the world's total fresh water whereas some 33 countries such as

Argentina, Azerbaijan, Bahrain, Bangladesh, Benin, Bolivia, Botswana, Cambodia, Chad, Congo, Djibouti, Egypt, Eritrea, Gambia, Iraq, Israel, etc., depend on other countries for 50% of their renewable water resources.

The variation in precipitation and availability of fresh water within a continent and a country in a year is significant. For example, major rainfall in most of the states of India occurs within 2–3 months of the year, causing rivers to overflow; in the dry months, during April to May vast regions remain under drought and are forcedto transport water from one area to another area for survival.

1.3 WATER-QUALITY STANDARDS AND SOURCES AND CLASSIFICATION OF POLLUTANTS

1.3.1 Drinking Water: Standards and Guiding Principles

Since naturally occurring fresh and safe drinking water is limited there is no alternative to treatment of water and wastewater from different sources to ensure supply of adequate and safe drinking water to the world's growing population. Thus scientifically it needs to be ascertained what quality of drinking water is actually safe for direct human consumption. Guidelines on the permissible limits of possible contaminants of water have been issued by organizations such as the World Health Organization (WHO) to ensure the safety of drinking water. These recommendations are listed in Tables 1.1–1.3.

To sustain life, water is essential, and a satisfactory (adequate, safe, and accessible) supply must be available to everyone. As noted by the guidelines, safe drinkingwater is water that poses no significant risk to health over a lifetime of consumption, including different sensitivities that may occur between life stages. The greatest risk of waterborne diseases is faced by infants and young children and the sick and elderly, especially when living under unsanitary conditions. Those at risk of waterborne illness may need to take additional steps to protect themselves against exposure to waterborne pathogens, such as boiling their drinking water. Safe drinking water is required for all domestic purposes, including drinking, food preparation, and personal hygiene. In general, these guidelines are intended to support the development and implementation of risk-management strategies that will ensure the safety of drinking water supplies through the control of hazardous contaminants. These tables will immensely benefit the planners, designers, and operators of water

Table 1.1 WHO standards (1993) for inorganic contaminants of safe drinking water

Element/substance	Symbol/formula	Safe limits
Aluminum	Al	0.2 mg/L
Ammonia	NH_4	No guideline
Antimony	Sb	0.005 mg/L
Arsenic	As	0.01 mg/L
Barium	Ba	0.3 mg/L
Beryllium	Be	No guideline
Boron	B	0.3 mg/L
Cadmium	Cd	0.003 mg/L
Chloride	Cl	250 mg/L
Chromium	Cr^{+3}, Cr^{+6}	0.05 mg/L
Copper	Cu	2 mg/L
Cyanide	CN^-	0.07 mg/L
Dissolved oxygen	O_2	No guideline
Fluoride	F	1.5 mg/L
Hardness	$CaCO_3$	No guideline
Hydrogen sulfide	H_2S	No guideline
Iron	Fe	No guideline, but 1−3 mg/L may be acceptable
Lead	Pb	0.01 mg/L
Manganese	Mn	0.5 mg/L
Mercury	Hg	0.001 mg/L
Molybdenum	Mb	0.07 mg/L
Nitrate and nitrite	NO_3, NO_2	50 mg/L total nitrogen
Turbidity		Not mentioned
pH		No guideline
Selenium	Se	0.01 mg/L
Silver	Ag	No guideline
Sodium	Na	200 mg/L
Sulfate	SO_4	500 mg/L
Inorganic tin	Sn	No guideline
TDS		No guideline
Uranium	U	1.4 mg/L
Zinc	Zn	3 mg/L

treatment plants in selection of appropriate strategy and type of plant as well as their design and operation. For some elements such as iron no specific guidelines have been provided. However, being a human nutritional element 10−50 mg/L of iron may be required daily. Maximum permissible daily intake is 0.8 mg/kg body weight. If anaerobic groundwater is used as drinking water, 1−3 mg/L may be acceptable.

Table 1.2 WHO standards for permissible limits of organic compounds in safe drinking water (set up in Geneva 1993)

(a) Substance		Formula	Health based guideline by the WHO
Carbon tetrachloride		CCl_4	2 μg/L
Dichloromethane		CH_2Cl_2	20 μg/L
1,1-Dichloroethane		$C_2H_4Cl_2$	No guideline
1,2-Dichloroethane		$ClCH_2CH_2Cl$	30 μg/L
1,1,1-Trichloroethane		CH_3CCl_3	2000 μg/L
1,1-Dichloroethene		$C_2H_2Cl_2$	30 μg/L
Benzene		C_6H_6	10 μg/L
Ethylbenzene		C_8H_{10}	300 μg/L
Styrene		C_8H_8	20 μg/L
Polynucleararomatic hydrocarbons (PAHs)		$C_2H_3N_1O_5P_{13}$	0.7 μg/L
Monochlorobenzene (MCB)		C_6H_5Cl	300 μg/L
Dichlorobenzenes (DCBs)	1,2-Dichlorobenzene (1,2-DCB)	$C_6H_4Cl_2$	1000 μg/L
	1,3-Dichlorobenzene (1,3-DCB)	$C_6H_4Cl_2$	No guideline
	1,4-Dichlorobenzene (1,4-DCB)	$C_6\ H_4\ Cl_2$	300 μg/L
Trichlorobenzenes (TCBs)		$C_6H_3Cl_3$	20 μg/L
Di(2-ethylhexyl)adipate (DEHA)		$C_{22}H_{42}O_4$	80 μg/L
Di(2-ethylhexyl)phthalate (DEHP)		$C_{24}H_{38}O_4$	8 μg/L
Acrylamide		C_3H_5NO	0.5 μg/L
Epichlorohydrin (ECH)		C_3H_5ClO	0.4 μg/L
Hexachlorobutadiene (HCBD)		C_4Cl_6	0.6 μg/L
Ethylenediaminetetraacetic acid (EDTA)		$C_{10}H_{12}N_2O_8$	200 μg/L
Nitrilotriacetic acid (NTA)		$N(CH_2COOH)_3$	200 μg/L
Organotins	Dialkyltins	R_2SnX_2	No guideline
	Tributil oxide (TBTO)	$C_{24}H_{54}OSn_2$	2 μg/L

(Continued)

Table 1.2 (Continued)

(b) Substance		Formula	Health based guideline by the WHO
Carbon tetrachloride		CCl_4	2 μg/L
Dichloromethane		CH_2Cl_2	20 μg/L
1,1-Dichloroethane		$C_2H_4Cl_2$	No guideline
1,2-Dichloroethane		$ClCH_2CH_2Cl$	30 μg/L
1,1,1-Trichloroethane		CH_3CCl_3	2000 μg/L
1,1-Dichloroethene		$C_2H_2Cl_2$	30 μg/L
Benzene		C_6H_6	10 μg/L
Ethylbenzene		C_8H_{10}	300 μg/L
Styrene		C_8H_8	20 μg/L
Polynuclear aromatic hydrocarbons (PAHs)		$C_2H_3N_1O_5P_{13}$	0.7 μg/L
Monochlorobenzene (MCB)		C_6H_5Cl	300 μg/L
Dichlorobenzenes (DCBs)	1,2-Dichlorobenzene (1,2-DCB)	$C_6H_4Cl_2$	1000 μg/L
	1,3-Dichlorobenzene (1,3-DCB)	$C_6H_4Cl_2$	No guideline
	1,4-Dichlorobenzene (1,4-DCB)	$C_6H_4Cl_2$	300 μg/L
Trichlorobenzenes (TCBs)		$C_6H_3Cl_3$	20 μg/L
Di(2-ethylhexyl)adipate (DEHA)		$C_{22}H_{42}O_4$	80 μg/L
Di(2-ethylhexyl)phthalate (DEHP)		$C_{24}H_{38}O_4$	8 μg/L
Acrylamide		C_3H_5NO	0.5 μg/L
Epichlorohydrin (ECH)		C_3H_5ClO	0.4 μg/L
Hexachlorobutadiene (HCBD)		C_4Cl_6	0.6 μg/L
Ethylenediaminetetraacetic acid (EDTA)		$C_{10}H_{12}N_2O_8$	200 μg/L
Nitrilotriacetic acid (NTA)		$N(CH_2COOH)_3$	200 μg/L
Organotins	Dialkyltins	R_2SnX_2	No guideline
	Tributil oxide (TBTO)	$C_{24}H_{54}OSn_2$	2 μg/L

Table 1.3 WHO standards (1993) for disinfectants and disinfectant byproducts

Group	Substance		Formula	WHO STD
Disinfectants	Chloramines		$NH_nCl^{(3-n)}$, where $n = 0$, 1 or 2	3 mg/L
	Chlorine		Cl_2	5 mg/L
	Chlorine dioxide		ClO_2	No guideline
	Iodine		I_2	No guideline
Disinfectant byproducts	Bromate		$Br\ O_3^-$	25 μg/L
	Chlorate		$Cl\ O_3^-$	No guideline
	Chlorite		$Cl\ O_2^-$	200 μg/L
	Chlorophenols	2-Chlorophenol (2-CP)	C_6H_5ClO	No guideline
		2,4-Dichlorophenol (2,4-DCP)	$C_6H_4Cl_2O$	No guideline
		2,4,6-Trichlorophenol (2,4,6-TCP)	$C_6H_3Cl_3O$	200 μg/L
	Formaldehyde		$HCHO$	900 μg/L
	MX (3-Chloro-4-dichloromethyl-5-hydroxy-2(5H)-furanone)		$C_5H_3Cl_3O_3$	No guideline
	Trihalomethanes	Bromoform	$CHBr_3$	100 μg/L
		Dibromochloromethane	$CHBr_2Cl$	100 μg/L
		Bromodichloromethane	$CHBrCl_2$	60 μg/L
		Chloroform	$CHCl_3$	200 μg/L
	Chlorinated acetic acids	Monochloroacetic acid	$C_2H_3ClO_2$	No guideline
		Dichloroacetic acid	$C_2H_2Cl_2O_2$	50 μg/L
		Trichloroacetic acid	$C_2HCl_3O_2$	100 μg/L
	Chloral hydrate (trichloroacetaldehyde)		$CCl_3CH(OH)_2$	10 μg/L
	Chloroacetones		$C_3\ H_5\ O\ Cl$	No guideline
	Halogenated acetonitriles	Dichloroacetonitrile	C_2HCl_2N	90 μg/L
		Dibromoacetonitrile	C_2HBr_2N	100 μg/L
		Bromochloroacetonitrile	$CHCl_2CN$	No guideline
		Trichloroacetonitrile	C_2Cl_3N	1 μg/L
	Cyanogen chloride		$ClCN$	70 μg/L
	Chloropicrin		CCl_3NO_2	No guideline

Data from: http://www.lenntech.nl/toepassingen/drinkwater/normen/who-s-drinking-water-standards.htm#ixzz45W3oEK92 [5].

1.3.2 Industrial Discharge Standards: MINAS

In addition to the guidelines recommended by the WHO, some countries have their own standards depending on the ground realities. For example, the Central Pollution Control Board of India (CPCB) issued a set of standards in regards to effluent concentration, called Minimum Acceptable Standards (MINAS), defined for each type of industry and for each type of medium of release. One such set of guidelines is presented in Table 1.4.

1.3.3 Sources of Water Pollution

Surface Water Pollution Sources

To effectively combat pollution of surface water and ground water, the sources of pollution need to be identified first. Surface water bodies (e.g., ponds, lakes, rivers, and oceans) are polluted from either domestic or industrial wastes when such wastes are discharged directly or indirectly into water bodies without adequate treatment to remove harmful compounds. Another major source of surface water pollution is industrial or agricultural runoff. Plants and living organisms present in water bodies are affected by pollutants present in the water bodies. Water pollution from farming according to OECD study (organization of Economic Cooperation and Development) is the maximum. In 2012, global chemical fertilizer use hit 180 million tons reflecting an increase of 500% in last 50 years. Erosion of topsoil and run-off results in massive water pollution. Agricultural runoff affects some 400 aquatic ecosystems worldwide, with the Gulf of Mexico and South China Sea being the most prominent among them. Just a quarter of cropland spread in India, China, USA leads to 50% of the global fertilizer waste generation during cultivation of rice, corn, wheat. The US alone spends $4.8 billion in nitrogen removal annually. Pesticides also heavily pollute both surface and groundwater. In 2007, it stands at 5.2 billion tons worldwide. Animal manure results in 1.2–1.3 billion tons of animal waste in the US alone. Thus millions of tons of nitrogen and phosphorus escapes into the rivers and water bodies affecting both surface and groundwater [6,7].

Particulate or gaseous air pollutants also indirectly contaminate surface water through rain or acid rain. When contaminated surface water seeps through soil to the underground aquifer it pollutes the groundwater. Surface water and groundwater are interrelated. Surface water seeps

Table 1.4 Maximum permissible limits for industrial effluent discharges (in mg/L) Indian standards (1974)

Parameter	Into inland surface water	Into public sewers	On the land for irrigation
pH	5.5–9.00	5.5–9.00	5.5–9.00
BOD (5 days at 20°C)	30	350.00	100.00
COD	250		
Suspended solids	100.00	600.00	200.00
Total dissolved solids (inorganic)	2100.00	2100.00	2100.00
Temperature (°C)	40	45.00	
Oil and grease	10.00	20.00	10.00
Phenolic compounds	1.00	5.00	
Cyanides	0.20	2.00	0.20
Sulfides	2.00		
Fluorides	2.00	15.00	
Total residual chlorine	1.00		
Pesticides			
Arsenic	0.20	0.20	0.20
Cadmium	2.00	1.00	
Chromium (hexavalent)	0.10	2.00	
Copper	3.00	3.00	
Lead	0.10	1.00	
Mercury	0.01	0.01	
Nickel	3.00	3.00	
Zinc	5.00	15.00	
Chlorides	1000.00	1000.00	1000.00
Boron	2.00	2.00	2.00
Sulfates	1000.00	1000.00	1000.00
Sodium (%)		60.00	60.00
Ammonical nitrogen	50.00	50.00	
Radioactive materials			
Alpha emitters (millicurie/mL)	10-7	10-7	10-8
Beta emitters (curie/mL)	10-6	10-6	10-7

through the soil and becomes groundwater, and conversely, groundwater can also feed surface water sources. Sources of surface water pollution are generally grouped into two categories based on their origin: point sources and nonpoint sources.

Point-source water pollution comes from discrete conveyances and alters the chemical, biological, and physical characteristics of water. Point-source water pollution refers to contaminants that enter a waterway from a single, identifiable source, such as a pipe or ditch carrying wastewater from a factory, municipal sewer system, or industrial storm water system.

Nonpoint pollution refers to pollution that does not originate from any single specific source but from a vast area such an agricultural field or a vast stretch of industrial area. Pollution in the form of agricultural runoff or industrial storm water falls into this category.

Groundwater Pollution Sources

Some of the major sources of groundwater pollution include storage vessels and reservoirs of petroleum products, storage vessels or chemicals, septic systems, hazardous waste sites, landfills, agricultural fields with high amounts of unabsorbed fertilizers, pesticides, and other chemicals.

Saline ingress following over drafting of aquifers or natural leaching from naturally occurring deposits are natural sources of groundwater pollution. Leaching of minerals like arsenic and fluoride from their crystal lattice due to geological disturbance causes heavy groundwater contamination. While concern over groundwater contamination has focused on pollution associated with human activities, in many cases, groundwater contamination is related to private sewage disposal systems, land disposal of solid waste, municipal wastewater, wastewater impoundments, land spreading of sludge, brine disposal from the petroleum industry, mine wastes, deep-well disposal of liquid wastes, animal feedlot wastes, and radioactive wastes. Interactions between groundwater and surface water are complex. Consequently, groundwater pollution, sometimes referred to as groundwater contamination, is not as easily classified as surface water pollution. By its very nature, groundwater aquifer are susceptible to contamination from sources that may not directly affect surface water bodies. A spill or ongoing release of chemical or radionuclide contaminants into soil can contaminate the aquifer below, defined as a toxin plume.

Classification of Major Water Pollutants

1. Suspended solids: Suspended solids can lead to the development of sludge deposits and anaerobic conditions when untreated wastewater is discharged into the aquatic environment.

2. Biodegradable organics: Composed principally of proteins, carbohydrates, and fats, biodegradable organics are measured most commonly in terms of BOD (biochemical oxygen demand) and COD (chemical oxygen demand). If discharged untreated to the environment, their biological stabilization can lead to the depletion of natural oxygen resources and to the development of septic conditions.
3. Heavy metals: Heavy metals are usually added to wastewater from industrial activities and may have to be removed if the wastewater is to be reused.
4. Pathogens: Communicable diseases can be transmitted by pathogenic organisms present in wastewater.
5. Nutrients: Both nitrogen and phosphorus along with carbon are essential nutrients for growth. When discharged in excessive amounts on land, they can also lead to pollution of groundwater.
6. Priority pollutants: Organic and inorganic compounds selected on the basis of their known or suspected carcinogenicity, teratogenicity, or high acute toxicity. Many of these compounds are found in wastewater.
7. Refractory organics: These organics tends to resist conventional methods of wastewater treatment. Typical examples include surfactants, phenols, and agricultural pesticides.
8. Dissolved inorganics: Inorganics constituents such as calcium, sodium, and sulfate are added to the original domestic water supply as a result of water use and may have to be removed if wastewater is to be reused.

1.4 INTRODUCTION TO WATER RESOURCE MANAGEMENT APPROACHES

According toestimates from the World Economic Forum (2015), three of the top ten risks confronting humanity are environmental risks, with the water crisis at the top of the list. Failure of climate-change adaptability and biodiversity loss are the other risks. The current water crisis is linked to the following [8,9]:

1. Lack of integrated water management and sound land-use practices has led to losses in natural ecosystems (destruction of watersheds and forests) of cleaning water, preserving water, and for managing temporal and spatial variation of precipitation and flow of water.
2. Higher temperatures due to greenhouse gas emissions have speeded up water cycles unleashing intense drought in some regions already facing challenging water strains and extreme rainfall in the other

regions resulting in huge water pollution through agricultural and industrial runoff.

3. Water has been polluted in large scale through industrial discharge, agricultural runoff, and discharge through municipality sewers.
4. Massive groundwater withdrawal: For irrigation, excessive withdrawal of groundwater continues amidst demand for increasing quantum of food grain for feeding the ever-growing world population. Though agricultural productivity in irrigated land is twice that in rain-fed agriculture, massive water use in irrigation has severely strained groundwater resources in many parts of the world threatening drinking water availability. Roughly 70% of the fresh water available is used in irrigation globally. Irrigated farming is increasing with growing population aggravating the situation further. In Africa rain-fed agriculture is 95%, in China it is 70%, and in India it is 60%. In rain-fed agriculture, poor water management and huge evaporation losses lead to enormous water losses.
5. Aging water infrastructures and lack of new investment to maintain and develop new infrastructure is causing severe strain in freshwater availability worldwide.
6. Poor regulations: Lack of regulations in water use, particularly groundwater use, has led to inefficient use and waste of fresh water in many parts of the world. However, under the emerging threats, water authorities and regulators are enacting new laws and regulations to ensure sustainability of water resources through controlled use of scarce water resources, relocation of water sources from agriculture for urban use and energy needs, etc., regulation of groundwater pumping, etc.

Both the qualitative and quantitative limitations of freshwater resources require an effective water resource management strategy to ensure sustainable use of water resources. While the strategy used will likely be, in part, location-specific, in general these approaches broadly include zero discharge, flow management, preservation, and purification as explained in the next section.

1.4.1 Pollution-Prevention Approach

Agricultural runoff, industrial wastewater discharge, stormwater runoff, and sewer line leaks pollute freshwater on a massive scale, resulting in significantly less safe freshwater for human use. Thus the best water resource management strategy focuses on preventing water pollution and maintaining the safety of freshwater resources. This approach is particularly very relevant to industrial pollution and implies switching over to green technologies. While total elimination of pollutants may be difficult, reducing the level of pollution close to zero has been possible in many cases through adoption of clean technologies. Switching over to clean technologies and adoption of the principles of process intensification has the potential of saving fresh water in a big way [10].

1.4.2 Flow Management Approach

Significant amounts of water flows out to the sea through river lines in many countries during the months of heavy precipitation in the catchment area. Without an adequate reservoir facility such water cannot be saved for the dry months. Diverting at least a part of this huge outflow from these rivers to big reservoirs or other areas of relatively low rainfall may help reduce the freshwater scarcity in many parts of the world. Dredging river beds help increase water-holding capacity thereby preventing flood during the periods of heavy rain and storing more water for effective utilization as well as recharging underground aquifers. Interlinking of rivers is often suggested to divert river water from one region to another dry region, but these are often very expensive routes and not without conflict among the involved regions.

1.4.3 Efficient Water-Use Approach

Considering the limited availability of freshwater, it is necessary to ensure efficient use of water that minimizes water use as well as saves on fresh water intake. Agriculture being the largest user of water (85% of all freshwater use) needs to urgently address this issue of efficient water use [11].

Efficiency in Irrigation and Water Distribution

Use of an efficient irrigation technology is another widely advised practice. For example, instead of flood irrigation, drip irrigation can save enormous quantity of fresh water. Uncontrolled withdrawal of groundwater for irrigation and industrial purposes has resulted in significant groundwater depletion in several parts of the world. A day may not be very far away when many current human habitats will have to be deserted for lack of drinking water as evident in the history of human civilization through centuries. There needs to be proper regulations on such groundwater use. Lack of proper monitoring and supervision leads to waste of fresh water in public water distribution networks through leaks and open water valves without manning.

Use of Water Footprint in Agriculture

The water footprint of a product is defined as the volume of freshwater used to produce the product [11]. This water footprint may be a green water footprint (rainwater consumed), blue water footprint (volume of surface and groundwater consumed (evaporated) as a result of production of the product and gray water footprint meaning the volume of freshwater required to assimilate the load of pollutants based on existing ambient water-quality standard. Table 1.5 shows the water footprint of various crops. Cropping pattern should be thoroughly examined to find out the

Table 1.5 Water footprint of various crops

Crop	Global average water footprint (m^3/ton)
Sugar crop	197
Fodder crop	253
Vegetables	322 (cauliflower, broccoli 285; tomatoes 214, onion 272)
Roots, tubes	387 (potatoes 287)
Fruits	967 (apple 822; mangoes 1800; strawberries 347; banana 790; orange 560; watermelon 235; avocados 1981; grape 506)
Cereals	1644 (rice 1647; wheat 1828; rye 1544; oats 1788; maize, corn 1222)
Pulses	4055 (lentils 5874)
Spices	7048
Rubber	13,748
Nuts	9063
Almond	8047
Onion	272
Cotton	4029
Coffee	15,897
Potatoes	287
Mustard seed	2809
Castor oil	24,740
Sunflower seed oil	3366
Soybean	2145
Strawberries	347
Mangoes	1800

Data Source: Food and Agricultural Organization, UN, Rome, Italy, 2008a.

best crop for a region particularly if it is a water stress region. Water intensity of crops needs to be checked and if other conditions permit then switching over to less water intensive crop is most advisable for a water stress area.

Global fresh water withdrawal has witnessed a sevenfold increase in the last century largely due to massive growth in population, change in dietary pattern, and change in lifestyle along with increase in the intensity of agricultural and industrial activities [12]. A cup of tea (based on 3 g tea leaves in a cup of tea) uses less water than a cup of coffee (7 g coffee powder in a cup) as a cup of tea involves consumption of only 27 L of fresh water compared to 130 L in case of a cup of coffee. Water intensities vary very widely with food items (15,415 L/kg of beef, 4325 L/kg chicken, 287 L/kg potato, and 560 L/kg orange).

1.4.4 Preservation Approach

Water preservation approaches have been in practice for centuries and include creating reservoirs, preventing outflow, and using ponds as a means of water preservation. This is a supply side water management approach. Since precipitation alone does not guarantee sustainable supply of fresh water throughout the year, water preservation is needed. Normally water shortage is not reported from Jaisalmer located in the desert of Rajasthan in India which receives a meager annual rainfall of only around 100 mm whereas people in Cherrapunji in the North-Eastern part of India faces acute water shortage despite heavy rainfall of 15,000 mm annually. This example illustrates the role and importance of water preservation. In any country, infrastructure for water storage is the most important requirement for sustainable water resource management, but variations in water storage or preservation infrastructure are observed in different countries, as seen in Table 1.6.

This implies that massive investment in water storage infrastructure is necessary in a country like India which has only 225 m^3 per capita annual water storage facility and where almost 88% of the rain water gets drifted to the sea. There may be manifold benefits in preserving water such as:

1. Hydel power generation (from dams).
2. More irrigation coverage (through canals and modern channels like pipelines).
3. Recharging of groundwater.
4. Prevention of saline ingress in the coastal areas.

Table 1.6 Variation of per capita water storage across the countries

Some selected countries across the world	**Annual per capita water storage based on dam based storage infrastructure (m^3)**
Brazil	3145
Russia	6103
India	225
China	1111
USA	1964
Spain	1410
Turkey	1739
Mexico	1245
South Africa	753

From: Food and Agriculture Organization of the United Nations, Water Development & Management Unit, 2015, Reproduced with permission.

5. Need based distribution to different areas.
6. Ensuring supply throughout the year.
7. Protection of cattle, human life and property from flood.

Water preservation approaches may be multifaceted as outlined below.

Rainwater Harvesting

Rainwater harvesting for immediate use and for recharging of groundwater is an efficient water preservation strategy. Rainwater can be preserved on rooftops of houses and free surface areas and in ponds and underground reservoirs. Without much treatment rainwater can be used for flushing toilets, gardening, car washing, and washing of pavements and for agriculture when preserved in ponds. With further treatment rainwater can be used of cleaning clothes and for drinking purpose.

Preventing Groundwater Depletion: Pond Management

Saline ingress and depletion of groundwater could be checked in many cases by using preserved rain water for recharging underground aquifers. Municipalities across the world in many cases have made it compulsory for the housing societies to have provisions of rain water preservation while constructing buildings. For example, in the Chennai city of Tamil Nadu, India, a 50% increase in groundwater could be achieved along with significant improvement in water quality within 5years of its implementation.

In the Irrawaddy Delta of Myanmar, the groundwater is saline and communities rely on mud-lined rainwater ponds to meet their drinking water needs throughout the dry season. Some of these ponds are centuries old and are treated with great reverence and respect. Ponds in the villages across many countries for centuries have served as the major source of drinking water and other water requirements for domestic purposes. But over the last few decades, many of these ponds have been dried up to make space for human habitation or cultivation, thus jeopardizing the associated ecology and closing an important channel for groundwater recharging and important water storage facility. However, recently policymakers have aimed at effective management of such ponds to ensure sustainability of ecosystems as well as water supplies across thousands of villages in many regions particularly in the hugely populated countries like India and China [13]. The most important task before the stakeholders is preventing uncontrolled and excessive withdrawal of groundwater and protection of the traditional surface water structures, which will significantly prevent groundwater depletion.

Check Dams

Generally, check dams are constructed across streams to enhance the percolation of surface water in to the subsoil strata. The water percolation in the water impounded area of the check dams can be enhanced artificially many folds by loosening the subsoil strata/overburden by using ANFO explosives as used inopen cast mining. Thus local aquifers can be recharged quickly by using the available surface water fully for use in the dry season.

Rejuvenation of Inland Waterways

Rejuvenation of inland water waterways can multiple benefits. For example, it will make bulk transport easier, cheaper, and ecofriendly while ensuring passage of excess water during the monsoon months of heavy rain preventing floods and recharging underground aquifers throughout the year.

Freshwater Flooded Forests

Rainwater preservation is also possible by growingfreshwater flooded forests. This will facilitate the use of locally available rain water to throughout the year without the need of huge capital expenditure. This would facilitate availability of uncontaminated water for domestic, industrial, and irrigation needs. Construction of check dams on small streams and planting of "flooded forest trees" will enhance water percolation into the ground. The underground moisture will help the forest during dry summer months, and will also ensure animals also get drinking water and shelter during summer months.

Educating School Children

If rainwater-harvesting systems are installed in school buildings, this will not only preserve water but will also go a long way toward educating school children about water-conservation principles and bridging divides between people of different religious and ethnic backgrounds while addressing the water-scarcity issues.

In general, rainwater harvesting is an independent and inexpensive water supply source and helps prevent groundwater depletion and saline ingress.

1.4.5 Purification Approach: Closing the Loop as Sustainable Solution

As pollution makes huge quantity of water unfit for consumption, volume of effectively usable fresh water goes down. There cannot be a better alternative to recovery and reuse of this water through appropriate treatment. This is an end of pipe approach. Through decades of research aimed at developing better water treatment technologies, it is now possible to close the water use cycles for many industrial operations as well for domestic purposes through adoption of advanced water purification approaches for eventual recovery and reuse of water.

Human civilizations grew and developed along the banks of many rivers across the world. From the 18th century onward, the world has been witnessing development of industries along the banks of the rivers. Today, thousands of such river bodies that have served as lifelines for centuries are now dirty and heavily polluted as a result of continuous discharge of mostly untreated and polluted wastewater. The only practical way of saving and protecting such river bodies is closing the loop of water use in the industries and the municipalities where modern membrane-based treatment technology will permit recycle and reuse of such water within the industry and the municipality [14]. Applicability of a particular technology depends on a number of criteria such as type and source of water, end use of treated water, desired degree of purification, cost of treatment, etc. In many cases, combination of major treatment technologies may be necessary. Treatment technologies broadly fall into the following categories:

1. Chemical treatment technology.
2. Biological treatment technology.

3. Physicochemical treatment technology.
4. Membrane separation technology.

In subsequent chapters, we will describe how these treatment methods and technologies and purification of raw water from different sources can be achieved for various purposes.

REFERENCES

[1] World Health Organization and UNICEF Joint Monitoring Programme (JMP). Progress on drinking water and sanitation. 2015 Update and MDG Assessment; 2015.

[2] United States Census Bureau Estimates. United States and world population clock; 2015.

[3] International Telecommunication Union (ITU). The world in 2015 ICT facts and figures, 2015.

[4] Lockwood, H., and Smits, S. (2011). Supporting rural water supply. Moving towards a service delivery approach.

[5] http://www.lenntech.nl/toepassingen/drinkwater/normen/who-s-drinking-water-standards.htm#ixzz45W3oEK92.

[6] Fonterra, Clarity: Fonterra annual review 2013.

[7] Tim Searchinger, T., Hanson, C., Ranganathan, J., Lipinski, B., Waite, R., Winterbottom, R., et al. Creating a sustainable food future: interim findings. A menu of solutions to sustainably feed more than 9 billion people by 2050, December 2013, Washington DC, USA.

[8] World Economic Forum (2015). Global risks 2015 report.

[9] United Nations' Environmental Program, Global Environmental Outlook Report GEO-4, 2007.

[10] Pal P, Nayak J. Development and analysis of a sustainable technology in manufacturing acetic acid and whey protein from waste cheese whey. J Clean Prod 2015;112 (1):59–70.

[11] Hoekstra AY, Chapagain AK, Addaya MM, Mekonnen MM. Water footprint manual: state of the art 2009. Enschede, The Netherlands: Water Footprint Network; 2009.

[12] Gleick PH. The changing water paradigm: a look at the twenty first century water resources development. Water Int 2000;25(1):127–38.

[13] Roy M. Managing the village level open access water resources in a region facing rapidly declining water availability. Environ Dev Sustain 2010;12(6):999–1012.

[14] Thakura R, Chakrabortty S, Pal P. Treating complex industrial wastewater in a new membrane-integrated closed loop system for recovery and reuse. Clean Technol Environ Policy 2015;17(8):2299–310.

CHAPTER 2
Chemical Treatment Technology

2.1 INTRODUCTION

To make ordinary surface water or groundwater suitable for drinking as per drinking water standards or to make industrial waste water suitable for discharge following effluent discharge standards as described in Chapter 1, a variety of chemical treatments may be adopted depending on the source of water, requirement of the end use, and cost. Such chemical treatments broadly include chemical coagulation, chemical precipitation, chemical disinfection, chemical oxidation, advanced oxidation, ion exchange, and chemical neutralization. Chemical treatment processes are remarkably fast but aren't considered ecofriendly processes due to the involvement of harsh chemicals, the creation of harmful byproducts in some cases, and the generation of huge amounts of sludge. But in many cases, as a pretreatment step, chemical treatment can help in removing certain specific contaminants with a high degree of efficiency while rendering subsequent downstream treatments easier. For example, the presence of certain toxic chemicals like cyanide in wastewater turns biological treatment to be very difficult but a chemical pretreatment for cyanide removal can make the biological treatment very successful. Chemical neutralization of wastewater streams is often the most important first step in the train of water-treatment units. Aeration is the least expensive option for improvement of water quality in terms of taste and odor. Table 2.1 some of the broad chemical treatment options and their applicability.

2.2 AERATION

Aeration is a simpler operation and has the potential to improve water quality significantly. The taste and odor problems frequently encountered in drinking water can be largely overcome by aeration particularly when it is due to dissolved gases like H_2S and volatile organic compounds (VOCs). The presence of algae, iron, manganese, and other impurities may contribute to taste and odor problems and in such cases no single

Industrial Water Treatment Process Technology.
DOI: http://dx.doi.org/10.1016/B978-0-12-810391-3.00002-3

Table 2.1 Broad chemical treatment options for various applications

Chemical treatment processes	Typical application
1. Aeration	Removal of odor-causing substances, VOCs
2. Chemical coagulation	Chemical destabilization of suspended particulates in water for aggregation during ortho and perikinetic flocculation
3. Neutralization	Controlling pH for chemical precipitation and downstream treatments
4. Chemical oxidation	Reduction of BOD, COD, microorganisms, odor-causing substances of water
5. Chemical precipitation	Removal of heavy metals and hardness
6. Ion exchange	Removal of hardness-causing ions or specific toxic ions and metals
7. Chemical disinfection	Elimination of pathogens to prevent water-borne diseases
8. Advanced oxidation	Removal of refractory organic compounds

treatment will suffice. Taste and odor of water is measured as TON (threshold odor number) and flavor profile.

2.2.1 Mechanism of Water-Quality Improvement by Aeration

Aeration or air stripping improves water quality through the following:

1. Aeration removes dissolved gases such as carbon dioxide and hydrogen sulfide mainly by a mechanical scrubbing mechanism but stripping off of the volatile reaction products and precipitation of the solid phase may also take place through chemical conversions such as those shown in Eqs. (2.1) and (2.2).
2. Removal of the dissolved gases eliminates the causes of taste and odor problems.
3. Water quality improves as VOCs are stripped out.
4. Additional oxygenation of the treated water improves its quality as potable water through high oxygen saturation.
5. Additional oxygenation helps in the oxidation of iron and manganese by making them insoluble and thus causing their separation from the aqueous phase through chemical precipitation.
6. Oxygenation through aeration prevents formation of reducing environments that exacerbate taste and odor problems.

2.2.2 Oxygen Mass Transfer in Aeration

Transfer of oxygen from air mass to water mass depends on the difference of concentration (C_s) of oxygen in water in saturation and the actual concentration of dissolved oxygen in water at a given temperature (C). The rate of oxygen mass transfer can be expressed as:

$$\frac{dc}{dc} = K_L a[C_s - C] \tag{2.1}$$

where $K_L a$ is the liquid phase oxygen mass transfer coefficient.

Aeration equipment is specified in terms of standard oxygen rate (SOR), which is defined as:

$$\frac{dc}{dt} = (K_L a)_{1,20}\left[C_s, 20, \mathrm{sp}\right] \tag{2.2}$$

$(K_L a)_{1,20}$ refers to the liquid-phase oxygen mass transfer coefficient at standard conditions.

C_s, 20, sp refers to the saturated dissolved oxygen concentration at 20° C and standard pressure. In rating aeration equipment, the standard condition implies temperature of 20°C, pressure of 1 atmosphere, and dissolved oxygen concentration of 0 mg/L for the water body. In Eq.(2.2), the third bracket includes only the term C_s, 20, sp as the DO at the standard conditions is 0 mg/L (i.e., $C = 0$).

The $K_L a$ under actual field conditions is determined using the Arrhenius relation:

$$(K_L a)_f = (K_L a)_{1,20}(\Theta^{T-20}) \tag{2.3}$$

The actual oxygenation rate (AOR) is related to the SOR by:

$$(\mathrm{AOR}/\mathrm{SOR}) = \alpha[(\beta[C_s] - [C_m])/C_{s,20,sp}]\Theta^{T-20} \tag{2.4}$$

where $\alpha = (K_L a)_f/(K_L a)_{1,20}$ and c_m is the mean oxygen concentration of water.

$$\beta = C_{sf}/C_s \tag{2.5}$$

where C_{sf} = saturated DO of field water and C_S = saturated DO of clean water at the field condition. The saturation DO values for clean as well as untreated water are determined through a jar test. A 1-L jar half-filled with the water sample is vigorously agitated to fully saturate the water with oxygen, and the dissolved oxygen concentrations are measured at different temperatures. The higher the temperatures the lower the saturated dissolved oxygen concentration. Similarly saturated dissolved oxygen concentration will be higher at higher barometric pressure and actual C_{sf} (saturated concentration of oxygen at field) needs to be corrected for pressures other than the standard pressure ($P_s = 760$ mm Hg) through the relation:

$$Cs_f = (C_s)(P_b/P_s) \tag{2.6}$$

where P_b is the barometric pressure of the field.

2.2.3 Methods of Aeration

There are several methods of aeration and the most appropriate method is generally based on contaminant-removal efficiency and cost. The most widely adopted aeration practices are based on the use of fountain-type devices, cascade aerators, cascade tray aerators, and diffused aerators.

Fountain or Spray-Nozzle Aerators

Spray nozzles have been used for many years in the water-treatment field. The most common application is fountain-type aeration in which water is sprayed under pressure into the open atmosphere. Fountain-type aerators are widely used to control taste and odor problems, to prevent anaerobic decay of natural organic matter accumulated in the reservoir, and to prevent the solubilization of iron and manganese present in reservoir bottom sediments Fig. 2.1.

Cascading-Tray Aerators

Cascading tray aerators have been used in water treatment for a long time. These aerators use multistage waterfalls that help control taste

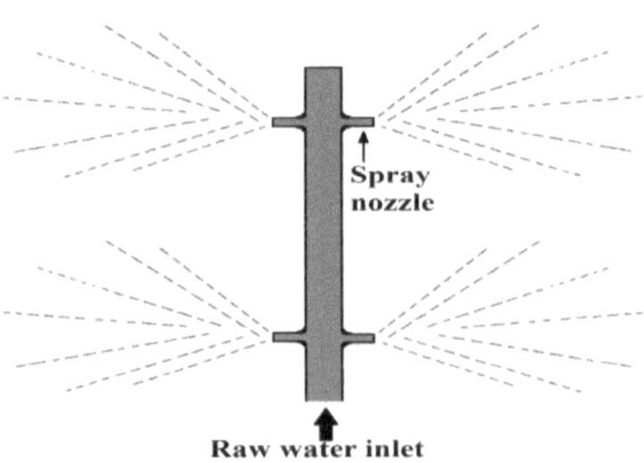

Figure 2.1 Spray-nozzle aerator.

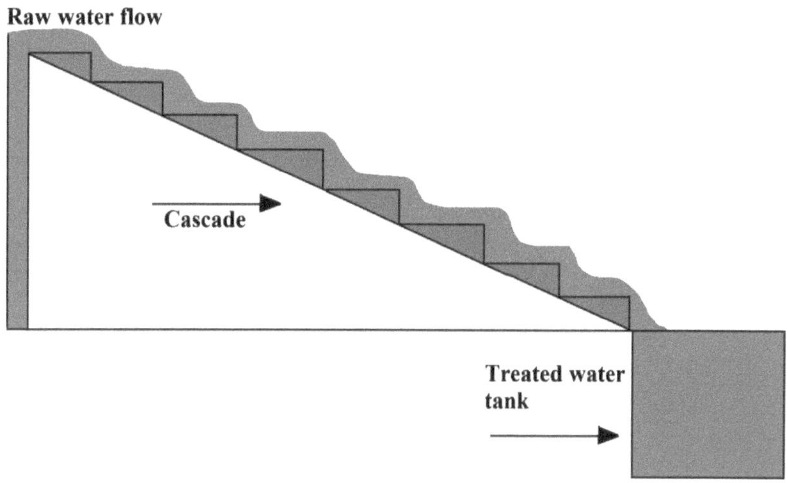

Figure 2.2 Cascade aerator.

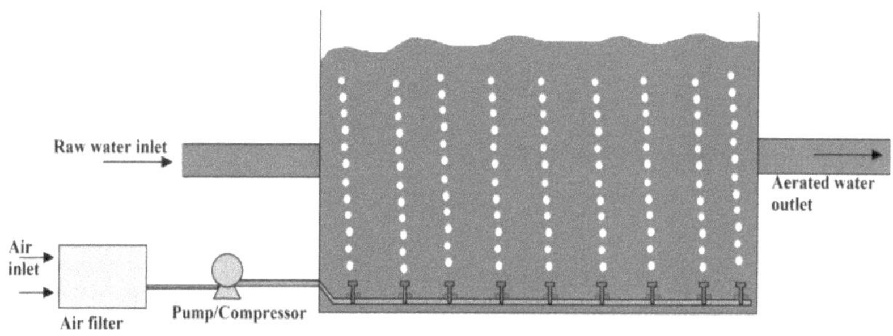

Figure 2.3 Diffused aerator.

and odor problems while precipitating soluble iron and manganese Fig. 2.2.

Diffused Aerators

Diffused aeration involves releasing compressed air bubbles from a diffuser element located at the bottom of a water column. Diffused aeration is sometimes used for VOC removal but is not generally cost-effective or as efficient as packed-tower air stripping. Diffused aeration is sometimes employed in source-water reservoirs to control taste and odor problems and to oxidize iron and manganese Fig. 2.3.

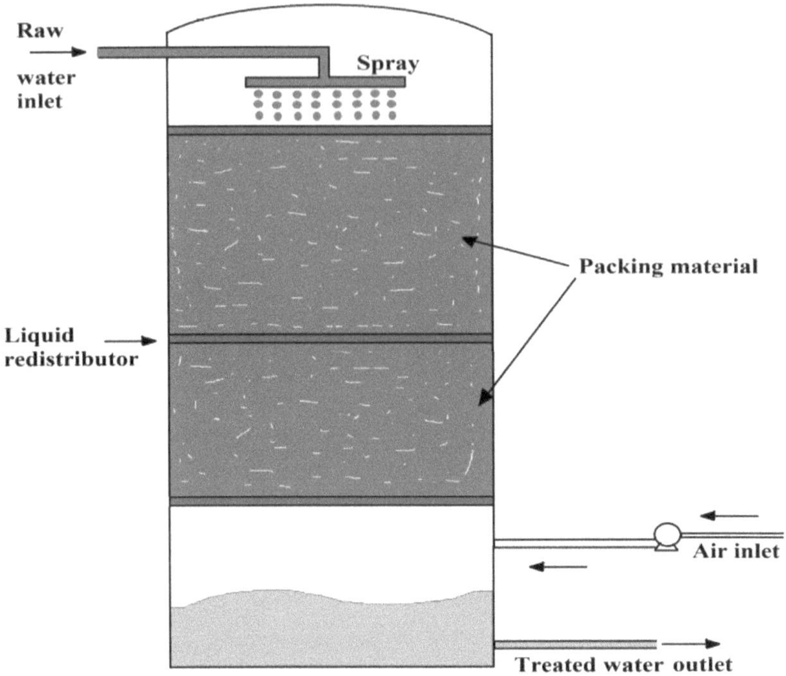

Figure 2.4 Packed-tower stripping aerator.

Packed-Tower Stripping Aerators

Packed-tower aeration is widely used in water-treatment fields because of increasing concern and regulation of VOCs in drinking water. Packed-tower stripping generally consists of packing materials with high-surface area supported and contained in a cylindrical shell. Water normally flows downward through the packing material with forced draft or induced draft upward airflow. The high specific surface area of the packing material provides a higher liquid–gas mass transfer area than provided by other aeration and stripping methods. Packed-tower stripping is more effective for removal of VOCs that are more volatile than traditional contaminants of concern such as carbon dioxide and hydrogen sulfide. Sometimes off-gas treatment of contaminants stripped from the water may be necessary depending on the characteristics and quantity of the gaseous contaminants, as well as on the location of the stripping column and the prevailing air-quality regulations of the area of operation Fig. 2.4.

Aeration in Odor Removal

While coagulation followed by filtration and adsorption are the major physical methods of removal of odor-causing substances, aeration can also be used. Oxygen of air mixed into water by aeration effectively removes tastes and odors caused by iron, manganese, and hydrogen sulfide. Aeration rings the offending material in to direct contact with oxygen in the atmosphere. The materials are then oxidized into stable, safe factors.

Aeration can be very effective for removing hydrogen sulfide (H_2S) primarily by mechanical scrubbing action of air as it through water. Apart from the scrubbing mechanism, oxidation also directly helps in stripping off dissolved H_2Sthrough the following reactions:

$$\text{Hydrogen sulfide} + 2H_2S/O_2 \text{ aeration} \rightarrow 2H_2O + 2S\downarrow(\text{sulfur}) \tag{2.7}$$

$$2S + 4O_2 \text{ aeration} \rightarrow 2SO_4^{-2}(\text{sulfate ion}) \tag{2.8}$$

2.2.4 Other Oxidizing Agents in Odor Removal

Chlorine (Cl)

In addition to oxygen, chlorine and potassium permanganate are also very successful odor- removing agents. Many substances that cause taste and odor can be effectively oxidized with the help of chlorination. For example, algae-caused odors described as fishy, grassy, or septic can be controlled by prechlorination of the water to a free residual of 0.25–5.0 mg/L. Chlorine can also intensify certain tastes and odors. For example, chlorine combines with phenols in water, producing extremely objectionable medicinal tastes. Chlorine also intensifies the earthy odors caused by certain types of algae and actinomycetes.

Potassium Permanganate ($KMnO_4$)

Potassium permanganate ($KMnO_4$), used either alone or in combination with other chemicals, is effective in removing iron and manganese and oxidizing organic and inorganic materials that cause taste and odor. When it is added to water containing taste–odor compounds, the reaction is:

$$\begin{aligned}&2KMnO_4 + H_2O + \text{taste-and-odor compound} \rightarrow 2MnO_2\downarrow + 2KOH\\&\quad + 3O^{-2} + \text{taste-and-odor compounds.}\end{aligned}$$

Dosages of $KMnO_4$ vary from 0.5 to 15 mg/L, although dosages in the range of 0.5–2.5 mg/L are usually adequate to oxidize most taste-and-odor-producing chemicals.

Aeration in Removal of Iron (Fe^{+2})

Both ferrous iron (Fe^{2+}) and manganous manganese (Mn^{+2}) can be removed by aeration as follows:

$$4Fe(HCO_3)_2 + O_2 + 10H_2O \xrightarrow{\text{aeration}} 4Fe(OH)_3\downarrow + 8H_2CO_3 \quad (2.9)$$

$$2MnSO_4 + O_2 + 4NaOH \xrightarrow{\text{aeration}} 2MnO_2 + 2Na_2SO_4\downarrow + 2H_2O \quad (2.10)$$

The removal of iron requires a two-step reaction process. In the first step, the soluble ferrous bicarbonate [Fe $(HCO_3)_2$] is converted into ferrous hydroxide [Fe $(OH)_3$]. Subsequent oxidation of this ferrous hydroxide results in formation of insoluble ferric hydroxide [Fe $(OH)_3$], which can be filtered out or settled out of the solution as fluffy, rust-colored sludge. The optimum pH range for the reaction is 7.5−8.0 that goes to completion in about 15 minutes. 0.14 mg/L of oxygen removes 1 mg/L of iron (Fe^{2+}) in this type of aeration.

Aeration in Removal of Mn^{+2}

In removal of manganese from water, the same mechanism works where the soluble salt of manganese [manganoussulfate ($MnSO_4$)] is oxidized to the insoluble manganese dioxide (MnO_2), which can be filtered out or settled out of solution. The optimum pH for this reaction is 9.0−10.0. As shown in Eq. (2.4), sodium hydroxide (NaOH) raises the pH to the desired level and provides the hydroxide alkalinity (hydroxide ions, OH^-) necessary to increase the pH for the reaction. 0.27 mg/L of oxygen removes 1 mg/L of manganese (Mn^{2+}). In this case, the reaction completion time is about 15 minutes.

2.3 CHEMICAL COAGULATION

To coagulate means to drive together. In coagulation, colloidal and finely divided suspended matters are driven together to help form flocks that can subsequently be separated out from the aqueous medium through sedimentation or filtration leaving a clear water stream. Chemical flocculating agents are added to the water to be treated. Destabilization of the colloidal matter takes place through a series of surface reactions depending on the type and dose of the flocculating reagent added, medium pH, types of impurities present and their concentration, agitation in the medium, duration of flocculation, zeta potential, and temperature. While there are several coagulants available for water treatment, in this section, the chemical changes that take place in presence of the two most widely used flocculants, namely alum and ferric sulfate, will be described.

2.3.1 Alum as Coagulant and the Chemical Reactions

Alum has been used in water treatment for centuries because of its effectiveness, low cost, availability, and ease of handling. Alum is available in two forms: filter alum and liquid alum. Filter alum is the solid or dry alum [$Al_2(SO_4)_3.14H_2O$] that is ivory-white in color and available in lump ground or powdered form. Liquid alum is alum in solution [$Al_2(SO_4)_3.xH_2O$]. Liquid alum is available in different strengths, the strongest being less than half the strength of the dry filter alum. There may be variation in color of liquid alum depending on strength, from a slight, white, iridescent-like color to a yellow-brown.

Depending on the form, filter alum varies in density. Liquid alum has a density considerably higher than that of water.

When alum is added to water, it reacts as follows:

$$Al_2(SO_4)_3 + 3Ca(HCO_3)_2 \rightarrow 2Al(OH)_3\downarrow + 3CaSO_4\downarrow + 6CO_2 \quad (2.11)$$

$$Al_2(SO_4)_3 + 3Na_2CO_3 + 3H_2O \rightarrow 2Al(OH)_3\downarrow + 3NaSO_4\downarrow + 3CO_2 \quad (2.12a)$$

$$Al_2(SO_4)_3 + 3Ca(OH)_2 \rightarrow 2Al(OH)_3\downarrow + 3CaSO_4\downarrow \quad (2.12b)$$

In alkaline pH only, the reactions take place and such alkaline pH may be already present as in case of Eq.(2.5) or the alkalinity may be added during coagulation as in Eqs. (2.6) and (2.7). Alkaline substances react with the alum to form aluminum hydroxide [$Al(OH)_3$], the sticky flock material. For each mg of alum added per liter about 0.5 mg/L of alkalinity is required. Without an adequate supply of alkalinity, alum will not form aluminum hydroxide and will pass through the filters. Such alkalinity added later for corrosion control may form flocks that settle out either in the clear well or in the distribution system causing serious problems. The mechanism of coagulation–flocculation can be explained through the following three steps:

Step-1

The positively charged aluminum ions (Al^{3+}) attract the negatively charged particles that cause color and turbidity, then they form tiny particles called microflocks. This marks the beginning of coagulation.

Step-2

As many of these microflock particles are now positively charged, they begin to attract and hold more negatively charged color-causing and turbidity-causing material in the second step. These first two coagulation steps occur very quickly, in a matter of microseconds.

Step-3

In the final step, the microflock particles grow into easily visible, mature flock particles. This growth occurs partly by the continual attraction of the color and turbidity-causing materials; partly by the adsorption of viruses, bacteria, and algae into the microflock; and partly by the random collision of the microflocks that cause particles to sticks together.

Later, during sedimentation, the large flock particles settle rapidly, leaving the water clear. The optimum alum dosage for the best results depends on the various factors described in the previous section. In general, alum dosage ranges from 15 to 100 mg/L, and the effective pH ranges between 5.5 and 8.5. The pH needs to be adjusted within this range (by increasing alkalinity) for the best performance of alum. The optimum pH for can be found through jar-test experiments. To address the fluctuations in water quality following rainstorm and high runoff periodic jar tests need to be conducted to determine the optimum conditions of coagulation and flocculation.

2.3.2 Ferric Sulfate as Coagulant and the Chemical Reactions

Ferric sulfate is a reddish-gray [commercially called Ferric-flock, $Fe_2(SO_4)_3.3H_2O$] or grayish-white [commercially called Ferriclear, $Fe(SO_4)_3 \cdot 2H_2O$] granular material. On addition of ferric sulfate to water the following reactions take place:

$$Fe_2(SO_4)_3 + 3Ca(HCO_3)_2 \rightarrow 2Al(OH)_3 \downarrow + 3CaSO_4 \downarrow + 6CO_2 \quad (2.13)$$

Like alum, ferric sulfate also requires alkalinity in the water in order to form the flock particles ferric hydroxide [$Fe(OH)_3$]. When natural alkalinity is not sufficient, alkaline chemicals (such as soluble salts containing HCO_3^-, CO^{-2} and OH^- ion) should be added. Compared to alum, ferric sulfate has some advantages. For example, the flock particles [$Fe(OH)_3$] of ferric hydroxides have much higher density than alum flocks and are more easily or quickly removed by sedimentation. The desired chemical reactions involving ferric sulfate take place favorably over a much wider range of pH (3.5–9.0). However, there are also disadvantages. For example, ferric sulfate can stain equipment and is difficult to dissolve, and its solution is corrosive and it may react with organics to form soluble iron (Fe^{+2}).

The best ferric-sulfate dosage to use in any coagulation application must be determined on a case-by-case basis based using jar tests as the optimum dosages in water treatment vary between 5 and 50 mg/L.

2.4 CHEMICAL NEUTRALIZATION

Chemical neutralization is done by adding acids or alkali to a water stock prior to the final treatment. This is particularly essential for protectingdownstream equipment from corrosion attack. Many downstream chemical or biological treatments need pH adjustment for optimum performance. Microbial treatment almost always demands an optimum pH. For many water-treatment plants, neutralization and equalization of the wastewater streams emerging from different units is required prior to their introduction to the main treatment units. Such neutralization and equalization address the flow and quality fluctuations of the wastewater streams, thereby preventing shock loads to the sensitive downstream treatment units (chemical or biological).

2.5 CHEMICAL OXIDATION

Oxygen (O_2), O_3, Cl_2, $KMnO_4$, and H_2O_2 are some of the strong oxidizing agents used in water treatment. These oxidizing agents improve water by removing dissolved odor- producing gases by converting soluble forms of some metals such as manganese and iron into insoluble forms thereby facilitating their removal. Some of these oxidizing agents help in disinfection of water as well. We have already seen how O_2 helps in removing odor-causing gases (H_2S), iron, and manganese from water. In this section, we will see how chlorine removes iron and manganese (the two most troublesome constituents affecting water quality). Iron and manganese are found predominantly in groundwater supplies and occasionally in the anaerobic bottom waters of deep lakes. In nature, iron and manganese occur in stable forms such ferric (Fe^{+3}) form and the manganic (Mn^{4+}), which are insoluble in water. The insoluble forms get converted into soluble forms under anaerobic conditions that may develop in groundwater aquifers as well as at the bottom of deepwater reservoirs or lakes. Thus the mechanism of removal of naturally occurring iron and manganese from water entails their conversion into insoluble forms through their oxidation using oxidizing agents like chlorine. Once oxidized by a strong oxidizing agent like chlorine these metals settle out as insoluble salts.

2.5.1 Oxidation Reactions of Chlorine During Iron Removal

Chlorine can very easily remove iron when present in bicarbonate form as ferric hydroxide precipitates through the following reaction:

$$2Fe(HCO_3)_2 + Cl_2 + Ca(HCO_3) \rightarrow 2Fe(OH)_3\downarrow + CaCl_2 + 6CO_2 \tag{2.14}$$

This instantaneous reaction occurs best at pH 7. However, a pH range of 4–10 allows the reaction at reasonably fast rate where removal of every milligram of iron per liter of water needs 0.64 mg/L of chlorine. The ferric-hydroxide precipitate obtained in the reaction is easily recognizable as fluffy, rust-colored sediment. The calcium bicarbonate [$Ca(HCO_3)_2$] in the reaction represents the bicarbonate alkalinity of water. Both free and combined chlorine residual can cause the precipitation of iron as ferric hydroxide.

2.5.2 Oxidation Reactions of Chlorine During Manganese Removal

Manganese removal reactions using chlorine proceed almost in the same way as in the case of iron. If manganese is present as manganoussulfate ($MnSO_4$), the corresponding reaction proceeds as follows:

$$MnSO_4 + Cl_2 + 4NaOH \rightarrow MnO_2\downarrow + NaCl + 2H_2O \tag{2.15}$$

Chlorine and added sodium hydroxide oxidize manganese producing the precipitate manganese dioxide (MnO_2). For removal of every milligram of manganese, 1.3 mg/L of free chlorine needs to be present in the reaction medium. However, chloramines (combined chlorine residuals) have little effect on manganese.

This reaction can take place in the pH range of 6–10 where added sodium hydroxide (NaOH) causes the hydroxyl alkalinity in the sample. However, the reaction rate varies sharply between acidic pH (6.0) and alkaline pH (10) involving just a few minutes (at pH 10) to even 12 hours at pH 6.0 for the same conversion.

2.5.3 Oxidation Reactions of Potassium Permanganate During Iron and Manganese Removal

Potassium permanganate ($KMnO_4$) is a powerful oxidizing agent and can cause rapid precipitation out of both iron and manganese present in water through the following reactions:

$$\begin{aligned} Fe(HCO_3)_2 + KMnO_4 + H_2O + 2H^+ \rightarrow MnO_2\downarrow + Fe(OH)_3\downarrow \\ + KHCO_3 + H_2CO_3 \end{aligned} \quad (2.16)$$

$$3Mn(HCO_3)_2 + 2KMnO_4 + 2H_2O \rightarrow 5MnO_2\downarrow + 2KHCO_3 + 4H_2CO_3 \quad (2.17)$$

$$3MnSO_4 + 2KMnO_4 + 2H_2O \rightarrow 5MnO_2\downarrow + K_2SO_4 + 2H_2SO_4 \quad (2.18)$$

The normal dose of $KMnO_4$ is about 0.6 mg/L for removal of 1 mg/L of iron (Fe^{2+}). For removal of 1 mg/L of manganese (Mn^{2+}) 2.5 mg/L of $KMnO_4$ is required.

2.6 CHEMICAL PRECIPITATION

The principles of chemical precipitation are widely exploited in the removal of heavy metals and hardness-causing metals. Removal of hardness of water using lime [Ca(OH)2] or soda ash[Na_2CO_3] is called water softening. Such metal-removal reactions are favored at high pH in the range of 10–11. Removal of iron and manganese also takes place during softening at high pH (10–11), although iron in this case is eliminated in the form of ferrous hydroxide $Fe(OH)_2$, instead of the familiar ferric hydroxide $Fe(OH)_3$ form.

2.6.1 Hardness of Water and Softening by Chemical Precipitation

Hardness of water is caused by the presence of multivalent ions (e.g., Ca^{2+}, Mg^{2+}) in high concentration. Aluminum, strontium, zinc, and iron may also contribute to hardness but normally these metals are not present in water in high enough concentration to cause hardness. Soft water may also contain multivalent ions (Ca^{2+}, Mg^{2+}) but in small concentration. Such hardness may be temporary when multivalent ions are present as bicarbonates (calcium bicarbonate, magnesium bicarbonate) as this hardness can be easily removed by boiling water, resulting in precipitation of the ions as carbonates. On the other hand, permanent hardness is caused by the presence of sulfate or chlorides of multivalent ions like calcium and magnesium that do not precipitate out on boiling the water. As such, hardness does not have any major health impact. The World Health Organization (WHO) suggests 40–80 mg/L Ca and 20–30 mg/L Mg in drinking water. Hard water in domestic use results in the formation of soap scum. Soap scum is a solid-white precipitate of calcium stearate produced from the reaction of hardness-causing ions with sodium stearate, the main component of soap. Hardness leads to more consumption of soap. However, scum does not form with synthetic detergents. Soap-scum formation occursdue to this reaction:

$$2\, C_{17}H_{35}COO^{-}(aq) + Ca^{2+}(aq) \rightarrow (C_{17}H_{35}COO)_2Ca(S) \tag{2.19}$$

The other problems that arise from use of hard water is the formation of scale, which can clog pipelines and damage heater and boiler. The scale may consist of $CaCO_3$, $Mg\ (OH)_2$, $CaSO_4$, and $MgCO_3$. Calcium (Ca^{2+}) and magnesium (Mg^{2+}) are the two major contributors to water hardness.

In general, there are broad two types of water hardness:

1. Carbonate hardness, caused primarily by calcium bicarbonate;
2. Noncarbonated hardness, caused primarily by calcium chloride [$CaCl_2$], magnesium chloride [$MgCl_2$], and magnesium sulfate [$MgSO_4$].

Lime is used to remove carbonate hardness and soda ash is suggested for removal of noncarbonated hardness.

2.6.2 Chemical Precipitation During Removal of Carbonate Hardness

During treatment for removal of carbonate hardness, added lime reacts as follows:

$$Ca(HCO_3)_2 + Ca(OH)_2 \rightarrow 2CaCO_3\downarrow + 2H_2O \tag{2.20}$$

Magnesium bicarbonate and carbonate are removed following these reactions:

$$Mg(HCO_3)_2 + Ca(OH)_2 \rightarrow CaCO_3\downarrow + MgCO_3 + 2H_2O \tag{2.21}$$

$$MgCO_3 + Ca(OH)_2 \rightarrow CaCO_3\downarrow + Mg(OH)_2\downarrow \tag{2.22}$$

From Eqs. (2.15), (2.16), and (2.17), we can see that the requirement of lime for removal of magnesium bicarbonate hardness is double the requirement for calcium bicarbonate hardness as $MgCO_3$ formed in the first reaction (2.16) is soluble unlike $CaCO_3$ of Eq. (2.15). In all cases, the insoluble calcium carbonate precipitates out.

2.6.3 Chemical Precipitation During Removal of Noncarbonated Hardness

When noncarbonated hardness is due to magnesium compounds, both lime ($Ca(OH)_2$ and soda ash (Na_2CO_3) need to be added as follows:

$$MgCl_2 + Ca(OH)_2 \rightarrow Mg(OH)_2\downarrow + CaCl_2 \tag{2.23}$$

$$CaCl_2 + Na_2CO_3 \rightarrow CaCO_3\downarrow + 2NaCl \tag{2.24}$$

$$MgSO_4 + Ca(OH)_2 \rightarrow Mg(OH)_2\downarrow + CaSO_4 \tag{2.25}$$

$$CaSO_4 + Na_2CO_3 \rightarrow CaCO_3\downarrow + 2NaSO_4 \tag{2.26}$$

The removal of magnesium noncarbonated hardness with lime forms calcium noncarbonated hardness, which are subsequently removed with soda ash.

2.6.4 Removal of Dissolved CO_2 Prior to Lime Softening

Prior to lime softening, dissolved CO_2must be removed which otherwise will lead to higher consumption of lime through the following reaction:

$$CO_2 + Ca(OH)_2 \rightarrow CaCO_3\downarrow + H_2O \tag{2.27}$$

To take care of this dissolved CO_2, excess lime should be added. All chemical precipitation reactions described here take place at high pH. For example, during removal of calcium carbonate hardness, a pH of 9.4 is necessary and for magnesium hydroxide precipitation the required pH is 10.6, which are maintained through addition of lime.

2.6.5 Addition of CO_2 After Lime Softening

Normally removal of metals from water requires high pH. For example, for removal of magnesium, a high pH has to be maintained during precipitation by adding more lime. A considerable amount of this lime remains in the water and can be removed after coagulation–flocculation and before filtration using CO_2 as follows:

$$CO_2 + Ca(OH)_2 + H_2O \rightarrow CaCO_3\downarrow + H_2O \tag{2.28}$$

This form of recarbonation is performed after coagulation and flocculation but before final settling. Carbon dioxide reacts with the excess lime. However, the correct amount of CO_2 needs to be introduced as excess CO_2 may lead to hardness through the following reaction:

$$Ca(OH)_2 + 2CO_2 \rightarrow Ca(HCO_3)_2 \quad (2.29)$$

2.6.6 Recarbonation After Water Softening: Removal of Excess $CaCO_3$

Lime softening makes water supersaturated with calcium carbonate with a pH of above 10.0. The very finely suspended calcium carbonate has high potential for deposition of scale on filter media and water distribution system piping. Recarbonation can prevent such scale deposition through the following reaction when carbon dioxide is bubbled into the water, lowering the pH and removing calcium carbonate:

$$CO_2 + CaCO_3 + H_2O \rightarrow Ca(HCO_3)_2 \quad (2.30)$$

This type of recarbonation should be done after the coagulated and flocculated waters are settled but before their filtration to prevent the suspended $CaCO_3$ from being carried out of the sedimentation basin and cementing the filter media.

2.7 ION EXCHANGE

While the major use of ion exchange is in water softening, it can also be used in general for selective removal any ion by another ion. The main material involved is ion-exchange bed in which selected cation or anion may remain embedded.

A cation-exchange material can be represented by the general symbol $(R^{-n})_r(C^{+m})_c$ where R^{-n} is the host and C^{+m} is the exchangeable cation. Similarly, anion-exchange material can be represented by $(R^{+O})(A^{-p})_a$ where R^{+O} is the host and A^{-p} is the exchangeable anion. The capability of an ion of displacing another from the exchange bed depends on the relative positions of the ions in the displacement series. For example, in the displacement series of cations, the positions of Ca^{2+}, Mg^{2+} and Na^{1+} are 5th, 6th and 9th respectively. With a higher position in the series, Ca^{2+} and Mg^{2+} can displace Na^{1+} from the cation-exchange bed. This implies that water containing hardness ions Ca^{2+} and Mg^{2+} can be softened by using cation- exchange bed with embedded displaceable Na^{1+} ions. Similarly, I^{-}, Cl^{1-}, F^{1-} being in the 7th, 8th and

Table 2.2 Displacement series of selected ions

Position	Cation	Anion
1	La^{3+}	SO_4^{2-}
2	Y^{3+}	CrO_4^{2-}
3	Ba^{2+}	NO_3^{2-}
4	Sr^{2+}	AsO_4^{3-}
5	Ca^{2+}	PO_4^{3-}
6	Mg^{2+}	MOO_4^{2-}
7	Rb+	I^{1-}
8	K^+	Cl^{1-}
9	Na^{1+}	F^{1-}
10	Li^+	OH^-
11	H^+	–

9th positions in the anion-exchange series, can be removed from water by using anion- exchange with OH^- embedded in the exchange material. A few cations and anions are shown in Table 2.2 with their relative positions in the displacement series.

Normally multivalent cations like Ca^{2+} and Mg^{2+}cause hardness of water. Ion-exchange materials can exchange such hardness-causing ions by nonhardness-causing cations such as sodium. Both natural as well as synthetic zeolites are widely used in water softening. Green sand or gluconate are natural zeolites while synthetic zeolites may be organic, inorganic, or resin types. Zeolites are compounds of aluminum, silica, and sodium where the sodium is the base. Artificial or synthetic zeolites are manufactured by mixing in definite proportions of feldspar, soda, and clay and fusing in furnace. After fusion, the material is cooled and crushed. Since zeolites have strong affinity for bivalent ions, when hard water is passed through a zeolite bed, the hardness-causing ions like Ca^{2+} and Mg^{2+} get embedded in the zeolite on replacement of the sodium ions. Zeolite material fails in water softening when concentrations of iron and manganese exceed 0.5 mg/L and total hardness of water is more than 850 mg/L.

Synthetic polystyrene resins are the most widely used cation-exchange materials. These resinous ion-exchange materials (normally in the shape of spheres) contain sodium ions that are released into the water in exchange for hardness-causing ions like calcium and magnesium. Properly designed and operated ion-exchange bed has the potential for complete removal of all hardness-causing ions, be it carbonate or noncarbonate.

The following two reactions show how a sodium-based cation-exchange resin can remove calcium and magnesium hardness of water:

$$Ca^{+2} + Na_2X \rightarrow CaX + 2Na^{+} \tag{2.31}$$

$$Mg^{+2} + Na_2X \rightarrow MgX + 2Na^{+} \tag{2.32}$$

"X" stands for the exchange resin of the material Na_2X, which though not a chemical compound somewhat behaves like one. The sodium cations are released into the water just as the sodium in Na_2SO_4 would be released on dissolution of the compound in water. Insoluble in water the resin "X" acts similar to an anion such as SO_4^{2-}, but does not react chemically as SO_4^{2-}. Rather, it functions more like a "parking lot" for exchangeable cations. The terms CaX and MgX represent the same resin after the exchange has been made. Calcium and magnesium hardness ions are removed from the water and get embedded onto the surface of the resin while replacing sodium ions. One hardness ion (Mg^{2+} or Ca^{2+}) with a charge of +2 is exchanged for two sodium ions, each having a charge of +1, implying that a total charge of +2 is exchanged for a +2 charge.

The following equations explain the removal of hardness-causing ions:

$$Ca(HCO_3)_2 + Na_2X \rightarrow CaX + 2NaHCO_3 \tag{2.33}$$

$$CaSO_4 + Na_2X \rightarrow CaX + 2Na_2SO_4 \tag{2.34}$$

$$CaCl_2 + Na_2X \rightarrow CaX + 2NaCl \tag{2.35}$$

$$Mg(HCO_3)_2 + Na_2X \rightarrow MgX + 2NaHCO_3 \tag{2.36}$$

$$MgSO_4 + Na_2X \rightarrow MgX + 2NaSO_4 \tag{2.37}$$

$$MgCl_2 + Na_2X \rightarrow MgX + 2NaCl_2 \tag{2.38}$$

The equation shows that anions originally associated with the hardness-causing cation stay in the softened water being attached to the sodium cations released by the resin. Hence, the softened water contains sodium bicarbonate ($NaCHO_3$), sodium sulfate (Na_2SO_4), and sodium chloride. These compounds do not cause taste problems and can remove both carbonate and noncarbonated hardness. Both are removed by the same exchange reaction.

2.7.1 Regeneration of Ion-Exchange Material

After some time, the exchange bed loses its power to soften water further. At this point, regeneration of the resin bed must be done to restore its softening capacity. This implies reversing the exchange process that forces out the embedded hardness cations such as calcium and magnesium from the resin while bringing back the sodium ions onto the bed. Regeneration by such reverse exchange is achieved by passing a strong brine solution through the resin bed. The associated ion-exchange regeneration reactions take place as follows:

$$CaX + 2NaCl \rightarrow CaCl_2 + Na_2X \quad (2.39)$$

$$MgX + 2NaCl \rightarrow MgCl_2 + Na_2X \quad (2.40)$$

As sodium is taken back into the exchange resin, the resin becomes ready again to be used for the softening while calcium and magnesium, released during regeneration, are carried to disposal by the spent brine solution. Properly maintained and operated, cation exchange removes all hardness.

Zero Hardness is not Desirable

Water with zero hardness is corrosive, so the final step in ion-exchange softening process is to mix a portion of the unsoftened water with the softened effluent to provide water that is still relatively soft, but that contains enough hardness to be noncorrosive (stable) Fig. 2.5.

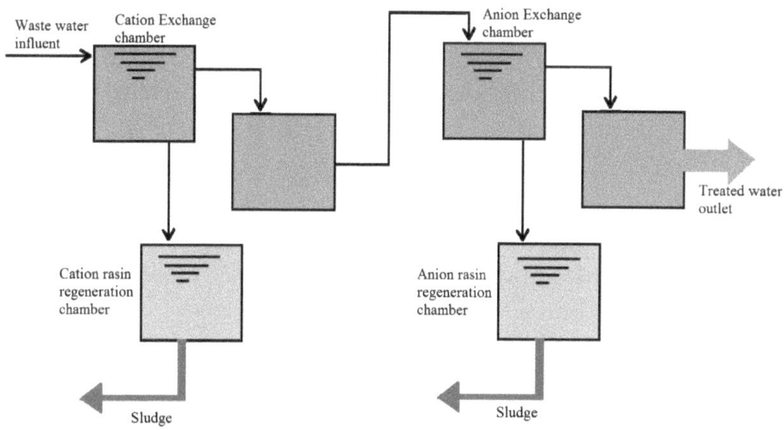

Figure 2.5 Water softening plant based on ion exchange-based technology.

2.8 DISINFECTION OF WATER

Disinfection refers to rendering pathogens ineffective and harmless while sterilization refers to killing pathogens. Disinfection of water is done following the coliform rule, which states that there should not be any *E. coli* present in a random water sample. *E. coli* represents pathogens. Disinfection of drinking water may be done by following any of the physical and chemical methods described as follows:

1. Physical methods consists of application of
   a. heat;
   b. ultrasonic wave;
   c. Electron beam or gamma irradiation;
   d. UV irradiation.
2. Chemical methods may include any of the oxidizing agents listed below:
   a. $KMnO_4$;
   b. Ozone (O_3);
   c. Halogens such as chlorine, bromine (used in swimming pool), iodine (used in tablet form);
   d. Acids and alkalis (highly acidic or alkaline conditions can make survival of pathogens difficult).

The use of electron beam or gamma irradiation is limited. The most widely adopted disinfection methods are based on ozone treatment, UV irradiation, or chlorine-based treatment.

2.8.1 Technology-based on Ozone Treatment

Ozone can effectively render pathogens ineffective. The typical ozone dose required for effective disinfection is 1.0–5.3 kg/1000 m^3 of water. Ozone-based disinfection is ecofriendly and leaves no harmful byproducts after reaction. But ozone has no residual effects to deal with microbial contamination during transport of water from one location to another, and needs to be used in treating water close to the point of use. Power consumption in ozone production is high, being of the order of 10–20 kW/kg ozone. Such ozone is produced by passing either pure oxygen or dry and clean air through high-voltage electric field.

2.8.2 Technology Based on UV Radiation

UV irradiation at 254–260 nm wavelength range is considered most effective for disinfection. Such UV light is readily absorbed by the nucleic acid

Table 2.3 Effective UV radiation dose for selected pathogens

Pathogens	**Essential Dose ($mW/\ s.cm^2$)**
E. coli	3.2
Mycobacterium tuberculosis	6.0
Salmonella typhii	2.1
Shigelladysenteriae	2.2
Bacillus anthracis	4.5
Influenza virus	3.6
Polio virus	7.5
Rotavirus	11.3

of microorganisms causing damage to the cells that fail to replicate and eventually die. UV radiation is generated using low-pressure mercury vapors lamps that may be suspended in stainless-steel channel. UV lamps usually last 3–4 years. The power input of 30 $\mu W/cm^2$ is normally applied on thin sheets of turbidity-free water. The dose of radiation is exponentially related to microbial decay or destruction. For example, if a particular dose has the capability of 98% reduction of microbial concentration then doubling the dose will reduce 98.9% microbial concentration. The lethal doses for various pathogens vary widely as shown in Table 2.3.

Like ozone, UV radiation has no residual effects to deal with microbial contamination during long-distance transport of water. Both ozone treatment and UV irradiation need to be done at the point of water use and are not considered economical for large-scale water treatment.

2.8.3 Chlorination Technology

Disinfection of water by chlorine or its compounds is widely practiced because of the ease of use, low cost, effectiveness, and residual effects of chlorine. However, the disadvantage of chlorine-based disinfection is the production of harmful disinfection byproducts that in many cases are either carcinogens or suspected carcinogens. Chlorine may be used in three forms, namely Cl_2, $Ca(OCl)_2$, and NaOCl, for disinfection of water.

Liquid Chlorine

Chlorine is available as both a liquid and gas. Liquid chlorine is a compressed amber-colored product containing 99.5% pure chlorine. At room temperature and pressure, the liquid can expand to approximately 500 times its volume to become a gas.

Chlorine Gas

Chlorine gas is greenish-yellow in color and is visible at high concentrations. It is highly toxic, even at concentrations as low as 0.1% by volume. Plain chlorine mixed with water produces hypochlorous acid:

$$Cl_2 + H_2O \rightarrow HOCl + HCl \tag{2.41}$$

Chlorine liquid or gas is not spontaneously combustible, but can support combustion. Dry chlorine is not corrosive but moist chlorine can be extremely corrosive. Chlorine liquid is approximately 1.5 times heavier than water while gaseous chlorine is approximately 2.5 times heavier than air.

Calcium Hypochlorite $Ca(OCl)_2$

Calcium hypochlorite is a dry, white, or yellow-white granular material available in tablet form. The granular material contains 65% available chlorine by weight.

$$Ca(OCl)_2 + 2H_2O \rightarrow 2HOCl + Ca(OH)_2 \tag{2.42}$$

Calcium hypochlorite needs to be stored carefully to avoid contact with easily oxidized organic material.

Sodium Hypochlorite (NaOCl)

Sodium hypochlorite is a clear, greenish-yellow liquid chlorine solution normally used in bleaching. Normal household bleach is an example of sodium hypochlorite. It contain 5% available chlorine, which is equivalent to 0.42 lb/gal. Commercial bleaches are stronger, containing 9–15% available chlorine in various NaOCl strengths.

Sodium hypochlorite reacts with water to produce HOCl as follows:

$$NaOCl + 2H_2O \rightarrow HOCl + NaOH \tag{2.43}$$

Sodium-hypochlorite solution needs to be stored in rubber-lined steel vessels Fig. 2.6 shows a chlorine-based disinfection unit.

2.8.4 Mechanisms of Improvement of Water Quality by Chlorine-based Treatment Technology

When introduced to water, chlorine or its compounds may react directly with the water or impurities such as ammonia, organic compounds, and

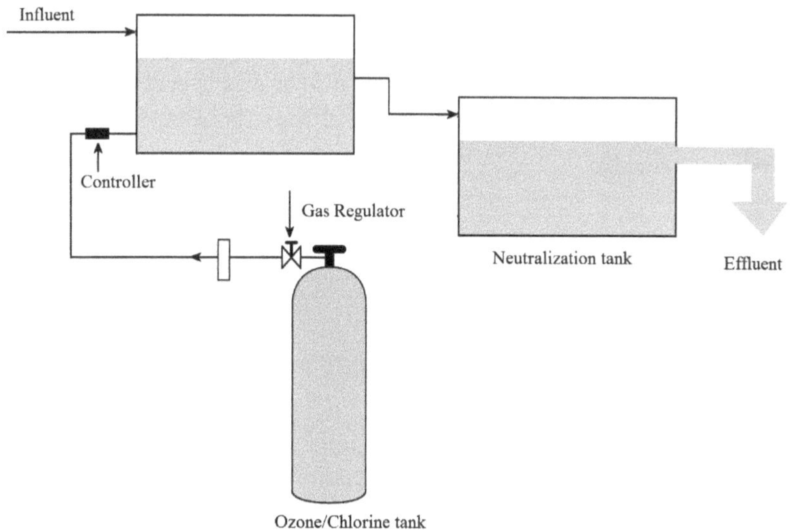

Figure 2.6 Typical chlorine-based disinfection scheme.

compounds of sulfur, manganese, and iron. These reactions and their products depend on pH, temperature, concentration, and impurities. While most of the reactions with water, ammonia, compounds of sulfur, manganese, and iron contribute to improvement of quality of water, some of the reactions of chlorine with organic compounds have the potential to form harmful disinfection byproducts (DBPs) that are either carcinogens or suspected carcinogens. We now describe the reactions through which improvement of quality of water occurs and under what reaction conditions. As chlorine or chlorine-based compounds are introduced to water they may react directly with water as shown in Eqs. (2.44)–(2.46).

Water-Quality Improvement through Disinfection by Chlorine and its Compounds

The most desired reaction of chlorine with water as shown in Eq. (2.44) produces strongly effective disinfectant hypochlorous acid (HOCl):

$$Cl_2 + H_2O \rightarrow HOCl + HCl \tag{2.44}$$

$$Ca(OCl)_2 \rightarrow Ca^{+2} + 2\ OCl^- \tag{2.45}$$

$$NaOCl \rightarrow Na^+ + OCl^- \tag{2.46}$$

HOCl is one of the two freely available chlorine residual forms and is considered the most effective form of chlorine-based disinfectant because of the ease with which HOCl penetrates into and kills bacteria. However, subsequent dissociation of a part of HOCl into the ions H^+ and OCl^- leads to loss of disinfection potential of total applied chlorine as ions such as OCl^- and H^+ do not have the same strong disinfection capability as HOCl.

$$HOCl \rightleftarrows H^+ + OCl^- \tag{2.47}$$

Hydrolysis constant $K_h = [HOCl][H^+][Cl^-]/[Cl_2] = 4.5 \times 10^{-4}\ (mol/L)^2$ at 25°C.

Ionization constant $K_i = [H^+][OCl^-]/[HOCl]$

pH plays a significant role in dissociation of HOCl, which in turn determines the effectiveness of disinfection. For pH <3, chlorine remains as Cl_2 and in the pH range 3–5, it remains as HOCl. In the pH range 5–10, chlorine remains as HOCl and OCl^- and beyond a pH of 10.0, chlorine remains in the form $OCl^{-\cdot}$ of all the different chloride species, HOCl is the strongest disinfectant.

Hydrogen and chlorine ions have no disinfection properties. The effectiveness of disinfection by chlorination depends on pH, temperature, contact time, concentration, and presence of impurities in water.

As HOCl has strong disinfection property while OCl^- is a weak disinfectant, higher effectiveness of chlorine-based disinfectant requires that the reaction conditions be so maintained that formation of HOCl is favored while that of OCl^- (being weak) is discouraged.

Eq. (2.47) indicates that high concentration of H^+ on the right side will push the equilibrium of the reaction (2.47) to the left ensuring high concentration of HOCl, which is the most effective disinfectant in the present case. This means that low pH will favor high HOCl reserve in the medium, which suggests that chlorine-based disinfection should be done at low pH. It has been observed that at around 5.5 pH and at temperature ranging from 0 to 20°C, concentration of HOCl reaches almost 100%, implying high effectiveness.

High pH values favor the formation of OCl^-, the less effective free residual form. As pH increases from 7.0 to 10.7, the OCl^- begins to dominate and the time required for the free residual to effectively disinfect increases. In the pH range 7.0–8.5 this effect is less strong but beyond a pH greater than 8.5 disinfection time drastically increases.

Improvement in Water Quality through Reaction of Chlorine with Iron (Fe)

Iron is a human nutritional element where depending on body weight, physiological conditions, age, and gender 10–50 mg/L of iron may be necessary daily. Maximum permissible daily intake is 0.8 mg/kg of body weight. However, drinking water from underground aquifers may contain 1–3 mg/L and thus excess iron is harmful and needs to be removed.

Iron is often found in groundwater usually in the form of ferrous bicarbonate [$Fe(HCO_3)_2$]. This iron can be removed by chlorination through the reaction (2.48) where iron is precipitated out from drinking water as ferric hydroxide.

$$2Fe(HCO_3)_2 + Cl_2 + Ca(HCO_3)_2 \rightarrow 2Fe(OH)_3\downarrow + CaCl_2 + 6CO_2 \quad (2.48)$$

Precipitation of ferric hydroxide [$Fe(OH)_3$] takes place instantly, forming a fluffy, rust-colored sludge. The calcium bicarbonate [$Ca(HCO_3)_2$] involved in reaction (2.48) represents the alkalinity in the water. Iron may be removed using either the free or combined forms of chlorine residual. Approximately 0.64 mg/L of chlorine is required to remove 1 mg/L of iron.

Improvement in Water Quality through Reaction with Manganese (Mn)

The maximum permissible limit of manganese in drinking water is 0.5 mg/L as set under the guidelines of the World Health Organization (WHO). Just as the presence of excess iron is indicated by red-colorwater, excess manganese may produce brown or black water. Manganese may normally be present as a salt such as sulfate salt and chlorine reacts with manganese sulfate to precipitate manganese as MnO_2 shown in Eq. (2.49):

$$MnSO_4 + Cl_2 + NaOH \rightarrow MnO_2\ 2NaCl + Na_2SO_4 + 2H_2O \quad (2.49)$$

The reaction completes in about 2 hours.

Improvement in Water Quality through Reaction with Hydrogen Sulfide

Hydrogen sulfide (H_2S) is normally found in groundwater, and rarely, in surface water. The presence of H_2S in water even in concentrations as low as 0.05 mg/L can cause an unpleasant taste. If inhaled with air in more than 0.1–0.2% by volume, it may be fatal. At high concentration above 4.3% by volume in air, H_2S is flammable. H_2S can be removed from water with chlorine by the following reactions:

$$Cl_2 + H_2S \rightarrow 2HCl + S \quad (2.50)$$

$$4Cl + H_2S + 4H_2O \rightarrow 8HCl + H_2SO_4 \quad (2.51)$$

Sulfur is formed as colloidal type fine particles that need to completely separate out from water by coagulation followed by filtration. Reaction (2.50) turns the water milky-blue. The necessary chlorine dose is 2.2 mg/L for 1 mg/L H_2S.

2.8.5 Reaction Conditions for Improvement of Water Quality by Chlorination

Effect of Temperature

At high temperature, microbes are killed faster but main formation of HOCl is not favored at high temperature. Retention of chlorine-based residuals in water is also not favored at high temperature. However, the overall effects of high temperature are positive for disinfection as rates of chemical and biochemical reactions increase at high temperature offsetting the negative effects.

Effect of Contact Time

The destruction of organisms is directly related to the contact time T and concentration of chlorine. Over a longer contact time, the chlorine concentration needed to accomplish the same killing is lower. Similarly, at higher dosage of disinfectant, the contact time needed for accomplishing the same degree of disinfection decreases.

Effect of Impurities

During chlorination undesirable side reactions of chlorine with ammonia, iron, manganese, hydrogen sulphide, and dissolved organic materials may reduce the effectiveness of chlorine-based disinfection.

Ammonia in water may be due to decay of natural vegetation or discharge of wastewater after domestic or industrial use. Chlorine reacts with ammonia to form chloramines, compounds containing both nitrogen and chloride ions as shown in the following equations:

$$NH_3 + HOCl \rightarrow NH_2Cl + H_2O \quad (2.52)$$

$$NH_2Cl + HOCl \rightarrow NHCl_2 + H_2O \quad (2.53)$$

$$NHCl_2 + HOCl \rightarrow NCl_3 + H_2O \quad (2.54)$$

Formation of mono, di, or other forms of chloramine compounds depends on the pH of water and on the concentration of ammonia. Monochloramine and dichloramine are far less effective than free chlorine or HOCl. During chlorination, free available chlorine reacts rapidly with any oxidizable substance. If water contains natural or added ammonia, the free available chlorine will react to form combined available chlorine, which is less effective than free chlorine. Monochloramine (NH_2Cl) has effectiveness of only 1/150 times the effectiveness of HOCl, whereas the strength of hypochlorite ion (OCl^-) is only 1/100 times that of HOCl.

2.8.6 Strength of Chlorine Disinfection

The strength of a chlorine-based disinfectant is measured in terms of available chlorine, which is defined as the ratio of mass of chlorine to the mass of the disinfectant with the same unit of oxidizing power as chlorine. The unit of oxidizing power of chlorine may be computed from the ionization reaction (2.48) as follows:

$$Cl_2 + 2e^- \leftrightarrow 2Cl^- \tag{2.55}$$

From this equation, the oxidizing power of chlorine in terms of 1 mol of electron may be computed as:

$$\text{Unit of oxidizing power of chlorine} = \text{Mass of chlorine}/2 = 35.5$$

Let's look at another example of chlorine-based disinfectant NaOCl that undergoes the ionization reaction:

$$NaOCl + 2e^- + 2H^+ \leftrightarrow Cl^- + Na^+ + H_2O \tag{2.56}$$

The unit of disinfection power of NaOCl from Eq. (2.49) is computed as mass of NaOCl/2 = 37.24.

The available chlorine in NaOCl equals the ratio of mass of chlorine to the mass of NaOCl that has the same oxidizing power as chlorine and may be computed as:

$$35.5/37.24 = 0.95 \text{ or } 95\%.$$

This implies that NaOCl is 95% as effective as chlorine itself.

2.8.7 Chlorine Residuals

When chlorine is added to water for disinfection, it reacts with ammonia if present forming chloramine-based residuals such as monochloramine (NH_2Cl), dichloramine ($NHCl_2$), and nitrogen trichloride (NCl_3).

These residuals are collectively called combined available chlorine. Chlorine also forms residuals like Cl^-, OCl^-, and HOCl where the total amount of these residuals (Cl^-, OCl^-, and HOCl) is called free available chlorine. The sum of the combined available chlorine and free available chlorine is called the total residual chlorine (TRC).

Residual chlorine plays a significant role in ensuring protection of water from any microbial contamination during pipeline transport of potable water from treatment plants to user points. However, concentration and effectiveness of residual chlorine depends on the chlorine dose applied, as described in the next section.

The chlorine dose-residual curve as shown in Fig. 2.7 indicates the consumption of chlorine and formation of chlorine residuals through a series of reactions represented by Eqs. (2.44)–(2.54) in Sections 2.7.4 and 2.7.5.

In Fig. 2.7, zone OA represents the chlorine-destruction zone when almost all applied chlorine is used for useful reactions (2.48–2.51) through which iron (Fe^{+2}), manganese (Mn^{+2}), and hydrogen sulphide (H_2S) are separated from water thereby improving the quality. Thus in zone OA, no residual chlorine is observed. Zone AB represents the zone of formation of combined chlorine residuals comprising organo-chloro compounds like chloramines such as monochloramine, dichloramine, nitrogen trichloride, and disinfection byproducts like trihalomethanes,

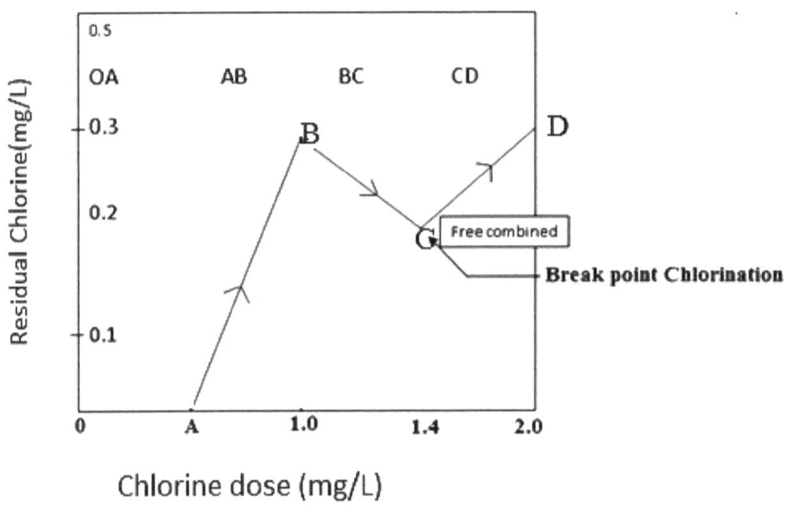

Figure 2.7 Chlorine dose-residual curve.

bromochloromethane, etc., through reactions (2.50)–(2.52). Zone BC represents the formation of free chlorine residuals through reactions (2.42)–(2.45). Zone CD represents formation and stockpiling of free and combined chlorine residuals as chlorine is continually added to water. Point C in the chlorine dose-residual curve represents the breakpoint chlorination dose at which all chloramines get decomposed to nitrogen trichloride, N_2, or N_2O, and free chlorine residuals (Cl^-, OCl^-, HOCl) start increasing sharply. Beyond this breakpoint chlorination, chlorine residuals in drinking water remain as free and combined residuals. The breakpoint chlorination dose is the dose of chlorine that satisfies all chloride demand of water so that further addition of chlorine will result only in an increase in free residual chlorine. The safe dosage of chlorine is 1–2 mg/L but during an outbreak of any epidemic it has to be increased by several fold (e.g., 10–15 mg/L) and is called superchlorination.

2.8.8 Conventional Technology of a Typical Municipal Water-treatment Plant

The water-treatment scheme using conventional technology is presented in Fig. 2.8. In a municipal water-treatment plant using conventional technology, chlorine-based disinfection plays a significant role in low-cost treatment of the millions of gallons of water required in a city. In most cases, the water is drawn from surface sources like river bodies by means of low-pressure lifting pumps via rough screening arrangement that arrests

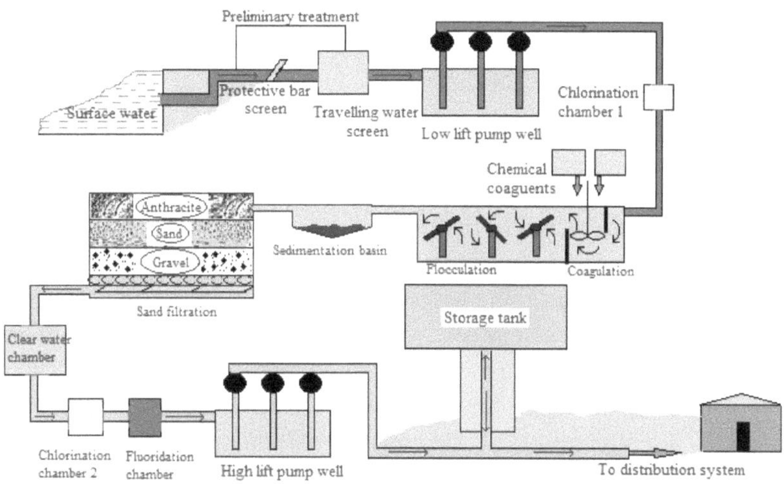

Figure 2.8 Typical municipal water-treatment plant using conventional technology.

floating bodies from water. After some primary chlorination that improves the water quality through a series of reactions as described in Section 2.7.4, this water is subjected to coagulation—flocculation—chemical precipitation with the addition of chemical reagents like alum. Such chemical precipitation removes hardness-causing substances and suspends other impurities in the settling tank. Overflow of clear water from the sedimentation unit is subsequently passed to the sand-filtration unit where microfiltration takes place. After final chlorination and in some cases, fluoridation (in essential doses for dental health), this water is pumped by high-pressure pumps to municipal overhead storage tanks from which households get supply by gravity flow. This water is not used entirely for the purpose of drinking but is also used for industrial and domestic use. Thus chlorine treatment is considered justified in view of its effectiveness, residual effects even after long-distance transport, and low cost to treatenormous quantities of water. However, the possible formation of disinfection byproducts (DBPs) many of which are carcinogens or suspected carcinogen very much remains. The best option maybe nanofiltration of only the amount of water expected to be used for drinking with prior ultrafiltration in small domestic treatment units to avoid drinking water from being contaminated with DBPs.

2.8.9 Harmful Effects of Chlorine-based Treatment Technology

The reactions of natural organics such as fulvic and humic acids with chlorine may produce trihalomethanes such as chloroform and bromochloromethane, which are carcinogens. While such trihalomethanes are found in high concentrations in water, a large variety of other DBPs may be produced in drinking water during chlorine-based treatment such as haloacetic acids, halonitriles, haloaldehydes, and chlorophenols. The maximum permissible limits of such DBPs are 0.1 mg/L in drinking water. Formation of such harmful DBPs should be avoided either by removal of the organic compounds from water prior to chlorination or removal of the DBPs by adsorption using activated carbon or similar adsorbents or by advanced treatments such as membrane-based filtration. Nanofiltration- or reverse osmosis-based membrane technology can produce safe drinking water without the need for additional disinfection such as chlorine treatment. Membrane-based treatment technologies are discussed in Chapter 5, but for the purposes of this discussion, a membrane-based advanced treatment technology is schematically presented in Fig. 2.9.

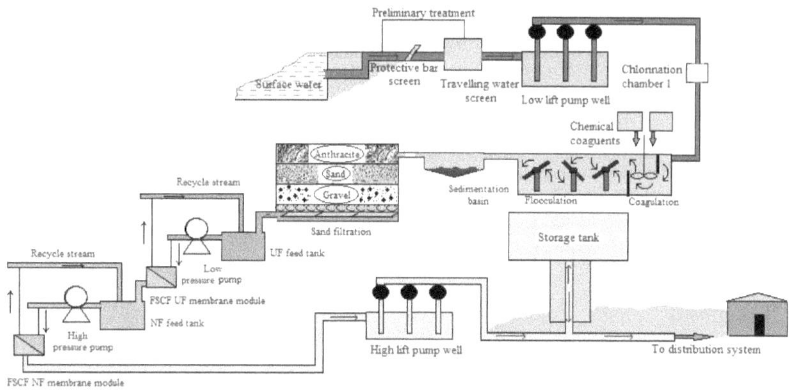

Figure 2.9 Treatment scheme with advanced membrane-based technology.

After sand filtration, water can be treated by ultrafiltration membrane followed by nanofiltration. Ultrafiltration will largely prepare feed for safe and long-term trouble-free filtration by a sensitive nanofiltration membrane that can separate all sorts of pathogens from water without the risk of formation of DBPs. Membrane in appropriate modules like flat sheet cross-flow type (FSCF) can ensure long-term filtration without significant membrane fouling. Today, with the emergence of tailor-made and durable/replaceable membranes, the cost of membrane-based treatment is gradually going down. Although in packaged membrane-based treatment plants reverse osmosis (RO) is used to produce safe potable water, the use of nanofiltration has the added advantage of filtration at much lower transmembrane pressure thereby reducing pumping cost. Moreover, unlike RO membranes, nanofiltration membranes do not retain all the useful minerals thus eliminating the need for postfiltration addition of minerals to water for general health. After RO, addition of calcium and magnesium minerals is required. Thus the membrane-based technology scheme presented in Fig. 2.9 is human health-friendly.

2.8.10 Determination of Chlorine Doses

Chlorine dose is determined by the amount of chlorine needed to kill pathogens and for oxidation of reducing agents like iron, manganese, and hydrogen sulfide. It may be determined by adding excess chlorine to water and then by measuring the difference between the added amount of chlorine and the residual chlorine after a specified time period.

A few case studies are discussed in this section to show methods of determination of doses of chlorine and other chemical reagents in water disinfection and water softening for the purpose of producing potable water.

The chlorine requirement, or chlorine dosage, is the sum of the chlorine demand and the desired chlorine residual. It can be expressed mathematically as

Chlorine dosage (mg/L) = Chlorine demand (mg/L) + Chlorine residual (mg/L)

Case study 1

Surface water drawn from a river is found to have a chlorine demand of 10 mg/L. The desired chlorine residual is 0.1 mg/L. We have to determine the daily chlorine dose for treating 10 kL of water per day.

Solution:

Chlorine dosages (mg/L) = chlorine demand (mg/L) + chlorine residual (mg/L)
Dosages = 10 mg/L + 0.1 mg/L;
= 10.1 mg/L
Daily chlorine dose = (10.1 mg/L) × (10 × 1000 L)/day = 101 kg/day

Case study 2

A disinfection chamber has an effective volume of 2000 L. River water at 100 L/min is passed through the chamber for disinfection. The tank has residual chlorine of 1.5 mg/L. C*T for the chlorinator has to be determined. C* stands for effective chlorine dose to which water needs to be exposed for the residence or contact time T to produce safe drinking water.

Effective contact or residence time = 2000 L/ 100 L/min = 20 minutes
C*T = effective contact time × residual concentration
= 20 min ×
1.5 mg/L = 30 min-mg/L

2.8.11 Determination Lime and Soda Ash Dosages in Water Softening

Case study 3

Water drawn from a surface source is tested in the laboratory for hardness and alkalinity. The test results indicate magnesium = 35 mg/ L as Mg, total hardness = 345 mg/L as $CaCO_3$, bicarbonate alkalinity = 160 mg/L as HCO_3^-, total alkalinity = 128 mg/L as $CaCO_3$, and carbon dioxide = 5 mg/L as CO_2. It has to be determined whether lime alone(CaO) or soda ash(Na_2CO_3) is needed to treat hardness of this water and at what dosages. No excess lime or soda ash need to be considered in this assessment.

Analysis

It is observed that the total hardness (345 mg/L as $CaCO_3$) of the water is greater than the total alkalinity (128 mg/L as $CaCO_3$). This indicates that the water will have both carbonate and noncarbonated hardness constituents ($Ca^{2+} + Mg^{2+}$) and so both lime (CaO) and soda ash (Na_2CO_3) will be necessary to remove the total hardness.

General Conversion Methods

All concentrations need to first be computed in terms of $CaCO_3$. To do this the following two steps are necessary:

Step-1: Equivalent weight calculation

Molar mass (CaCO3) = M(Ca) + M(C) + 3 × M(O) = 40 + 12 + 3 × 16 = **100 g**/mole. Divide the molar mass by the ion charge or oxidation number (for CaCO3) to determine equivalent weights. For example, Eq. weight of (HCO3(−)) = 61/1 (charge) = 61 g/Eq. Equivalent weight of (CO3(2−)) = 60/2 (charge) = **30 g**/Eq. Equivalent weight of $CaCO_3$ = molar mass/ion charge number or oxidation number = 100/2 = 50 (CO_3^{2-} has ion charge number = 2).

Step-2: Expressing all concentrations in terms of $CaCO_3$

Multiply the dosages by the equivalent weight of the chemical being changed to, and divide that result by the equivalent weight of the chemical being changed from.

Thus to express Mg hardness in terms of $CaCO_3$ the following equation is used:

$$(\text{Dose of Mg}) \times (\text{Equivalent weight of } CaCO_3)/(\text{Equivalent weight of Mg}) = (35\ \text{mg/L})(50/12) = 145.833\ \text{mg/L as } CaCO_3,$$

where equivalent weight of $CaCO_3$ = 50.

Total hardness = (already given in terms of $CaCO_3$) = 345 mg/L as $CaCO_3$

Bicarbonate alkalinity = (160 mg/L)(50/61) = 131.1475 mg/L as $CaCO_3$

Total alkalinity = (already given in terms of $CaCO_3$) = 128 mg/L as $CaCO_3$

Carbon dioxide = (5 mg/L)(50/22) = 11.36 mg/L as $CaCO_3$

Lime dose calculation (CaO)

$$\begin{aligned}\text{Lime dosages} &= [CO2] + [HCO - 3] + [Mg] + [\text{excess}] \\ &= 11.36\ \text{mg/L} + 138\ \text{mg/L} + 145.833\ \text{mg/L} + 0 \\ &= 295.193\ \text{mg/L as } CaCO_3\end{aligned}$$

Soda ash dosages (Na_2CO_3) calculation

$$\text{Soda ash dosage} = [\text{total hardness}] + [HCO_3^-] + [\text{excess}]$$

$$\begin{aligned}\text{Soda ash dosage} &= 345\ \text{mg/L} + 138\ \text{mg/L} + 0 \\ &= 483\ \text{mg/L as } CaCO_3\end{aligned}$$

If, instead of a magnesium ion concentration, a calcium ion concentration is given in the test results, the magnesium ion concentration can still be calculated as total hardness = [Ca] + [Mg]. The lime dosages in this case can also be determined in the usual manner.

Case Study 4

Surface water drawn from a river was tested in a laboratory for necessary softening before supplying it to city households. The test results are as follows:

Calcium = 50 mg/L; Total hardness = 150 mg/L as $CaCo_3$; Bicarbonate alkalinity = 148 mg/L

Total alkalinity = 130 mg/L; Carbon dioxide = 12 mg/L

Dosages (in terms of $CaCO_3$) of lime and soda need to be determined for this purpose.

Analysis

Since the total alkalinity (130 mg/L) is less than the total hardness, the water has both carbonate and noncarbonated hardness constituents necessitating both lime (CaO) and soda ash (Na_2CO_3) dosing.

Computation

All the concentrations need to be expressed in terms of calcium carbonate.

Step1: Expressing all concentration in terms of $CaCO_3$ as below following the same procedure of case 3.

Calcium = 50 mg/L (50/20) = 125 mg/L as $CaCO_3$

Total hardness = 150 mg/L as $CaCO_3$

Bicarbonate alkalinity = 148 mg/L (50/61) = 121.31 g/L as $CaCO_3$

Total alkalinity = 130 mg/L as $CaCO_3$

Carbon dioxide = 12 mg/L (50/22) = 27.27 mg/L as $CaCO_3$

Total hardness is the sum of calcium and magnesium ion concentrations: [Ca] + [Mg].

$$= 150\ \text{mg/L} = 125\ \text{mg/L} + \text{Mg}$$
$$\Rightarrow \text{Mg} = 25\ \text{mg/L}$$

Calculation of lime dose (CaO)

$$\begin{aligned}\text{Lime dosages (CaO)} &= CO_2 + HCO_3^- + Mg^+ + \text{excess}\\ &= 27.27\ \text{mg/L} + 121.31\ \text{mg/L} + 25\ \text{mg/L} + 0\\ &= 174.4173.58\ \text{mg/L as } CaCO_3\end{aligned}$$

Calculation of soda ash dosage (Na_2CO_3)

$$\begin{aligned}\text{Soda ash dosages } (Na_2CO_3) &= [\text{total hardness}] + [HCO_3^-] + [\text{excess}]\\ &= 150\ \text{mg/L} - 121.31\ \text{mg/L} + 0\\ &= 28.69\ \text{mg/L as } CaCO_3\end{aligned}$$

2.9 ADVANCED OXIDATION TECHNOLOGY

2.9.1 Wet-Air Oxidation Technology Using Bubble Column Reactor

Conventional treatment technologies often fail to effectively treat water that contains refractory organic compounds or toxic compounds. Direct biological treatment methods are unsuccessful because the toxic environment does not allow the microbes to sustain and grow. Incineration is also not advisable for such cases due to additional fuel cost or low heat recovery resulting in only marginal savings and shifting of pollution from water to air. The combustion takes place on a catalyst usually at temperatures several degrees below those required for thermal incineration. This is where advanced treatment technologies such as those based on wet-air or thermal liquid-phase oxidation (WAO) processes have been gradually emerging.

Wet-air oxidation technology using a bubble column reactor is shown in Fig. 2.10. The main components of the treatment technology are the heat-exchanger unit and a reactor to destroy pollutants present in the feed. In such advanced oxidation technology, generation of active oxygen species, such as hydroxyl radicals, takes place at high temperatures and pressures. These active radicals are known to have great potential for the treatment of effluents with high concentrations of organic matter (COD, 10–100 g/L), or toxic contaminants for which direct biological treatment

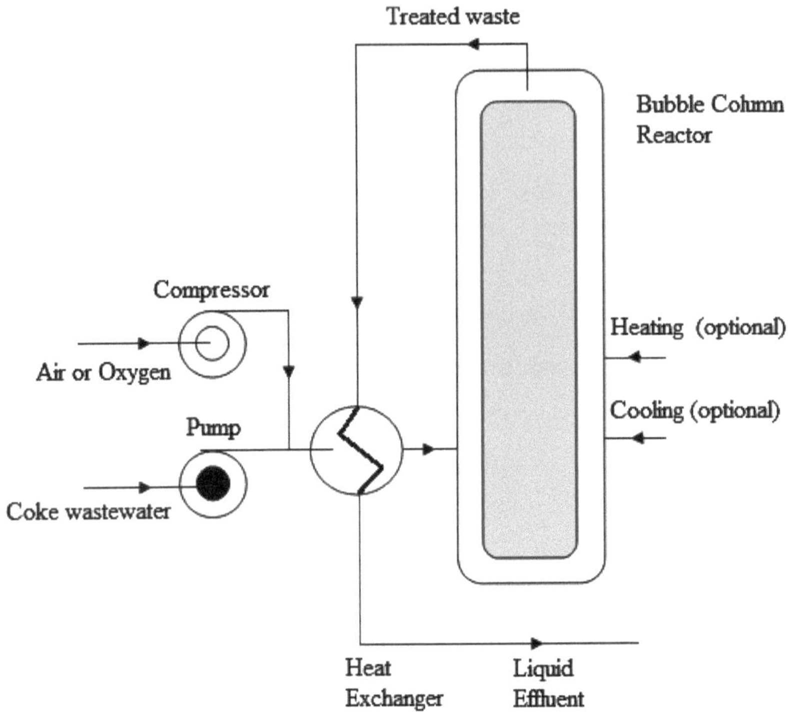

Figure 2.10 Scheme of wet-air oxidation bubble column reactor technology [1].

is not feasible. In this process, molecular oxygen dissolved in the wastewater reacts with the organic and inorganic pollutants. The oxidizing power of the process is based on the high solubility of oxygen at these conditions and the high temperature, which increases the reaction rates and production of free radicals. Technology based on catalytic wet-air oxidation (CWAO) is one such technology that has already established itself as attractive and useful for the treatment of effluents with low concentrations of organic pollutants The temperature that is necessary to initiate the reaction depends on the type of pollutant present.

2.9.2 Supercritical Wet-Air Oxidation Technology (SCWO)

Supercritical wet-air oxidation technology (SCWO) is considered a green technology in which oxidation reactions occur in superheated water at a temperature above the normal boiling point of water (100°C), but below the critical point (374°C). Supercritical water oxidation or SCWO is a process that occurs in water at temperature and pressure above the mixture's thermodynamic critical point. Under these conditions, water becomes a fluid with unique properties that can cause destruction of hazardous waste such as phenolic and other refractory compounds.

Ruthenium (Ru)-based eggshell catalysts are effective in treating coke wastewater by wet oxidation in a bubble-bed reactor. In a bubble-bed reactor filled with eggshell and supported with Ru catalysts, removal of COD and ammonia/ammonium compounds (NH_3-N) at temperature of 250°C and pressure of 4.8 MPa has been successfully achieved. The catalytic activity of uniform catalyst depended strongly on the distribution of active sites of Ru on catalyst [2]. Supercritical water has successfully treated high-strength coking wastewater by oxidation [3]. The SCWO process is particularly successful in treating hazardous industrial wastes and wastewaters. Supercritical water acquires a state in which water becomes a nonpolar solvent above its critical point ($T_C = 374°C$, $P_C = 22.1$ MPa). Supercritical water is an excellent medium for the rapid destruction of organic wastes by oxidation, because organics and oxygen easily dissolve in supercritical water to form a homogeneous phase. Fig. 2.11 shows such a SCWO technology scheme for the treatment of coke wastewater in a continuous-flow reactor system consisting of preheater, reactor, salt separator, and gas liquid separator. SCWO can readily achieve high destruction efficiencies (>99.99%) of organics in short residence time (<1 min) under an operating temperature of 450–700°C and operating pressure of 23–30 MPa.

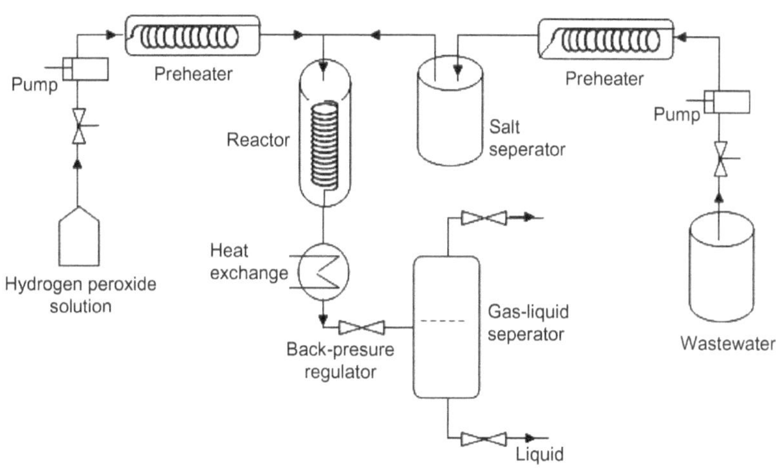

Figure 2.11 Schematic of SCWO technology for the treatment of coke wastewater in continuous mode [2].

The finally formed products are simple molecules like water, carbon dioxide, and molecular nitrogen instead of Nitrogen Oxides (NOX) as the temperature is much lower than that of incineration.

While SCWO technology is widely considered a very viable technology, its application still remains limited due to the formation of salt. In supercritical water, the solubility of inorganic salts decreases dramatically, causing the dissolved salts to deposit and accumulate on the wall of the reactor and lines. The precipitated salts in turn result in plugging and corrosion of the reactor. New reactor designs and operating techniques have been suggested to overcome the difficulties that arise due to salt precipitation. However, complicated reactor designs and complex operating techniques are often difficult to carry out in long-term operations. Using SCWO technology to dilute coking wastewater with MnO_2 as catalyst at a temperature of 500°C, 90% or higher conversions of both NH_3 and TOC (total organic carbon) within 2 s of contact time has been achieved. However, destruction of NH_3 was inhibited by both phenol and other coexisting compounds at short contact times. The requirement for high temperature (to the tune of around 500°C) during wet-air oxidation makes the technology expensive compared to technologies based on pure chemical and UV oxidation techniques [4].

2.9.3 Purification Technology based on Fenton and Photo-Fenton Oxidation

Fenton's reaction is based on the catalyzed decomposition of hydrogen peroxide by iron (II) to produce very reactive hydroxyl radicals, according to the following mechanism [5]:

$$Fe^{2+} + H_2O_2 \rightarrow Fe^{3+} + OH^{\bullet} + OH^- \quad (2.57)$$

$$Fe^{3+} + H_2O_2 \rightarrow Fe^{2+} + OOH^{\bullet} + H^+ \quad (2.58)$$

$$^{\bullet}OH + H_2O_2 \rightarrow HO_2^{\bullet} + H_2O \quad (2.59)$$

$$Fe^{2+} + {}^{\bullet}OH \rightarrow Fe^{3+} + OH^- \quad (2.60)$$

$$^{\bullet}OOH + H_2O_2 \rightarrow O_2 + H_2O + {}^{\bullet}OH \quad (2.61)$$

$$RH + {}^{\bullet}OH \rightarrow R^{\bullet} + H_2O \quad (2.62)$$

The Fe^{3+} produced in Fenton oxidation is a powerful coagulant for removing high-molecular-weight constituents of the feed by adsorption on the new Fe $(OH)_3$ flocks formed at a neutral pH [6]. Another modification over the Fenton's method by combining it (Fenton's method) with a biologically activated carbon process has successfully treated coke wastewater contaminated with phenolic and cyanide compounds [7]. Success of Fenton's oxidation largely depends on parameters such as H_2O_2 to Fe^{2+} molar ratio, dosage of Fe^{2+} reagent, initial pH, reaction time, and initial COD strength.

Fenton's oxidation breaks down most of the persistent organic pollutants and complex cyanides present in the waste. The hydroxyl radicals, which are second only to fluorine among common oxidants, react rapidly and nonselectively with nearly all-organic pollutants.

Biologically treated effluents still contain a considerable amount of phenol, cyanide, and thiocyanide, which must be removed below the standard effluent permissible limits. Thermodynamically, free cyanide can easily form stable complexes with metals such as nickel, iron, and cobalt as illustrated below:

$$Fe^{3+} + 6CN^{-} \Leftrightarrow Fe(CN)_6^{3-} \quad \log k = 43.9 \tag{2.63}$$

Considering the economic advantage of changing ferric-chloride solution with ferrous-sulfate solution, cyanide in some cases has been treated by ferrous sulfate. When the ferrous ion is added to the solution a blue precipitate is formed and the residual cyanide concentration decreases. During the reaction, the solution pH decreases from 7.15 to 4.98 with increasing amount of ferrous ion. The removal efficiency of cyanide by ferric ion ($FeCl_3$) at low concentration of cyanide is very low. When ferric-ion concentration increases up to 4.0 mmol/L, the color of the precipitate turns slightly dark-blue, and the removal efficiency of the cyanides increases sharply [8]. Cyanides are largely present in coke wastewater in the form of free cyanide and ferricyanide, as this wastewater contains only iron in the form of ferric ion and other metals at concentrations below 0.1 mg/l. Iron cyanide solids are complex coordination compounds that are produced and used in various commercial products and processes like Prussian blue, Prussian brown, Turnbull's blue, etc. Mixing solutions of soluble ferric ion and ferricyanide will lead to formation of a blue precipitate through:

$$3Fe(II)^{2+}2Fe(III)(CN)_6^{3-} \rightarrow Fe(II)_3[Fe(III)(CN)_6]_2 \tag{2.64}$$

This ferrous ferricyanide precipitate is unstable as the reducing $Fe(II)^{2+}$ cation and the oxidizing $Fe(III)(CN)_6{}^{3-}$ anion have an open path for electron transfer through the CN bridge. This internal process is very fast and results in a mixed valence ferrous-ferric ferricyanide system through:

$$Fe(II)_3[Fe(III)(CN)_6]_2 \rightarrow Fe(II)Fe(III)_2[Fe(II)(CN)_6]_2 \tag{2.65}$$

The mixed valence species can be oxidized to a charged ferric ferricyanide species by dissolved air or by ferricyanide in solution following the reaction below:

$$Fe(II)Fe(III)_2[Fe(II)(CN)_6]_2 \xrightarrow{O_2 \text{ or } Fe(III)(CN_6^{3-})} \{[Fe(III)_3[Fe(II)(CN)_6]_2\}^+ \tag{2.66}$$

This positively charged ferric ferricyanide species on acquiring a negatively charged species like Cl^-, OH^- or $(SO_4)^{2-}$ to neutralize the charge form Turnbull's blue (TB) $Fe(III)_3A\left[Fe(II)(CN)_6\right]_2$.

While free cyanide is very toxic to microorganisms, ferricyanide is essentially nontoxic except under UV-irradiating conditions. Despite its acute toxicity, various aerobes and anaerobes can easily and rapidly degrade free cyanide. Ferricyanide is resistant to biodegradation due to its thermodynamic stability, which explains the presence of only ferricyanide in the effluent of the predenitrification process. Residual cyanides and fluorides can be removed by reacting with ferric chloride and polyaluminum chloride (PAC) solutions in a chemical treatment process, after the biological treatment [9].

2.9.4 Ozone-based Oxidation Technology

Ozone-based oxidation technology belongs to the special class of advanced oxidation technologies that perform the oxidation at near ambient temperature and pressure through generation of $OH^{\bullet}$ radicals. The production of hydroxyl radicals can be achieved through homogeneous photochemical reaction (UV-irradiated H_2O_2), or alternatively by catalysts like Fe^{2+}, Cu^{2+}, TiO_2. In photochemical oxidation, oxygen is the most common electron accepter and is relatively efficient. But toward more efficient inhibition of the electron–hole recombination effect, alternatives like hydrogen peroxide as an electron acceptor has also been attempted for its higher potential [10]. Due to its high oxidation capability, ozone has been successfully applied in water and wastewater treatment to

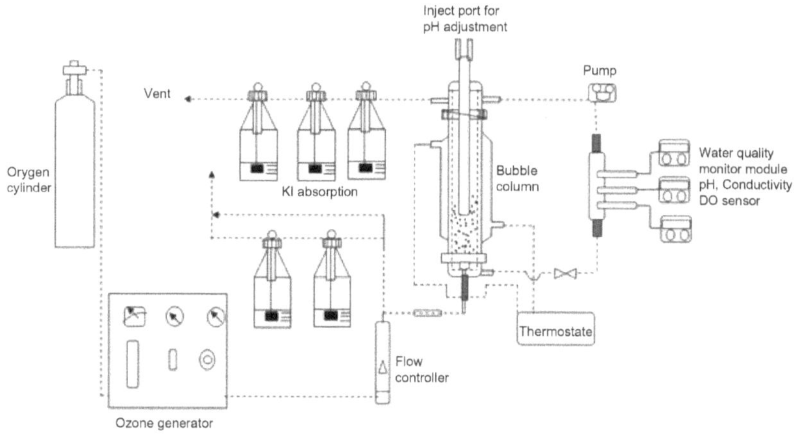

Figure 2.12 Scheme of ozone-based treatment technology [11].

remove color and reduce the chemical oxygen demand (COD) and total organic (TOC) of wastewater. Ozonation-based technology has been used to treat wastewater-containing refractory compounds and has been applied in the petrochemical, coke wastewater, textile, leather tanning, and paper and pulp industries, and is considered promising.

Fig. 2.12 shows aschematic diagram of the ozone-contacting bubble column [11]. Ozone-containing gas and raw wastewater flow countercurrently in the column, with wastewater flow in the downward direction. KI solution is used in measuring the ozone concentration in the offgas. An ozone generator is used to generate ozone from pure oxygen. CN^- and $SCN^-$are the important pollutants in coke wastewater and oxidized by ozone to form sulfate, nitrogen, and other nontoxic compounds.

$$SCN^- + 3O_3 + 2OH^- \rightarrow CN^- + SO_4^{2-} + 3O_2 + H_2O \quad (2.67)$$

$$SCN^- + O_3 + H_2O \rightarrow CN^- + H_2SO_4 \quad (2.68)$$

$$3CN^- + O_3 \rightarrow 3OCN^- \quad (2.69)$$

$$2OCN^- + O_3 + H_2O \rightarrow 2HCO_3^- + N_2 \quad (2.70)$$

Ozonation inhibits toxicity by degrading these pollutants, suggesting that preozonation benefits the subsequent biological treatment unit.

2.9.5 Electrochemical Oxidation Technology

Electrochemical oxidation is another promising technology for effective treatment of biorefractory and toxic compounds that cannot be taken care of by conventional biological or chemical treatments. Being environmentally benign, this technology has high efficiency in organic degradation, is simple and easy to design and operate, and has been used for degradation of various refractory substances such as phenol, chlorophenols, nitrophenols, dyes, surfactants, landfill leachate, and herbicides. Boron-doped diamond (BDD) electrodes are normally used for the electrochemical oxidation of biologically pretreated wastewater. Complete mineralization of the organic pollutants is almost achieved, and surplus ammonia–nitrogen (NH_3-N) is also removed by using BDD as an anode. However, the high energy consumption of electrochemical oxidation limits its industrial application. It is well known that the current efficiency of an electrochemical oxidation process strongly depends on anode material. In recent decades, a lot of electrode materials have been examined to improve the effectiveness of oxidation and current efficiency, such as graphite electrodes. This different kind of electrodes is having some advantages and disadvantages. Low current efficiency is found in the graphite and platinum electrodes and they are easy to foul. The IrO_2 and RuO_2 electrodes have low reactivity with organic oxidation. The high oxidation capacities for organic pollutants and high current efficiencies for organic oxidation are found in SnO_2,PbO_2, and BDD due to production of hydroxyl radicals by water discharge at these electrodes.

REFERENCES

[1] Debellefontaine H, Foussard JN. Wet air oxidation for the treatment of industrial wastes. Chemical aspects, reactor design and industrial applications in Europe. Waste Manag 2000;20:15–25.

[2] Yang M, Sun Y, Xu AH, Lu XY, Du HZ, Sun CL, et al. Catalytic wet air oxidation of coke-plant wastewater on ruthenium-based eggshell catalysts in a bubbling bed reactor. Bull Environ Contamin Toxicol 2007;79(1):66–70.

[3] Du X, Zhang R, Gan Z, Bi J. (2010). Treatment of high strength coking wastewater by supercritical water oxidation. Fuel, article in press.

[4] Oshima Y, Inaba K, Koda S. Catalytic supercritical water oxidation of coke waste with manganese oxide. Sekiyu Gakkaishi 2001;44(6):343–50.

[5] Pignatello JJ. Dark and photo assisted iron (3+)-catalyzed degradation of chlorophenoxy herbicides by hydrogen peroxide. Environ Sci Technol 1992;26:944–51.

[6] Ying W, Duffy J, Tucker E. Removal of humic acid and toxic organic compounds by iron Precipitation. Environ Prog 1998;7:262–9.

[7] Jiang W, Zhang W, Li B, Duan J, Lv Y, Liu W, et al. Combined fenton oxidation and biological activated carbon process for recycling of coking plant effluent. J Hazardous Mater 2011;189:308–14.
[8] Park D, Kim YM, Lee DS, Park JM. Chemical treatment for treating cyanides-containing effluent from biological cokes wastewater treatment process. Chem Eng J 2008;143:141–6.
[9] Akcil A. Destruction of cyanide in gold mill effluents: biological versus chemical treatments. Biotechnol Adv 2003;21:501–11.
[10] Dionysiou DD, Suidan MT, Baudin I, Lainé JM. Effect of hydrogen peroxide on the destruction of organic contaminants-synergism and inhibition in a continuous-mode photo catalytic reactor. Appl Catal B: Environ 2004;50:259–69.
[11] Chang EE, Hsing HJ, Chiang CP, Chen MY, Shyng JY. The chemical and biological characteristics of coke-oven wastewater by ozonation. J Hazardous Mater 2008;156: 560–7.

CHAPTER 3

Biological Treatment Technology

3.1 INTRODUCTION TO BIOLOGICAL TREATMENT TECHNOLOGIES

Microorganisms are the agents in biological treatment of wastewater. The primary goal of any microorganism species is to maintain its genetic heritage within the community. Different species of microorganism in a microbial community compete with each other for food, energy, and nutrients and through the process called selection the best fit microorganism species generates the greatest number of descendants thereby establishing a foothold in the community. And in doing so, the microorganisms try to find a suitable multidimensional space called a *niche* in terms of nutrients, energy, temperature, pH, and other environmental conditions. The best biological wastewater treatment system should, therefore, be so designed and operated that the most desired microorganism species get selected. The pollutants in wastewater serve as either electron donors or acceptors for prokaryotes. For example, biochemical oxygen demand (BOD) represents organic electron donor. Thus to remove BOD from water, heterotrophic bacteria needs to be selected by ensuring adequate supply of oxygen to serve as electron acceptor from this species. Oxygen (electron acceptor) in this case needs to be supplied at a rate commensurate with the expected rate of removal of BOD. The microorganisms that lose the competition get selected against. For their survival, growth, and reproduction these microorganisms mainly depend on the potential energy stored in the organic and inorganic molecules and sometimes partly depend on radiation energy such as solar energy. As the microorganisms transform these molecules through oxidation and reduction reactions, energy released is used for maintenance and or growth of new biomass. Thus the organic and inorganic molecules, which are pollutants of wastewater, get consumed and transformed into new biomass. However, there are real limitations to which microbes can decompose the polluting substances present in a water stream. In general, microbes are found to be very successful in degrading the organic molecules and thus for wastewater having high BOD, such microbial treatment is often

Industrial Water Treatment Process Technology.
DOI: http://dx.doi.org/10.1016/B978-0-12-810391-3.00003-5

suggested. BOD stands for the amount of oxygen used by the microorganisms to biochemically oxidize the organic matter present in a given sample of water at a given temperature and over a given period. It is not a precise quantitative test, although it is widely used as an indication of the quality of water. It broadly indicates the presence of organic pollutants. In a BOD_5 test, 300 mL of water sample saturated with oxygen is kept in incubation at a given temperature (normally 20°C) for five days and the difference in the dissolved oxygen (DO) concentration over this five-day period is taken as the BOD_5 of the water. Sometimes the water sample may have to be seeded.

Wastewater with high chemical oxygen demand (COD) indicates the presence of high concentrations of toxic chemicals or inorganic chemicals that are refractory in nature and pose a challenge in decomposing by biological treatment. COD refers to the amount of oxygen required to chemically break down or oxidize the compounds present in a given sample of water using the chemical-oxidizing agent potassium dichromate ($K_2Cr_2O_7$). COD may indicate the presence of both organic and inorganic compounds in a water sample. A high BOD/COD ratio indicates easy biodegradability. During this biochemical oxidation of the waste, initially a small fraction is oxidized, releasing energy for maintenance of the existing cells and for synthesis of new cells as shown in Eq. (3.1). Using this released energy, a part of the substance is then converted into new cells following Eq. (3.2). Finally, when the entire amount of the substrate is consumed, the cells undergo endogenous respiration when the new cells consume their own tissue to obtain energy for their maintenance following Eq. (3.3) as proposed by Hoover and Porges (1952).

First-stage oxidation

$$\begin{aligned}\text{COHNS} + O_2 + \text{microorganisms} &\rightarrow CO_2 + H_2O + NH_3 \\ &+ \text{other end products} + \text{energy}\end{aligned} \tag{3.1}$$

where COHNS represents carbon, oxygen, hydrogen, nitrogen, and sulfur or the organic waste, and $C_5H_7NO_2$ represents new cells.

New cell synthesis

$$\text{COHNS} + O_2 + \text{microorganisms} + \text{Energy} \rightarrow C_5H_2NO_2 \text{ (New Cells)} \tag{3.2}$$

Endogenous respiration

$$C_5H_7NO_2 + 5O_2 \rightarrow 5CO_2 + NH_3 + 2H_2O \tag{3.3}$$

The ultimate BOD or BOD_L is the total amount of oxygen consumed in the three reactions as represented by the Eqs. (3.1)–(3.3) if it is assumed that the water sample contains only organic carbon. Biodegradable COD is represented as bCOD or UBOD and includes soluble or dissolved, colloidal and particulate, or suspended biodegradable solids. The mixture of solids that result from mixing influent stream with recycled stream may be called mixed liquor suspended solids (MLSS) and mixed liquor volatile suspended solids (MLVSS), which broadly include microbial mass, inert inorganic suspended solids (SS), nonbiodegradable volatile suspended solids (nbVSS), or cell debris that may be produced from endogenous respiration of cell mass.

3.2 WASTEWATER BIODEGRADABILITY: SELECTION OF TREATMENT TECHNOLOGY

Before making a decision on the use of a biological treatment technology for wastewater, biodegradability needs to be assessed. The COD/BOD ratio is a useful tool for making such a decision. For example, if the COD/BOD ratio is ≤ 2.5 then the wastewater may be considered suitable for biodegradation and a treatment technology based on a process like activated sludge can be adopted. A COD/BOD ratio above 2.5 but below 5.0 suggests that some molecules are refractory to biodegradation and biological treatment may still be adopted but with provision of much longer residence time of the wastewater in the biotreatment unit. A COD/BOD ratio exceeding 5.0 may be an indicator of the presence of toxic substances that are highly likely to reduce the metabolic activity of the microbes in the biomass. Thus in this case, direct biological treatment should be avoided as the microbes will not survive in the environment. A pretreatment using chemical reagents or adsorbent additive may reduce the level of toxicity thereby permitting downstream biological treatment albeit with reduced effectiveness. Otherwise, adoption of a nonbiological oxidation process such as ultraviolet oxidation process may be more advisable.

3.3 MICROBIAL GROWTH KINETICS: UNSTRUCTURED MODEL

In an unstructured growth modeling approach, microbial cells as a whole are considered an entity and the interactions of this entity with the environment are captured. Instead of segregating the microbial population, they are viewed as a lumped biophase called species. On the other hand, the structured models divide the biomass into a number of components like DNA, RNA, proteins, etc., and the reactions taking place within these individual components are considered. Structured models are quite complex and difficult to develop and implement. We consider here the deterministic models only where the outputs are determined completely by the inputs against the consideration of distributions of cell characteristics (i.e., time of generation of a cell population) across the whole cell population in the stochastic models leading to a very high degree of complexity in the modeling.

The primary requirements of the microbes for their growth are carbon and energy. In addition, they need nutrients for supply of elements like N, S, P, Cl, Na, K, Ca, Mg, and Fe. Apart from these major elements, elements like Mn, Zn, Mo, Cu, Co, and Ni may also be required in small quantities. Organisms that draw their required carbon from organic compounds are called heterotrophs, while those receiving carbon from atmospheric carbon dioxide are called autotrophs. Thus autotrophs need more energy than the heterotrophs for their cell synthesis resulting in their lower growth rate. Based on the source of energy, organisms may be classified as phototrophs or chemotrophs. Phototrophs meet their synthesis energy requirement from light whereas chemotrophs derive their cell synthesis energy from chemical reactions that may be oxidation of both inorganic and organic compounds. Chemoautotrophs are those organisms that derive their energy from oxidation of inorganic compounds like ammonia, nitrite, etc., whereas chemoheterotrophs receive their energy from oxidation of organic compounds only.

Microbes employed in wastewater treatment grow rapidly through a high rate of cell division and the rate of growth is proportional to the concentration of the microbial mass in the medium (X). Thus the growth rate can be expressed as a first-order kinetic expression:

$$\frac{dX}{dt} = \mu X \tag{3.4}$$

where μ = specific growth rate.

3.3.1 Monod Kinetic Equation

Microbes grow at the cost of substrate and therefore the microbial growth rate is directly proportional to the substrate concentration (S). If the substrate is limited, the growth rate of microbes (for pure culture) according to French microbiologist Jacques Monod (Monod, 1940) may be expressed as:

$$\mu = (\mu_{max}).(S)/(K_s + S) \tag{3.5}$$

Thus the Monod equation shows that under substrate-limited growth environment (i.e., when S is very low), microbial growth follows first-order kinetics as represented by Eq. (3.4) and in an environment of excess or high substrate concentration, the growth follows zero-order kinetics as the denominator of the right-hand term of Eq. (3.5), which may be taken as equal to S (i.e., $K_s + S \approx S$).

μ_{max} is the maximum growth rate and K_s is the substrate concentration at which the microbial growth rate reaches half the maximum reaction velocity or $\mu = \mu_{max}/2$ and is called the Monod half-velocity rate constant. The substrate dependence of the microbial growth rate may be expressed graphically as shown in Fig. 3.1.

K_s = Monod half-velocity constant, i.e., the substrate concentration at which

$$\mu = \frac{\mu_{max}}{2}$$

The decrease of substrate is proportional to the increase of microbial mass and hence the rate of the decrease of substrate may be expressed as:

$$-\frac{dS}{dT} = U \cdot \frac{\mu_{max}.S}{K_s + S} X = \frac{1}{Y} \mu_{max} \frac{S}{K_s + S} X \tag{3.6}$$

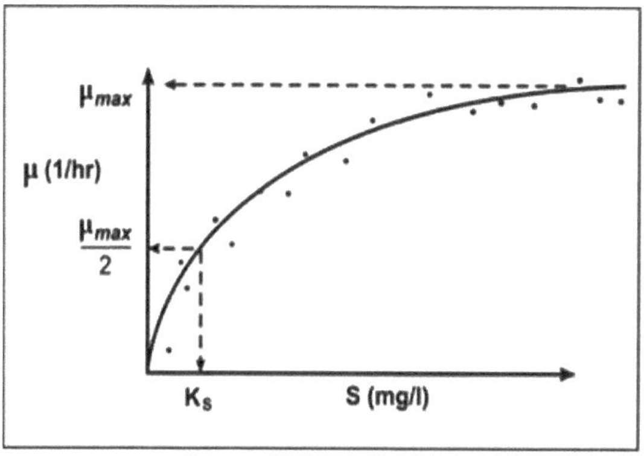

Figure 3.1 Variation of microbial growth rate with substrate concentration.

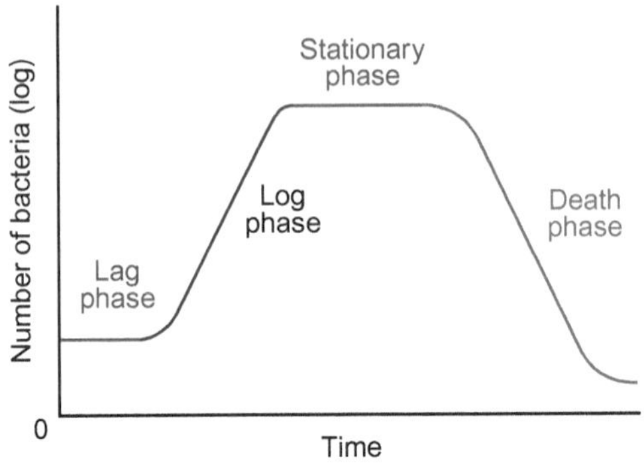

Figure 3.2 Microbial growth characteristics.

where U is the proportionality constant (the specific substrate utilization rate) and the reciprocal of Y the specific yields of organisms.

The lifecycle of bacteria includes four stages as depicted in Fig. 3.2. The first phase, called the lag phase, indicates the period in the lifecycle when microbes adjust or acclimatize themselves to a new environment. When cells are placed in a new medium they need to catabolize the new carbon source for intracellular activities prior to cell division; during this period, transportation of various cofactors like vitamins, amino acids, ions, etc., takes place through cell membrane. In this phase there is zero growth of microbes. If the inoculum is grown in a medium with the same composition as the reactor medium, then the lag phase in the reactor can be substantially reduced. The lag phase can also be reduced by adding adequate intermediates like amino acids, ions, nitrogen, and essential intermediate nutrients to the reactor. The next phase is the log phase during which microbes grow exponentially. Cell division takes place during this phase and cells increase in geometric progression (2^0, 2^1, 2^2,....2^n after n divisions). This is also called the logarithmic phase as the nature of the plot (exponential) is obtained by plotting the log of the number of microbes against time. The microbial growth rate for a batch reactor in this phase may be expressed by Eq. (3.4) (already presented at the beginning of this section):

$$\frac{dX}{dt} = \mu X$$

On integration of Eq. (3.4) between the limits ($X = X_0$ to X, $t = t$ to t_{lag}), we arrive at the relation:

$$\ln(X/X_0) = \mu(t - t_{lag}) \tag{3.7}$$

The log phase follows a very short decline phase when microbial growth starts taking a downward direction and then enters a stationary or a stagnant growth phase. After the stationary phase, microbes eventually enter the decay or death phase that takes place outside the reactor such as in the downstream-settling unit and recycle line.

3.3.2 Diauxic Microbial Growth

A growth medium containing multiple carbon sources induces more than one lag phase in the microbial lifecycle. Microbial growth with more than one lag phase is called *diauxic growth*. For example, a medium containing glucose and lactose will have two lag phases as microbes in this case will first consume glucose following a normal lag phase and then there will be another lag phase during which microbes will synthesize β-galactosidase, which is essential to subsequent utilization of lactose. The best biological treatment system should be so designed that the log phase (i.e., exponential growth phase) of the lifecycle of the microbes is fully utilized within the reactor to ensure a high rate of conversion of the substrate.

3.4 BIOREACTOR CONFIGURATIONS OF BIOLOGICAL TREATMENT TECHNOLOGIES

Selection of type of reactor depends on type of reaction, type of wastewater, environment, possible capital investment and affordable cost. For example, for heterogeneous reactions involving solid-phase catalyst or ion-exchange resin, packed bed or fluidized bed reactor is normally suggested. For homogenous reactions, CSTR (continuous flow stirred-tank reactor) or plug-flow reactors are normally used. However, compared to the use of a plug-flow reactor, the use of CSTR type is extensive.

3.4.1 Biological Treatment Using Plug-flow Reactor Technology

In an ideal plug-flow reactor, there should not be any longitudinal mixing (i.e., mixing in the direction of flow) and thus the particles should emerge from the reactor outlet point in the same sequence in which they enter. There will only be radial mixing. According to the Monod Eq. (3.1), the rate of conversion of substrate decreases with a decrease in substrate concentration. In a CSTR, this substrate concentration drops immediately to the level of exit point because of agitation and uniform dispersion of substrate within the reactor. The reaction rate is determined by this reduced concentration (the rate being proportional to the substrate concentration). In a plug-flow reactor, on the other hand, the substrate concentration will not drop to this low level but rather decreases in the axial direction only. Thus a plug-flow reactor will have a higher rate of substrate conversion than a CSTR. Despite offering a high rate of substrate conversion, plug-flow reactors are seldom suggested for biological treatment of water demanding cell growth because of inherent difficulties in providing continuous inoculum of cells as well as in maintaining desired pH and oxygen supply along the length of the reactor. But a plug-flow reactor may be used successfully in bioprocesses involving immobilized cells or enzymes as the constraints of maintaining pH and supplying oxygen do not affect immobilized systems.

Efficient performance of a reactor is ensured by the absence of dead zones, short-circuiting, or inadequate mixing. Ideal reactors can avoid these undesired phenomena but practical reactors are hardly ideal reactors. Short-circuiting results from temperature difference while dead zones (i.e., a part of the reactor remaining unutilized) largely result from flow pattern and poor reactor space design. Plug-flow reactors may be either tubular types with closed channels or in the form of long rectangular open channels. A tubular plug-flow reactor is presented in Fig. 3.3.

In an ideal plug-flow reactor, the theoretical detention time (V/Q) is the same as the measured or actual detention time where V is the volume of the reactor and Q is the volumetric flow rate.

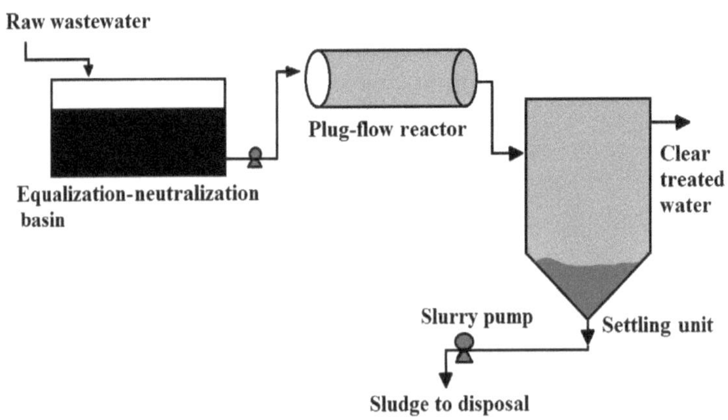

Figure 3.3 Biological treatment using plug-flow reactor technology.

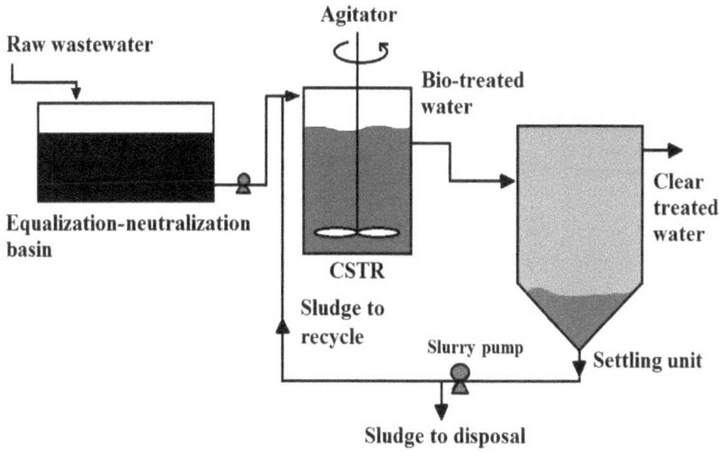

Figure 3.4 Biological treatment using continuous stirred-tank reactor technology.

3.4.2 Biological Treatment Using Continuous Stirred-Tank Reactor Technology

Stirred-tank reactors may be operated either in batch mode or in continuous mode. In large biological water treatment plants, such bioreactors are normally operated in continuous mode and are called continuous stirred-tank reactors (CSTRs). Fig. 3.4 presents a schematic diagram of a biological treatment plant using CSTR technology.

In a CSTR, the reactants are uniformly distributed instantly on entering the reactor. The extent of biodegradation, however, depends on microbial population and on optimum conditions of microbial growth including food to microorganism ratio, pH, temperature, mean cell residence time (MCRT), and other environmental conditions. CSTR technology finds wide application in activated sludge processes where a part of the sludge is continuously recycled back to the reactor as shown in Fig. 3.4.

3.4.3 Biological Treatment Using Pack Bed Reactor Technology

In a packed-bed reactor plastic or ceramic material forms the core of the reactor and wastewater passes through the packed bed. Microbes grow while remaining attached to the solid support. A packed-bed reactor is best illustrated by a trickling filter which will be described later (Fig. 3.5).

Packed-bed reactors may also be used successfully for immobilized cells or enzymes. High fructose corn syrup (HFCS) is commercially

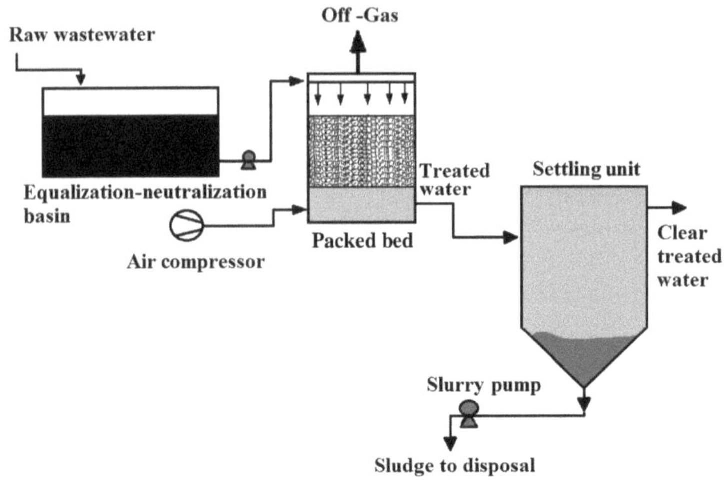

Figure 3.5 Biological treatment plant using packed-bed reactor technology.

produced in packed-bed reactors using immobilized enzyme. In this conversion of glucose to fructose, the enzyme glucose isomerase is kept immobilized on a solid support like alumina. A packed-bed reactor is well illustrated in submerged aerated filter which will be described later in this chapter. The carbon particles form the packed bed and serve as both medium for attached microbial growth as well as filter for wastewater.

3.5 BIOLOGICAL TREATMENT USING FLUIDIZED-BED REACTOR TECHNOLOGY

A fluidized-bed reactor works almost on the same principle as a packed-bed reactor the only difference being the expanded or fluidized form of the packing material, which is maintained by upward flow of the air and the wastewater. The degree of fluidization or total pore volume or reactor volume can be controlled by controlling the fluid flow in the reactor Fig. 3.6.

3.6 CONVENTIONAL BIOLOGICAL TREATMENT TECHNOLOGIES

Several conventional biological treatment technologies have been developed based on two basic microbial growth mechanisms: suspended growth

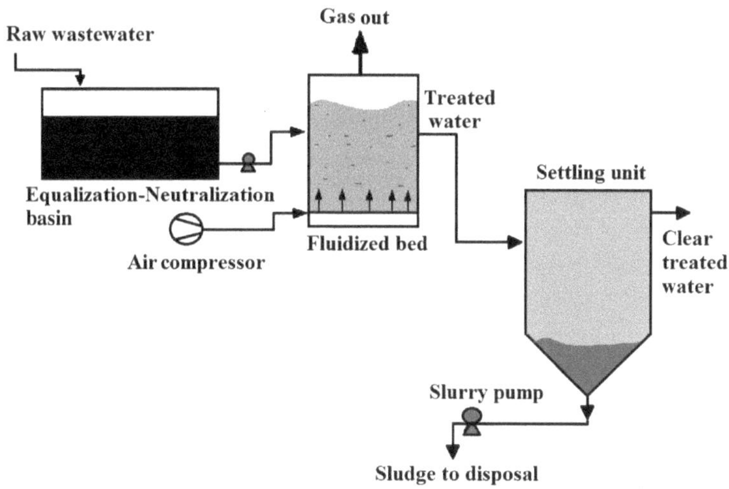

Figure 3.6 Biological treatment plant using fluidized-bed reactor technology.

mechanism and attached growth mechanism. In the suspended growth mechanism, microbes grow while remaining suspended in an aqueous medium. Such a process is well illustrated by the activated sludge process. In the attached growth mechanism, microbes grow while remaining attached to some solid support. A trickling filter or rotating-disc contactor represents a treatment technology based on attached growth. Conventional microbial treatment technologies have been developed based on the type of reactor device where the microbes are grown for degradation of the waste materials present in water. However, here we describe the most widely used technologies.

3.6.1 Activated Sludge-based Treatment Technology and Advances

An activated sludge process (ASP) is a biological wastewater treatment process that speeds up waste decomposition by maintaining activated microbial mass within the bioreactor through continuous return sludge, fresh feeding, and oxygenation. A part of the microbial sludge-bearing wastewater is continuously recycled to the bioreactor from the downstream settling unit when the dormant, inactive microbes under the settled sludge layer get reinvigorated on returning to the bioreactor as they get fresh food and adequate supply of oxygen in the agitated reactor. Adequate return sludge flow to the aeration tank is essential to sustaining bioconversion at a high rate. Return sludge pumping needs to be increased during peak flow hours to avoid formation of excessive sludge blanket in the settling unit. Normally 50%–75% of the design wastewater flow rate is the rate of the return sludge flow rate. Sometimes, depending on settling characteristics such return sludge flow rate is adjusted. For example, if on 30 min settling of 1000 mL aeration tank effluent, the volume of the settled sludge is 250 mL, then 100 × (250/750) or 33.33% of the peak wastewater flow rate to the bioreactor should be taken as the return sludge flow rate. If the peak flow rate is 4 m^3/s, then return sludge flow rate = (4/) × 33.33% = 4 × 33.33/100 = 1.33 m^3/s. The concentration of biological solids in the return sludge is 4000–10,000 mg/L. Construction of a rectangular tank is easier than a circular type. In the case of rectangular tanks, the width-to-length ratio is 1:2 to 1:4 with baffled inlet and the tank bottom sloping toward the discharge end facilitating sludge transport to a discharge trough. Circular tanks (with diameter-to-depth ratio of 3:1) are very common with bottom sloping (30°) from periphery to the center facilitating sludge collection in a central hopper. Frequent sludge removal is easier in circular tanks. The settled sludge is mainly taken out of the system for subsequent dewatering and disposal while allowing a small fraction for recycling through the system. The most basic configuration involves an aerated bioreactor with a downstream settling unit as illustrated in Fig. 3.7.

The three basic components of an activated sludge process are the complete mix bioreactor, sedimentation unit, and a sludge-recycle system.

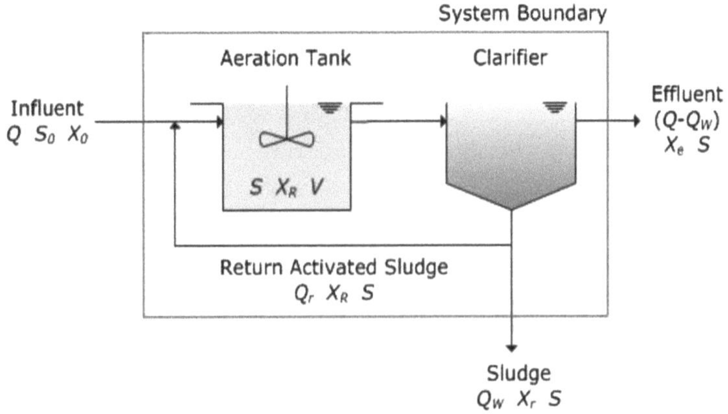

Figure 3.7 Schematic diagram of an activated sludge process.

Advances in ASP technology

A modern ASP-based plant functions with automatically operated valves, program logic controllers, level sensors, and integrated membrane-filtration modules. When a microfiltration module with membrane with 0.1–0.4 μm pores is integrated with the ASP unit, the treated water is separated out from the microbial mass continuously. Such an ASP reactor is called a membrane bioreactor (MBR). The membrane module may be placed within the bioreactor where through vacuum pump water is sucked through the membrane and separated from the microbial mass. The deposited solids from the membrane surface are cleaned out by using compressed air at the bottom of the reactor and the membrane module. In another modification, the membrane module tubular, hollow fiber or flat sheet cross-flow type may be integrated with the bioreactor or ASP reactor without being submerged in it but being attached to it from outside. The cross-flow pump drives the wastewater from the ASP unit through the membrane module outside the bioreactor causing pressure-driven microfiltration. The treated clear water is collected at the permeate side of the membrane module while the retentate is recycled back to the reactor. This type of integrated membrane module makes the treatment plant more compact and permits plant operation at higher microbial concentration resulting in much better quality of water. The downstream sedimentation unit is redundant in modern treatment plants and the quality of treated effluent is much better than that obtained in a conventional activated sludge plant with a downstream sedimentation unit. Recycling of activated sludge is always required for plants sensitive to a large number of process parameters, since the plant generates sludge that needs safe disposal.

Modeling Activated Sludge Process

Microbes grow at the expenses of the substrate. As microbes grow, substrate concentration decreases. To capture this microbial growth rate, we perform material balance around the control volume (shown as the box in Fig. 3.7). The box containing the reactor and the downstream settling or sedimentation unit represents the control volume in Fig. 3.7. With reference to this figure, the microbial cell growth rate and hence the accumulation rate within the control volume may be expressed mathematically as:

$$\frac{dX}{dt}V = Q_0X_0 + \mu_{\max}\frac{S}{K_s + S}.X.V - K_dX.V - (Q_0 - Q_w)X_e - Q_wX_u \tag{3.8}$$

The cell accumulation rate within reactor = rate of cell inflow + rate new cell mass growth from substrate consumption − rate of cell decay (or death) − rate of cell exit with treated effluent − rate of cell exit with underflow sludge (which is wasted).

X is the microbial cell concentration within the reactor with volume V; X_u is the cell concentration in the leaving underflow sludge; and X_e is the concentration of cell (escaping from the reactor).

$\frac{dX}{dt}$ = rate of microbial cell growth per unit volume of the reactor where $V.\frac{dX}{dt}$ is the total accumulation rate within reactor volume V.

Q_0 is the feed inflow rate; Q_e and Q_u are flow rates of treated effluent and the sludge underflow; K_d the microbial decay constant; μ_{max} is the maximum substrate conversion rate; S is the substrate concentration in the reactor; and S_0 is the substrate concentration in the influent. K_s is the Monod half-velocity rate constant, which stands for the concentration of the substrate at which reaction velocity reaches half the maximum value.

At steady state, $\frac{dX}{dt} = 0$.

This results in:

$$\mu_{max}\frac{S}{K_s + S} = \frac{Q_0X_u + (Q_0 - Q_w)X_e - Q_0X_0}{V.X} + K_d \qquad (3.9)$$

where μ = the specific cell growth rate.

The equation for the rate of substrate conversion and hence depletion or decomposition within control volume may be expressed as:

$$-V\frac{dS}{dt} = Q_0.S_0 - \frac{\mu_{max}.S}{Y(K_s + S)}(X.V) - (Q_0 - Q_w)S - Q_wS \qquad (3.10)$$

Depletion rate (substrate) within control volume = Inflow rate − conversion rate (degradation) − rate of discharge with effluent − rate of discharge with underflow sludge where the effluent flow rate $Q_e = Q_0 - Q_w$. The negative sign with the left-hand symbol indicates depletion or decrease.

At steady state:

$$\frac{ds}{dt} = 0$$

Therefore from Eq. (3.10), we get:

$$\mu_{max}\frac{S}{k_s + S} = \frac{Q_0Y}{V.X}(S_0 - S) \qquad (3.11)$$

Thus the expression for MCRT θ_c or sludge retention time (SRT) or sludge age may be derived from Eqs. (3.9) and (3.11) by equating the right-hand sides of these two equations as:

$$\frac{Q_w X_u + (Q_o - Q_w)X_e - Q_o X_o}{V.X} = \frac{Q_o Y}{V.X}(S_o - S) - K_d \qquad (3.12)$$

If we now define the MCRT θ_c as the ratio of the active biomass in the reactor to the rate of production of biomass then this term may be expressed in terms of reactor volume (V), the associated flow rates, and the corresponding microbial cell concentrations as:

$$\frac{V.X}{Q_w.X_u + (Q_o - Q_w)X_e - Q_o X_o} = \theta_c \qquad (3.13a)$$

where

$Q_w X_u =$ biomass wastage rate with underflow sludge

$(Q_0 - Q_w)\ X_e =$ microbial mass leaving with the effluent

$Q_0.X_0 =$ mass entering the system

Normally $X_0 = 0$ and then Eqs. (3.13a,b) can be rewritten as:

$$\frac{V.X}{Q_w.X_u + (Q_0 - Q_w)X_e} = \theta_c = SRT \qquad (3.13b)$$

$SRT = 1/\mu$ where μ is specific biomass growth rate.

The specific substrate utilization rate U can be related to X as:

$$U = S_{utR}/X = \frac{Q(S_0 - S)}{VX} = \frac{S_0 - S}{HRT.X}$$

where

HRT = hydraulic retention time

$S_0 =$ the influent soluble substrate concentration gBOD or COD/gVSS.d

$S =$ effluent soluble substrate concentration gBOD or bsCOD/m^3

$X =$ biomass concentration kg/m^3

SRT refers to the solids retention time and represents the mass of solids in the bioreactor divided by the mass of solids wasted ($Q_w.X_u$) intentionally through underflow sludge plus solids escaping the system through clarifier effluent $(Q_0 - Q_w)X_e$.

Thus the net rate of microbial mass production = net cell output − net cell input

$$= Q_w X_u + (Q_0 - Q_w)X_e - Q_0.X_0$$

The physical significance of the term θ_c is that it stands for the average residence time of the biomass in the reactor and is sometimes called SRT or sludge age. The MCRT is always greater than the nominal hydraulic residence time (NHRT) or θ, which is defined as

$$\frac{V}{Q_0} = \theta = \text{NHRT(Nominal Hydraulic Retention Time)} \tag{3.14}$$
$$= 1/D, \text{ where } D \text{ is the dilution rate } (Q_0/V).$$

The term "nominal" is used to differentiate it from the actual time (MCRT or θ_c) the microbial mass spends within the reactor for effective substrate conversion during operation of a biological treatment system with provision for recirculation resulting in a magnitude of θ_c much higher than that of θ.

Concentration of Microbial Cell (X) and Substrate (S) in Terms of MCRT and HRT

Using the definitions of MCRT (θ_c) and HRT (θ) in Eqs. (3.13a,b) and (3.14), we arrive at the new expressions for X and S as follows:

$$X = \frac{\theta_c Y(S_o - S)}{\theta(1 + K_d\theta_c)} \tag{3.15}$$

$$S = S_o - \frac{X.\theta(1 + k_d\theta_c)}{\theta_c Y} \tag{3.16}$$

Operating Parameters of Activated Sludge Process

Net Specific Substrate Utilization Rate

The net specific substrate utilization rate may be expressed as:

$$U_{net} = \frac{S_o - S}{\varphi.X} = \frac{Q_o(S_o - S)}{V.X} = \frac{1 + K_d\varphi_c}{\varphi_c Y} \tag{3.17}$$

Food-to-microorganism Ratio (F/M)

The food-to-microorganism ratio, which indicates the availability of substrate to the microorganism, is a very important process design parameter.

F/M $= Q_0.S_0/V.X$ is the food-to-microorganism ratio:

$$= \frac{(S_o - S)/t}{X} = \frac{(\text{BOD}_{in} - \text{BOD}_{out})/t}{X}$$

Recirculation Ratio

The recirculation ratio may be expressed as:

$$R = \frac{Q_R}{Q_o}$$

Performing material balance around the clarifier and assuming $dX/dt = 0$ in the clarifier (as there is no aeration):

$$(Q_o + Q_R)X = (Q_o - Q_w)X_e + (Q_R + Q_W)X_u$$

(volumetric inflow rate to the clarifier) × (associated cell concentration) = (volumetric effluent flow rate) × (associated cell concentration) + (volumetric outflow of waste that splits into recycle and underflow sludge waste stream) × (associated cell concentration). Thus the recirculation rate may be expressed as:

$$R = \frac{Q_R}{Q_o} = \frac{Q_o(X - X_e) - Q_w(X_u - X_e)}{Q_o(X_u - X)} \tag{3.18}$$

Minimum Mean Cell Residence Time (θ_c^{min})

In deriving the expression for minimum MCRT, we consider activated sludge reactor as shown in Fig. 3.8 to behave as an ideal chemostat involving the assumptions below:

1. reactor functions under steady-state conditions;
2. no microbial concentration in the feed;
3. constant inflow rate to the reactor (Q);
4. concentrations of substrate and cell in the effluent are the same as in the reactor.

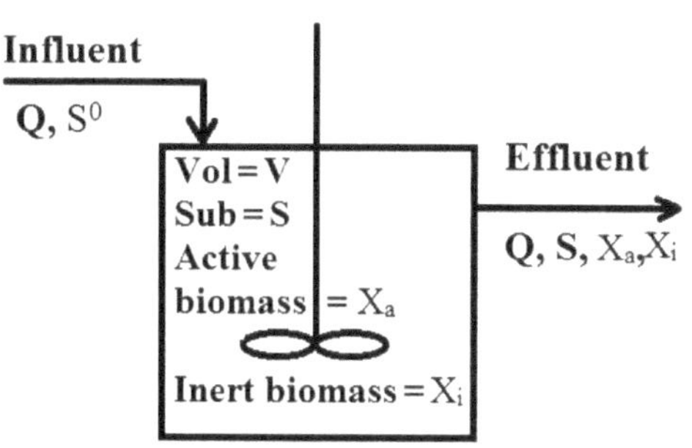

Figure 3.8 Activated sludge reactor as a steady-state chemostat. Activated sludge reactor as chemostat.

Now applying the Monod equation to express the microbial synthesis rate, we arrive at:

$$\mu_{\text{syn}} = \left(\frac{1}{x_a}\frac{dx_a}{dt}\right)_{\text{syn}} = \mu_{\max}\cdot\frac{S}{K+S} \tag{3.19}$$

where:

μ_{syn} = specific microbial growth rate due to synthesis (T^{-1})
$\mu_{\max}$ = maximum specific microbial growth rate (T^{-1})
t = time
K = half-velocity constant or substrate concentration giving half of the maximum specific growth rate (Mx/L^3)
S = substrate concentration

Microbes need flow of electrons and energy for their maintenance (resynthesis, repair, osmotic regulations, and transport) and considering loss of heat to the environment. This energy requirement is met by oxidation of decayed microbes, although the total mass of the decayed microbes does not undergo such oxidation (part remaining as inert mass). Cells oxidize themselves to meet this energy demand. This decay is called endogenous decay and is expressed mathematically as:

$$\mu_{\text{decay}} = \left(\frac{1}{x_a}\frac{dx_a}{dt}\right)_{\text{decay}} = -K_{\text{d}} \tag{3.20}$$

where

K_{d} = Endogenous decay coefficient (T^{-1})

Thus the net specific growth rate can be written as:

$$\mu = \mu_{\text{syn}} + \mu_{\text{decay}} = \frac{\mu_{\max} S}{K+S} - K_{\text{d}} \tag{3.21}$$

Substrate Utilization Rate

Substrate utilization results in microbial growth. The Monod equation can also be written in terms of substrate utilization rate as:

$$S_{\text{utR}} = \frac{-S_{\text{utRmax}} S X_{\text{a}}}{K+S} \tag{3.22}$$

where

S_{utRmax} = maximum specific substrate utilization rate ($M_sM_X{}^{-1}T^{-1}$)
S_{utR} = substrate utilization rate ($M_sL^{-3}T^{-1}$)

Substrate utilization and biomass growth are related by:

$$\mu_{\max} = S_{\text{utRmax}}.Y$$

where

Y = true yield of cells synthesis or yield coefficient ($M_sM_X{}^{-1}$)

Thus μ = net specific growth rate = $\frac{S_{\text{utRnet}}}{X_a}$:

$$S_{\text{utRnet}} = Y.\frac{S_{\text{utRmax}}SX_a}{K+S} - K_dX_a$$

and (3.23)

$$\mu = Y.\frac{S_{\text{utRmax}}S}{K+S} - K_d$$

Material Balance of the Activated Sludge Reactor as Steady-state Chemostat

With reference to the steady-state chemostat of Fig. 3.8, we now perform material balance around this chemostat to mathematically capture the minimum MCRT of an activated sludge reactor.

The steady-state chemostat is a completely mixed, continuous stirred-tank reactor containing uniformly distributed active biomass (X_a), inert biomass (X_i), and substrate S that includes both soluble and suspended substrate that can be hydrolyzed to soluble material. Substrate is the electron donor. We assume constant inflow (Q) with substrate concentration S^0 and zero microbial cell concentration.

The mass balance (under steady-state condition) for microbial cells is:

$$V\frac{dX_a}{dt} = 0 = \mu X_a V - QX_a \tag{3.24}$$

The mass balance for substrate under steady state is:

$$\frac{ds}{dt} = S_{\text{utR}} + Q(S^0 - S) = 0 \tag{3.25}$$

Now in the steady-state equations, using the values for μ and S_{utR} from Eqs. (3.21) and (3.22), we get:

$$0 = Y.\frac{S_{\text{utRmax}}.S}{(K+S)}.X_a.V - K_d.X_a.V - Q.X_a \tag{3.26}$$

$$0 = -\frac{S_{\text{utRmax}}.S}{(K+S)}.X_a.V - Q(S^0 - S)$$

$$\frac{S_{\text{utRmax}}.S}{(K+S)}.X_a.V = \frac{K_dX_a(V+Q)}{Y} \tag{3.27}$$

From Eq. (3.27), we get:

$$S = K\frac{1 + K_d(V/Q)}{[Y.S_{utRmax}.(V/Q)] - [1 + K_d(V/Q)]} \quad (3.28)$$

From Eq. (3.27) we also get:

$$\frac{S_{utRmax}.S}{(K + S)}.X_a.V = \frac{K_d X_a(V + Q)}{Y}$$

Using this value of $\frac{S_{utRmax}.S}{(K + S)}.X_a.V$ in Eq. (3.28) we get:

$$X_a = Y(S^0 - S)\frac{1}{1 + K_d(V/Q)} \quad (3.29)$$

We know that V/Q = HRT θ and MCRT or

$$\theta_c = \frac{\text{Active biomass in the system}}{\text{Production or wastage rate of active biomass(under steady state)}}$$

$$= \frac{V.X_a}{Q.X_a} = \frac{V}{Q} = \theta, \quad \text{i.e. under steady state,} \quad \theta = \theta_c$$

This implies that under steady-state the HRT is the same as the MCRT.

Thus Eqs. (3.28) and (3.29), can be rewritten as (3.30) and (3.31a,b) by replacing (V/Q) by θ_c:

$$S = \frac{K(1 + K_d\theta_c)}{YS_{utRmax}\theta_c - (1 + K_d\theta_c)} \quad (3.30)$$

$$X_a = Y\left(\frac{S^0 - S}{1 + K_d\theta_c}\right) \quad (3.31a)$$

When $\theta_c = \theta_c^{min}$, the system experiences washout resulting in $S = S^0$ and $X_a = 0$.

Thus from Eq. (3.30):

$$\theta_c^{min} = \frac{(K + S^0)}{S^0(Y.S_{utRmax} - K_d) - K.K_d} \quad (3.31b)$$

where

K is the Monod half-velocity rate constant

K_d is the decay constant

Y is the microbial cell yield coefficient

θ_c^{min} may also be expressed as:

$$\theta_c^{min} = \frac{K + S^o}{S^o(Y\mu_{max} - K_d) - K_d K} \tag{3.32}$$

The absolute minimum θ_c, or the limiting or boundary value of mean cell retention time for steady-state biomass, is defined as:

$$[\theta_c^{min}]_{lim} = \frac{1}{Y\mu_{max} - K_d} \tag{3.33}$$

In all practical designs, a safety factor is considered when deciding the MCRT.

Safety Factor Consideration

In general, a safety factor (S.F.), defined as $\frac{\theta}{\theta_c^{min}}$, is considered in design to address variation in temperature, washout characteristics, flow, operator skill, efficiency, reliability, and possible presence of inhibitory substances. Although typically a value of 20 is considered as S.F., its value may vary from 5 to 100.

$$\text{The design } \theta_c \text{ or } \theta_c^{min} = S.F[\theta_c^{min}]\text{lim}$$

Industrial Operations: Practical Considerations and Troubleshooting

The activated sludge process is almost temperature-independent and yields effluent of consistent quality. A microbial flock consisting of unicellular bacteria (rod and spherical shaped), fungi, nematode worms, metazoan, and protozoa remains in action in a diverse ecosystem where the HRT varies between 6 and 8 h. But imbalance in population among these diverse groups may result in poor functioning and poor effluent quality. Commissioning of a new plant is done using microbial seeding from a successfully operating plant or from settled sewage sludge and the process of stabilization may take 1 to 2 months' time during which adequate aeration and mixing is necessary.

During stable operation, MLSS should be maintained at a level of 2500–3000 mg/L. As an activated sludge plant generates a high quantity of sludge, there must be provision for continuous sludge removal and scientific handling of it. An activated sludge plant has the potential to be self-sufficient in terms of power supply where anaerobic digestion of the sludge can produce biogas that may either be directly marketed or used in power generation. Activated sludge tanks are usually up to 3 m deep of varying length and width. The inlet is fixed at one end but the outflow from the aeration tank is via a weir the height of which may be adjusted providing a scope for controlling the level of liquid in the tank. Such adjustable liquid height determines the degree of submergence of the mechanical aerators and hence DO concentration of the medium, which is maintained at an average of 2 mg/L (considered as the optimum level for the desired metabolism of the microbes and the settling characteristics of the sludge). The DO level needs to be monitored on a 24-h basis using large and robust DO probes that may be calibrated by immersion in sodium sulphate solution for 2 min (0% saturation) and by swinging in air (100% saturation). A DO level higher than 4 mg/L may be detrimental to balanced growth of a microbial population leading to rising sludge volume and difficulty in sludge settling. Mechanical aerators rotate at an average of 60 rpm. Aeration can also be done by diffused aerators, which are plastic-made fine bubbling domes installed at the bottom part of the aeration tank with plastic piping network having flexibility of occasional cleaning. DO probes may be synchronized with the functioning speed of the aerators for automatic control of DO level in the tank. Air is supplied by air blowers. An energy-efficient aerator should dissolve at least 2 kg of O_2/kWh energy consumption. The standard rating of such aerators may be 2.4 kg oxygen supply/m^3/h at 125 mm immersion to 3.7 kg oxygen supply/m^3/h at 200 mm immersion. Along with the desired level of oxygen, a supply of food at the rate of around 0.3 kg BOD/kgMLSS/day (i.e., the F/M) needs to be maintained for efficient running of the plant. Volumetric loading is typically 0.5 kg BOD/m^3/day, but a loading rate up to 1.0 may sometimes be allowed. Provision for aeration should ensure 1.0 to 3.5 kg O_2 /m^3/day.

Bulking Sludge—Problem of High Filamentous Growth

Most of the operational trouble in activated sludge plants is due to microbial flock characteristics. Unless these flocks are well settleable, clear water at the effluent outlet cannot be ensured and recycling of sludge from settling unit to the aeration unit will be difficult.

Bulking sludge refers to the poorly settling sludge resulting from huge filamentous growth of bacteria. While sufficient filamentous bacteria are required to form a strong and compact microstructure to which zoogleal microorganisms get attached and form a strong and compact filamentous macrostructure suitable for formation of a microbial flock with good settling characteristics, the presence of too many filamentous bacteria can have a disastrous effect on flock settling. Some of these filamentous bacteria have extended growth that serves as bridges between flocks and join them. This works against compaction of the flocks by preventing them from coming closer. Regular microscopic observation by trained personnel or measurement of the sludge volume index (SVI) can provide clues to the bulking sludge. The SVI should be around 100 mL/g or less, and should be monitored continuously. Sludge particles should be regular shaped, and with few filamentous bacteria should be compact so that it settles well.

Causes of Rising Sludge

Rising sludge may result from low activated sludge return, excessive aeration, low DO, enhanced denitrification at temperature above 20°C, low F/M ratio, and high presence of reduced sulfur (sulfides). Reduced sulfur encourages growth of filamentous *Thiothrix* as sulfur-oxidizing species. Under such conditions, fungal filamentous growth dominates, which drastically reduces the settleability of the sludge resulting in a significant loss of bacterial mass through weir outlet. This is called bulking and occurs when the stirred sludge volume index (SSVI) exceeds 150 mL/g. An SSVI value of 200 mL/g is an indicator of a serious bulking sludge situation. The SSVI is determined by measuring the sludge length in mm that settles during half an hour slow stirring of the sludge in a 4 L cylinder and by dividing this sludge height by the solids concentrations, which is typically 3.5 g/L for an activated sludge feed water. Bacterial filamentous growth resulting in low-density sludge and poor settling characteristics may also occur. Another problem that is encountered in aeration or settling is foam formation resulting from the presence of long-chain fatty acids or surfactants at the air–sludge interface and the presence of two foam-forming bacteria associated with the presence of fats and edible oils. *Nocardia and Microthrix parvicella* are two bacterial species with hydrophobic surfaces that attach to the air-bubble surfaces causing foam. These bacteria have filamentous structure. Foam present to a moderate degree in the aeration tank may not pose a big threat to stable operation, but it should not escape to the downstream settling unit through the weir.

Troubleshooting

Chlorination (1–3 mg/L) with return sludge may sometimes reduce such filamentous growth but may also invite another problem of formation of carcinogenic organo-chlorine compounds. Hydrogen peroxide (H_2O_2) is an alternate solution. In foam control, chlorination is often adopted, but oil and grease discharge should also be controlled. In the case of sludge bulking from reduced sulfur, the input reduced sulfur should be eliminated before passing the wastewater to the aeration tank. Chemical oxidizing agents such as H_2O_2 may be used to oxidize reduced sulfur through the following reaction:

$$4H_2O_2 + HS^- \rightarrow SO_4^{2-} + 4H_2O + H^+ \tag{3.34}$$

In the case of sludge bulking from low DO, adequate oxygenation following this relation is necessary:

$$\text{D.O.}(\text{mg/L}) > (R_{\text{COD/MLVSS}} - 0.1)/0.22 \tag{3.35}$$

where $R_{\text{COD/MLVSS}}$ = ratio of COD to MLVSS.

In this case enhanced oxygenation or reduced BOD loading may be the solution.

Use of Selector Technology in Addressing Sludge Bulking

For low F/M bulking, use of a selector contact tank provides a solution to the sludge-bulking problem. The selector tank is an additional contact tank placed before the entry point of the aeration tank where return sludge and feed wastewater are mixed without provision for aeration. Under anaerobic or anoxic conditions, growth of floc-forming bacteria is favored while growth of filamentous bacteria is limited. The filamentous bacteria cannot use nitrite or nitrate as electron acceptor, thus leaving space for growth of floc-forming bacteria, which can, however, use nitrite or nitrate as electron acceptor in the absence of oxygen.

General Monitoring for Stable Operation

For stable operation, maintenance of proper aeration, recycling of sludge, proper level of sludge withdrawal, F/M ratio, balance composition of the nutrients (carbohydrate, N, P), temperature, pH, and concentration of toxic substances need to be closely monitored and controlled. Continuous monitoring of growth of filamentous bacteria through SVI measurement or microscopic observation is absolutely essential to successful functioning of an activated sludge plant along with monitoring and exclusion of undesirable reduced sulfur and toxic compounds.

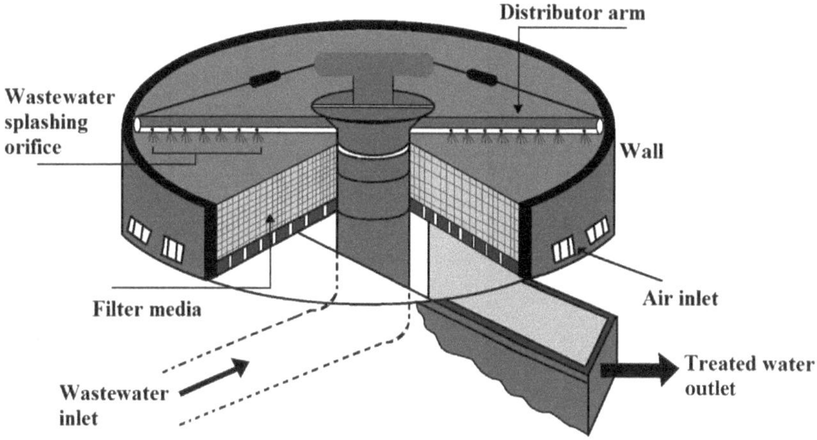

Figure 3.9 A typical trickling filter configuration.

3.6.2 Trickling Filter

A trickling filter operates on an attached growth mechanism of the microbes that grow on solid support forming filter bed. In the standard design, perforated distributor arms horizontally pivoted to a common wastewater header pipe slowly rotate as water is sprinkled out of the perforations and trickles down through the solid-filter bed covered with microbial film as shown in Fig. 3.9. A trickling filter offers secondary treatment where microbes degrade the pollutants, which are primarily organic. Prior to introduction of the wastewater to the trickling filter unit, primary treatment removes settleable and floatable solids. By the time the wastewater reaches the trickling filter it contains mostly colloidal and dissolved solids that are removed through the aerobic biological decomposition process.

Operation of a Trickling Filter

For successful operation of a trickling filter, hydraulic loading and rotation of the wastewater distributor arms play important roles.

Hydraulic loading is defined as:

$$\text{H.L} = (Q + Q^{r})/A_{pv}$$

where Q and Q_r are wastewater inflow rate and water recycling rate, respectively. A_{pv} is the plan view surface area of the filter bed. Hydraulic loading has to be controlled because it determines the water-layer thickness. Higher hydraulic loading results in greater thickness of this water layer, which in turn also increases the liquid hold-up and detention time. Hydraulic loading also determines the wetted surface area of the filter medium where microbes grow for biodegradation. Greater hydraulic load tends to distribute the biofilm deeper into the bed facilitating more surface area to be active in biodegradation. Higher hydraulic loading also contributes to reduce clogging potential as the resulting shear stress causes increased detachment of the excess biomass, which otherwise might clog the filter bed. Increased hydraulic loading also increases the rate of oxygen mass transfer from the air phase.

Massive detachment of biofilm from the solid surface is called sloughing and is not desirable as the filter bed surface gets totally depleted of the active biomass, which is essential to biodegradation. Anaerobic conditions at the solid surface-biomass interface cause weakening of the biomass resulting in drastic reduction in the capacity of the microbes to hold on to the solid surface. Thus under anaerobic conditions, following structural weakness, large chunks of biomass get detached from the filter-bed surface. To prevent sloughing or massive detachment of microbial mass, adequate air ventilation is needed. Slowing down of the rotary distributor to cause pulsed hydraulic loading is another option as pulsation prevents excessive biomass buildup. Maintaining a flushing intensity of the order of 0.1 to 0.5 m/(arm revolution) is suggested for stable operation. Another very crucial operational control on the process performance is exercised through controlled use of recycled effluent. Recycled effluent increases hydraulic loading, dilutes influent, and increases DO concentration. Thus recycling of the effluent stream partially helps achieve a favorable BOD/DO ratio, which is particularly necessary when handling high-strength wastewater.

Fig. 3.9 shows the basic structure of a trickling filter unit. The complete treatment plant needs arrangement for pumping the wastewater through the common header, a settling unit for clarification of the filter-treated water, pumping arrangement for recycling a part of the treated water from the downstream settling unit to the filter bed. For adequate oxygenation, an air blower or compressor should be attached to the filter bed at the bottom so that the bed surface where microbes grow as attached film get a supply of oxygen.

Trickling filters may be operated at low-, moderate-, or super-flow rates depending on the wastewater quality and desired degree of purification. A low-rate filter is a relatively simple, highly dependable device that

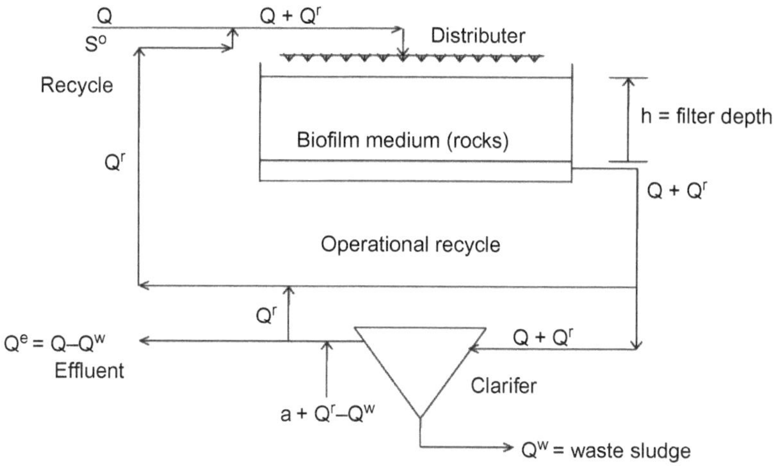

Figure 3.10 Material flow scheme of a trickling filter.

produces an effluent of consistent quality with an influent of varying strength. Moderate-rate filters operate with recirculation of the filter effluent or final effluent permitting higher organic loadings, whereas super-rate filters are used for wastewater with high concentration of the polluting substances. Super-rate filter-treated effluents will naturally be of poor quality and these filters are suggested as preliminary roughing units only Fig. 3.10.

A trickling filter involves moderate operating costs compared to an activated sludge process but initial investment is high. This system of attached growth can remove BOD up to 80–90%, and a two-stage system can reach 95% removal. A trickling filter can withstand shock load better than an activated sludge process. However, clogging of the bed is a problem often encountered, and can sometimes create a breeding ground for mosquitoes.

3.6.3 Lagoon: the Low-cost Bioremediation Technology

Anaerobic Lagoons

Anaerobic lagoons are most often used to treat animal wastes from dairy and pig farms, commercial or industrial wastes, or as the first treatment step in systems using two or more lagoons in series. Typically, anaerobic lagoons are designed to hold and treat wastewater from 20 to 150 days. They are relatively deep (usually 8 to 15 feet) and work much like septic tanks where anaerobic bacteria degrade pollutants in the absence of oxygen. Inside an anaerobic lagoon, solids in the wastewater separate and settle into layers. The top layer consists of grease, scum, and other floating materials. If not preceded with septic tanks, the layer of sludge that settles at the bottom of an anaerobic lagoon eventually accumulates and must be removed. The wastewater that leaves an anaerobic lagoon will require further treatment.

Natural Aerobic Lagoon

DO is present throughout much of the depth of aerobic lagoons in this class. They tend to be much shallower than other lagoons, so sunlight and oxygen from air and wind can better penetrate the wastewater. In general, they are better suited for warm, sunny climates, where they are less likely to freeze. Wastewater usually must remain in aerobic lagoons from 3 to 50 days to receive adequate treatment.

Wastewater treatment takes place naturally in many aerobic lagoons with the aid of aerobic bacteria and algae. Because they are so shallow, their bottoms need to be paved or lined with materials to prevent weeds from growing on them.

Facultative Stabilization Lagoon

Facultative stabilization lagoons are deep lagoons in which the upper part behaves as an aerobic lagoon and the lower part is deprived of oxygen supply under the cover of sedimentation layer of the sludge functions an anaerobic reactor. Fig. 3.11 shows a facultative lagoon.

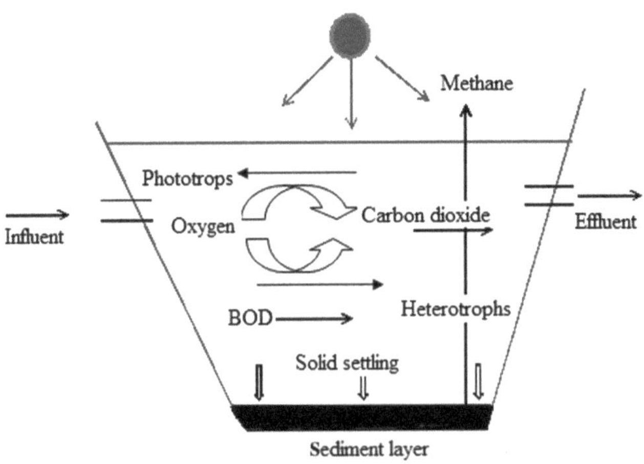

Figure 3.11 Facultative stabilization lagoon.

In an aerobic lagoon, phototrophs decompose polluting substrate utilizing atmospheric oxygen and sunlight where the main product of oxidation is CO_2. In an anaerobic lagoon (the bottom part of the lagoon) in the absence of oxygen under the layers of sedimentation, anaerobic decomposition of the substrates results in production of new cells and methane (CH_4) as biogas.

Advantages of Lagoons

- Lagoon systems can be cost-effective to design, construct, and operate in areas where land is inexpensive.
- The energy requirement of lagoons is almost negligible compared to the requirement of most other wastewater-treatment methods.
- They are simple to operate and maintain and generally require only part-time staff.
- Lagoons can take care of fluctuating flow and characteristics of wastewater. They can absorb shock loadings better than many systems, making them a good option for campgrounds, resorts, and other seasonal properties.
- They are very effective at removing disease-causing organisms (pathogens) from wastewater.

Disadvantages of Lagoons

- Lagoon systems require more land than other treatment methods and thus may be suitable for remote areas where land is not a problem.
- In cold climates efficiency of lagoons goes down demanding possibly larger land and longer time.
- In anaerobic lagoons and in lagoons that are inadequately maintained, odor can become a nuisance during algae blooms.
- Unless they are property maintained, lagoons can provide a breeding area for mosquitoes and other insects.
- They may not be successful in removing heavy metals from wastewater.

3.6.4 Submerged Aerated Filter Technology

Fig. 3.12 shows a submerged aerated filter (SAF) system that combines biological treatment with physical filtration within the same reactor making it compact. Wastewater is made to percolate at a high rate down through a fixed bed of granular material like activated carbon. Air is supplied counter-current via a grid of pipes and diffusers arranged 20−30 cm above the media base. The provision of highly porous bed material allows

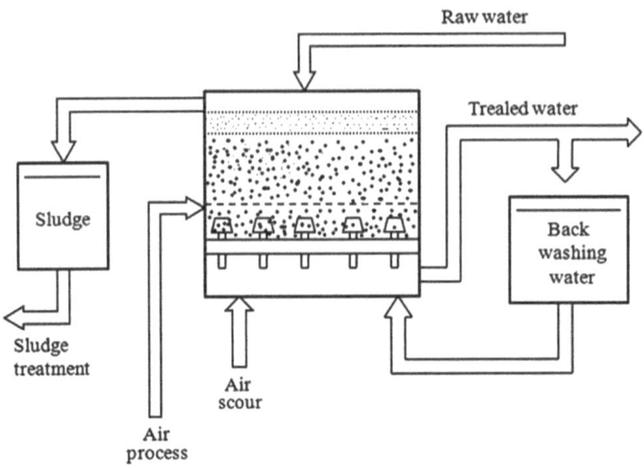

Figure 3.12 Bicarbone process flow diagram.

a concentration of biomass four times greater than in an activated sludge plant. While the whole porous media itself acts as a filter, the nonaerated portion of the medium below the diffusers in particular provides the final physical filter for retaining the SS. A part of the treated water is used in backwashing for cleaning the clogged bed material.

Like lagoons, submerged aerated filters can also accommodate large variations in flow and load. These systems do not involve high capital or operating costs and leave a relatively small footprint. The major advantage of this treatment technology is operational robustness coupled with simplicity.

3.6.5 Upward Flow Anaerobic Sludge Blanket Reactor Technology

Upward flow anaerobic sludge blanket reactor (UASB) technology is effective mainly for organic-loaded wastewater such as sugar industry wastewater where the whole process is completed in three phases involving a total time of 10–30 days depending on the waste strength. In the first phase of bacterial action, which takes around 10–15 days, the complex organics are completely hydrolyzed or solubilized to be fit for absorption by the bacteria in the subsequent phase. In the second stage, another group of bacteria converts solubilized substrates into organic acids through an action called acidogenesis. Then the methane-producing (methanogenic) anaerobic bacteria use the acid products, which causes complete decomposition of the substrates producing methane gas. The process thus produces a three-phase material comprising solid-sludge

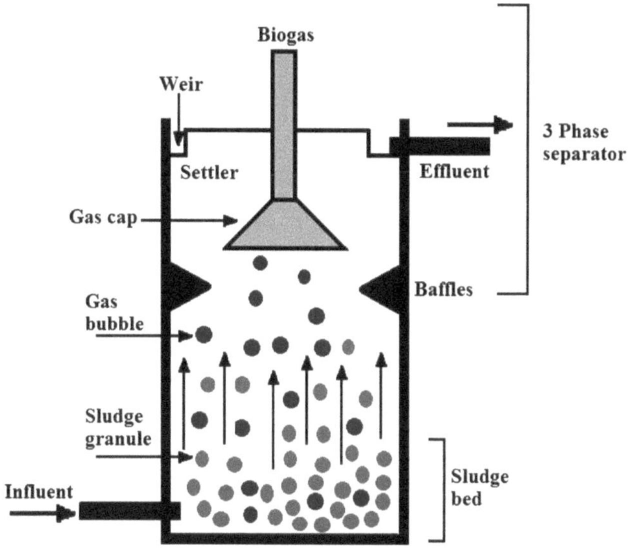

Figure 3.13 Upward flow anaerobic sludge blanket reactor.

particles, treated water, and generated methane gas. At the top of the reactor, the water outlet via a V-notch weir, the conical gas cap, and the horizontal baffles effect three-phase separation as shown in Fig. 3.13. The bottom sludge bed is about 2 m deep and contains thick material while the top sludge blanket is about 2% Dry Matter (DM). The technology is similar to submerged aerated filter technology as the sludge in this case serves the purpose of a filter. As incoming wastewater enters the reactor at the bottom and moves up through sludge blanket, which is located slightly above the feedwater entry point, the suspended particles get largely filtered out by the sludge. Settling of the sludge particles is also aided by the actions of the baffles that change the direction of the uprising particles.

The process works best at a pH in the vicinity of 7.0 with a COD: N:P ratio of 350:5:1. Trace elements like Fe, Ni, Se, Mo, Cu, Mn, Cr, and Co need to be added if not present in the wastewater. One to 3 days of HRT and around 25–30 days of SRT are normally required for the UASB process.

UASB is considered a low-energy consuming technology demanding less land and low overall cost compared to other biological anaerobic digestion systems. However, disadvantages include long start-up period and the need for sufficient seed sludge. Around 60% COD reduction using this technology is considered reasonable.

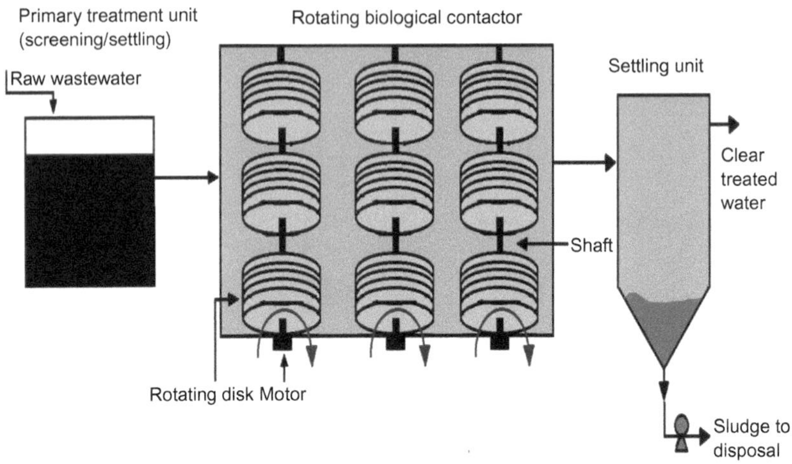

Figure 3.14 Biological treatment scheme using rotating-disc biological contactor.

3.6.6 Rotating-Disc Biological Contactor Technology

Rotating-disc biological contactors (RBC) function on the same principles as trickling filters where the microbes grow on the surface of the slowly rotating (1–2 rpm) circular-disc surfaces (0.5 m to 5.0 m in diameter). A typical wastewater treatment scheme using RBC is presented in Fig. 3.14.

A series of discs made primarily of polymer (PE, PVC, etc.) rotate in a vertical plane on a common horizontal motor-driven shaft while remaining partially (50%–80%) submerged in the wastewater to be treated. Alternate exposure of the rotating discs to the waterpool and the atmosphere help maintain both an aerobic (when exposed to atmosphere) and anaerobic environment (when submerged in wastewater) for the microbes, although RBCs are primarily used for aerobic treatment of wastewater. The dual-function capability is particularly useful for nitrogen removal when both the aerobic environment for nitrification and anaerobic environment for denitrification leading to conversion of ammonia to nitrogen are essential. Both aerobic and anaerobic microbes can grow on the disc surfaces. RBCs are considered compact designs making them suitable for installation even within urban sites. The presence of high microbial populations results in a high (8–10 times that of trickling filters) degree of biodegradation of the polluting substances. A well-operated RBC can ensure close to 90% biodegradation of polluting substances. The need for continuous power supply and the plant operation and maintenance (washing of the surfaces, lubrication of the rotating parts and motors) required by skilled staff, however, results in higher cost of treatment compared to ACS or trickling filters. Anaerobic digestion of the produced sludge leading to biogas generation and subsequent power generation can substantially offset the power cost of RBCs and can even make these units self-sustaining. If proper operating conditions (rotation of discs, oxygenation, nutrient supply, good hydraulics) are maintained, RBCs can be used to treat a range of wastewater from domestic (300−500 mg/L BOD loading) to high-strength industrial wastewater (BOD loading of 4000−5000 mg/L). However, both pretreatment of influent (for removal of heavy particles, suspended materials) and post-treatment of the effluent (for pathogen removal by sand or membrane filtration and disinfection) are necessary. Covering of the RBCs with transparent material may be necessary for their protection against rain, sunlight, and fluctuating weather that affect the efficiency of RBCs. However, such cover may necessitate additional oxygenation. Normally 2−3 months' time is required is to stabilize a biological treatment plant using RBC where seeding is not required whereas an activated sludge process is stabilized within one month with seeding from a successfully operating plant.

3.7 ADVANCES IN BIOLOGICAL TREATMENT TECHNOLOGIES

3.7.1 Introduction

Conventional biological treatment technologies have been in use for a very long time for large-scale treatment of huge quantities of wastewater both sewage water as well as industrial wastewater largely because of low cost. However, over time and with large-scale industrialization, regulations on industrial and municipal waste discharge have become more and more stringent in light of findings on the disastrous health effects of such waste discharges to natural water bodies. In many developing countries, thousands of kilometers of river bodies considered as vital life lines have been severely polluted by such untreated or improperly treated waste discharges. Large amounts of discharged waste, high concentration of polluting substances, and thousands of varieties of discharged compounds has made water management complex and tough amidst ever-rising demand for safe and clean water. These developments in many cases are resulting in limited success of conventional treatment technologies.

For example, coke-making units, which are often located within steel-making industries or coal-based power plants, generate enormous quantities of wastewater containing a number of highly hazardous compounds like cyanide, phenol, ammonia, thiocyanate, and other toxic and pathogenic contaminants represented by high COD and BOD values. Effective treatment of the organic and inorganic compounds present in coke wastewater has remained a longstanding problem in the industry. Failure of treatment plants often leads to discharge of heavily polluted water into natural water bodies like rivers and lakes causing serious environmental pollution. Many coke-making industries use hazardous wastewater to quench hot coke resulting in serious air pollution as carcinogenic aromatics, phenolic compounds, cyanide compounds and ammonia eventually pollute air bodies. The contaminants present in coke wastewater often oppose effective treatment of each other. Considering the enormous volumes of coke wastewater, biological treatment is often a low-cost option for major contaminants like phenol, ammonia, and similar compounds, but the presence of cyanide compounds makes the environment difficult for microbes to survive, which often leads to plant failure. In this context, development of integrated process technologies has gained attention. Phenol is the organic contaminant that contributes the most to the total COD in coke wastewater. Nitrogen compounds (NH_4^+-N, organic bound N and NO_3^-) are other major contaminants in coke wastewater. In recent years, attempts have been made to treat such complex and hazardous wastewater using membranes [1–3]. Membrane-integrated advanced treatment has the potential to treat wastewater up to the reusable criteria level leading to twofold benefits. On the one hand, hazardous wastewater will be prevented from polluting otherwise clean surface water bodies where wastewater is normally discharged, and on the other hand, it will reduce freshwater consumption by the industry.

Advanced development of biological wastewater treatment has focused on integration of conventional biological treatment units with chemical and membrane-based separation units [4]. In Chapter 6, development of a membrane-integrated biochemical treatment technology is described in detail. This technology is sustainable and promises to increase the amount of reusable reclaimed water and valuable byproducts and removes toxic compounds such as cyanide from a hazardous wastewater stream in a prechemical treatment unit using Fenton's reagents under optimized conditions. More than 95% of NH_4^+-N is recovered as a valuable byproduct called struvite through addition of appropriate doses of magnesium and phosphate salts. Wastewater eventually becomes reusable through a polishing treatment of nanofiltration (NF) membranes in a largely fouling-free membrane module following a biodegradation step. In this section, we describe some important aspects of membrane-integrated hybrid bioremediation technology [5]. These aspects include the fundamental principles involved in the chemical, biological, and membrane separation steps, process optimization, analysis of plant performance under varying operating conditions, and issues of sustainability of the technology.

3.7.2 Membrane-integrated Hybrid Treatment Technology

Principles of Membrane-integrated Hybrid Treatment Technology

Chemical Treatment

Cyanide oxidation by Fenton's reagent is highly dependent on pH. At high pH cyanide is present as CN^- ions and reacts easily with H_2O_2 and Fe^{2+} ions, but in acidic conditions, cyanide is present as HCN gas, which is difficult to oxidize. Fenton's treatment has two distinct stages, namely Fenton's oxidation (OH^- generation) and Fenton's coagulation, which is mainly simple ferric coagulation following the oxidation stage and resulting in sludge. Degradation of cyanide using Fenton's reagent follows first-order kinetics [6]:

$$-\frac{d[CN^-]}{dt} = k_{[CN^-]}[CN^-] \tag{3.36}$$

It has been observed that [7] Fe^{2+} added in the form of $FeSO_4 7H_2O$ acts as catalyst and the peroxide radicals ($OH_2^{\bullet}$) produced are capable of further oxidizing other species including Fe^{2+} present in the reaction medium as follows:

$$HO_2^{\bullet} + Fe^{2+} \rightarrow O_2 + Fe^{3+} + H^+,\ k = 1.26 \times 10^6\ M^{-1}S^{-1} \tag{3.37}$$

There is also the possibility of autoregeneration of Fe^{2+} in this system that may act as catalyst.

$$Fe^{+3} + H_2O_2 \rightarrow HO^{2\bullet} + Fe^{2+} + H^+ \tag{3.38}$$

Biological Treatment

The degradation kinetics as observed during experimental investigation [5] are presented here. The Monod model is applied. In the biological treatment process, the relationship between the rate of growth of microorganisms and the rate of substrate utilization is expressed by:

$$\frac{dX}{dt} = Y\frac{ds}{dt} - K_d.X \tag{3.39}$$

where

X = microbial concentration

Y = growth coefficient, mass of microorganisms produced per unit mass of substrate utilized

S = concentration of organic food substrate utilized by microorganisms

K_d = microbial decay coefficient, time^{-1}

Dividing both sides of Eq. (4) by X, we get:

$$\frac{dX/dt}{X} = Y.\frac{dX/dt}{X} - K_d \tag{3.40}$$

In the above equation, $\frac{dx/dt}{X}$ is the specific growth rate, often represented by μ. The inverse of μ is referred to as the solids retention time or mean cell retention time, θ_c.

$$\theta_c = \frac{X}{\frac{dX/dt}{X}} \tag{3.41}$$

The term $\frac{ds/dt}{s}$ is the substrate utilization rate per unit amount of biomass and is called the specific substrate utilization rate. It can be approximated by the following expression:

$$\frac{ds/dt}{X} = \frac{k_s}{k_s + s} \tag{3.42}$$

where k_s = maximum specific substrate utilization rate, time$^{-1.}$

Substituting Eqs.(3.41) and (3.42) in Eq.(3.40), we get Eq.(3.43) below

$$\frac{1}{\theta_c} = Y\left[\frac{k.s}{k_s + s}\right] - K_d \tag{3.43}$$

In an activated sludge system, it is assumed that the contents in the aeration tank are completely mixed and that there are no microbial solids in the raw wastewater influent. It is further assumed that the influent substrate concentration, S_0, remains constant and that the system operates under steady-state conditions. The solids are wasted from the sludge-recycle line, although they may also be wasted from the aeration tank.

For a completely mixed system, Eq. (3.43) can be written in terms of the system parameters. Thus the solids retention time can be expressed as:

$$\theta_c = \frac{V.X}{[Q_w.X_r + (Q - Q_w)X_e]} \tag{3.44}$$

An expression for S_e can be obtained by rewriting Eq. (3.43) as:

$$\frac{1}{\theta_c} = Y\left[\frac{k.S_e}{k_s + S_e}\right] - k_d \tag{3.45}$$

Rearranging this expression, we get,

$$S_e = \left[\frac{k_s(1 + \theta_c.k_d)}{\theta_c Y.k - (1 + \theta_c k_d)}\right] \tag{3.46}$$

The relationship between X and S_e in all tanks can be obtained by first considering a substrate material balance around the tanks. This step gives the amount of substrate utilized per unit time and per unit volume of the aeration tank as:

$$\frac{ds}{dt} = \frac{Q(S_o - S_e)}{V} \tag{3.47}$$

A loading parameter that has been developed over the years, called the HRT, q

$$q = \frac{V}{Q} \tag{3.48}$$

where

V = volume of aeration tank, m^3, and Q = sewage inflow, m^3/d

where

S_o = influent substrate organic matter, pollutants (g/m^3)

S_e = effluent substrate organic matter, pollutants (g/m^3)

A similar loading parameter is MCRT or SRT, q_c, expressed as:

$$q_c = \frac{V.X}{Q_w.X_r + (Q - Q_w)X_e} \tag{3.49}$$

Under steady-state operation the mass of waste activated sludge is given by:

$$Q_w.X_r = YQ(S_o - S_e) - k_d.V.X \tag{3.50}$$

Q = influent flow rate, m^3/d

Q_w = waste sludge flow rate, m^3/d

X_e = concentration of biomass in effluent, g VSS/m^3
X_r = concentration of biomass in the return line from clarifier, g VSS/m^3
where Y = maximum yield coefficient

Principles of Membrane Separation

NF is a liquid-separation membrane technology positioned between reverse osmosis (RO) and ultrafiltration membranes and its performance is predicted in two separate components: the pure water flux and the solute flux that their relationships are independent from each other; in each of the models the pure water flux can be related to pressure (ΔP). The separation of the solutes from the solution depends on two ways, steric (sieving) and Donnan (electrostatic) mechanisms based on whether solutes are charged or uncharged, which can be explained by continuum hydrodynamic models such as that originally proposed by Ferry (1936) and the extended Nernst–Planck model [8], respectively. In the hydrodynamic models, porous membranes are represented as a bundle of straight cylindrical pores and solute transport is corrected for hindered conversion and diffusion due to solute–membrane interactions. The solvent velocity through the pores of the NF membranes may be expressed using the Hagen–Poiseuille equation as [9]:

$$J_W = \frac{r_p^2 \Delta p}{8\,\mu\delta} \tag{3.51}$$

where

J_w = pure water permeability
r_p = pore radius
ΔP = difference in applied pressure across the membrane
δ = thickness of the membrane
μ = viscosity of fluid

According to Eq. (3.51), increasing the pressure will increase the pure water flux. The solute flux is proportionally related to the solute concentration gradient across the membrane. The osmotic pressure difference $\Delta\pi$ may be calculated using Van't Hoff equation:

$$\Delta\pi = RT\sum(C_{is} - C_{ip}) \tag{3.52}$$

where C_{is}, C_{ip} are, respectively, feed- and permeate-side concentrations of the solute.

Fluxes of permeate and solute may be computed respectively as:

$$J_v = L_w(\Delta P - \Delta\pi)/\eta \tag{3.53}$$

$$J_i = VC_{ip} \tag{3.54}$$

where L_w is pure water permeability determined experimentally.

Solute flux is defined using the extended Nernst–Planck equation as follows:

$$J_i = \left[-D\frac{dc_i}{dx}\right] - \left[D_i z_i c_i \frac{F}{RT}\frac{d\psi}{dx}\right] + \left[K_i c_i J_v\right] \tag{3.55}$$

where

J_i = solute flux i

D_i = diffusivity of solute i

c_i = concentration of solute i at the surface of the membrane

x = mole fraction for solute i

z_i = the valance of solute i

F = Faraday's constant

R = gas constant

T = temperature

Ψ = electric potential

K_i = distribution coefficient of solute i

J_v = volume flux and can be estimated based on the membrane area

The three different terms of the Nernst–Planck equation describe a different component of the solute flux. The first term describes (as a function of concentration gradient across the membrane), the second term quantifies the flux due to electrostatic forces (as a function of charge gradient), and the last represents the convection of solute i at the surface of the membrane. Ionic transport through NF membrane has widely been explained by models based on Donnan equilibrium theory, which states the equality of electrochemical potential (Ψ) of the solutions on either side of the solution-membrane interface. Charged molecule transport through NF membrane has been explained by models based on Donnan equilibrium theory, which states the equality of electrochemical potential (Ψ) of the solutions on either side of the solution-membrane interface.

The concentration at the surface of the membrane can be estimated using the Donnan equilibrium theory as follows:

$$\psi_D = \psi_m - \psi_s \text{ or } c_i = C_i \exp\left[-z_i \frac{F}{RT}\Delta\psi_D\right] \tag{3.56}$$

where

C_i = bulk (or feed) concentration of solute i

Donnan potential (Ψ_D) is the difference between the electrical potential of the solution (Ψ_s) and the electrical potential of the membrane (Ψ_m).

The Nernst–Plank equation coupled with the Donnan equilibrium theory has been shown to accurately predict the rejection of various salts by NF and RO membrane. Eqs. (3.55) and (3.56) explain why solute concentrations at the surface of the membrane determine its passage. If the charge (or electrical potential) of the membrane increases, the concentration of the counter ions (ions with the opposite charge) also increases. For neutral species, the concentration at the membrane surface is unaffected by the charge of the membrane [9]. Most of the polyamide composite NF membranes possess negative zeta potential at pH values greater than 7.0. Due to the charged nature of the NF membrane, solutes with an opposite charge compared to the membrane (counterions) are attracted, while solutes with a similar charge (coions) are repelled. In addition, distribution of co- and counterions will occur, thereby causing more separation.

Functioning of the Treatment Plant

Materials

The treatment plant may be made of high-grade stainless steel (SS-316) to avoid rusting during long operation. Thin-film composite polyamide NF membranes of 165 μm thickness and average pore size of 0.5–1.5 nm were used in flat sheet cross-flow membrane module may be used. The characteristics of such membranes are presented in Table 3.1.

Operation

The membrane-integrated bioremediation system used for experimental investigations is shown in Fig. 3.15.

Table 3.2 shows progressive improvement in the quality of treated water as it is subjected to chemical, biological, and membrane treatments.

Table 3.1 Major characteristics of some flat-sheet, polyamide composite NF membranes [5]

	Membranes			
Characteristics	**NF1**	**NF2**	**NF3**	**NF20**
Solute rejection, %				
$MgSO_4$	99.5	97	98	98
NaCl	90.0	50	60	35
pH	2–11	2–11	2–11	2–11
Maxi^{m} temp (°C)	50	50	50	50
Maxi^{m} pressure (bar)	83	83	83	83
Pore radius (nm)	0.53	0.57	0.55	0.54
Thickness (μm)	165	165	165	165

Reprinted from Journal of Water Reuse and Desalination, Kumar, R., Pal, P., 2013B. Membrane-integrated Hybrid Bioremediation of Industrial wastewater: A Continuous Treatment and Recycling Approach, Desalination and Water Reuse, 1, (3), 26–38, with permission from the copyright holders, IWA Publishing.

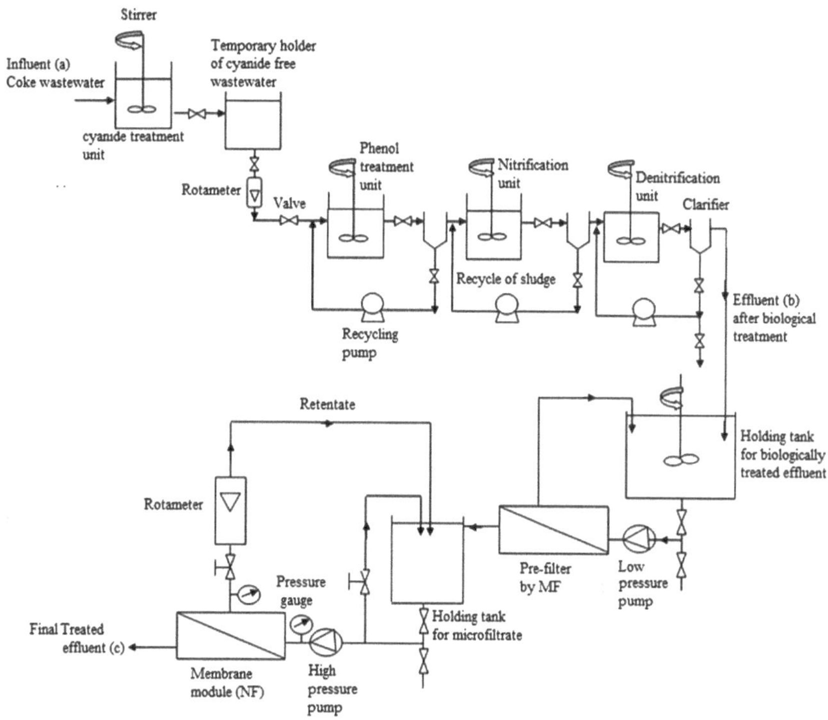

Figure 3.15 Scheme of the hybrid treatment technology integrating biological and chemical treatments with membrane-separation units [5]. *Reprinted from Journal of Water Reuse and Desalination, Kumar, R., Pal, P., 2013B. Membrane-integrated Hybrid Bioremediation of Industrial wastewater: A Continuous Treatment and Recycling Approach, Desalination and Water Reuse, 1, (3), 26–38, with permission from the copyright holders, IWA Publishing.*

Table 3.2 also shows that the membrane-integrated treatment eventually succeeds in producing a final effluent that meets the industrial effluent = discharge limits.

After three stages of treatment, which includes chemical treatment for cyanide, biological treatment for phenol and ammonia, and finally polishing for removal of trace elements mostly in ionic forms by NF membrane wastewater becomes reusable. Fenton's reagent ($FeSO_4$ $7H_2O$ and H_2O_2) is used in the chemical treatment unit for removal of cyanide considered toxic to the microbes, even for removal of cyanide in the first reactor of the series. pH is maintained in the range of 7.0–8.0 by adding 10M NaOH or concentrated HCl solution. Clear solution from this unit overflows to the temporary holding tank. Microbial treatments are carried out with well-known microbial strains (*Pseudomonas species*, for phenol, *Nitrosomonas* (NCIM 5076), and *Nitrobacter* (NCIM 5078) for nitrification, and *Pseudomonas aeruginosa* for denitrification after acclimatization for a

Table 3.2 Major characteristics of water before and after treatment [5]

Parameter	Influent (mg/L)	Effluent 1 after chemical and biological treatment (mg/L)	Effluent 2 after NF (mg/L)	Removal (%)	Permissible limit (mg/L)
Cyanide	121	0.105	ND	100	<0.1
Phenols	159	0.1	ND	100	<0.5
Ammoniacal-N	2720	35	15	99.44	<30
COD	1813	980	11	98.87	<250
TDS	16470	12458	880	92.93	–
Conductivity (mS/cm)	14.44	9.25	0.883	92.79	–
BOD	2445	1345	6	99.95	<30
Salinity	8.5	5.85	0.2	95.58	–
Oil and grease	51	11.5	ND	100	<10

Reprinted from Journal of Water Reuse and Desalination, Kumar, R., Pal, P., 2013B. Membrane-integrated Hybrid Bioremediation of Industrial wastewater: A Continuous Treatment and Recycling Approach, Desalination and Water Reuse, 1, (3), 26–38, with permission from the copyright holders, IWA Publishing.

sufficiently long time are used). Recycling is done using centrifugal pumps between the reactors and the settlers. pH is maintained at 7.5–8.5 range while reactions are carried out at 30–35°C temperature by circulating water through the reactor jackets from a thermostatic bath. During nitrification a food-to-microorganism ratio (F:M) in the range of 0.10 to 0.125 kgCOD/(kg MLVSSd) needs to be maintained. The nitrification unit is provided with a mechanical stirrer along with an air sparger from a compressor for aeration. Wastewater containing nitrate (1250–1300 mg/L) is treated in the denitrification unit containing facultative heterotrophic bacteria *Pseudomonas aeruginosa* (NCIM5032), which reduces nitrate to nitrogen gas at ambient temperature (30–35°C). Denitrification is carried out with a methanol dosage of 2.4 L/m^3 (equivalent to consumption of 7 mg COD/mg NO_3^--N). Mixed liquor from each biological treatment unit is passed to the clarifier units for settling of biomass, which is partially recycled back to the respective reactors. The effluent after chemical and biological treatments from the integrated pilot plant is directed to the cross-flow microfiltration membrane module followed by the NF membrane module of the same type.

Monitoring Plant Performance

Cyanide, ammonia, and pH concentrations are determined using a pH-ion meter with respective electrodes. The phenol content is determined by

HPLC with a Zorbax SB-Phenyl column with mobile phase methanol: water (70:30) at flow rate 1 mL/min, residence time of 3.567 min, and injection volume of 5 μL. The COD and BOD are measured using a COD analyzer, while the total dissolved solids (TDS), conductivity, and salinity are measured using a conductivity meter. DO needs to be monitored using a DO probe. The influents and effluents are periodically analyzed for residual contaminants (i.e., COD) to monitor the performance of the chemical, biological, and membrane-based processes. The samples are always kept under refrigeration at 4°C when immediate analysis cannot be done. Nitrate, oil, and grease content can be determined using a cadmium reduction method (4500 NO_3^-E) and a partition-method (5520B) as described in the standard methods of the APHA (1998). During NF the percentage of removal of pollutants is calculated using as follows:

$$\%\text{ removal of COD} = \left(1 - \frac{C_f}{C_i}\right) \times 100 \tag{3.57}$$

where C_i and C_f are the COD in the feed and permeate streams, respectively.

Plant Performance Analysis

Response Surface Optimization of Chemical Degradation Process of Cyanide using Design Expert Software

The minimum and maximum levels for H_2O_2 (1.50–5.50 g/L), iron salt (1.00–3.75 g/L), and pH (3.0–10.0) optimization of the cyanide degradation process is done using composite design (CCD) of Design Expert Software. In determining the interrelationships of variables, a second-order polynomial equation is fitted to the experimental data. From fit summary section in design, the model *F*-values as obtained for cyanide removal (145.93) implies that models are significant Table 3.3.

The value of *P* (0.0001) in this case being less than 0.0500 also indicates that the model terms are significant. The final regression equation made by analysis of variance (ANOVA) shows the empirical relationship among the target variables (cyanide removal) and the three operating conditions or variables, and the statistical parameters obtained from the ANOVA for the cyanide removal are listed in Table 3.4. The equation in terms of coded factors is represented by:

$$\begin{aligned}\sqrt{(\text{cyanide removal }\%)} &= +\,9.94 + 1.28 \times H_2O_2 + 1.31 \times \text{iron salt} \\ &+ 1.23 \times \text{pH} - 0.48 \times H_2O_2 \times \text{iron salt} - 0.14 \times H_2O_2 \times \text{pH} + 0.43 \\ &\times \text{iron salt} \times \text{pH} - 0.78(H_2O_2)^2 - 0.70(\text{iron salt})^2 - 2.07(\text{pH})^2\end{aligned} \tag{3.58}$$

Table 3.3 Experimental response under suggested operating conditions [5]

STD	Run	H_2O_2 dose (g/L)	Iron salt dose (g/L)	pH	Response cyanide removal (%)
14	1	3.50	2.38	12.39	36
18	2	3.50	2.38	6.50	99
1	3	1.50	1.00	3.00	7
15	4	3.50	2.38	6.50	99
19	5	3.50	2.38	6.50	99
3	6	5.50	1.00	3.00	26
6	7	1.50	3.75	10.00	59
13	8	3.50	2.38	0.61	4
16	9	3.50	2.38	6.50	99
12	10	6.86	2.38	6.50	100
9	11	3.50	0.06	6.50	28
5	12	1.00	1.50	10.0	18
4	13	5.50	3.75	3.00	51
11	14	0.14	2.38	6.50	33
20	15	3.50	2.38	6.50	99
10	16	3.50	4.69	6.50	100
8	17	5.50	3.75	10.0	98
17	18	3.50	2.38	6.50	99
7	19	5.50	1.00	10.00	83
2	20	1.50	3.75	3.00	35

Table 3.4 Statistical parameters obtained from the ANOVA for the regression models [5]

Response	R^2	Adj. R^2	CV (%)	S.D.	A.P.
Cyanide removal	0.9962	0.9928	3.06	0.23	50.17

A.P.: adequate precision; S.D.: standard deviation; CV: coefficient of variance.

It is clear from Eq. (3.58) that the percentage removal of cyanide is linear with respect to H_2O_2, iron salt, and pH and also quadratic with respect to the same parameters. The quality of the model is based on the correlation coefficient R^2 and standard deviation value. The closer the R^2 value to unity the smaller the standard deviation and the more accurate the response can be predicted by the model. The R^2 value for Eq. (3.58) is found to be 0.9940, which shows that 99.40% of cyanide removal is

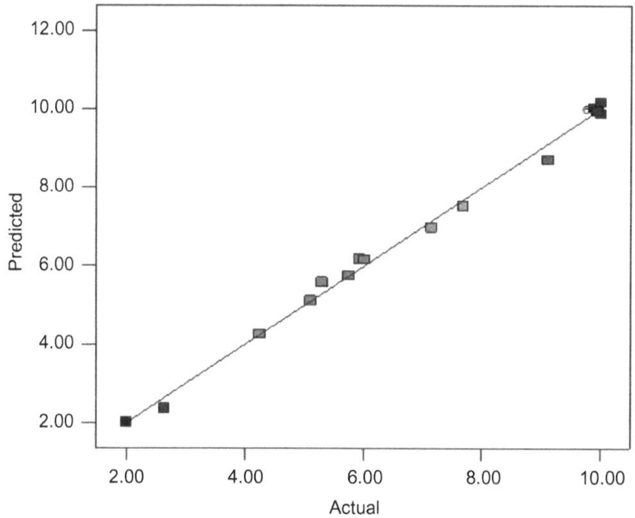

Figure 3.16 Distribution of experimentally determined values versus statistically predicted values of cyanide removal (%) [5]. *Reprinted from Journal of Water Reuse and Desalination, Kumar, R., Pal, P., 2013B. Membrane-integrated Hybrid Bioremediation of Industrial wastewater: A Continuous Treatment and Recycling Approach, Desalination and Water Reuse, 1, (3), 26–38, with permission from the copyright holders, IWA Publishing.*

attributed to the governing parameters. The value of the adjusted determination coefficient (adj R^2 for cyanide and removal % = 0.9828) is in reasonable agreement with the predicted R^2 0.9685 (cyanide removal %), which implies that the model is significant. The effects of H_2O_2, iron salt, and pH are highly significant as the P values are $<.0001$ in both cases.

The high significance of the model is also established in the plot of calculated values against the experimental values of cyanide removal % (Fig. 3.16). Clustering of points around the diagonal line indicate capability of the model to predict the actual performance in both cases. Fig. 3.17 presents the response surface modeling in 3D reflecting the effects of H_2O_2, iron salt, and pH on the cyanide removal after one hour of reaction. As a general trend, it is observed that the effects of H_2O_2 and iron salt on the removal of cyanide are pH-dependent. At low pH, cyanide degradation by $H_2O_2$and iron salt is very difficult due to the presence of cyanide mainly as an HCN gas. At high pH, cyanide is present as CN^- ions, so it reacts easily with H_2O_2 and iron salt. H_2O_2 is a powerful oxidizing agent (oxidation potential 1.77 V) and cyanide is converted to cyanate and ammonia as intermediate products.

During design of experiments, criteria are selected for optimization of H_2O_2 (in the range), pH (7.5) at minimum concentration of iron salt for maximum removal of cyanide. Some optimized solutions with different

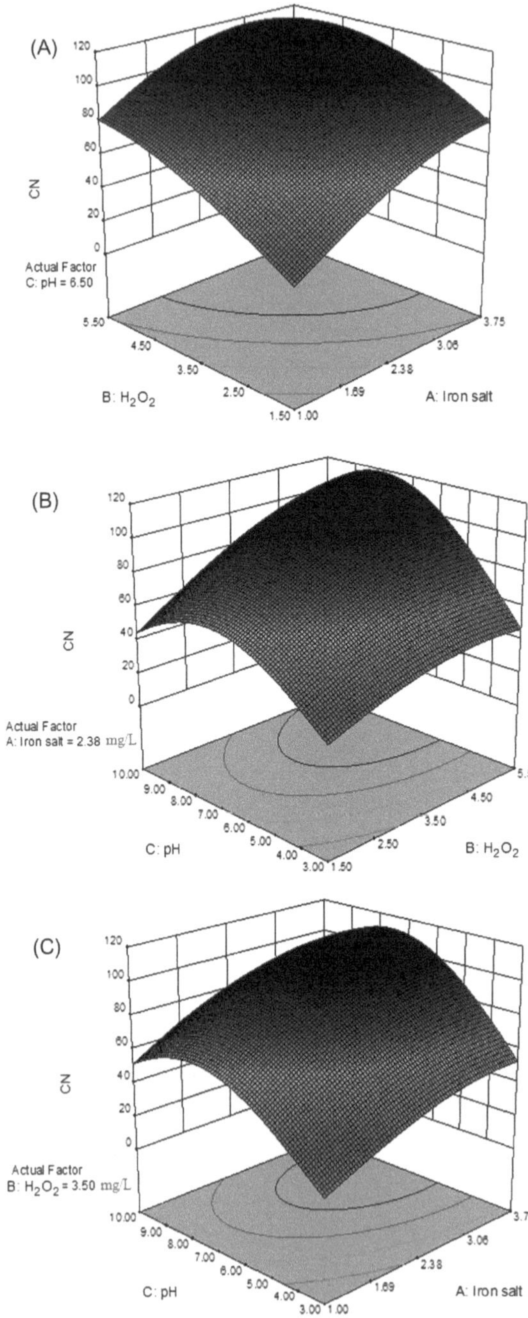

Figure 3.17 Response surface plot showing the removal % of cyanide (a, b, c) with variable parameters pH, iron salt, and H_2O_2 [5]. *Reprinted from Journal of Water Reuse and Desalination, Kumar, R., Pal, P., 2013B. Membrane-integrated Hybrid Bioremediation of Industrial wastewater: A Continuous Treatment and Recycling Approach, Desalination and Water Reuse, 1, (3), 26–38, with permission from the copyright holders, IWA Publishing.*

criteria are suggested by the software. From those suggested solutions, one acceptable solution is taken as follows: at pH 7.50, H_2O_2 5.10 g/L, and iron salt 1.52, while the excepted cyanide removal is 100%. As the selected optimum criteria for variables are not among the 20 investigations previously designed by CCD and are just assumptions, experiment with selected criteria is performed in shaker flask level and is found that the predicted response is in close agreement with the actual performance. The solution is accepted as the complete removal of cyanide with a minimum concentration of iron salt, which reduced the sludge generation during treatment and avoided iron contamination in final effluent. Fig. 3.18 shows the result of cyanide concentration before the Fenton's treatment and after the treatment with optimized concentration. Cyanide first oxidized to cyanate, which further oxidized to ammonium, and carbonate ions as shown by the following reactions:

$$CN^- + H_2O_2 \rightarrow CNO^- + H_2O \quad (3.59)$$

$$CNO^- + 2H_2O \xrightarrow{H_2O_2} NH_4^- + CO_3^{-2} \quad (3.60)$$

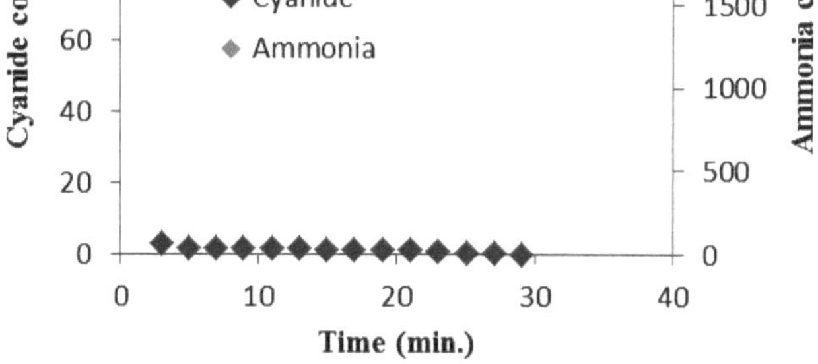

Figure 3.18 Cyanide and ammonia concentration profile during chemical pretreatment. Experimental conditions: influent cyanide concentration 121 mg/L; optimized concentrations of $FeSO_4.7H_2O = 1.52$ g/L; $H_2O_2 = 5.1$ g/L; pH = 7.5; and temperature: 308 K [5]. *Reprinted from Journal of Water Reuse and Desalination, Kumar, R., Pal, P., 2013B. Membrane-integrated Hybrid Bioremediation of Industrial wastewater: A Continuous Treatment and Recycling Approach, Desalination and Water Reuse, 1, (3), 26–38, with permission from the copyright holders, IWA Publishing.*

Cyanide may also be mineralized to bicarbonate and ammonia following the reaction:

$$CN^- + 2H_2O \xrightarrow{H_2O_2} NH_4^- + HCOO^- \quad (3.61)$$

Biological Degradation of Phenol and Ammonia

The HRT is optimized for the successful treatment of phenol and it was found that in 17.5 h, phenol is degraded up to a minimum detection level as shown in Fig. 3.19.

Phenol is almost completely removed regardless of loading variation, and the removal efficiency is always higher than 99%. For an influent ammonia concentration of 2720 mg/L its removal increases with increasing HRT. After 35 h of retention time, removal is almost stabilized. The highest ammonia removal (98.7%) is achieved on operation of the system for 70 h of HRT. The following equations describe the nitrification process:

$$NH_4^+ + 1.5\ O_2 \rightarrow 2H^+ + 2H_2O + NO_2^- \text{ by Nitrosomonas} \quad (3.62)$$

$$NO_2^+ + 0.5\ O_2 \rightarrow NO_3^- \text{ by Nitrobacter} \quad (3.63)$$

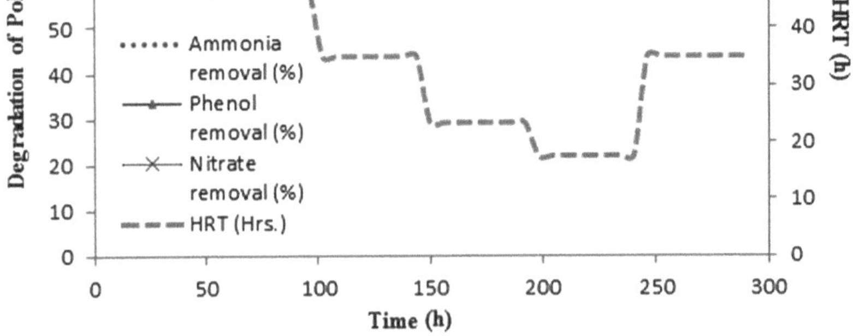

Figure 3.19 Phenol-, ammonia-, and nitrate-removal efficiencies obtained in the biological treatment of coke wastewater in a separate tank with respective microorganisms. Operating condition: Influent concentration of phenol = 159 mg/L and ammonia = 2720 mg/L [5]. *Reprinted from Journal of Water Reuse and Desalination, Kumar, R., Pal, P., 2013B. Membrane-integrated Hybrid Bioremediation of Industrial wastewater: A Continuous Treatment and Recycling Approach, Desalination and Water Reuse, 1, (3), 26–38, with permission from the copyright holders, IWA Publishing.*

During nitrification the pH of the medium comes down due to acid produced during nitrification. Hence the pH is adjusted by addition of 5N NaOH. For proper nitrification, the ratio of mass of $CaCO_3$/mass of NH_4^+-N is very important. Alkalinity in the nitrification process is maintained at around 150–200 mg/L of the medium by adding $CaCO_3$ at a rate of 5 g/L to achieve adequate buffering, as shown by:

$$H_2O + CO_2 \Leftrightarrow H_2CO_3 \Leftrightarrow HCO_3^- + H^+ \Leftrightarrow CO_3 + 2H^+ \qquad (3.64)$$

The initial concentration of nitrate in the nitrification is found to be 1254 mg/L due to conversion of ammonia to nitrate by two subsequent biological reactions. Nitrate is reduced to free nitrogen in the denitrification unit, which is the last step, at a methanol dosage of 2.4 L/m^3 (equivalent to a consumption of 7 mg COD/mg NO_3^--N) as external organic or inorganic carbon sources are necessary for stable operation of the denitrification unit. By maintaining almost an anoxic condition (0.5 mg/L DO) nitrate is completely converted to free nitrogen at a minimum HRT of 17.5 h.

Nanofiltration of Biologically Treated Coke Wastewater

The best possible membrane can be found by investigation and screening. Effects of transmembrane pressure and cross-flow velocity on flux as well as on removal of COD, BOD, TDS, salinity, and conductivity can be found. Microfiltration with 0.45 μm Polyvinylidene Difluride (PVDF) membrane in the cross-flow membrane module is done prior to NF. SS and microbes are largely screened out during microfiltration, paving the way for better NF under a more conducive environment that ensures long hours of largely fouling-free NF.

Effect of Cross-flow Rate and Pressure on Flux

When pressure is increased from 5 to 15 bars flux at a volumetric cross-flow rate of 750 liters/h(L/H), flux through NF-2 membrane increases from 158 to 260 liters/m^2 h (LMH). A similar tendency of increase of flux with increase of cross-flow rate is also shown by other membranes. The highest flux is exhibited by NF-2, since it has the highest porosity among the four investigated membranes, followed by NF-3, NF-20, and NF-1, respectively. NF-1 is found to be the tightest, i.e., it has the lowest porosity among the four types and so the flux is the lowest. Adequate cross-flow and transmembrane pressure help in reducing membrane fouling, optimization of membrane area requirement, and maximization of removal of the hazardous compounds.

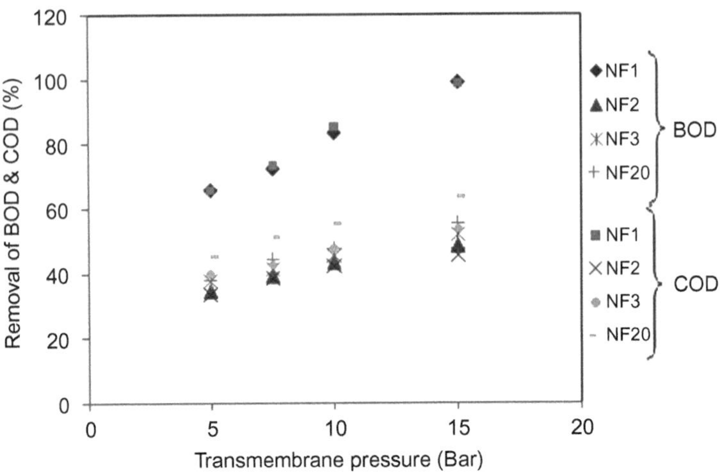

Figure 3.20 BOD and COD removal performance by different NF membranes with varying pressure. Experimental conditions: cross-flow rate = 750 LPH; pressure = 5–15 bars; pH = 7.6; and temperature = 308 K [5]. *Reprinted from Journal of Water Reuse and Desalination, Kumar, R., Pal, P., 2013B. Membrane-integrated Hybrid Bioremediation of Industrial wastewater: A Continuous Treatment and Recycling Approach, Desalination and Water Reuse, 1, (3), 26–38, with permission from the copyright holders, IWA Publishing.*

Effect of Transmembrane Pressure on the Rejection of COD and BOD

Four different types of NF membranes are investigated at different pressures to select the best possible membrane and the optimum operating pressure in removing COD loading as shown in Fig. 3.20.

The first stage of treatment removes around 45–46% of the influent BOD and COD. In all four membranes, rejection of BOD and COD increases with increase in applied pressure. The Sepro-made NF-1 membrane with a flux of 79–80 LMH is a well-performing membrane in terms of COD reduction (99%) as well as BOD reduction (98%) at a pressure of 15 bars and cross-flow rate of 750 L per hour. COD and BOD are used to measure the oxygen equivalence of the organic-matter content of a sample. When negatively charged organic-matter ions come into contact with the negatively charged membrane surface of NF1 membrane, charge repulsion takes place causing rejection of the organic matter. The entire separation is effected by the combined action of Donnan exclusion and sieving mechanism.

Effect of Nanofiltration (NF1) on TDS, Salinity, and Conductivity

NF membrane efficiently removes TDS, salinity, and conductivity (>90%). Microfiltration (0.45 μm) membrane prefilters the suspended materials prior to NF thus reducing load on the NF membrane.

The charged inorganic and organic molecules are then removed by negatively charged NF1 membrane up to 93% due to repulsion and sieving mechanism. High salinity in water and soil could negatively affect crop yields, degrade land, and pollute groundwater. NF1 removes the salinity by 95%–96% at a pressure of 15 bars and cross-flow rate of 750 LMH. During chemical and microbial treatment, chemicals and microbes are decomposed into ions in addition to generation of such ions through microbial metabolism thus raising the conductivity. In aerobic treatment, oxygen is consumed releasing $CO_{2,}$ which is subsequently converted into carbonic acid (H_2CO_3), bicarbonate ions (HCO_3^-), and carbonate ions (CO_3^-) decreasing the pH and increasing the conductivity. These charged ions are removed by the charged NF1 membranes by up to 93%.

Economic Evaluation of the Treatment Scheme

Economic assessment of treatment cost for treatment of around 20 m^3 wastewater/h shows that the total annual cost (investment + operating cost) amounts to around 0.46$/$m^{3,}$ which indicates economic viability. Novelty of membrane-integrated hybrid treatment plant

Membrane-integrated hybrid treatment technology succeeds in making highly hazardous wastewater reusable, and protects both air and surface water bodies from toxic contaminants like ammonia, phenol, cyanide, thiocyanate, and other carcinogenic aromatic compounds that are normally released into the environment during discharge of coke wastewater and during quenching of coke by wastewater. Apart from these hazardous substances, oil, grease, other organics, and even trace elements can also be very effectively removed from wastewater by logical sequencing of chemical, biological, and finally nanomembrane-based treatments in an integrated hybrid plant. After almost 99% removal of highly toxic cyanide compounds in a well-optimized Fenton's treatment unit, subsequent biological treatment units could be made very effective. All these pretreatments help achieve microbial nitrification and denitrification of more than 98% of ammonia. Composite NF membranes selected through investigation could separate ionic trace contaminants from water with a high degree of purification permitting recycling and reuse of the treated water. The cross-flow membrane module employed in this case allows long hours of largely fouling-free operation under a reasonably low transmembrane pressure of only 15 bars while yielding an industrially acceptable flux of 80 L of pure water per hour per square meter of membrane surface.

This technology represents a sustainable technology that closes the industrial water-use loop effectively protecting surface water bodies from the onslaught of hazardous waste discharge while reducing freshwater consumption through recycle of reusable water in the loop.

3.7.3 Anaerobic Anoxic Oxic process

The ANANOX process (anaerobic anoxic oxic) is a two-stage biological process. In the first stage, operating in anaerobic conditions, a three-series chamber ABR (anaerobic baffled reactor) is used, while in the second stage, aerobic activated sludge with a settler is used [10].

Removal of nitrogen from wastewater before discharge to surface water bodies is always suggested as nitrogen represents one of the main causes of the eutrophication of the aquatic environment. Moreover, high nitrate content in drinking water can cause methemoglobinemia, especially in kids. Thus it is necessary to pretreat drinking water for the removal of nitrates.

As described in the scheme above, wastewater is introduced to the first chamber of the ABR reactor following a microscreening process. Under this anaerobic condition, the organic matter is transformed to methane instead the nitrogen compounds are transformed mainly to ammonia. The three chambers guarantee good removal of organics and well-clarified effluent. In the second stage, the ammonia and the remaining part of the organic matter together with the sulfide compounds are completely oxidized. From the settling tank, part of the supernatant rich with nitrates is sent back to the third chamber of the ABR for the denitrification process. This third chamber is used as an anoxic chamber where only the supernatant is introduced with the advantage that the process is not limited by the availability of reducing substances from the previous anaerobic process.

The main advantages of this process are the possibility of the recovery of methane from wastewater as bioenergy and the use of the same anaerobic reactor for the removal of nitrates from effluents. The presence of an anaerobic process before the aerobic treatment reduces sludge production and can be used for the digestion of the excess sludge produced by the second stage of the process. These contribute to reduce the quantity of biological sludge as well as disposal costs. The three ABR chambers provide a large volume whose capacity can be settled to reduce the flow peaks that can occur during the process.

The use of an anaerobic process in the treatment line of wastewater offers several advantages including reduction of the energy required for the treatment of wastewater and generation of bioenergy in the form of methane. This energy can be used to make the whole plant self-sufficient in power consumption.

3.7.4 Anaerobic Ammonium Oxidation

The recently developed Anammox (anaerobic ammonium oxidation) process [11] has been successful at removing nitrogen from wastewater. In this process, ammonium is oxidized to di-nitrogen gas with nitrite as the electron acceptor. The overall reaction for the Anammox process (Eq. (3.65)) is exothermic as indicated by the negative value of the standard Gibbs free energy ($DG^0 = -358$ kJ/mol NH^{4+}). The biochemical process can, therefore, supply sufficient energy for growth as follows:

$$NH^{4+} + NO^{2-} \rightarrow N_2 + 2H_2O \quad (3.65)$$

The Anammox process is a biological, autotrophic process that is strictly anaerobic and is unable to convert ammonia into nitrite. This implies that the Anammox process requires a source of nitrite in order to remove ammonia from the system. The bacteria responsible for the Anammox process have been identified as *planctomycete* [12].

3.7.5 Chemical-biological Integrated Treatment Process

Treatment of wastewater containing complex components with high levels of toxicity and biorefractory character is quite challenging. For effective treatment of such biorefractory yet toxic wastewater, an integrated treatment scheme has evolved [13]. The scheme combines interior microelectrolysis (IME) and Fenton oxidation—coagulation (FOC) as pretreatment processes in combination with biological treatments using a hydrolysis acidification (HA) unit and two-stage biological contact oxidation (BCO) processes as shown in Fig. 3.21. The treatment scheme is effective treatment for complex wastewater such as pharmaceutical wastewater containing a vast variety of complex components high in organic concentrations, and variable concentrations of salts, toxins, and biorefractory compounds like pyridine, hydroxylamine hydrochloride, cyclohexanone, toluene, tetrahydrofuran, methylene chloride, and metals, which are difficult to manage in pharmaceutical plants.

Wastewater is pumped into an equalization tank to obtain a stable influent. During the FOC, supernatant from chemical sludge dewatering is added to the equalization tank. After coagulation, the effluent's pH is adjusted to 7—7.5 using an acid solution in a buffer tank. Prior to the continuous run, biofilm formation is activated in the biological tank with inoculated sludge from aerobic sediments of a successfully running

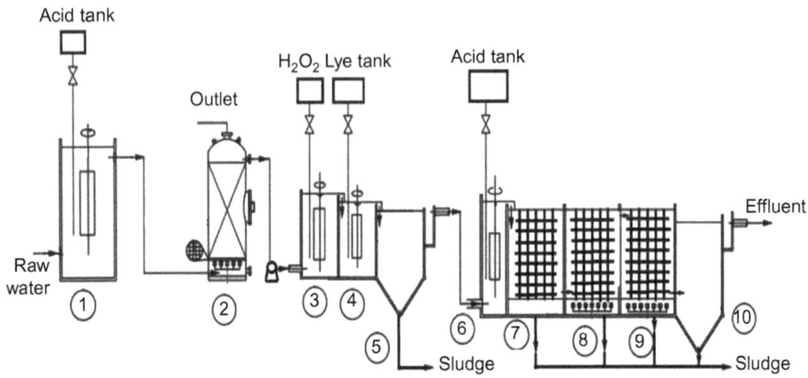

Figure 3.21 Chemical-biological integrated treatment scheme [13]. 1. Equalization Tank, 2. IME Reactor, 3. Fenton Tank, 4. pH Regulation, 5. Settling Tank, 6. Buffer Tank, 7. HA Tank, 8. BCO Tank, 9. BCO Tank, 10. Settling tank.

wastewater treatment plant. After nearly a month of inoculation, the biofilm gets firmly attached to the carrier surface and the biomass concentration in the BCO tank is made to grow to at least 2–3 mg/L. The average DO concentration in the HA and BCO tanks reaches 1.0 mg/L to 3 mg/L, respectively. The COD in the effluent becomes relatively stable after the successful completion of the acclimation stage. The plant performs at a temperature ranging from 25°C to 40°C.

As toxicants and macromolecules in wastewater may inhibit microbial activity to some extent and also cause process failure, it is necessary to develop proper pretreatment methods to improve the biodegradability of sewage before it enters biological treatment units. Physicochemical pretreatment units combined with biological treatments have been widely applied to treat this type of wastewater, because of their high removal efficiency and cost effectiveness. One such method is IME-FOC, which is the integration of interior microelectrolysis (IME) and Fenton's oxidation–coagulation (FOC) process with biological treatment process. IME-FO technologies are effective methods for the pretreatment of biorefractory organic wastewater due to their high efficiency, simple operation, and low cost. It is known that during the IME process, which uses iron as a sacrificial anode and carbon as a cathodic catalyst [14], rhR macromolecular organic contaminants can be broken down into small organic matter molecules, which tend to biodegrade during IME due to the strong reducibility of Fe, Fe^{2+}, and [H]. A portion of these compounds can also be adsorbed at the surface of Fe-C fillings by adsorption and electrophoresis, after which they can then be separated from wastewater. Oxygen can compete as an electron acceptor under aerated conditions, thus generating H_2O_2 in the process of IME. The reactions are as follows [15,16].

At cathode (under aerobic and acidic conditions) [15,16]:

$$O_2(g) + 4H^+(aq) + 4e^- \rightarrow 2H_2O, \quad E^o\left(O_2/H^+\right) = +1.23\ V \quad (3.66)$$

$$O_2(g) + 2H^+(aq) + 2e^- \rightarrow H_2O_2, \quad E^o\left(O_2/H_2O_2\right) = +0.68\ V \quad (3.67)$$

With the Fenton process, hydroxyl radicals, which have strong potential to oxidize organic compounds, can be procured in the presence of H_2O_2 and Fe^{2+} under acidic conditions. After FO, the pH of the solution should be balanced to around 9.0. Under these conditions, the Fe^{n+} (i.e., Fe^{2+}, Fe^{3+}) compounds produce nascent $Fe(OH)_n$ compounds, which have a good flocculation effect and may contribute to further contaminant removal (i.e., organics and inorganics). IME pretreatment can achieve high COD removal efficiency and improve the BOD5/COD (B/C) values of refractory wastewater with high organic matter concentration. However, economical and effective treatment for pharmaceutical wastewater using single FO or IME processes is difficult. As Fe^{2+}-rich IME-treated effluent is generally suitable for successive FO treatment without Fe^{2+} addition or pH adjustment, the combined process of IME-FOC has been used on specific industrial wastewater, like landfill leachate and electroplating wastewater, with better treatment efficiency and lower reagent dosage.

3.8 CASE STUDIES

3.8.1 Activated Sludge Process

Design Problem 1

An activated sludge process is set to operate at a MCRT of 20 days and at a nominal hydraulic retention time (NHRT) of 4 h. The bioreactor volume is 2000 m^3 and underflow concentration X_u is 10,000 mg/L. If the MCRT and NHRT are to be maintained, what should the sludge-wasting rate for microbial mass X = 4000 mg/L of MLVSS be? Assume negligible influent and effluent biomass concentrations.

By definition of MCRT we know that

MCRT =

$$\theta_c = \frac{V.X}{Q_w.X_u + (Q_o - Q_w)X_e - Q_o.X_o}$$

In this case both the influent and effluent do not contain microbial cells, therefore, we can rewrite the expression for MCRT in this case as (putting X_0 and X_e equal to zero):

$$\theta_C = \frac{V.X}{Q_w X_u} \Rightarrow 20 = \frac{2000(4000)}{Q_w(10{,}000)} \Rightarrow Q_w = 40\ m^3/day$$

This means (Q_w) the sludge-wasting rate is 40 m^3/day and considering a solids concentration of 10,000 mg/L (X_u) of the sludge, the solids mass wasting rate is $Q_w X_u = 40\ m^3/day \times 10{,}000\ mg/L$.

$(40\ m^3/day) \times 10{,}000\ mg/(10^{-3})m^3$

$4 \times 10^8\ mg\ X/day/\ (10^6\ mg/kg) = 400\ kg/day$

3.8.2 Case Study 2

A pilot plant study established the following data:

Yield coefficient (Y) = 0.5 kg MLVSS/kgBOD$_5$

Decay constant (K_d) = 0.05 per day

Q_0 (inflow rate) = 10,000 m^3/day

Q_w (sludge-wasting rate) = 40 m^3/day

X_u (underflow concentration of microbes) = 10,000 mg/L

The plant has to operate for an MCRT of 10 days and NHRT of 4 h.

a. If the influent to the bioreactor contains 150 mg/L of BOD$_5$, what MLVSS should be maintained in the bioreactor to meet an effluent limit of 5 mg/L of BOD$_5$?

b. At what recirculation ratio should the plant be operated?

$$a)\quad X = \frac{\theta_c y(S_o - S)}{\theta(1 + k_d.\theta_c)} = \frac{10(0.05)(0.15 - 0.005)}{(4/24)[1 + 0.05(10)]} = 0.29\ kg/m^3 = 2900\ mg/L$$

c. X_e is very small compared to X and X_u, so can be neglected:

$$R = \frac{Q_R}{Q_o} = \frac{Q_o(X - X_e) - Q(X_u - X_e)}{Q_o(X_u - X)}$$

$$= \frac{Q_o X - Q_w X_u}{Q_o(X_u - X)} = \frac{10{,}000(2900) - 40(10{,}000)}{10{,}000(10{,}000 - 2900)} = 0.40$$

3.8.3 Case Study 3

An influent of 10,000 m^3/day to a secondary reactor (activated sludge reactor) has a BOD$_5$ of 150 mg/L. It is desired to have an effluent BOD$_5$ of 5 mg/L, an MLVSS of 3000 mg/L, and an underflow concentration of 10,000 mg/L. Using the kinetic constants $y = 0.5$ kg MLVSS of BOD_5, $k_d = 0.05$ per day, design for the volume of the reactor. What are the volume and mass flows of sludge wasted per day?

Assume $\theta_c = 20$ days

Solution:

$$a)\quad X = \frac{\theta_c y(S_o - S)}{\theta(1 + k_d.\theta_c)} = \frac{Q_o\theta_c y(S_o - S)}{V(1 + k_d.\theta_c)}$$

$$3 = \frac{10,000(20)(0.5)(0.15 - 0.005)}{V[(1 + 0.05(10))]} = V = 3222\ m^3$$

$$\theta_c = \frac{V.X}{Q_w X_u}, \quad 20 = \frac{3222(3)}{Q_w X_u}$$

$$Q_w X_u = 483.3\ kg/day \Rightarrow \text{mass flow}$$

$$Q_w = \frac{483.3}{10} = 48.33\ m^3/day \Rightarrow \text{vol.flow}$$

3.8.4 Detailed Design of Activated-sludge Process

Before we get into designing biological treatment systems, let us first define the relevant design parameters and explain the meaning of the involved terms. *X* stands for concentration of active microbial mass in the reactor, which is often referred as MLVSS in the design. This may be expressed as mass per unit volume. VSS stands for volatile suspended solids that include active biomass and cell debris. As microbial cells enter the decay or death phase, cell lysis occurs releasing cellular materials in the medium that may be consumed by other bacteria. A part of the released material (e.g., the cell wall material) is not biodegradable and remains as cell debris.

Case 3.7.4

The process design has to be done for an activated sludge plant (CSTR type) to treat 25,920 m^3/day of wastewater having BOD_5 loading of 400 mg/L so that the treated effluent will not have more than 40 mg/L BOD_5. The operating temperature may be assumed to be 20°C where microbial concentration of the inflow is negligible. The decay constant may be assumed to be 0.06 d^{-1}. The other conditions of design are presented below. Determine the efficiency of plant operation based on soluble BOD_5 (i.e., overall plant efficiency) in the effluent as well as soluble substrate in the effluent.

1. Ratio of MLVSS to MLSS = 0.8
2. Return-sludge concentration = 10,000 mg/L of SS
3. MLVSS = 3500 mg/L
4. Design MCRT $\theta_c^d = 10$ days
5. Effluent contains 22 mg/L of biological solids, of which 65% is biodegradable.
6. The value of BOD_5 can be obtained by multiplying the value of BOD_L by the factor of 0.68 (corresponds to a K value of 0.1 d^{-1} in the BOD equation).
7. Waste contains adequate nitrogen, phosphorus, and other trace nutrients for biological growth.

SOLUTION

Estimate the concentration of soluble BOD_5 in the effluent using the following relationship:

Effluent BOD_5 = soluble component of BOD_5 escaping treatment + SS component of BOD_5 in effluent

MLVSS and MLSS

MLVSS is generally defined as the microbiological suspension in the aeration tank of an activated-sludge biological wastewater treatment plant. The biomass solids in a biological wastewater reactor are usually indicated as total suspended solids (TSS) and volatile suspended solids (VSS). TSS can be determined by filtering wastewater by standard glass fiber filter and by drying at 103–105°C to constant weight. The weight gain of the filter paper indicates the TSS. The mixture of solids resulting from combining recycled sludge with influent wastewater in the bioreactor is called MLSS and MLVSS. MLVSS can be determined by burning the volatile matter of the filter paper solids at 550°C. The difference in weight after burning will indicate the MLVSS. Solids are comprised of biomass, nonbiodegradable volatile suspended solids (nbVSS or inert solids), and inert inorganic total suspended solids (iTSS).

Measure BOD using a laboratory test that lasts a certain number of days: results based on tests lasting 5 days are denoted as BOD_5. The ultimate BOD is the BOD that would be consumed given an infinite amount of time (i.e., the total amount of oxygen consumed if the biochemical reaction was allowed to proceed to completion following the reactions in Eqs.(3.1), (3.2), and (3.3) in Section 3.1). We know the total amount of oxygen consumed for these three reactions is the BOD_L or ultimate BOD.

The ultimate BOD is too time consuming, so BOD_5 has almost universally been adopted as a measure of relative pollution effect, and we will use it here.

Net Biomass Yield

This is defined as the ratio of net biomass growth rate to substrate utilization rate where net biomass refers to active biomass:

$$Y_{net} = X_{GR}/S_{utR}$$

Observed Biomass Yield

This is the ratio of actual solid production rate to substrate utilization rate:

$$\mathrm{Y_{obs}} = \mathrm{VSS_{PR}}/\mathrm{S_{utR}}$$

$\mathrm{Y_{obs}}$ refers to the sludge production rate.

3.8.5 Calculation of COD

For biological reactions, it is practically impossible to know the exact stoichiometry as substrates are comprised of a mixture of compounds. However, following COD mass balance we can observe the changes in substrate concentrations with time. For example, assuming $C_6H_{12}O_6$ as the representative organic pollutant and $C_5H_7NO_2$ as the new cells, we can present the oxidation of substrate as follows considering only NH_3 as the present nutrient:

COD of Glucose

$$C_6H_{12}O_6 + 6O_2 \rightarrow 6CO_2 + 6H_2O$$

COD of glucose = 6 × 32 g of oxygen/mole/ (180 g of glucose/mole) = 1.07 g oxygen/g of glucose

Yield of Cells

$$3C_6H_{12}O_6 + 8O_2 + 2NH_3 \rightarrow 2C_5H_7NO_2 + 8CO_2 + 14H_2O$$

Now we can express the yield of cells per g of substrate as:
Y = 2(113 g/mole)/ 3(180 g/mole) = 0.42 g of cells/g of glucose

$$C_5H_7NO_2 + 5O \rightarrow 5CO_2 + 2H_2O + NH_2$$

$$C_5H_7NO_2(\text{Cell}) = 113; \quad 5O_2 = 5 \times 32 = 160$$

$$\text{BOD of Cell} = \frac{\text{KgO}_2}{\text{Kg Cells}} = \frac{160}{113} = 1.42$$

Determine the Treatment Efficiency E

$$E = \frac{S_0 - S}{S_0} \times 100$$

where
E = process efficiency, percent
S_0 = influent substrate concentration
S = effluent substrate concentration (soluble effluent BOD = 12.8)

Determination of SS of the Effluent BOD_5

SS of the effluent BOD_5 = (40 mg/L) × (0.68) = 27.2 mg/L

Biodegradable portion of the effluent biological solids = 0.65(22 mg/L) = 14.3 mg/L

Ultimate BOD_L of the biodegradable effluent solids = [0.65(22 mg/L)] (1.42 mg/mg) = 20.3 mg/L

Soluble BOD_5 of the Influent that Escapes Treatment

Total escaping BOD = 40 mg/L (as mentioned in the problem) = S (soluble BOD) + 27.20 mg/L(suspended BOD); So, S = 12.80 mg/L.

Determination of the Overall Plant Efficiency

The overall plant efficiency is:

$$E_{\text{overall}} = \frac{(400 - 40)\text{mg/L}}{400\ \text{mg/L}} \times 100 = 90\%$$

where
Effluent substrate concentration S = 40 mg/L
Efficiency based on soluble BOD_5 = (400 − 12.80)/400 = 96.8%

Designing for the Reactor Volume

The volume of the reactor can be determined using:

$$X = \frac{\theta_c^d Y(S_o - S)}{\theta(1 + K_d\theta_c^d)} = \frac{\theta_c^d (Y)(S_0 - S)}{(V/Q)(1 + K_d\theta_c^d)}$$

where
$\theta = \frac{V}{Q}$
X = mass concentration of microorganism
θ_c = MCRT, i.e., the average time period a unit of biomass is retained in biological reactors, including the activated sludge process, also known as the SRT

$$\theta_c = \frac{VX}{QX} = \frac{\text{mass of cells in reactor}}{\text{mass of cells wasted per day}}$$

S_0 = influent substrate concentration
S = effluent substrate concentration
Y = yield coefficient
K_d = decay coefficient
Substituting for θ in the above equation and solving for V yields:

$$V = \frac{\theta_c^d YQ(S_o - S)}{X(1 + K_d\theta_c^d)}$$

Designing the Reactor Volume

$$\theta_c^d = 10\ d$$

$Q = 25920\ m^3/d$
$Y = 0.50$ mg/mg (assumed)
$S_0 = 400$ mg/L
$S = 12.80$ mg/L
$X = 3500$ mg/L
$K_d = 0.06\ d^{-1}$ (assumed)

$$V = \frac{(10)(0.5)(25920)[(400 - 12.8)\text{mg/L}]}{(3500\ \text{mg/L})[(1 + (0.06\ d^{-1} \times 10))]}$$

$8961\ m^3$

Determining Sludge-wasting Rate Per Day

$$\theta_c = \frac{V.X}{Q_W.X_u + (Q_o - Q_W)X_e - Q_o.X_o}$$

$$\frac{VX}{\theta_c^d} = Q_W.X_u =$$

Or $(8961\ m^3)(3500\ mg/L)/(10\ d) = Q_W.(10{,}000\ mg/L)$

$$\begin{aligned} Q_W &= 313.6\ m^3/day = 313.6 \times 10000(m^3/day) \times (mg/L) \\ &= 313.6 \times 10000 \times 10^{-6}(kg/mg)/10^{-3}\ m^3 = 3136\ kg/day \end{aligned}$$

$mg/L = 10^{-3}\ kg\ /m^3$

Sludge wasting is Q_W in volumetric rate and mass wastage rate is $Q_W.X_u$

For negligible influent and effluent cell concentration, we already know that Y_{obs} represents sludge generation. So we use the equation below to arrive at Y_{obs} first:

$$Y_{obs} = \frac{Y}{1 + K_d\theta_c^d} = \frac{0.5}{(1 + (0.06\ d^{-1} \times 10))} = 0.3125$$

$$Y = \frac{R_g}{S_{utR}}$$

R_g = rate of bacterial growth, (mass)/(unit volume)(time)

S_{utR} = substrate utilization rate in (mass)/(unit volume)(time)

Y = maximum yield coefficient measured during any finite period of logarithmic growth, and defined as the ratio of the mass of cells formed to the mass of substrate consumed, (mass)/(unit volume)(time)

Strip

Estimate the recirculation ratio by writing a mass balance around the inlet to the reactor.

Aerator VSS concentration = 3500 mg/L

Return VSS concentration = 8000 mg/L

$3500(Q + Q_r) = 8000\ (Q_r)$

$$\frac{Q_r}{Q} = \alpha = 0.78$$

Compute the HRT for the reactor.

$$\theta = \frac{V}{Q} = \frac{8961\ m^3}{25{,}920\ m/d} = 0.346\ d = 8.3\ h$$

The HRT (hydraulic retention time; residence time) θ is a measure of the average length of time a soluble compound remains in a constructed bioreactor.

Compute oxygen requirements based on the ultimate carbonaceous demand, BOD_L.

Compute the mass of the ultimate BOD_5 of the incoming wastewater converted in the process, assuming the BOD_5 is equal to 0.68 BOD_L.

$$\text{Mass of BOD}_L \text{ utilized} = \frac{Q(S_0 - S) \times 10^{-6}}{0.68} \text{kg/day}$$
$$= \frac{(25,920\,\text{m}^3/\text{d})(400 - 12.8) \times 10^{-6}\,\text{kg}/10^{-3}\,\text{m}^3}{0.68}$$
$$= 14759\ \text{kg/d}$$

Compute the oxygen requirement using:

$$\text{kg, O}_2/\text{d} = (\text{total mass of BOD}_L \text{ utilized, kg/d}) - 1.42(\text{mass of organism wasted, kg/d})$$

$$\text{kg, O}_2/\text{d} = \frac{Q(S_0 - S)(10^3\ \text{g/Kg})}{\text{f}} = 1.42(\text{P}_\text{x})$$

where f = conversion factor for converting BOD_5 to BOD_L

$$\text{Kg, O}_2/\text{d} = 14759\ \text{kg/d} - 1.42(3136\ \text{kg/d}) = 10,306\ \text{kg/d}$$

The air supply must be adequate to (1) satisfy the BOD of the waste, (2) satisfy the endogenous respiration by the sludge organisms, (3) provide adequate mixing, and (4) maintain a minimum dissolved-oxygen concentration of 1 to 2 mg/L throughout the aeration tank.

Determine the F/M ratio and the volumetric loading factor.

To determine the F/M ratio:

$$\text{F/M} = \frac{\text{S}_\text{o}}{\theta X} = \frac{400\ \text{mg/L}}{(0.346\ \text{d})(3500\ \text{mg/L})} = 0.3303\ \text{d}^{-1}$$

To determine the volumetric loading:

$$\text{Volumetric loading} = \frac{\text{S}_\text{o}\text{Q}}{\text{V}}$$
$$= \frac{(400)(25920) \times 10^{-3}\ \text{kg BOD/m}^3}{8961}$$
$$= 1.157\ \text{kg BOD}_5/\text{m}^3$$

The aeration requirement assuming an oxygen-transfer efficiency for the aeration equipment of 15%. A safety factor of 2 should be used to determine the actual design volume for the sizing the blowers.

The theoretical air requirement, assuming that air contains 23.2% oxygen by weight, is:

$$= \frac{10306 \text{ kg/d}}{(1.201 \text{ kg/m}^3)(0.232)} = 36988 \text{ m}^3/\text{d}$$

Actual air requirement (assuming 15% transfer efficiency):

$$= \frac{36988 \text{ m}^3/\text{d}}{0.15 \times 144 \text{ d/min}} = 1712 \text{ m}^3/\text{min}$$

Design air requirement:

$$2(1712 \text{ m}^3/\text{min}) = 3424 \text{ m}^3/\text{min}$$

Determine the air–volume requirement per unit volume of wastewater treated.

Air required per unit volume wastewater:

$$\frac{3424 \text{ m}^3 \times 1440/\text{d}}{25920 \text{ m}^3/\text{d}} = 190 \text{ m}^3 \text{ air/m}^3 \text{ water}$$

3.8.6 Designing Trickling Filters

Major Governing Parameters

- Hydraulic loading
- Flushing intensity
- Volumetric loading
- Overflow
- Aeration
- Sloughing or flushing
- Use of recycled effluent
- Surface loading

Major design parameters in standard symbol

S^0 = influent BOD

Q^r = recycle flow rate

a = medium sp. Surface

A_{pv} = plan view surface area of the filter = πr^2
A_s = plan view surface area of the settler/clarifier
O/F = overflow rate = $Q^e/A = (Q + Q^r - Q^w)/A_s$
H.L. = hydraulic loading = $(Q + Q^r)/A_{pv}$
S.L. = surface loading = $(QS^0)/(A_{pv}$ ha)
V.L. = $(QS^0)/(A_{pv}$ h) = volumetric loading
SK = flushing intensity = $(Q + Q^r)/$ $(A_{pv}$ nw) = m/arm −revn
n = number of distributor arms
w = rotation speed = rpm

Flushing Intensity

S.K. = $[(Q + Q^r)/$ $(A_{pv}$ nw)].[d/1440 min] = m/(arm. revolution)
Q, Q^r in m^3/h
A_{pv} in m^2
n = number of arms of wastewater distributor
W = revolution of distributor arm per min

Volumetric Loading

$$\begin{aligned} \text{V.L.} &= (Q.S^0)/(A_{pv}.h) \\ &= 0.3 - 1.0 \text{ kg BOD/m}^3\text{.day} \end{aligned}$$

Overflow

$$O/F = (Q + Q^r - Q^w)/A_s$$

Q = influent rate (m^3/h)
Q^w = sludge-wasting rate
Q^r = recycle rate (m^3/h)
A_s = plan view surface area of the settler (m^2)
Recommended value of O/F = 48 m/day.

Aeration

For a natural draft trickling filter:

$$\Delta P_{draft} = 0.33(1/T_{am} - 1/T_{pore}), \quad \text{T = temperature in K}$$

- If $T_{am} < T_{pore}$, then $\Delta P_{draft} > 0$ and air flows upward
- Conversely air flows downward if $T_{am} > T_{pore}$
- Under drain needs careful design for air flow
- Maintain flushing intensity of 0.1 to 0.5 m/arm rev.

Surface Loading

S.L. = $(Q.S^0)/(A_{pv}\ h\ A)$

Q = influent (m^3/h)

S^0 = influent substrate concentration (kg/m^3)

h = filter depth(m)

a = medium specific surface (m^{-1})

A_{pv} = plan view surface area of filter (m^2)

S.L. = 2.0– 7.0 kg BOD/1000 m^2.day

3.8.7 Case Study on Trickling Filter design [17]

A trickling filter system needs to be designed to treat wastewater with the flow and pollutant characteristics given in the table below [17].

Design Conditions

Item	Unit	Primary effluent	Target effluent
Flow	m^3/d	15,000	
BOD	g/m^3	125	20
TSS	g/m^3	65	20
Minimum temp.	22°C	14	

Note: $g/m^3 = mg/L = 10^{-3}\ kg/m^3$.

The following assumptions are made:

Towers: Assume two towers at 6.1 m depth.

Medium: Cross-flow plastic packing with a specific surface area of 90 m^2/m^3 and a packing coefficient n value of 0.5

Water distributor: A two-arm distributor system.

Wetting requirement: The required minimum wetting rate = 0.5 L/m^2 s.

Clarifier: Assume a secondary clarifier depth of 4.2 m.

Based on the above assumptions and the treatment requirements, we design for the following:

I. Diameter of tower trickling filter, m
II. Volume of packing required, m^3
III. Recirculation rate required, if any
IV. Total pumping rate, m^3/h
V. Flushing and normal dose rate, mm/pass
VI. Flushing and normal distributor speeds, min/rev
VII. Clarifier diameter, m (assume the ratio of the peak to average flow rate is 1.5)

Solution:

Determine k_{20} for the design conditions using the following empirical relation:

$$k_2 = k_1\left(\frac{D_1}{D_2}\right)^{0.5}\left(\frac{S_1}{S_2}\right)^{0.5}$$

where

k_2 = normalized value of k for the site-specific packing depth and influent BOD concentration

k_1 = k value at depth of 6.1 m and influent BOD of 150 mg/L (g/m^3)

S_1 = 150 gBOd/m^3

S_2 = site-specific influent BOD concentration, g BOD/m^3

D_1 = 6.1 m packing depth, m

D_2 = site-specific packing depth, m

Solution for k_2.

From empirical relation $k_1 = 0.210\ (\text{L/s})^{0.5}/\text{m}^3$

Trickling filter depth = 6.1 m

$$k_2 = 0.210\left(\frac{6.1\ \text{m}}{6.1\ \text{m}}\right)^{0.5}\left(\frac{150\ \text{g/m}^3}{125\ \text{g/m}^3}\right)^{0.5} = 0.230(L/s)^{0.5}/\text{m}^2$$

Correction for k_2 for temperature effect:

i. $k_T = k_{20}(1.035)^{T-20}$

ii. $k_{14} = 0.230(1.035)^{14-20} = 0.187$

where

K_{20} = filter-treatability constant at 20°C, (L/s)$^{0.5}$/m^3

Determine the hydraulic application rate and filter area, volume, and diameter.

Determine the hydraulic application rate:

$$\frac{S_e}{S_0} = e^{-kD/q^n}$$

$$q = [kD/\ln S_o/S_e]^{1/0.5}$$

$$q = [0.187 \times 6.1/\ln(125/20)]^{1/0.5}$$

$$q = 0.3875\ \text{L/m}^2.\text{s}$$

where

S_e = BOD concentration of settled filter effluent, mg/L (g/m^3)

S_o = influent BOD concentration to the filter, mg/L (g/m^3)

K = wastewater treatability and packing coefficient, (L/s)$^{0.5}$/m^2 (based on $n = 0.5$)

D = packing depth, m

q = hydraulic application rate of primary effluent, excluding recirculation, L/m^2 s

n = constant characteristics of packing used

Determine the tower area:

$Q = 15{,}000$ m^3/d = 173 L/s

Filter area = Q/q = (173 L/s)/(0.3875 L/m^2 s) = 448 m^2

Packing volume determination:

Packing volume = (448 m^2) (6.1 m) = 2732.8 m^3

Tower diameter:

Area/tower = (448 m^2)/2 = 224 m^2

Diameter = 17 m each for 2 filters

Determine the recirculation rate and recirculation ratio:

The minimum wetting rate = 0.5 L/m^2 s

$$q + q_r = 0.5\ \text{L/m}^2.\text{s}$$
$$q_r = 0.5 - 0.39 = 0.11\ \text{L/m}^2.\text{s}$$

Determine the recirculation rate:

$$R = q_r/q = 0.11/0.39 = 0.28$$

Determine the pumping rate:

$$q + q_r = 0.5\ \text{L/m}^2.\text{s}$$

Total pumping rate = (0.5 L/m^2 s) (448 m^2)

$$= 224\ \text{L/s} = 806.4\ \text{m}^3/\text{h}$$

Determine the flushing and normal dose rate using industrial data.

$$\begin{aligned}\text{BOD loading} &= QS_o/V \\ &= (15{,}000\ \text{m}^3/\text{d})(125\ \text{mg/L})(1\ \text{kg}/10^3\ \text{g})/2732.8\ \text{m}^3 \\ &= 0.68\ \text{kg/m}^3.\text{d}\end{aligned}$$

Determine the dosing rates.

From practical experience the estimated flushing and operation dose rates are:

Flushing dose = 300 mm/pass

Operating dose = 500 mm/pass

Determine the distributor speed using empirical relation.

For flushing:

$$n = \frac{(1 + R)q(1000 \text{ mm/min})}{(A)(DR)(60 \text{ min/h})}$$

A = 2

$$q = \frac{(15,140 \text{ m}^3/\text{d})}{(452.2 \text{ m}^2)(24 \text{ h/d})} = \frac{1.4 \text{ m}^3}{\text{m}^2}.\text{h}$$

R = 0.28

$$n = \frac{(1 + 0.28)(1.4)(1000 \text{ mm/min})}{(2)(300)(60 \text{ min/h})} = 0.0498 \text{ rev/min(i.e., } 20 \text{ min/rev)}$$

For normal operation:

$$n = \frac{(1 + 0.28)(1.4)(1000 \text{ mm/min})}{(2)(50)(60 \text{ min}/h)} = 0.30 \text{ rev/min(i.e., } 3.33 \text{ min/rev)}$$

To take care of different speed requirements for normal and flushing operations, a distributor drive with variable speed capability is suggested.

Determine the clarifier diameter:

Clarifier depth = 4.2 m

From Fig. 9.7, the recommended overflow rate for peak and average flow rates is 1.1 and 2.4 m/h, respectively. Because the ratio of the peak to average flow rate is 1.5, the average overflow rate controls the design.

Flow rate = (15,140 m^3/d)/(24 h/d) = 630.8 m^3/h

Clarifier area = (630.8 m^3/h)/(1.1 m/h)

Use two clarifiers.

Area of each = 573.5 m^2/2 = 286.7 m^2

Diameter of each = 14.1 m

Design summary [17]

Parameter	Unit	Value
Number of filters	Number	2
Diameter	M	17
Depth	M	6.1
Total packing value	m^3	2732.8
BOD loading	kg/m^3 d	0.68
Hydraulic application rate	L/m^2 s	0.39
Total pumping rate	m^3/h	806.4
Recirculation ratio	–	0.28
Distributor arms	Number	2
Normal distributor speed	Min/rev	3.33
Flushing distributor speed	Min/rev	20
Clarifiers	Number	2
Clarifier depth	M	4.2
Clarifier diameter	M	14.1

3.8.8 Case of simultaneous BOD Removal and Nitrification

Trickling filter with Plastic Packing [17]

A trickling filter needs to be designed for simultaneous removal of BOD and nitrification for wastewater whose characteristics are as follows:

Parameter	Unit	Value
Flow	L/s	100
BOD	g/m^3	160
TKN	g/m^3	25
TSS	g/m^3	70

The volume of plastic-packing medium with specific surface area of 100 m^2/m^3 with a depth of 6.1 m will have to be designed for 90% removal of Total Kjeldahl Nitrogen (TKN) in a trickling filter. Assume the specific surface area of plastic packing material is 100 m^2/m^3. Also design the relevant hydraulics.

Solution

Determine the specific TKN removal rate.

$$R_n = 0.82\left(\frac{\text{BOD}}{\text{TKN}}\right)^{-0.44}$$

BOD/TKN = 160/25 = 6.4

$$R_n = 0.82(6.4)^{-0.44}$$
$$= 0.36\ g/m^2.d.$$

where

R_n = nitrification rate, $g/N/m^2$ d

BOD/TKN = influent BOD to TKN ratio, g/g

Determine the TKN removal.

$Q = 100\ L/s = 8640\ m^3/d$

TKN removal = $0.90(8640\ m^3/d)(25\ g/\ m^3)$

$$= 194,400\ g/d$$

Surface area of packing material:

$$A_s = \frac{194,400\ g/d}{R_n} = \frac{194,400\ g/d}{(0.36\ g/m^2.d)} = 540,000\ m^2$$

Volume of packing material:

$$\text{Vol} = \frac{540,000\ m^2}{100\ m^2/m^3} = 5400\ m^3$$

Hydraulic application rate:

$$\text{Filter area} = \frac{\text{Volume}}{\text{depth}} = \frac{5400\ m^3}{6.1\ m} = 885\ m^2$$

hydraulic application rate, q

$$q = \frac{q}{A} = \frac{(100\ L/s)}{885\ m^2} = 0.11\ L/m^2.s$$

To meet the minimum hydraulic application rate given previously as $0.5\ L/m^2.s$ recirculation will be required.

BOD loading based on volume and surface area

Loading based on volume:

$$\text{BOD loading} = \frac{(8640\ m^3/d)(160\ g/m^3)(1\ kg/1000\ g)}{5400\ m^3}$$

Loading based on area:
BOD loading $= (0.26\ \text{Kg/m}^3\ \text{d})[1/(100\ \text{m}^2/\text{m}^3)](10^3\ \text{g/kg})$

$$= 2.6\ \text{g/m}^2.\text{d}$$

Volumetric oxidation rate:

$$\text{VOR} = \frac{[S_o + 4.6(\text{NO}_x)]Q}{V(10^3\ \text{g/kg})}$$

where

VOR = volumetric oxidation rate, Kg/m^3 d
S_o = influent BOD concentration, g/m^3.
NO_x = amount of ammonia-nitrogen oxidized, g/m^3
Q = influent flow rate, m^3/d
V = packing volume, m^3
$S_o = 160\ \text{g/m}^3$
$\text{NO}_x = 0.90(25) = 22.5\ \text{g/m}^3$

$$\text{VOR} = \frac{[160\ \text{g/m}^3 + 4.6(22.5\ \text{g/m}^3)](8640\ \text{m}^3/\text{d})}{(5400\ \text{m}^3)(10^3\ \text{g/kg})} = 0.42\ \text{kg/m}^3.\text{d}$$

3.8.9 Design of a Flow-through Aerated Lagoon [17]

A 4.3 m deep flow-through aerated lagoon assuming a 7 day SRT time has to be designed to treat wastewater having characteristics as presented below. Daily 3800 m^3 of wastewater needs to be treated. The design should specify the requirement of surface aerators along with their power rating. Design should be done assuming a 2-day detention time of the treated water in the settling basin. Assume the following flow characteristics.

(1) Influent TSS $= 200\ \text{g/m}^3$, (2) influent TSS are not degraded biologically, (3) influent sBOD $= 200\ \text{g/m}^3$, (4) effluent sBOD $= 30\ \text{g/m}^3$, (5) effluent SS after settling 20 g/m^3, (6) kinetic coefficients: $Y = 0.65$ g/g, $K_S = 100\ \text{g/m}^3$, $k = 6.0$ g/gd, $k_d = 0.07$ g/gd for $T = 20$ to 25°C, (7) total solids produced are equal to computed volatile SS divided by 0.85, (8) first–order observed soluble BOD removal-rate constant $k_{20} = 2.5\ \text{d}^{-1}$ at 20°C, (9) summer air temperature = 30°C, (10) winter air temperature during coldest month 6°C, (11) wastewater temperature during winter = 16°C, (12) wastewater temperature during summer = 22°C, (13) temperature coefficient = 1.06, (14) aeration constants: $\alpha = 0.85$, $\beta = 1$, (15) aeration oxygen rate = 1.8 kg O_2/kWh, (16) elevation = 500 m,

(17) oxygen concentration to be maintained in liquid = 1.5 g/m^3, (18) power requirement for mixing is 8 kW/10^3/m^3

Process Design Considerations

Factors that must be considered in the process design of aerated lagoons include:

1. BOD removal
2. Effluent characteristics
3. Oxygen requirements
4. Temperature effects
5. Solids separation

BOD Removal

- The pertinent equation for a single aerated lagoon is:

  where

  S = effluent BOD_5 concentration, mg/L

  S_o = influent BOD_5 concentration, mg/L

  k = overall first-order BOD_5 removal-rate constant

  V = volume, m^3

  Q = flowrate, m^3/day

Effluent Characteristics

The important characteristics of the effluent from an aerated lagoon include the BOD_5 and the SS concentration.

The effluent BOD_5 may occasionally contain small amounts of algae.

Oxygen Requirement

The oxygen requirement refers to the activated-sludge process design based on operating results obtained from a number of industrial and domestic installations.

The amount of oxygen required has been found to vary from 0.7 to 1.4 times the amount of BOD_5 removed.

Temperature Effects

The two most important effects of temperature are:

1. Reduced biological activity and treatment efficiency
2. The formation of ice

The effect of temperature on biological activity, i.e., on the influent wastewater temperature, air temperature, surface area of the pond, and wastewater flow rate.

The resulting temperature in the aerated lagoon can be estimated using the Mancini and Barnhart equation:

$$(T_i - T_w) = \frac{(T_w - T_a)fA}{Q}$$

or

$$T_w = \frac{AfT_a + QT_i}{Af + Q}$$

where

T_i = influent waste temperature, °C (°F)
f = friction factor, 0.5
T_w = lagoon water temperature, °C (°F)
A = surface area, $m^2(ft^2)$
T_a = ambient air temperature, °C (°F)
Q = wastewater flow rate, m^3/d

Solid Separation

1. The detention time must be adequate to achieve the desire degree of SS removal.
2. Sufficient volume must be provided for sludge storage.
3. Algal growth must be minimized.
4. Odors that must be controlled.

The expression can be used to estimate the decay of VSS:

$$W_t = W_o e^{-k_d t}$$

where

W_t = mass of volatile SS that have not degraded after time t, kg
W_0 = mass of solids deposited initially, kg
k_d = decay of coefficient, d^{-1} or yr^{-1}
t = time, d or yr

Determine Surface Area of the Lagoon based on 7-day SRT

Volume = Q*SRT = (3800 m^3/d) (7 d) = 26,600 m^3
Surface area of lagoon = V/D = 26,600 m^3/4.3 m = 6186 m^2
D = Lagoon depth = 4.3 m

Consideration of Temperature Fluctuations between Summer and Winter

We use the empirical relations below to arrive at temperature considerations in summer:

$$T_w = \frac{AfT_a + QT_i}{Af + Q} = \frac{(6186\ m^2)(0.5)(30) + (6186\ m^3/d)(22)}{(6186\ m^2)(0.5) + (6186\ m^2/d)} = 24.67^\circ C$$

For winter we use similarly the relation:

$$T_w = \frac{AfT_a + QT_i}{Af + Q} = \frac{(6186\ m^2)(0.5)(30) + (3800\ m^3/d)(16)}{(6186\ m^2)(0.5) + (3800\ m^2/d)} = 4.78^\circ C$$

where

T_i = influent waste temperature °C
T_w = lagoon water temperature °C
T_a = ambient air temperature °C
f = proportionality factor
A = surface area, m^2
Q = wastewater flow rate, m^3/d

The proportionality factor incorporates the appropriate heat-transfer coefficients and includes the effect of surface area increase due to aeration, wind, and humidity. A typical value of "f" in the United States is 0.5 in SI units.

Estimate the soluble effluent BOD at the lagoon outlet during the summer using:

$$S = \frac{K_s[1 + (K_d)\mathrm{SRT}]}{\mathrm{SRT}(Yk - K_d) - 1}$$

$$= \frac{(100\ g/m^3)[(1 + 0.07\ g/g.d)(0.5\ d)]}{(0.5\ d)[(0.65\ g/g)(6.0\ g/g.d) - (0.07\ g/g.d)] - 1} = 7.4\ g/m^3$$

where

S = the effluent dissolved substrate concentration, g BOD or bs COD/m^3
k_d = endogenous decay coefficient, gVSS/Gvss d
k_s = half-velocity constant, substrate concentration at one-half the maximum specific substrate utilization rate, g/m^3
k = maximum rate of substrate utilization, T^{-1}

SRT = average time the activated-sludge solids are in the system. The SRT is determined by dividing the mass of solids in the aeration tank by the solids removed via the effluent and by wasting for process control.

Observation and Special Considerations for Temperature Fluctuation

The BOD value in the effluent from the settling lagoon will be essentially the same.

The effluent sBOD is computed using kinetic-growth constants derived for the temperature in the range from 20 to 25°C. Thus, during the summer months, the effluent requirement of 20 g/m^3 or less is achieved easily. Because there is no reliable information on how to correct these constants for the winter temperature of 10°C, an estimate of the effect of temperature can be obtained using the first-order soluble BOD removal rate constant.

Estimating Effluent BOD

Correct the removal-rate constant for temperature effects using [17]:

Summer (24.7°C)

$$\frac{k_2}{k_1} = \theta^{T_1 - T_2}; \quad k_2 = k_1 \theta^{T_1 - T_2}$$

where

θ = temperature coefficient. The temperature dependence of the rate and equilibrium constants can be estimated using the van't Hoff–Arrhenius relationship.

T = temperature of water is a very important parameter because of its effect on chemical reactions and reaction rates, aquatic life, and the suitability of the water for beneficial uses. In addition, oxygen is less soluble in warm water than in cold water.

$$k_{25.4^\circ C} = 2.5(1.06)^{24.7-20} = 3.3/\text{d}$$

Winter (4.78°C):

$$k_{11.7^\circ C} = 2.5(1.06)^{4.78-20} = 1.03/\text{d}$$

Determine the effluent BOD values using [17]:

Summer (24.7°C)

$$S = \frac{S_o}{[1 + (k)\tau]} = \frac{200\ \text{g/m}^3}{[1 + (3.42/\text{d})(5\text{d})]} = 11\ \text{g/m}^3$$

Winter (11.7°C)

$$S = \frac{S_o}{[1 + (k)\tau]} = \frac{200\ \text{g/m}^3}{[1 + (1.54/\text{d})(5\text{d})]} = 22.9\ \text{g/m}^3$$

where

S = effluent BOD concentration, g/m^3

S_o = influent BOD concentration, g/m^3

k = overall first-order BOD removal rate constant, d^{-1}

τ = hydraulic retention time (V/Q), d

Based on the above analysis using removal-rate constants, it appears that the effluent requirement of 30 g/m^3 or less will be met during both the summer and winter. (Note: The foregoing calculations were presented only to illustrate the method. The removal rate constants need to be determined through pilot plant study.)

Estimate the concentration of biological solids produced using the empirical relation (SRT = τ).

$$X = \frac{Y(S_o - S)}{(1 + k_d)\text{SRT}} = \frac{(0.65\ \text{g/g})[(200 - 7.4)\text{g/m}^3]}{[1 + (0.07\ \text{g/gd})(5\text{d})]} = 92.7\ \text{g/m}^3$$

where

X = biomass concentration, ML^{-3}

Y = biomass yield, M of cell formed per M of substrate consumed

k_d = endogenous decay coefficient, T^{-1}

S = effluent BOD concentration, g/m^3

S_o = influent concentration, g/m^3

SRT = solid retention time

An approximate estimate of the biological solids produced can be obtained by multiplying the assumed growth yield constant (BOD basis) by the BOD removed.

Suspended Solids in the Lagoon Effluent before Settling

$$\text{TSS} = 200\ \text{g/m}^3 + (92.7\ \text{g/m}^3)/0.85 = 309.1\ \text{g/m}^3$$

With the extremely low overflow rate provided in a holding basin with a detention time of 2 d, an effluent containing less than 20 g/m^3 of SS should be attainable.

Determine the Oxygen Requirement

The oxygen required for the biodegradation of carbonaceous material is determined from a mass balance using the bCOD concentration of the wastewater treated and the amount of biomass wasted from the system per day. Thus, for a suspended growth process, the oxygen used is:

Oxygen used = bCOD removed − COD of waste sludge

$$R_o = Q(S_o - S) - 1.42\ W_{x.bio}$$

where

R_o = oxygen required, kg/d

$W_{x,bio}$ = biomass as VSS wasted per day, kg/d, including active biomass and cell debris from cell growth

Determine $W_{x \cdot bio}$, the amount of biological solids wasted per day.

$$W_x = (92.7\ g/m^3)(3800\ m^3/d)(1\ kg/10^3\ g) = 352.3\ kg/d$$

Assuming the conversion factor for BOD to COD is 1/1.16 (=0.625), determine the oxygen requirements.

$$R_o = \frac{(3800\ m^3/d)[(200 - 7.4)g/m^3]}{(0.625)(10^3\ g/kg)} - (1.42)(352.3\ kg/d) = 670.7\ kg/d$$

Ratio of oxygen required for bBOD removal:

$$\frac{O_2\ \text{required}}{\text{BOD removal}} = \frac{(670.7\ kg/d)}{(3800\ m^3/d)[(200-7.4)g/m^3](1\ kg/10^3\ g)} = \frac{0.92\ kg\ O_2}{kg\ BOD}$$

Surface aerator power requirement, assuming aerator rating at 1.8 kg O_2/kWh.

Correction for the surface aerators for summer conditions:

i. Oxygen saturation concentration at 25.4°C = 8.18 g/m^3

ii. Oxygen saturation concentration at 25.4°C corrected for altitude DO = (0.94)(8.18 g/m^3) = 7.69 g/m^3

Determine the required AOTR.

$$\text{SOTR} = \text{AOTR}\left[\frac{C_{o,\ 20}}{\alpha(\beta C_{s,TH} - C)}\right](1.024^{20-T})$$

$$\frac{(670.7\ \mathrm{kg/d})(9.08\ \mathrm{g/m^3})(1.024^{20-25.4})}{(0.85)[(1.0)(7.69\ \mathrm{g/m^3})-(1.5\ \mathrm{g/m^3})]}=1018.3\ \mathrm{kg/d}$$

Determine the energy required to supply the needed oxygen. The amount of O_2 transferred per day per aerator unit is equal to 1.8 kg O_2/kWh. The total power needed to meet the oxygen requirements is:

$$\mathrm{Energy}=\frac{(1018\ \mathrm{kg/d})}{(1.8\ \mathrm{kg/kWh})(24\ \mathrm{h/d})}23.6\ \mathrm{kW}$$

Energy requirement for mixing.

Power requirement

Lagoon volume = 19,000 m^3

Power required = (19,000 m^3)(8 kW/10^3 m^3) = 152 kW

The selection of an aerator option depends on the geometry of the lagoon site, the type and design of the aerator, and a lifecycle cost estimate of the lagoon construction and aerator installation with special attention to the zones of mixing and oxygen dispersion (obtained from manufacturers' technical information) to minimize solids deposition and to ensure that aerobic conditions are maintained throughout the lagoon. It has been shown that the energy requirement for mixing is over six times the energy requirement needed to satisfy the oxygen demand. For installations designed to treat domestic wastewater, the energy requirement for mixing is almost always the controlling factor in sizing the aerators for treating high-strength industrial wastes.

REFERENCES

[1] Kumar R, Bhakta P, Chakraborty S, Pal P. Separating cyanide from coke wastewater by cross flow nanofiltration. Sep. Sci. Technol. 2011;46:2119–27.

[2] Korzenowski C, Minhalma M, Bernandes AM, Ferreira JZ, Pinho MN. Nanofiltration for treatment of coke plant ammoniacal wastewaters. Sep. Purif. Technol. 2011;76:303–7.

[3] Zhao W, Huang X, Lee D. Enhanced treatment of coke plant wastewater using an anaerobic–anoxic–oxic membrane bioreactor system. Sep. Purif. Technol. 2009;66: 279–86.

[4] Kumar R, Pal P. A membrane-integrated advanced scheme for treatment of industrial wastewater: dynamic modeling towards scale up. Chemosphere 2013;92(10):1375–82.

[5] Kumar R, Pal P. Membrane-integrated hybrid bioremediation of industrial wastewater: a continuous treatment and recycling approach. Desalination and Water Reuse 2013;1(3):26–38.

[6] Sarla M, Pandit M, Tyagi DK, Kapoor JC. Oxidation of cyanide in aqueous solution by chemical and photochemical process. J. Hazard Mater. 2004;116:49–56.

[7] Bae W, Lee SH, Ko GB. Evaluation of predominant reaction mechanisms for the Fenton's process in textile dying wastewater treatment. Wat. Sci. Technol. 2004;49 (4):91–6.
[8] Ferry JD. Ultrafilter membranes and ultrafiltration. Chem. Rev. 1936;18:373–455.
[9] Bandini S, Vezzani D. Nanofiltration modelling: the role of dielectric exclusion in membrane characterization. Chem. Eng. Sci. 2003;58(15):3303–26.
[10] Garuti G, Giordano A, Pirozzi F. Full-scale ANANOX® system performance. Water SA 2001;27(2):189–97. Available on website <http://www.wrc.org.za>
[11] Crowe SA, Canfield DE, Mucci A, Sundby B, Maranger R. Anammox, denitrification and fixed-nitrogen removal in sediments from the Lower St. Lawrence Estuary. Biogeosciences 2012;9:4309–21.
[12] Zhang L, Zheng P, Tang C-j, Jin R-c. Anaerobic ammonium oxidation for treatment of ammonium-rich wastewaters. J. Zhejiang Univ. Sci. B 2008;9(5):416–26.
[13] Xu X, Cheng Y, Zhang T, Ji F. Treatment of pharmaceutical wastewater using interior microelectrolysis/Fenton oxidation-coagulation and biological degradation. Chemosphere 2016;152:23–30.
[14] Fan L, Ni J, Wu Y, Zhang Y. Treatment of bromoamine acid wastewater using combined process of micro-electrolysis and biological aerobic filter. J. Hazard Mater. 2009;162:1204–10.
[15] Lan S, Ju F, Wu X. Treatment of wastewater containing EDTA-Cu(II) using the combined process of interior microelectrolysis and Fenton oxidation coagulation. Sep. Purif. Technol 2012;89:117–24.
[16] Ying D, Peng J, Xu X, Li K, Wang Y, Jia J. Treatment of mature landfill leachate by internal micro-electrolysis integrated with coagulation: a comparative study on a novel sequencing batch reactor based on zero valent iron. J. Hazard Mater. 2012; 229-230:426–33.
[17] Metcalfe & Eddy, Wastewater Engineering: Treatment and Reuse, Tata McGraw-Hill Edition 2003.

CHAPTER 4

Physicochemical Treatment Technology

4.1 COAGULATION–FLOCCULATION–PRECIPITATION–FILTRATION

If present in aqueous form a contaminant in water can be separated out through filtration by transferring the contaminant to solid phase. Fine solids in the100–1000 μm size range and coarse solids greater than 1000 μm can be easily separated out from water by direct application filtration or sedimentation techniques. Solids of dimensions less than 1 μm (micrometer), which are called molecular dispersions, while those falling in the range of 1–100 μm, which are called colloidal suspensions, are difficult to separate by direct filtration or sedimentation techniques. However, colloidal suspensions exhibit certain properties or characteristics such as very high surface area-to-mass ratio, electrical charge on the surface, Brownian movement, Tyndall effect, and high adsorption capacity exploiting which such fine solids can be effectively separated out from water. Water-treatment technologies have accordingly been developed by directly exploiting the characteristics of the fine solids present in water or indirectly by using the potential of conversion of the aqueous contaminants into solid phase. The high specific surface of fine particles results in dominance of the surface phenomena in the behavior and treatment of such fine solids. Such large surface area in turn imparts high adsorption capability to the solids.

Colloidal particles acquire either positive or negative electrical charge with respect to the surrounding medium and the magnitude of such charge depends on the nature of the solids and the medium. Thus when placed in an electrical field, such charged particles migrate to the pole of opposite charge in a phenomenon known as electrophoresis. The rate of migration is proportional to the potential difference of the electrical field. Colloidal particles are always bombarded by the molecules of the dispersion medium because of their very small size resulting in their Brownian movement. Another very significant property of colloidal particles is the Tyndall effect. Since such solids have dimensions greater than the average

Industrial Water Treatment Process Technology.
DOI: http://dx.doi.org/10.1016/B978-0-12-810391-3.00004-7

wavelength of 400–640 nm of white light they interfere with its passage. This is the reason why an observer at right angles to the beam of light can see the same as it gets reflected by the particles. This is called the Tyndall effect, which is used to measure the turbidity of water.

4.1.1 Chemical Precipitation Technology for Mobilization to Solid Phase

By exploiting the insolubility of certain contaminants in an aqueous system, mobilization to solid phase can be done through chemical precipitation. For example, most dominant arsenic compounds that are precipitated out using insolubility property are arsenic sulphides, ferric arsenate, and calcium arsenate. In such precipitation, the role of pH is significant. Under neutral pH conditions, the inorganic arsenic compounds of Cu(II), Zn(II), Pb(II), and Fe(II) are very stable and for their precipitation out from aqueous medium, the pH needs to be increased where the dominant form of arsenic is As(V).

Chemical precipitation is generally considered to be a permanent, efficient, and easy-to-monitor method with immediate and fast results that allows simultaneous removal of many metal contaminants from water. Although chemical precipitation may be very useful for large-scale treatment of contaminated water, deep elimination of contaminants is often not possible by this method. The major disadvantage of chemical coagulation–precipitation is the simultaneous generation of a large amount of sludge coupled with relatively high maintenance and operational costs. Calcium, aluminum, and ferric ions are widely used for precipitation. In most precipitation methods, coagulation, flocculation is the associated steps through which eventual precipitation or mobilization to solid phase takes place. The term "coagulation" literally means driving together in Latin and refers to the chemical process of destabilization of the colloidal suspensions or charged particles. The next logical step in mobilization toward solid phase is flocculation or flock formation, which facilitates and enhances settling of the fine particles through agglomeration of the destabilized particles. For example, when it is alum precipitation, added aluminum ions reduce absolute values of zeta potential of the particles resulting in coagulation–flocculation of the fine particles. Thus arsenic ions precipitate with aluminum ions enmeshed in the coagulates–precipitates. Finally separation of arsenic from water is effected through downstream sedimentation and filtration. The coagulation–precipitation process is pH-dependent. For example, alum precipitation of arsenic is best in the

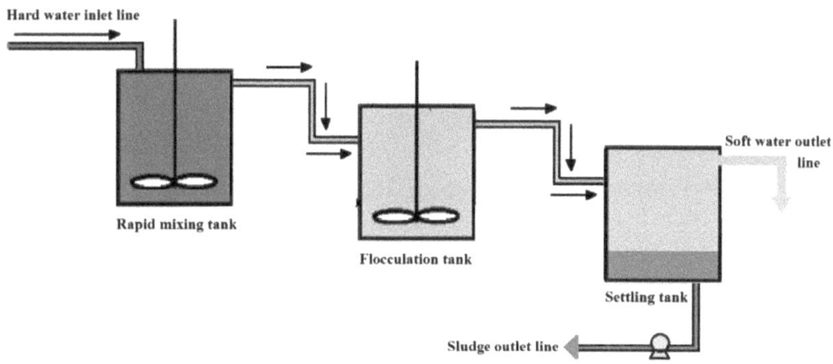

Figure 4.1 Schematic of coagulation–flocculation–precipitation–settling system.

pH range of 5–7. However, As(V) is more effectively mobilized to solid phase than As(III). Thus for efficient removal of arsenic from water where both forms of arsenic are present or where only As(III) is present, conversion of As(III) to As(V) prior to precipitation is important. Enmeshment precipitation is often followed in coagulation–precipitation. For example, in arsenic separation using iron-precipitation process, ferric ions are added to arsenic-bearing water where arsenic coprecipitates with ferric hydroxides are enmeshed in the coagulate–precipitate. In all coagulation–flocculation–settling processes, three basic units as shown in Figure 4.1 are essential; these are the rapid mixing, flocculation, and settling units. The rapid mixing unit is meant for rapid mixing of the coagulation reagents with the help of high-speed mechanical agitation. This facilitates fast dispersion of the chemical reagents throughout the aqueous medium resulting in efficient utilization of the consumable chemicals. The second unit facilitates flock formation where mild agitation helps particle–particle interaction and flock formation. The third unit should avoid agitation as this is the unit where formed flocks settle at the bottom leaving clear supernatant to overflow from the top of the settling unit.

The efficiency of a rapid mixing unit is measured in terms of velocity gradient G where

$$G = \frac{\partial u}{\partial y} \text{ in } X \text{ direction}, \quad u \text{ is velocity} \tag{4.1}$$

If fluid power is designated by P, then

$$P = (T)(\omega)$$

where T is the torque and ω is the fluid angular velocity. We may now express fluid power as:

$$P = (T)(G) = T\frac{\Delta u}{\Delta y} \tag{4.2}$$

Again, torque T = product of shear stress and area of shear or

$$T = (\tau)(A_s) \tag{4.3}$$

From Newton's second law, we get:

$$\tau = \mu\frac{\Delta u}{\Delta y} \tag{4.4}$$

where moment arm length in the y direction is Δy.

Therefore we may now express fluid power in terms of velocity gradient G as:

$$P = \mu\frac{\Delta u}{\Delta y}A_S\Delta y\frac{\Delta u}{\Delta y} = \mu V_T G^2 \tag{4.5}$$

where V_T stands for the rapid mixing tank volume and P is the fluid power. To arrive at the input power requirement, we have to take into account the efficiency of the blades (η_b) of the rapid mixing impeller and the motor efficiency (η_m), or in other words, we may express input power as:

$$P_{input} = P/(\eta_b \cdot \eta_m) \tag{4.6}$$

In the rapid mixing flocculation tank, G stays high whereas the final flocculation tank needs to stay low. A low value of G implies larger flock size. The time (t) allowed for flocculation is important in flocculation, and in fact, the dimensionless parameter Gt is considered a flocculation index. Thus both the parameters G and Gt assume significance in determining rate and size of formation of flocks.

4.1.2 Enhanced Coagulation Technology

Sometimes separation by chemical coagulation—precipitation can be enhanced by adding coarse particles like calcite. Smaller calcite particles (30—45 μm) can be more effective than larger ones due to larger specific surface area. During arsenic separation, for example, arsenic-borne coagulates get coated on the surface of such calcite particles for eventual removal as precipitates. After being coated onto the surface of the calcite particles, the arsenic-borne coagulates attain bigger size and greater density facilitating easy and quick settling in downstream sedimentation and filtration.

4.1.3 Flocculation

Perikinetic Flocculation

As described earlier, flocculation needs particle–particle interaction and when such interaction is facilitated by Brownian movement (without the aid of external mechanical stirring of the aqueous medium) it is called Perikinetic flocculation. The random movement of the colloidal particles following bombardment by the fluid molecules results in particle–particle interaction and hence flocculation of the fine particles.

The rate of change of concentration of the particles in suspension due to Perikinetic flocculation is expressed as [1]:

$$\frac{dN}{dt} = -\frac{4}{3}\frac{E}{\eta}KTN^2 \tag{4.7}$$

where "N" represents the total number of fine particles at any instant t in the medium

"E" is collision efficiency factor

K is Boltzmann's constant

T is absolute temperature

η is fluid absolute viscosity

Analysis of the terms of Eq. (4.1) shows that the rate of change of concentration of the fine particles in the medium becomes faster with increase in the values of the terms E, K, T, and N and only fluid viscosity impedes the change. Again, a faster rate implies that the number of particles N that remain at any instant t will decrease with increase in the values of E, K, and T.

If Eq. (4.1) is integrated between the limits $t = 0$ to $t = \mathrm{t}$ and $N = N_\mathrm{I}$ to $N = N$, then N, the number of particles at $t = t$, can be expressed in terms of initial number of particles (at $t = 0$) N_I as:

$$N = \frac{N_\mathrm{I}}{1 + (4/3)(EKTN_\mathrm{I}/\eta)t} = \frac{N_\mathrm{I}}{1 + (t/t_{0.5})} \tag{4.8}$$

where we define the term $1/t_{0.5}$ as equal to $(4/3)(EKTN_\mathrm{I}/\eta)$ and we see that when time t equals $t_{0.5}$ Eq. (4.2) reduces to $N = N_\mathrm{I}/(1 + 1) = N_\mathrm{I}/2$.

This implies that $t_{0.5}$ is the time required to reduce the number of particles in suspension to half the initial number of particles in colloidal suspension.

Orthokinetic Flocculation

The flocculation that results from particle–particle interaction–collision following external mechanical stirring of the medium is called orthokinetic flocculation. When bulk fluid is mechanically agitated, fluid molecules undergo both spatial as well as temporal variation. If the spatial variation is designated by a term "*G*," then the rate of change of total concentration of the particles assuming uniform size "*d*" in suspension may be expressed as [1]:

$$\frac{\mathrm{d}N}{\mathrm{d}t} = -(2/3)EGd^3N^2 \tag{4.9}$$

where the volume of the colloidal particles is V_c and can be expressed as:

$$\frac{4}{3}\prod r^3N = \frac{4}{3}\prod\left(\frac{d}{2}\right)^3N = \frac{\prod d^3}{6}N = V_c \tag{4.10}$$

Thus the average size (d) of the particles can be expressed as:

$$d^3 = \frac{V_c 6}{\prod N} \tag{4.11}$$

Now substituting d^3 by $V_c 6/\prod N$ in Eq. (4.3), we arrive at:

$$\frac{\mathrm{d}N}{\mathrm{d}t} = -\frac{4}{\prod}EGV_cN \tag{4.12}$$

As in the case of Perikinetic flocculation, if we integrate Eq. (4.6) with respect to the limits $t = 0$ to t and $N = N_I$ to N, we arrive at the expression:

$$\ln\frac{N}{N_I} = -\frac{4}{\prod}EV_cGt \tag{4.13}$$

Now to find the time t_{destab} required for complete destabilization when $E = 1$, we put $E = 1$ in Eq. (4.7) and arrive at:

$$t_{destab} = -\frac{\prod}{4GV_c}\ln\frac{N}{N_I} \tag{4.14}$$

The total destabilization time so arrived is generally very long suggesting that additional measures such as interparticle bridging and enmeshment precipitation in metal hydroxides should be adopted to bring down the time for complete destabilization of colloidal particles to the order of a few hours. In the next section, we discuss such destabilization-enhancing techniques.

4.1.4 Understanding Diffuse-Double-Layer Theory to Destabilize Colloidal Suspension

A colloidal suspension is basically a stable phase showing little tendency to aggregate and separate from the aqueous phase. For separation of chemical precipitates effectively from water phase, destabilization of colloidal state is necessary. As colloidal particles have large surface-to-mass ratio, the behavior of the colloidal suspension basically represents surface phenomena. Colloidal particles assume electrical charges with respect to the surrounding environment. Thus destabilization of colloids means neutralization of such surface charge. The effective way to destabilize is to aggregate the fine particles through coagulation. Driving particles together for destabilization is called coagulation. The stability of colloids has been explained most explicitly by diffuse-double-layer theory. We now look at understanding this diffuse-double-layer theory for effective separation of solids from aqueous phase.

According to diffuse-double-layer theory colloidal particles assume positive or negative charge in the presence of charged groups within or due to the presence of an adsorbed charged layer from the surrounding medium. An electrical double layer of opposite charge is thus formed at the solid—liquid interface consisting of charged colloidal particles and oppositely charged counterions that accumulate in water in the immediate vicinity of the charged colloidal particles. Formation of this double layer ensures electroneutrality and hence stability of the overall colloidal system as it can't have a net electrical charge. Diffusion of counterions is induced by thermal agitation or replacement of ions by some reaction. In the presence of high concentration of counterions, the diffuse double layer gets compacted. The fixed layer of positive charge adjacent to the diffuse double layer is called the Stern layer as depicted in Figure 4.2.

A colloid surface attracts the counterions of the aqueous phase with opposite charge. In the immediate vicinity of the colloidal particles, concentration of the counterions is naturally high. This concentration diffuses

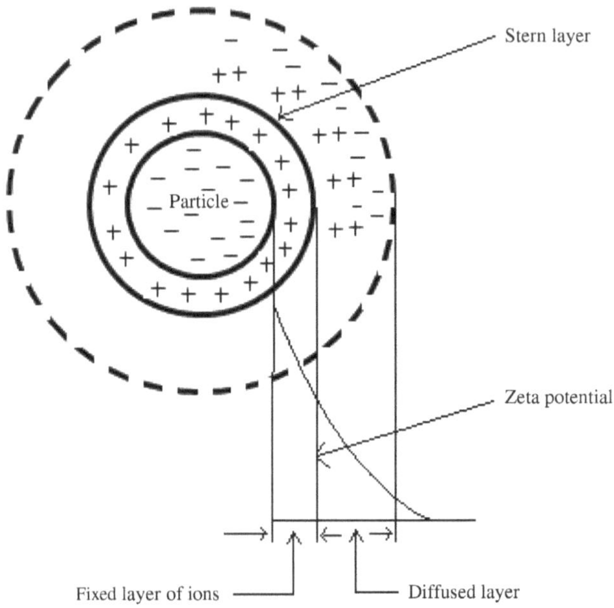

Figure 4.2 Diffuse-double-layer diagram (Source: *P. Pal, Groundwater Arsenic Remediation—Treatment Technology and Scale Up, Elsevier Science; 2015*).

out as the distance between the surface of the colloidal particle and the bulk solution increases. The magnitude of the charge at the surface of shear (at the interface of diffuse layer and stern layer) is called the zeta potential and is measured through electrophoresis where migration of colloidal particles toward opposite pole of charge is captured. This zeta potential ψ may be defined in terms of charge q on the colloidal particle and its zone of influence δ as:

$$\psi = 4\pi\, \delta q / D$$

where D is the dielectric constant of the medium.

A repulsive force and an attractive van der Waal's force remain active between two similarly charged colloidal particles. If we designate these repulsive and attractive forces as V_R and V_A, respectively, then the net interactive force or energy may expressed as $V_R - V_A$. This difference between the repulsive and attractive forces is called the energy barrier of double-layer interaction. This net-energy barrier determines the stability of the colloidal suspension. Sufficient kinetic energy needs to be added to the colloidal system to overcome this energy barrier leading to its destabilization. Mathematically we can express the potential repulsive energy of electrical double-layer interaction in a suspension of heterogeneous particles, between two heterogeneous spheres of radii a_1 and a_2 with ψ_1 and ψ_2 as the outer Helmholtz plane (OHP) potentials or zeta potentials, respectively, as [2]:

$$V_R = 32\pi\varepsilon_0\varepsilon_r \frac{2a_1a_2}{(a_1+a_2)}\left(\frac{K_B T}{ev}\right)^2 \tanh\left(\frac{ev\psi_1}{4K_B T}\right)\tanh\left(\frac{ev\psi_2}{4K_B T}\right)\exp(-kh) \tag{4.15}$$

where ε_0 = vacuum dielectric permittivity;
ε_r = relative dielectric permittivity of the medium;
T = absolute temperature;
K_B = Boltzmann constant.

In Eq. (4.9), the elementary charge and ionic valence of the electrolyte have been designated as e and ν, respectively. The Debye reciprocal length is k and the shortest separation between the two particles is denoted by h. The potential energy (V_A) of van der Waal's interaction between two heterogeneous particles 1 and 2 in medium 3 is expressed as [3]:

$$V_A = -\frac{a_1a_2}{(a_1+a_2)}\left(\frac{A_{132}}{6h}\right) \tag{4.16}$$

where
A_{132} = Hamaker constant of particles 1 and 2 in the medium 3.

To compute the same we need to compute the Hamaker constants of the particles 1 (A_{11}), 2 (A_{22}), and medium 3 (A_{33}) in vacuum using Eq. (4.11) [4]:

$$A_{132} = \left(\sqrt{A_{11}} - \sqrt{A_{33}}\right)\left(\sqrt{A_{22}} - \sqrt{A_{33}}\right) \tag{4.17}$$

To illustrate this we can consider the case of removal of arsenic (1) from medium water (3) by using precipitation reagent (2) ferric hydroxide [$Fe(OH)_3$].

Let us consider that A_{11} represents the Hamaker constant of the arsenic-borne coagulates, A_{22} refers to the Hamaker constant of the precipitation reagent $Fe(OH)_3$, and water is represented by A_{33}. In this case, A_{11} is greater than A_{33} and A_{22} is greater than A_{33}. Thus A_{123} is greater than 0, and V_A is less than 0 according to Eqs. (4.10) and (4.11). This indicates an attractive interaction of van der Waals' between the two heterogeneous particles. In the pH range of around 5–9, $\Psi_1 < 0$ (arsenic-borne coagulates), and $\Psi_2 > 0$ (ferric hydroxide), thus $\tanh(ev\psi_1/4kT) < 0$ and $\tanh(ev\psi_2/4kT) > 0$, implying $V_R < 0$ according to Eq. (4.9). This means the electrical double-layer interaction is attractive between the two heterogeneous particles involved. Therefore, in the above pH range, the total potential energy of interaction between the arsenic-borne coagulate and the ferric hydroxide particle will be attractive at every distance. This in turn implies that there is no potential energy barrier between the two particles represented by arsenic-bearing fine coagulates and ferric hydroxide or coarse calcite particles, which might prevent them from coming together. This explains the mechanism that facilitates coating of the arsenic-borne fine coagulates onto the surface of the precipitation reagent ferric hydroxide or coarse calcite particles to eventually coprecipitate with ferric hydroxide. Such coprecipitation is called enmeshment precipitation in which the process of settling becomes much faster than in the precipitation without enmeshment.

4.1.5 Treatment Strategies for Fast Settling of Particles

Now that we understand diffuse double layer in colloidal systems coupled with the mechanisms of enmeshment coprecipitation, we can now devise a strategy for making the settling process fast during coagulation–flocculation–precipitation for purification of water. Fast-settling technology could be more successful with the adoption of double-layer compression techniques, adsorption and neutralization of charge, enmeshment precipitation, and interparticle bridging. In one such approach, coagulant ions in high concentration can be added to the system. As similarly charged ions of the coagulants are repelled, the oppositely charged ions are attracted by the primary charge of the colloidal particles, thereby causing compression of the diffuse double layer. The decrease of the diffuse double layer with high concentration of the counterions of the coagulants will help overcome the energy barrier of a colloidal system while destabilizing it. The higher the charge of the coagulant ion the higher the degree of coagulation. Thus the coagulation effect of Al^{+3} will be greater than that of Ca^{+2}.

Another effective strategy is adsorption and neutralization of charge where the charge of the colloidal particles is neutralized by the addition of molecules of opposite charge that adsorb onto the surface of the colloidal particles. But caution has to be exercised as overdosing of such charged molecules may lead to restabilization of the system by the residual charges of the added molecules after neutralization of the primary charges of the colloidal particles.

In the enmeshment-precipitation approach, metal salts such as aluminum sulfate, ferric chloride, and calcium oxide are added to the colloidal system to precipitate them as hydroxides in which the colloidal particles get enmeshed and coprecipitate.

In interparticle bridging, as proposed by Lamer (1963), some long-chain charged polymeric molecules are added to a colloidal system where one charged end of the polymer molecule attaches to a site of the colloid and the other end extends to the bulk solution. When the other end attaches to another colloidal particle, an effective bridging takes place between two colloidal particles resulting in their settling together.

4.1.6 Particle Settling

Successful operation of the settling unit is important in the overall coagulation–flocculation–settling process of water treatment. Settling of a particle depends on the net force that results from three forces, drag force F_D, buoyant force F_B, and gravity force F_G, as depicted in Figure 4.3.

Discrete particles that fall through still liquid accelerate downward due to gravity force (F_G) until the frictional resistance, which manifests itself as drag force (F_D), equals the driving force. When the driving force equals the drag force, a particle starts settling with uniform velocity called terminal settling velocity.

According to Newton's second law, for a particle of settling velocity V_s, mass m, characteristic diameter d, and density ρ_s the acting forces on a settling particle can be expressed as:

$$m\frac{dv_s}{dt} = F_G - F_B - F_D \tag{4.18}$$

$$F_G = \rho_s \cdot V_s g; \quad F_B = \rho \cdot V_s \cdot g$$

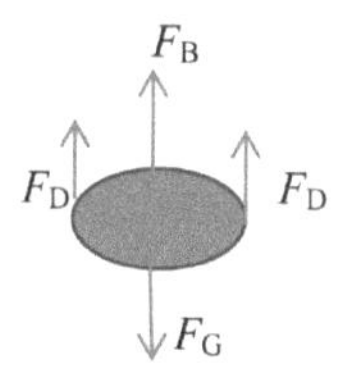

Figure 4.3 Action of the forces acting on a settling particle.

where the density of the particle and water are ρ_s and ρ, respectively, and V_s stands for the particle volume or volume of water displaced by the particle volume V_s, which is equal to:

$$\frac{4}{3}\prod r^3 = \frac{4}{3}\prod\left(\frac{d}{2}\right)^3 = \frac{\prod}{6}d^3 = V_s$$

The projected area of a particle $= A_s = 4/\prod d^2$.
If C_d = Newton's drag coefficient, then $F_d = (1/2)C_D \rho A_s V_s^2$.
As $V_s/A_s = (2/3)d$, we get:

$$V_s = \sqrt{\frac{4}{3}\left(\frac{g}{C_D}\right)(\rho_s - 1)d} \tag{4.19}$$

C_D as a function of Reynolds number $= dV_s\rho/\mu$.
For $Re < 1.0$, $C_D = 24/R_e$. Thus:

$$V_s = \sqrt{\frac{4}{3}\left(\frac{g}{24}\right)(\rho_s - 1)(R_e)(d)} \tag{4.20}$$

This is Stokes law.

At $R_e > 1000$, $C_D = 0.4$, so

$$\begin{aligned} V_s &= \sqrt{\frac{4}{3}\left(\frac{g}{0.4}\right)(\rho_s - 1)(d)} \\ &= 1.82\sqrt{g(\rho_s - 1)d} \end{aligned} \tag{4.21}$$

In designing such settling systems, the parameters that need to be considered are flow rate of the water through the settling system, residence time of water in the system, size of the system, and ability of the settling unit to remove the sludge.

The rate of flow of water should be slow enough to allow particles to settle out. If the water flow is too rapid particles will escape the settling process. A properly designed tank or baffle system will reduce the water flow and allow particles to settle out. However, a flow that is too slow implies low capacity of the system, suggesting a tradeoff in deciding on the flow rate commensurate with the desired treatment volume and quality of the treated water.

Adequate residence time of water must be ensured in the settling unit to allow the particles to settle. A properly designed settling tank or baffle system reduces water flow, but additional turbulence in turn reduces particle settling and may even disturb settled sludge.

Let us examine the dimensions of a settling tank that ensure effective settling that is commensurate with flow rate. Let the effective length of the settling zone, breadth, volume, surface area of settling, depth of settling zone, volumetric flow rate of water, and velocity of water be represented by L, B, V, A_s, H, Q, and U, respectively.

Thus the detention time of a particle at a height H from the sludge zone is:

$$\theta = \frac{V}{Q} \quad \text{and} \quad \text{flow velocity } U = \frac{Q}{HB}$$

The settling velocity of this particle, which has traversed a height H, may be expressed as $V_{SH.}$ We can write settling velocity V_{SH} = depth of the tank/detention time = H/θ and

$$V_{SH} = \frac{H}{V/Q} = \frac{H}{LB} \cdot \frac{Q}{H} = \frac{Q}{A_S} \tag{4.22}$$

where

Q/A_S = the overflow rate or surface-loading rate (SOR).

Let the settling velocity of the particle, which was initially at a height H, be V_{sh}.

A particle initially at a height H will be removed by the time it traverses the settling zone. Obviously all particles initially at a height less than H will also settle, whereas the particles at greater heights will not reach the bottom of the settling tank before reaching the outlet zone from where particles escape settling. Thus all particles with settling velocity $V_s > V_{SH}$ will be removed partly depending on their position at a height from the top of the sludge zone. The efficiency of removal of such particles may be expressed as:

$$\eta_{settling} = \frac{h}{H} = \frac{V_{Sh} \cdot \theta}{V_{SH} \cdot \theta} = \frac{V_{Sh}}{V_{SH}} = \frac{V_{Sh}}{Q/A_S} \tag{4.23}$$

where Q/A_S = surface-loading rate or overflow rate.

Therefore it may be concluded that the rate of settling of particles for a given velocity depends only on the surface-loading rate and the surface area of settling rather than the depth of settling.

The size of the system needs to ensure the desired water flow rate and the desired residence time. A system that is too small size will force water to flow at high rate allowing inadequate residence time. Proper design minimizes sludge disturbance. When determining the settling tank size, the size of the particles is also very important to consider as finer particles need longer time to settle. The length of the tank must be long enough to allow longer settling time to the finest particles. Width and depth should consider requirement of total holding capacity and appropriate so as not to disturb sludge. Baffling often helps in fast settling but the design must ensure easy removal of sludge. A regular sludge removal schedule is needed to keep the system running. Heavy buildup of sludge reduces the effectiveness of the settling system. To avoid corrosion, settling tanks can also be made of polymers.

4.1.7 Filtration

After settling of the contaminants the next step in the processing is filtration of the precipitated solids from the water. While the large flocks mostly settle in the settling tanks, the fine particles escape with the overflow and need to be separated out by physical filtration. While fabric, ceramic, or polymeric microfiltration devices can be used at this stage for separation of the fine solids, sand filtration is considered a fast and low-cost option and is used for large-scale filtration where sand or silica are abundantly available. Fabric, polymeric, or ceramic filtration are covered in the membrane-filtration chapter and therefore, in this section, we only deal with some aspects of sand filtration. Water percolates through large sand beds down to the clear water well. Sand is first properly sieved and used as per the desired criteria of effective size and uniformity coefficient. If D_{10} and D_{60} are the sieve sizes that do not allow more than 10% and 60% by weight of sand, respectively, to pass through, then D_{10} is considered an effective size while D_{60}/D_{10} is the uniformity coefficient. The usable stock sand percentage is expressed as:

$$P_{usable} = 2(P_{60} - P_{10}) \tag{4.24}$$

where P_{10} and P_{60} are the percentages of stock sand smaller than D_{10} and D_{60}, respectively, and sand of grain size between D_{10} and D_{60} forms half of the specified sand.

Some of the major problems encountered during sand filtration include air binding resulting from release of dissolved air or gases of water that occupy the sand pores thereby effectively preventing water flow and resulting in rapid loss of head. By avoiding air saturation of the water and preventing algal growth these problems can be largely overcome.

The filter bed needs to be cleaned by backwashing occasionally with water jet in the reverse direction, which transforms the entire sand bed into a fluidized bed. Formation of mud balls (mud-coated ball-shaped sand) during backwashing is another problem frequently encountered. As the mud balls sink due to increased weight they prevent effective backwashing. The problem of mud balls can be overcome by replacing the sand or by adding 2%–5% caustic soda solution to the system or eventually breaking the balls.

4.2 PHYSICOCHEMICAL TREATMENT TECHNOLOGY BASED ON COAGULATION–FLOCCULATION–SETTLING

Now that the basic unit operations and the principles of physicochemical treatment of water have been discussed we focus on a technology developed recently for large-scale treatment of arsenic-contaminated groundwater by [5]. The technology is based on coagulation, flocculation, and precipitation where sand filtration at the final stage produces potable water. For arsenic-affected regions where sources of alternate safe drinking water are limited and expensive, this technology can be promising as a low-cost and reasonably efficient technology.

The technology uses ferric chloride as the coagulant and potassium permanganate ($KMnO_4$) as the oxidizing agent for converting trivalent arsenic to pentavalent form as the latter form of arsenic precipitates out more easily than the trivalent form.

This oxidation is fast and follows first-order kinetics as shown in Eq. (4.12):

$$\frac{\mathrm{d}}{\mathrm{d}t}\left[\mathrm{As}^{\mathrm{III}}\right] = -K\left[\mathrm{As}^{\mathrm{III}}\right] \tag{4.25}$$

where K = first-order rate constant (s^{-1}).

Oxidation of iron follows similar first-order kinetics. In the presence of oxidizing agent $KMnO_4$ the following reactions take place through which As(III) gets converted into As(V) and Fe^{+2} gets converted into Fe^{+3}:

$$KMnO_4 \rightarrow K^+ + MnO_4^- \tag{4.26}$$

$$H_2O \rightarrow H^+ + OH^- \tag{4.27}$$

$$3Fe^{2+} + MnO_4^- + 4H^- \rightarrow 3Fe^{3+} + MnO_2 + 2H_2O \tag{4.28}$$

$$As^{3+} + MnO_4^- + 4H^+ \rightarrow As^{5+} + MnO_2 + 2H_2O \tag{4.29}$$

Concentration of $KMnO_4$ strongly influences oxidation of As^{+3} to As^{+5} and of Fe^{+2} to Fe^{+3}. The oxidation reactor is a Continuous Stirred Tank Reactor (CSTR) where reaction ingredients need to be quickly and thoroughly mixed.

After coagulation–precipitation, up to 98% removal of arsenic is possible following subsequent sedimentation and sand filtration. The pH needs to be controlled in both the oxidizing as well as the coagulation–precipitation units. In the oxidization unit, the optimum pH is 5–6 whereas in the coagulation–flocculation unit it is in the range of 7–8. The treatment plant is shown schematically in Figure 4.4.

In this physicochemical treatment technology, a preoxidation reactor equipped with a mechanical high-speed stirrer is the first unit followed by a high-mixing coagulator, a flocculation unit with provision for mild agitation, a settling unit, and a sand-filtration unit. The cylindrical oxidizer unit is provided with a three-impeller mechanical agitator in addition to some baffles.

The filter medium of the sand-filtration unit consists of granular sands with an average diameter of 0.00001 m. In the continuously operated

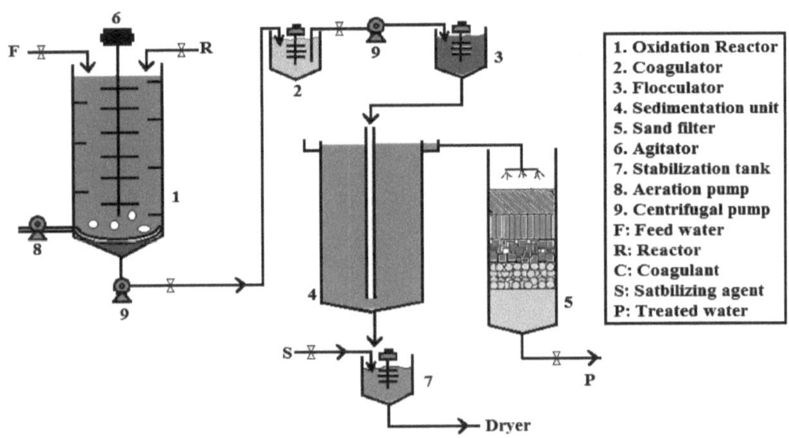

Figure 4.4 A physicochemical water treatment plant for arsenic removal [5].

plant, $KMnO_4$ solution (15 mg/L) is added to the oxidation unit along with the feed solution in measured doses. The coagulant ferric chloride solution (30 mg/L) is also added to the coagulation unit in the same way. Caustic soda (NaOH) is added to adjust the pH to around 7.6.

The aqueous stream from the coagulator passes to the flocculation unit where mild stirring facilitates particle–particle interaction for larger flock formation and subsequent settling. The well-flocculated stream then passes to the settling unit for separation of the precipitates from the aqueous phase through sedimentation. The suspended solids that do not settle in the sedimentation are eventually separated out from the aqueous stream by sand filter. Further details on this technology can be found in *Groundwater Arsenic Remediation—Treatment Technology and Scale Up* (Elsevier Science, 2015, by Parimal Pal) [6].

4.3 ADSORPTION PRINCIPLES

4.3.1 Introduction

In water purification, the role of adsorption is significant. The fact that we can get almost clean and pure groundwater (barring cases of some specific contamination) is largely because of the natural adsorbents of soil. Beds of clays, sands, and minerals of soil through adsorption purify this groundwater that we extract for various purposes. Such adsorption of contaminants or impurities on a solid surface from an aqueous system is driven by the unbalanced forces of attraction of the atoms at the free surface of the solid. At the interior of the solid mass, such attractive forces of the atoms get balanced through interactions with forces of other atoms in the lattice. When such adsorption or concentration of solutes at the solid surface is driven by van der Waal's forces of attraction only, we call it physical adsorption. However, in concentrating solutes on the free solid surfaces, chemical interactions between the solutes and the solid surface may also be decisive and we then call it chemical adsorption.

Adsorption is a very general phenomenon where the availability of free surface of the adsorbent largely determines how long the process of adsorption will effectively continue for practical purposes. As the process of adsorption goes on, a time is reached when the surface of the adsorbent material gets completely exhausted or saturated. At this stage, even a highly efficient or effective adsorbent fails to perform any further purification of the contaminated fluid. Certain specific characteristics such as selectivity, capacity, and life of the adsorbent material largely determine

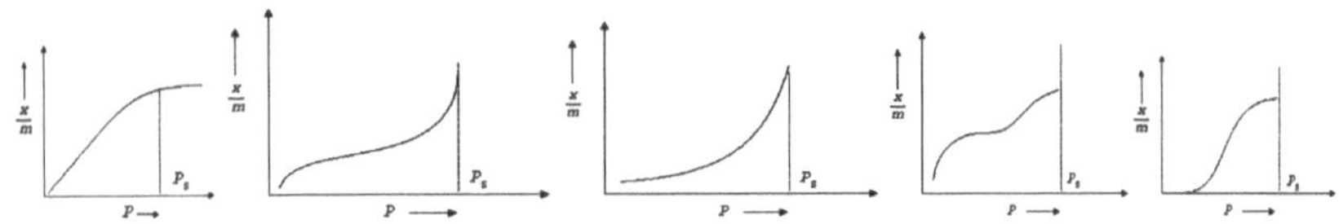

Figure 4.5 BET adsorption isotherms [8].

the effectiveness of adsorbents for economical and practical use. Henry's constants or directly measured chromatographic retention time can give an indication as to the suitability of an adsorbent for a practical situation.

Adsorption isotherms are the other useful tools for understanding the mechanism of adsorption and for quantitative assessment of partition or distribution of the solutes of interest at equilibrium between the involved solid and fluid phases.

Out of the adsorption isotherms developed over the years, the most widely referred isotherms are those of Langmuir [7] and Brunauer, Emmett, and Teller (BET) [8]. Langmuir isotherms are based on the concept of monolayer adsorption and apply to mainly low-pressure conditions. BET isotherms are based on the concept of multilayer adsorption and accurately describe physical adsorption under high-pressure conditions.

Figure 4.5 shows five basic types of BET isotherms where the x-axis represents the mass of adsorbate per mass (m) of adsorbent against pressure P. P_s represents the maximum saturation pressure.

The relative sizes of the adsorbate molecules and the pores of the adsorbent largely determine the degree of adsorption and hence the types of the adsorption isotherms. For example, isotherm takes the shape of Type I when the size difference between the micropore of the adsorbent and the sorbate molecules is very negligible. Complete filling of the micropore by the adsorbate molecules results in a saturation limit up to which maximum adsorption can take place. Thus every adsorption system under a set of conditions has a saturation limit. Type II and type III isotherms are exhibited by adsorbents with a wide range of pores in the structures. The adsorbents of Type II and III exhibit a continuous increase in adsorption capacity with increase in pressure due to capillary condensation under such pressure. Where the pore diameters of the adsorbents are much larger than the size of adsorbate molecules, Type IV isotherms are observed representing adsorption on two distinct surfaces—one on the plane surface and the other on the walls of the pores.

4.3.2 Adsorption Kinetic Models

There have been several models developed on adsorption kinetics. But four models, namely those of Langmuir [7], BET [8], Freundlich [9], and Redlich–Peterson [10], are the most widely applied.

The kinetics of the Langmuir model, which is valid for monolayer adsorption on a surface with a finite number of adsorption sites of equal energy, is expressed as:

$$A_e = \frac{A_{m,e} k_L C_e}{1 + k_L C_e} \tag{4.30a}$$

where A_e and $A_{m,e}$ are the amount of adsorption and the maximum amount of adsorption, respectively, per unit of adsorbent at equilibrium concentration of the solute C_e. The equilibrium coefficient of the Langmuir model is k_L. Thus in linearized form, the model may be expressed as:

$$\frac{C_e}{A_e} = \frac{1}{k_L A_{m,e}} + \frac{C_e}{A_{m,e}} \tag{4.30b}$$

For adsorbents with a heterogeneous surface comprised of different classes of adsorption sites, the Freundlich empirical model is used. The model is based on the assumption that each site can be modeled by:

$$A_e = k_F C_e^{\frac{1}{n}} \tag{4.31}$$

where k_F is the equilibrium coefficient of the Freundlich isotherm and n represents the intensity constants of the adsorbents.

The linearized form of this model is:

$$\ln A_e = \ln k_F + \frac{1}{n} \ln C_e \tag{4.32}$$

In the Redlich–Peterson mode, the models of Freundlich and Langmuir are combined to describe the number of adsorption sites with the same adsorption potential in addition to heterogeneity of the sorbent surface. With this approach, we arrive at the Redlich–Peterson model described by:

$$A_e = \frac{k_R C_e}{1 + a C_e^b} \tag{4.33}$$

where k_R is the equilibrium coefficient of the Redlich–Peterson isotherm and a, b represent equation constants. The linearized form of this model may be presented as:

$$\ln\left(\frac{k_R C_e}{A_e} - 1\right) = \ln a + b \ln C_e \quad (4.34)$$

High-pressure, low-temperature conditions permit a large number of the adsorbate molecules to remain in contact with the surface of the adsorbents because of their low thermal energy resulting in their multilayer adsorption. The BET kinetic model as presented in Eq. (4.35) describes such multilayer adsorption where active sites have different energies. Multilayers of adsorbate are formed in different parts of the surface. The equilibrium coefficient k_{BET} of such adsorption and the concentration of the adsorbate at the filled layer C_{Fl} are expressed as:

$$A_e = \frac{A_m k_{BET} C_e C_{Fl}}{(C_{Fl} - C_e) + (C_{Fl} + k_{BET} C_e - C_e)} \quad (4.35)$$

The BET isotherm in linearized form is:

$$\frac{C_e}{A_e(C_{Fl} - C_e)} = \frac{1}{A_m k_{BET}} + \frac{k_{BET} C_e - C_e}{A_m k_{BET} C_{Fl}} \quad (4.36)$$

where the equilibrium constants (k_L, k_F, k_R, and k_{BET}) and different constants (a, b, and n) of the equations can be calculated from the slope and the intersection of the line graph by plotting C_e/A_e vs. C_e of Eq. (4.35), $\ln A_e$ vs. $\ln C_e$ of Eq. (4.32), $\ln\left((k_R C_e/A_e) - 1\right)$ vs. $\ln C_e$ of Eq. (4.34) and $C_e/A_e(C_e - C_{Fl})$ vs. C_e/C_{Fl} of Eq. (4.36).

4.3.3 Adsorbent Materials in Water Purification

Activated Carbon

Activated carbon is one of the most widely used adsorbents with a removal efficiency of 90%–99%.The major characteristics of activated carbon include its large active surface area, inert nature, and stability over a wide pH range. Activated carbon is prepared by thermal decomposition of carbonaceous material followed by activation with steam at around 800–1100°C. During activation, highly volatile tarry carbonization products are released from the carbonaceous mass thereby opening the pores and resulting in a highly porous structure that consists of elementary microcrystallites of graphite stacked in random orientation. The spaces between the crystals constitute the micropores. Size and distribution of the actual pores, however, largely depend on the conditions of pyrolysis and activation leaving the possibility open for modifying porosity and enhancing adsorption capacity by controlling such conditions. For example, during adsorption of impurities from aqueous phase, pores can be made bigger to reduce mass-transfer resistance, whereas such pores can be made still smaller in gas-phase purification. While the surface of carbon is essentially nonpolar, slight polarity may, however, be induced on surface oxidation. Activated carbon is thus hydrophobic and organophillic and can be effectively used in the removal of organic impurities from water. In decolorizing sugar, activated carbon adsorption is widely practiced. However, the neutral nature of activated carbon does not allow it to be very selective for high purification application. Such selectivity of activated carbon can be further enhanced by making the pore-size distribution extremely narrow during activation, thereby permitting it to act as a molecular sieve.

The major disadvantage of activated carbon is that regeneration and recovery of the adsorbent on exhaustion is difficult and expensive. The powdered form in particular poses a challenge in recovery from aqueous stream. The recent trend is developing composite adsorbent incorporating activated carbon [11].

Zeolite

Zeolites are another widely used mineral adsorbent consisting of SiO_4 and AlO_4 tetrahedral units joined in a regular fashion through shared oxygen atoms. The Si/Al ratio can have any value ≥ 1.0. The vertex of each structure made up of several polyhedral of SiO_4 and AlO_4 is occupied by Si or Al atom. Each Al atom induces one negative charge that is balanced by one exchangeable cation present in some preferred locations within the framework, which eventually plays a significant role in determining the adsorptive capacity of the zeolite crystal. By changing this exchangeable cation through ionexchange, modifications of the adsorptive capacities of adsorbents can be enhanced. While aluminum-rich adsorbents will show high water adsorption capacity, silica-rich adsorbents will show hydrophobic character and can be used for adsorbing nonpolar substances like hydrocarbons. The transition from hydrophilic to hydrophobic occurs at a Si/Al ratio of around 8–10. Thus we observe that zeolite-based adsorbents can be tailor-made with tremendous potential for enhancing selectivity in adsorption for any specific separation-purification application. The most striking feature of this aluminosilicate adsorbent is that there is no pore-size distribution. The uniform pore size of molecular dimension resulting from definite crystalline structure makes this class of adsorbent useful molecular sieves different from other adsorbents. More than 30 varieties of natural as well as synthetic zeolite-based adsorbents have been reported [12,13]. The windows in the intracrystalline channel structures are constricted by oxygen rings made up of different numbers of oxygen atoms such as six, eight, or ten with different free diameters. The sieving property is determined by this free diameter. For example, in the case of six-membered oxygen rings, the free diameter is 2.8 Å, and in case of eight-member oxygen rings, the free diameter is 4.2 Å. The diameter of oxygen atom is 1.4 Å. Thus zeolites are accordingly called small-port or large-port adsorbents. The free diameters and hence kinetic selectivity of the zeolite-based adsorbents can be modified by using different cations that can block the windows to different extents depending on their number, nature, and affinity for different sites of the windows. There are different types of zeolites such as ling zeolite, clinoptilolite, etc. Clinoptilolite can effectively remove Pd^{+2}, Cd^{+2}, Cu^{+2}, and Zn^{+2} from water by adsorption and such adsorption can be improved at higher temperature. Zeolite cation-exchange capacity can be improved by pretreatment of the adsorbent. For example, NaOH treatment of zeolite improves its cation-exchange capacity. But a disadvantage of zeolite adsorbent is its low permeability.

Natural Clay

Clay is one of the cheapest and abundantly available materials used as an adsorbent. The four major clay groups are kaolinite, montmorillonite-smectite, illite, and chlorite. Bentonite is one such clay adsorbent comprised of montmorillonite used in the removal of dye from water. A recent trend is modification/activation of clay adsorbent by incorporating nanoparticles like TiO_2 to increase efficiency of adsorption.

Activated Alumina

Alumina has been in use as adsorbent since the dawn of the twentieth century though its commercial production and use began in 1932 by the Alcoa Company [14] for adsorption of water. While traditionally it was used as a desiccant, in recent years diversified use of activated alumina as catalyst in chemical production systems and as adsorbent for removal of many impurities in water treatment has been seen. Due to its abundance, high mechanical strength, large specific surface area, and ease of regeneration, activated alumina has been used widely in the removal of arsenic and fluoride from water. Sen and Pal [15] developed a low-cost activated alumina with large specific surface area of 335–340 m^2/g from gibbsite precursor through partial thermal dehydration. The continuous treatment system using activated aluminum as adsorbent proved very successful in the removal of arsenic from contaminated water up to the WHO-prescribed limit of 10 μg/L.

Activated aluminum is prepared from bauxite ($Al_2O_3 \cdot 3H_2O$) or from monohydrate of aluminum by thermal dehydration and recrystallization. This adsorbent is polar and exhibits amphoteric character (both acidic and basic).

When the alumina surface gets saturated with the solute of interest or the contaminant, it is necessary to regenerate the adsorbent for further use. The adsorbed contaminant or pollutant may be removed from the alumina surface by treating the saturated adsorbent with caustic soda in this case. Subsequent neutralization with sulfuric acid should be done to reuse alumina. However, regeneration is also limited since after two to three regenerations adsorption capacity is lost by more than almost 25%. However, the main advantage of using activated alumina is that it is a relatively low-cost material with reasonably high adsorption capacity. The rate and degree of arsenic removal from water by alumina depends on the oxidation states of the arsenic and pH of the feed water. For example, removal of As(V) from water is done more efficiently than the removal of As(III). Thus As(III) needs to be converted to As(V) by oxidation prior to adsorption. A pH in the range of 5.5–6.0 is found to be very effective in arsenic removal. Activated carbon and activated alumina are the two most widely used adsorbents in arsenic removal. However, considering the cost of raw material and the cost of regeneration such adsorbents can be made more attractive through cutting costs in the synthesis process.

Gel precipitation of gibbsite powder with sodium hydroxide and sulfuric acid is the most widely practiced method of producing activated alumina. This method produces chemically pure activated alumina with well-defined physical parameters that are essential to its use as catalyst. However, this gel precipitation method is expensive. An alternative approach is partial thermal dehydration of gibbsite powder at a suitable temperature over an appropriate time thereby removing 28%–31% of water of crystallization from the gibbsite, leaving a loss on ignition of 4–7 wt%. This gives rise to high active surface area. Parameters like dehydration temperature, rate of increase of temperature, residence time, and particle size are found to have significant effects on generation of specific surface area of the adsorbent.

Composite Adsorbent

Composite adsorbents seek to exploit the useful properties of zeolites, activated carbon, chitosan, activated clay, and other mineral-based precursors. The main aim of composite adsorbents is to enhance adsorption of contaminants with improved chemical strength and mechanical ability. Among the composite adsorbent materials, mineral carbon-based adsorbent has been found to be very effective in general due to its mosaic characteristic, which gives it the capability to adsorb both inorganic and organic substances.

4.4 ADSORPTION-BASED TECHNOLOGY

Adsorption-based technology has in general been applied in water treatment for removal of dyes and for removal of heavy metals, phenolic compounds, pesticides, removal of fluoride, and arsenic. Heavy metals are found naturally in earth. They get concentrated due to human activities and contaminate aquatic and terrestrial life. Common contamination sources are mining and industrial waste, vehicle emission, lead-acid batteries, fertilizers, paints, treated woods, aging water supply infrastructures, etc. The four most toxic heavy metals are cadmium, mercury, lead, and arsenic. Other examples include chromium, manganese, cobalt, nickel, copper, zinc, selenium, silver, antimony, and thalium. These are toxic as well as carcinogenic in nature. In removal of such toxic heavy metals adsorbents have been used successfully where the process of adsorption has been found to be enhanced at higher temperature. Over the years, thousands of adsorbent materials have been tested in water purification. Even very low-cost ordinary material like fly ash has been used in the removal of many water contaminants such as phenolic compounds.

We now touch on activated alumina-based treatment technology for removal of arsenic and fluoride from contaminated groundwater. Figure 4.6 depicts the complete flow diagram of the plant. The adsorbent activated alumina 0.5−0.9 mm in size is packed in a column on stainless-steel mess and graded gravel. Contaminated water is continuously passed through the column, which may be made polycarbonate material or stainless steel. The treatment plant is very simple consisting of the column. The plant initially works with a high degree of removal efficiency

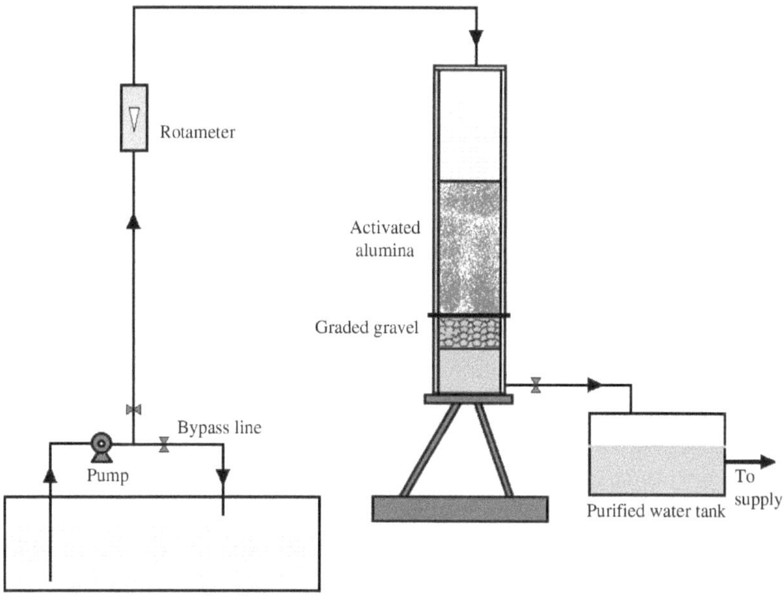

Figure 4.6 Adsorption plant using activated alumina-packed column for continuous treatment.

that gradually goes down with the progress of time. When the bed is completely saturated the spent adsorbent needs to be replaced by fresh material.

In any adsorption column employed in the removal of a water contaminant, the issue of backwashing of the bed needs to be addressed properly as beds often get clogged by fine sands, colloidal silvery sands, and the precipitated materials. In many field applications of adsorption columns installed for providing safe drinking water to a community, this problem of bed clogging has been encountered and has led to malfunctioning or closure of the units. At regular intervals (e.g., two to three times per week, depending on the rate of sedimentation on the adsorption bed) provision for backwashing needs to be made. During backwashing, the valves in the normal well-to-filter direction should be closed and the valves that allow flow in the reverse direction should be opened and water pumped under pressure. Continuous monitoring of the water quality must be done to ensure that the adsorption bed has not been exhausted and the water is still pure. For all adsorption materials, experimentally the break through point and the exhaustion point should be determined experimentally. This is done by continuously measuring the treated effluent volume and the concentration of the contaminant in the influent as well as in the effluent. The cumulative volume of water passed through the column up to the time of sudden rise of concentration of the contaminant in the effluent is the breakthrough volume and the cumulative volume of water up to the point when concentration of the target contaminant in the effluent is same as that in the influent is called the exhaustion volume of water. The adsorbent needs to be replaced at the point of breakthrough. Two major drawbacks of adsorption-based technology are the necessity of continuous monitoring of the effluent quality and necessity of regeneration or replacement of adsorbent. Hundreds of adsorbent materials have been investigated for their adsorption efficiency and the applicable adsorption isotherms. The general observation is that adsorption isotherms largely converge into the standard ones and adsorption efficiency though initially high eventually goes down. In gas-phase separation such as in air separation, or in chromatic separation and recovery of high-value components from a mixture, adsorption-based technology has been used in large-scale production system. But in aqueous-phase separation, particularly in water purification, the role of adsorption-based technology remains limited to community water-filtration systems. As an exothermic process adsorption suggests that desorption can be done for regeneration and reuse of adsorbent at elevated temperature. For gas-phase application, there are quite a few options like subjecting the saturated adsorbent to vacuum, to high temperature, or to purging by other gases. But in water-purification applications, backwashing, high-temperature treatment, or replacement by new adsorbent material are the only options.

REFERENCES

[1] O'Melia CR. Coagulation and flocculation. In: Weber Jr WJ, editor. Physicochemical processes for water quality control. New York: Wiley-Interscience; 1972.
[2] Wang Q. Theoretical analysis of Brownian heterocoagulation of fine particles at secondary minimum. J Colloid Interface Sci 1991;145:305–13.
[3] Hiemenz PC, Rajagopalan R. Principles of colloid and surface chemistry. 3rd ed. New York: Marcel Dekker; 1997.
[4] Israelachvili JN. Intermolecular and surface forces. 2nd ed. New York: Academic Press, Inc; 1992.
[5] Pal P, Ahamad Z, Pattanayak A, Bhattacharya P. Removal of arsenic from drinking water by chemical precipitation—a modeling and simulation study of the Physical-chemical processes. Water Environ Res 2007;79(4):357–66.

[6] Pal Parimal. Groundwater arsenic remediation treatment technology and scale up. Amsterdam, The Netherlands: Elsevier Science; 2015.
[7] Langmuir I. The constitution and fundamental properties of solids and liquids. J Am Chem Soc 1916;38:2221.
[8] Brunauer S, Emmett PH, Teller E. Adsorption of gases in multimolecular layers. J Am Chem Soc 1940;62:1723.
[9] Freundlich HMF. Over the adsorption in solution. J Phys Chem 1906;57:385.
[10] Redlich O, Peterson DL. A useful adsorption isotherm. J Phys Chem 1959;63:1024.
[11] Yanagisawa, et al. Adsorption of Zn(II) and Cd(II) ions onto magnesium and activated carbon composite in aqueous solution. Appl Surf Sci 2010;256:1619–23.
[12] Smith JV. In: Rabo J, editor. Zeolite chemistry and catalysis. ACS monograph no. 171. London: American Chemical Society; 1968. p. 10.
[13] Meier WM, Oslon DH. Atlas of zeolite structure types. Zurich: Juris Druck and Verlag AG; 1978.
[14] J.B. Barnitt, U.S. patent no.1 868 869, July 26, 1932.
[15] Sen M, Pal P. Treatment of arsenic-contaminated groundwater by a low cost activated alumina adsorbent prepared by partial thermal dehydration. Desalination Water Treat 2009;11:275–82.

CHAPTER 5

Water Treatment by Membrane-Separation Technology

5.1 INTRODUCTION

The fact that membrane as a semipermeable barrier is highly efficient in separation and purification of components is amply evident throughout nature. No living cell could survive without the screening service of membranes. This knowledge has led to numerous applications of membrane-based technology in separation and purification. For example, in artificial kidney dialysis, membranes are used in removing low molecular weight urea from blood to a stripping solution thereby purifying blood very efficiently. In equally significant scale, ultrafiltration (UF) membranes have been used in the pharmaceutical industry to produce injection-grade water.

Since the 1970s, reverse osmosis (RO) membrane has been in use in large-scale seawater desalination. Despite clear evidence of purification capability, the high cost of synthesis of membranes coupled with their short life limited the development of membrane-based technology. Over time, however, membranes with increased selectivity, mechanical strength, and durability at reduced cost have been developed, paving the way for large-scale integration of membrane-based technology in industrial water treatment.

Today's membranes can be tailor-made for a multitude of applications, and membrane-based technologies offer several advantages over conventional separation–purification technologies. For example, this technology is inherently energy-saving as no phase change is involved in the purification process.

Membrane technology based on modular design also offers operational flexibility in terms of capacity utilization. High selectivity of tailor-made membranes can ensure a high degree of purification of a system in a simplified plant configuration compared to conventional technology. A production technology based on membrane-integrated separation–purification is

Industrial Water Treatment Process Technology.
DOI: http://dx.doi.org/10.1016/B978-0-12-810391-3.00005-9

also ecofriendly as no chemical reagents or harsh chemicals need to be used and often multistep operations can be simplified into one step. In other words, it can safely be said that the use of membrane technology culminates in extensive process intensification, which implies energy-efficient, low cost, and environmentally friendly production with a high degree of purity. While versatile materials from polymerics to ceramics are being used in membrane synthesis, incompatibility of polymeric membranes in many solvent systems, low-temperature resistance of most of the polymeric membranes, fabrication limitations of ceramic membranes beyond the MF range, fouling of membranes in most of the available modules, limited scale-up confidence, and reluctance to embrace a new technology are some of the current challenges. Despite these limitations, membrane technologies are rapidly making progress in process intensification and water-treatment schemes. Such applications are determined by the requirement of degree of separation of the target components, relative pore sizes, and surface and physicochemical properties of the membranes as classified in the following section.

5.2 CLASSIFICATION OF MEMBRANE-BASED PROCESSES

Membrane-based separation processes are classified by the size of the materials or the molecules being separated, although size alone does not determine degree of separation. In addition to the relative sizes of the materials and membrane pores, the surface and physicochemical properties of the membranes and solute–membrane interaction often determine the degree of purification. Table 5.1 shows the classification along with the separation mechanism and transport regime.

5.3 MEMBRANE-SEPARATION TERMINOLOGY

Volume flux
The volume of the solvent collected on the permeate side per unit area of the membrane per unit time is called the volumetric flux and is expressed as $L/(m^2 \cdot h)$ or LMH.
Mass flux of solute
Mass flux = (mass of permeated solute)/(membrane surface area) × (permeation time).
Solvent flux
Flux is calculated based only on solvent permeation.
Total flux = solute flux + solvent flux

Table 5.1 Membrane Process Classification

Membrane Process	Pore Size	Separation Mechanism	Transport Regime	Membrane	Driving Force	Operating Pressure, bar
Microfiltration	0.2–10 μm >50 nm	Sieving	Macroporous	Porous	Hydrostatic pressure	2–5
Ultrafiltration	2–50 nm	Sieving	Mesoporous	Porous	Hydrostatic pressure	5–10
Nanofiltration	<2 nm	Sieving + Donnan exclusion	Microporous	Partly porous, partly dense	Hydrostatic pressure	10–20
Reverse osmosis	<5 Å	Solution diffusion	Molecular	Dense	Hydrostatic pressure	20–50
Forward osmosis	<5 Å	Solution diffusion	Molecular	Dense	Pressure gradient	1–2
Pervaporation	<5 Å	Solution diffusion	Molecular	Dense	Vapor pressure gradient	–
Gas separation	<5 Å	Knudsen diffusion	Molecular	Porous	Pressure gradient	15–130
		Solution diffusion	Molecular	Dense	Pressure gradient	
Electrodialysis	<5 Å	Electromigration	Ionic	Dense with electrical charge	Electrical potential gradient	–
Dialysis	<5 Å	Diffusion	Molecular	Dense	Concentration gradient	–

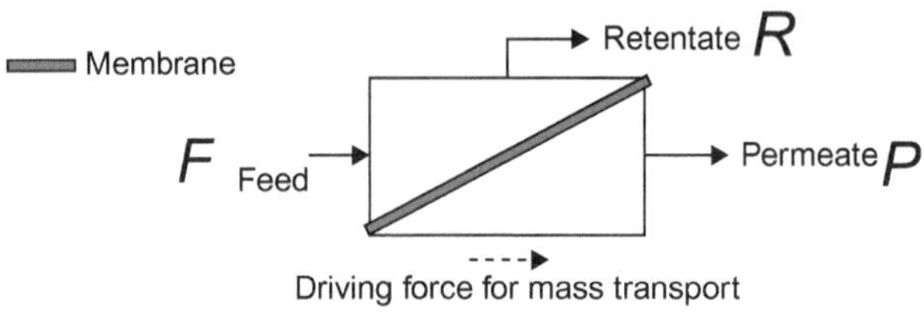

Figure 5.1 A typical membrane module showing the feed, permeate, and retentate streams.

5.3.1 Solute Retention

Solvent flux and solute retention are the two most important membrane-performance indicators.

The solute retention, or rejection *R*, is defined as the concentration difference across the membrane divided by the bulk concentration on the feed or concentrate side (fraction of solute remaining in the feed stream). Fig. 5.1 shows the feed, permeate, and retentate streams of a typical membrane module. The retention of solute can be expressed as:

$$R = \frac{c_1 - c_2}{c_1} = 1 - \frac{c_2}{c_1} \tag{5.1}$$

where c_1 = concentration of the diffusing solute in the bulk liquid phase (*F*); c_2 = concentration of the solute in the permeate (*P*)

Fouling

Fouling is accumulation of the solute on the membrane surface in the retentate side after certain operation and blockage of the pores.

Reversible fouling: When washing the membrane, fouling is removed making the membrane reusable, the fouling is called reversible fouling.

Irreversible fouling: When even after adequate rinsing and washing the membrane, original flux cannot be restored, we call it irreversible fouling.

5.3.2 Concentration Polarization

CP occurs when solutes on the membrane surface build up thereby reducing flux, and is considered a major hindrance in the sustainable use of membrane.

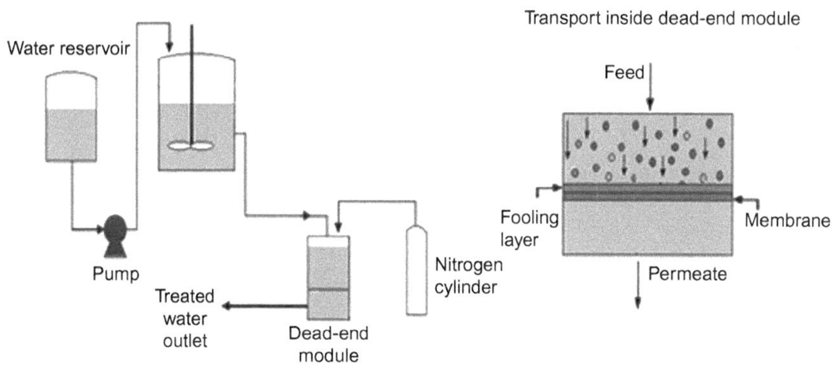

Figure 5.2 A membrane module operated in dead-end mode with high fouling potential.

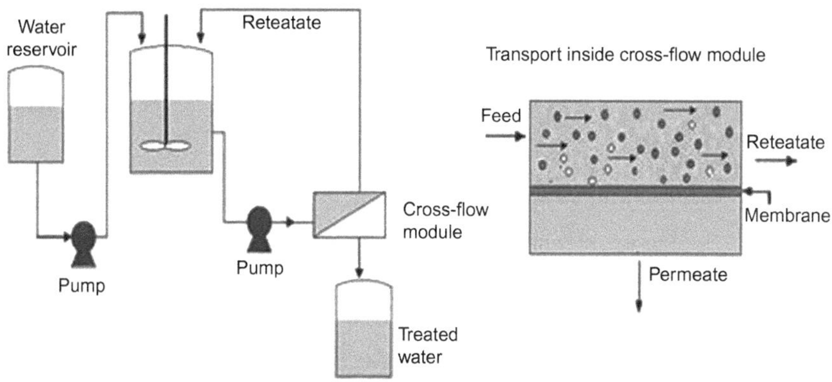

Figure 5.3 A membrane module operated in cross-flow mode with limited fouling possibility.

5.4 FLOW MODES

In any membrane module, fluid flow may be in dead-end mode as depicted in Fig. 5.2 or in cross-flow mode as depicted in Fig. 5.3. Dead-end mode is fouling-prone due to rapid concentration polarization (CP) whereas cross-flow mode is largely free from fouling due to the sweeping action of fluid on the membrane surface.

5.4.1 Cross-Flow Model and Pressure Drop

The flow of bulk solution is parallel to the membrane surface in cross-flow mode. A relatively high flow rate tangential to the surface sweeps the deposited particles toward the filter exit leaving a relatively thin

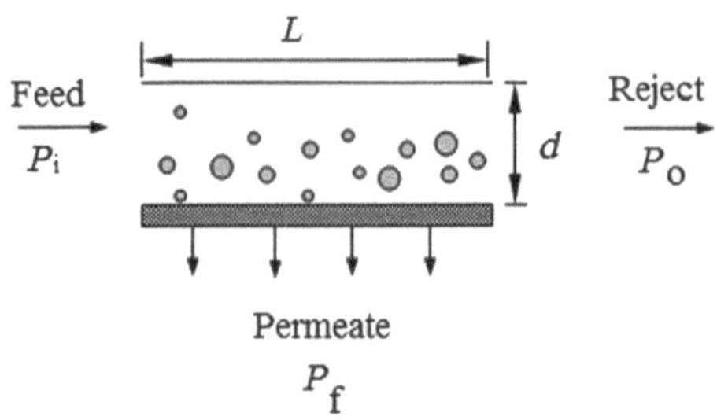

Figure 5.4 Pressure drop in a cross-flow module.

deposited cake layer, which is similar to the gel layer formed in UF. This cross-flow effectively controls the CP.

From the flow pattern and pressure drop with respect to Fig. 5.4, the expressions for transmembrane pressure drop are as follows:

$$\Delta P = P_i - P_o \tag{5.2}$$

Transmembrane pressure:

$$\Delta P_{TM} = \left(\frac{P_i - P_o}{2}\right) - P_f \tag{5.3}$$

ΔP_{TM} = Average driving force

P_f is normally assumed zero as the permeate side is open to atmosphere.

For laminar flow (Poiseuille flow)

$$\begin{aligned} \Delta P &= \frac{(C_1 \mu L v)}{d^2} \\ &= \frac{(C_2 \mu L Q)}{d^4} \end{aligned} \tag{5.4}$$

For turbulent flow (Fanning equation)

$$\begin{aligned} \Delta P &= \frac{(C_3 f L v^2)}{d} \\ &= \frac{(C_4 f L Q^2)}{d^5} \end{aligned} \tag{5.5}$$

where C_1, C_2, C_3, C_4 = constants based on channel geometry; f = factor based on Reynolds's number; Q = volumetric flow rate; v = velocity; L = filter length; d = fluid channel height above the membrane; μ = viscosity.

5.4.2 Dead-End Flow Model

In many applications, the batch process is run as dead-end flow. The particles build up over time forming cake and the clarified permeate is forced through the membrane. Membrane resistance is constant, whereas cake resistance increases with time due to cake buildup.

5.4.3 Permeate or Solvent Flux (Nw)

$$N_{\mathrm{w}} = \frac{\Delta P}{\mu(R_{\mathrm{m}} + R_{\mathrm{c}})} \tag{5.6}$$

where N_{w} = solvent flux (kg/(m$^2 \cdot$ s)); ΔP = pressure difference, Pa; R_{m} = membrane resistance, m^2/kg; R_{c} = cake resistance (m^2/kg); μ = viscosity of the solvent (Pa · s).

5.5 MEMBRANE MATERIALS

Membranes may be porous or nonporous and made of polymer, glass, metal, liquid, ceramic, zeolite, or composite materials. Nonporous membranes may be dense homogeneous or dense heterogeneous carrying electrical charges, functional groups, or catalysts. The polymers most widely used in membrane synthesis are cellulose acetate, polyamide, polysulfones, Polyvinylidene difluride (PVDF), polyacrylonitrile (PAN), polypropylene (PP), and polyethersulfone (PES). The use of zeolites is a recent development. The 3D crystalline material with lattice structure consisting of aluminum, silicon, and oxygen has well-defined pores and channels and offers both molecular sieving as well as catalytic services. Such zeolite-based membranes have recently been used in gas separation and in high-temperature environments. Liquid membranes are dynamically formed membranes mainly made of organic liquid, surface stabilizers, and embedded carrier molecules, which facilitate transport. The composite membranes consist of two distinct layers: the top thin layer does the screening or selection of materials that permeate through the membrane and the thick bottom layer (100–300 μm) provides mechanical support to the thin top layer (0.1–5.0 μm). The RO and nanofiltration (NF) membranes are thin-film composite (TFC) membranes in which the top layer may be made of polyamide or polyimide and the support layer made of polysulfones. A NF membrane may be synthesized by depositing a polyamide thin layer on a polysulfone UF membrane.

Membranes may be symmetric (isotropic) or asymmetric (anisotropic) in structure. In symmetric membranes, the structure of the membrane throughout the cross-section is the same, whereas it varies in asymmetric membranes. In symmetric membranes, which are mostly used in dialysis, electrodialysis, or MF, the flow characteristics are the same along the cross-section and the flux is almost proportional to the thickness of the membrane (30–500 μm). The mechanical stability of this class is generally low. Asymmetric membranes with high mechanical stability are pressure-driven membranes offering high flux. Symmetric membranes are synthesized from the same material. When asymmetric membranes are prepared by a single step-phase inversion process, the same material forms the skin layer and the support layer differently. Alternately, in a two-step process, a thin skin layer is deposited on a support layer of different material. Membrane modules may be flat-sheet, hollow-fiber, tubular, or spiral-wound types. Flat-sheet membranes can be both symmetric and asymmetric but hollow-fiber and capillary types are primarily asymmetric membranes.

5.6 MEMBRANE MODULES

Industrial membrane plants often require hundreds to thousands of square meters of membrane surface to perform the separation required to be useful. For large-scale application, membranes have to be packaged economically and efficiently. These packages, called membrane modules, are classified broadly as plate-and-frame, tubular, spiral wound, hollow-fiber, and flat-sheet cross-flow types.

5.6.1 Plate-and-Frame Membrane Module

Membrane, feed spacers, and product spacers are layered together between two end plates in a plate-and-frame module as shown in Fig. 5.5. The feed mixture is forced across the surface of the membrane. The fluid passes through the membrane, enters the permeate channel, and makes its way to a central permeate collection manifold. This type of module can be used in electrodialysis, pervaporation, and in rough-water purification prior to UF, NF, and RO applications with highly fouling feeds.

Cleaning of the membrane surface and replacement of the membrane can be done easily. Fine control over flow can be exercised on both the permeate and feed side of the membrane. However, this module is only suitable for small-scale applications, since it is expensive and requires a

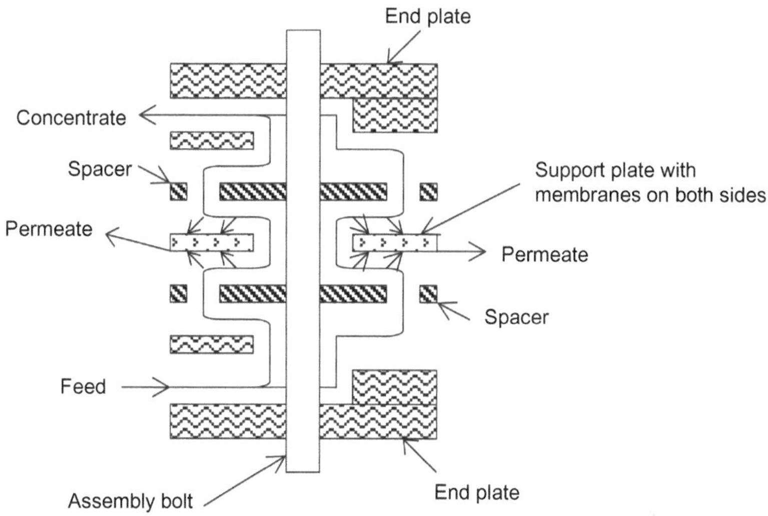

Figure 5.5 Schematic of a plate-and-frame module [1].

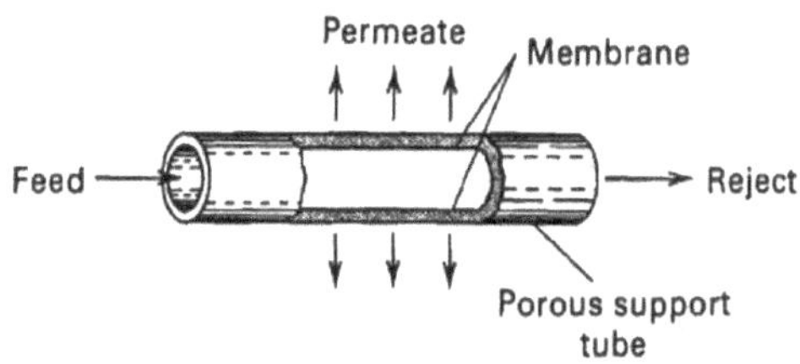

Figure 5.6 Tube of a shell and tube membrane module.

large number of spacer plates and seals. This module also suffers from maintenance problems as gaskets are used in large number between the plates and leaks can develop.

5.6.2 Shell and Tube or Tubular Membrane Modules

In a tubular or shell and tube membrane module, membrane tubes 5–15 nm in diameter are arranged in a shell similar to the fashion of a shell and tube heat exchanger. A large number of tubes (Figs. 5.6 and 5.7) may be connected in series. The feed solution (FS) is pumped through all the tubes connected in series, and the permeate is removed from each tube and sent to a permeate collection common header. The tubes consist of a porous paper or fiberglass support with the membrane formed on the inside of the tubes.

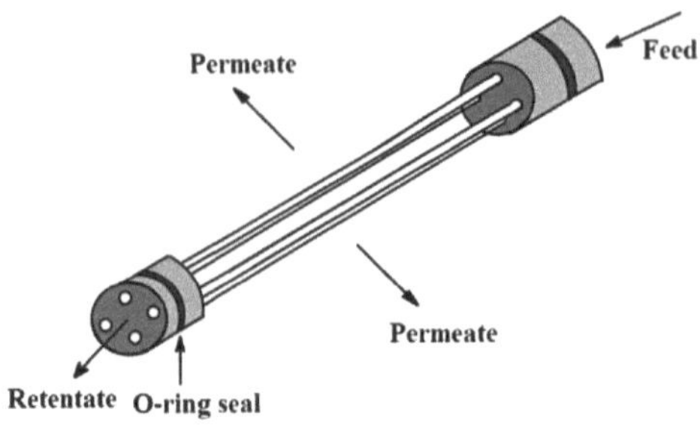

Figure 5.7 Shell and tube membrane module.

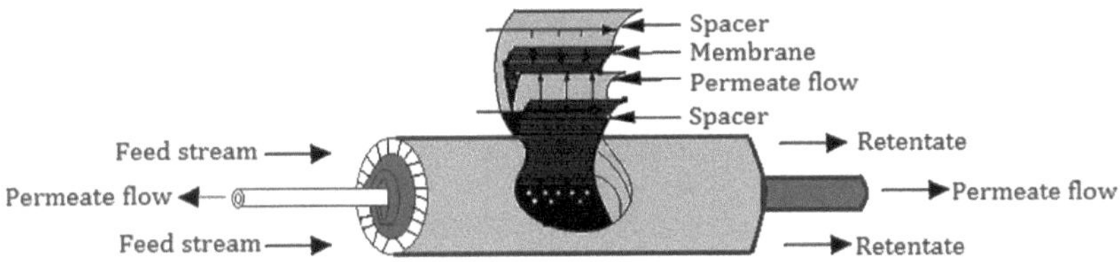

Figure 5.8 Spiral-wound membrane module [1].

The shell and tube design provides a large membrane area per unit volume but membrane formation is complex as the supporting and selecting layers are formed as integral cylindrical units during spinning. High surface area results in high fouling and high volumetric holdup.

Tubular modules are now generally limited to UF applications, for which the benefit of resistance to membrane fouling due to good fluid hydrodynamics compensates for their high cost. This means that this type of module should be used when a turbulent flow regime is preferred due to high solid concentration of the fluid (Re >10,000).

5.6.3 Spiral-Wound Membrane Modules

This design, as shown in Fig. 5.8, consists of a membrane envelope of spacers and a membrane wound around a perforated central collection tube. The module is placed inside a tubular pressure vessel. Feed passes axially down the module across the membrane envelope. A portion of the feed permeates into the membrane envelope, where it spirals toward the center and exits through the collection tube. The module is constructed easily and inexpensively and the hydrodynamic (flow rate, transmembrane pressure) can be adjusted by changing the spacer thickness to overcome for the CP and fouling. However, the membrane surface area per unit volume is low in this module. The system may also be expensive for high-pressure

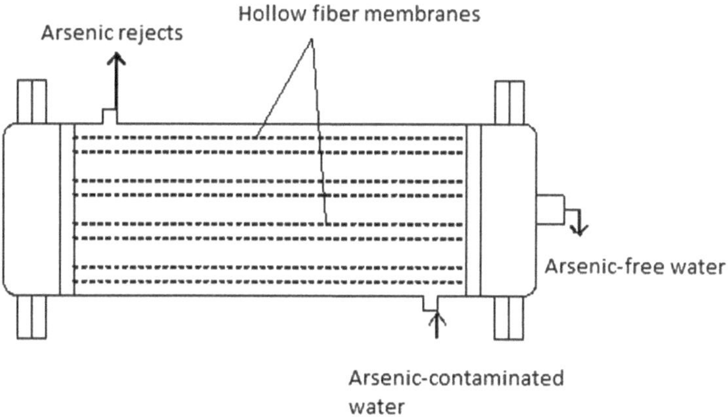

Figure 5.9 Hollow-fiber modules—shell-side feed design [1].

applications because of need for a high-pressure vessel for housing. Bypassing of feed may also occur due to nonuniform wrapping of spacers.

Spiral-wound modules are the most common module design for RO and UF as well as for high-pressure, gas-separation applications in the natural gas industry.

5.6.4 Hollow-Fiber Membrane Modules

Hollow-fiber membrane modules come in two designs: the shell-side feed design and the bore-side feed design. The shell-side feed design is shown in Fig. 5.9. In the shell and tube design, shell-side feed modules are generally used for high-pressure applications up to 70 bars. In such a module, a loop or a closed bundle of fibers is contained in a pressure vessel. The system is pressurized from the shell side as the permeate passes through the fiber wall and exits through the open fiber ends. As the fiber wall needs to support considerable hydrostatic pressure, the fibers usually have thick walls with small diameters, typically with a 50-μm internal diameter and 100–200-μm outer diameter.

Bore-Side Feed Design

In bore-side design, the fibers are open at both ends, and the feed fluid is circulated through the bore of the fibers. To minimize pressure drop inside the fibers, the diameters are usually larger than those of the fine fibers used in a shell-side feed system and are generally made by solution spinning. Feed pressures are normally limited to below 10 bars. The bore-side feed types are suggested in UF, pervaporation, and some low-to-medium-pressure applications.

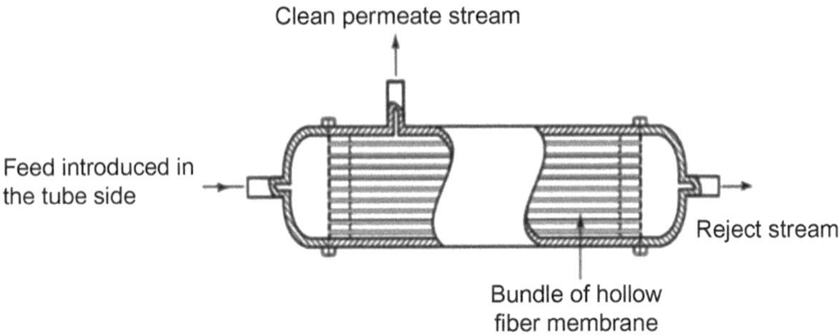

Figure 5.10 Hollow-fiber modules—bore-side feed design.

A hollow-fiber module with bore-side feed design as shown in Fig. 5.10 offers a very compact design by packing a very large membrane area into a single module. However, in the shell-side feed module, fouling on the feed side of the membrane is often a problem requiring pretreatment of the feed stream prior to introduction to this module. Bore-side feed modules are generally used for medium-pressure feed streams up to 10 bars where fine flow control is essential to minimize fouling and CP on the feed side of the membrane.

5.6.5 Flat-Sheet Cross-Flow Membrane Module

Flat-sheet cross-flow membrane modules (Fig. 5.11) are very similar to the plate-and-frame type but the fluid dynamics in these two cases are

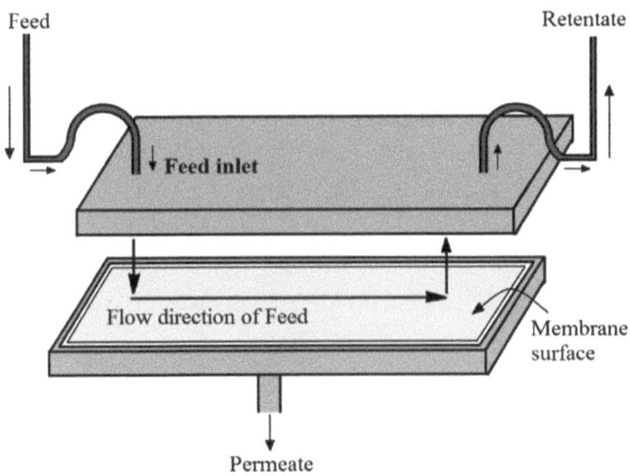

Figure 5.11 Flat-sheet cross-flow membrane module.

quite different. Upon entering the membrane module from its top surface at one end of the membrane box the fluid flows tangentially over the membrane surface horizontally and leaves from the bottom of the module at the other end after traversing the whole membrane surface. The sweeping action of the tangentially flowing fluid keeps the membrane surface largely free from fouling.

5.7 TRANSPORT MECHANISMS IN THE MEMBRANE-SEPARATION PROCESS

In biological membranes, transport is active with active participation of the membrane in the transport process through contribution of energy resulting from some sort of chemical reactions in the membrane itself. In contrast to biological membranes, all transport of contaminants through synthetic membranes is passive without any contribution of energy or driving force from the membrane in material transport across it.

Membrane acts as a semipermeable barrier between two phases, and mass transport between the two phases takes place as a result of the difference in the thermodynamic states of the two phases where the membrane is permeable to mass, heat, and electrical charge. The thermodynamic state of a system depends on a number of state functions such as internal energy, enthalpy, and Gibbs free energy. State functions in turn are dependent on a number of intensive and extensive state parameters. Mass, volume, and total energy are some examples of extensive state parameters whereas temperature, pressure, density, and molar properties are some examples of intensive parameters independent of the size of the system. The state function internal energy U and enthalpy H are defined as:

$$U = q + w$$

$$H = U + pV$$

where q is the total heat of the system and w is the work done by or to the system.

The change of the heat (q) of a system at constant temperature (T) is defined as entropy.

As in a complete reversible process, where no change of entropy takes place, entropy can be expressed as:

$$\mathrm{d}S = \frac{\mathrm{d}q_{\mathrm{rev}}}{T} = 0$$

and for an irreversible or spontaneous process, entropy may be expressed as:

$$dS = \frac{dq_{irrev}}{T} \geq 0$$

The state function enthalpy H is related to the Gibbs free energy as:

$$G = H - TS.$$

The Gibbs free energy change plays the most significant role in the membrane-separation process. The change in Gibbs free energy in a reversible process in equilibrium is given by:

$$(dG)_{p,T} = 0$$

implying that the total change in the Gibbs free energy is used to bring the system to equilibrium.

The Gibbs free energy change for an irreversible process is given by:

$$-(dG)_{p,T} = T ds$$

implying that the total energy in the Gibbs free energy change is not used to bring the system to equilibrium.

In transport of any specific component from one phase to the other through membrane, the driving force is the difference in Gibbs free energy of the component between the two phases, which in turn is the difference in electrochemical potentials of the component between the two phases. The electrochemical potential differences result from differences in concentration, hydrostatic pressure, temperature, and electric potential of the specific components between the two phases. In the case of dense membranes, transport is mainly through diffusion whereas in porous membrane it is the viscous flow that constitutes the transport regime. The rate of transport depends on the mobility of the components, which in turn is determined by the relative size of the component coupled with its interaction with the membrane-phase material. Membrane-separation mechanisms thus can be broadly classified into five types described as follows.

5.7.1 Size-Exclusion or Sieving Mechanism

In a size-exclusion mechanism, the relative sizes of the pores and the components of the mixture or solution determine the permeation of some components and the retention of others. This mechanism applies to porous membranes like microfiltration (MF) and UF. In this case, convective transport of solute takes place. Solute and solvent fluxes are coupled in this mechanism, implying that an increase in solvent flux will lead to an increase in solute flux.

5.7.2 Knudsen-Diffusion Mechanism

In the transport of liquid molecules through the pores of a membrane, the diameter of the permeating species or molecules is always significantly smaller than the pore diameter resulting in more interaction of the permeating components with themselves than with the wall of the pore. The flow is thus always viscous flow. In the case of transport of gas molecules, the relative size of not only the gas molecules and pore diameter but also the mean-free path length of the gas molecules determine the flow regime and rate. The transport may be by viscous flow or diffusion depending on whether the mean-free path length of the gas molecules is smaller or larger than the pore diameter. In the case of large pore diameter, where the mean-free path length of the gas molecules is much smaller than the pore diameter, the interaction of the gas molecules with themselves dominates interaction with the pore wall resulting in viscous flow. But when the mean-free path of the gas molecules is larger than the pore diameter in the case of narrow pores, the interaction between the permeating components and the pore wall dominates and the transport takes place by Knudsen diffusion. The deciding factor mean-free path length (λ) of the gas molecules is defined as the distance traveled by a permeating gas molecule between two successive collisions with other gas molecules. The mean-free path length (λ) depends on the absolute temperature (T), pressure (p), and diameter of the molecule (d_{gas}) and may be expressed mathematically by:

$$\lambda = \frac{kT}{\pi d_{gas}^2 p\sqrt{2}} \tag{5.7}$$

where k is the Boltzmann constant.

The flux (J_i) of the permeating component (i) in Knudsen diffusion through cylindrical membrane pore depends on the pressure difference (Δp) across the membrane, the pore radius (r), number of pores (n), thickness (Δz) of the membrane, and tortuosity factor (τ), which are determined as follows:

$$J_i = \frac{n\pi r^2 D_i^k \Delta p}{RT\tau\Delta z} \tag{5.8}$$

where D_i^k is the Knudsen diffusion coefficient of the component i and is related to the molecular weight of the permeating component (M_i) as:

$$D_i^k = 0.66r\sqrt{\frac{8RT}{\pi M_i}} \tag{5.9}$$

Eq. (5.9) shows that diffusion of the component i is inversely proportional to the square root of its molecular weight. From this relation it follows that separation ($\alpha_{i,j}$) of the two components i and j present in a mixture is determined by the ratio of their molecular weights (M_i and M_j) and can be expressed as:

$$\alpha_{i,j}\alpha\sqrt{\frac{M_j}{M_i}} \tag{5.10}$$

In Knudsen diffusion, selective retardation of the components takes place by the pores of the membranes when the involved pore diameters are close to molecular size.

5.7.3 Solution-Diffusion Mechanism

The solutes of interest dissolve in the membrane-phase material and get transported to the other side through diffusion. Such a transport is obviously a phenomenon of dense membranes involved in RO type, pervaporation, or gas separation. NF membranes also partly follow solution diffusion and partly charge repulsion of the Donnan principle. In solution diffusion, the membrane is assumed to be in equilibrium with its adjacent phases where concentration and mobility of the permeating components in the membrane phase determine transport. The permeability of a component "i" is expressed as:

$$P_i = S_i \times D_i \tag{5.11}$$

where P_i = permeability of the membrane; S_i = solubility; D_i = diffusivity.

The flux (J_i) of the component "i" is related to its concentration C_i, mobility m_i, and chemical potential μ_i in the membrane phase as:

$$J_i = -C_i^m m_i^m \frac{d\mu_i^m}{dz} \tag{5.12}$$

where $d\mu_i/dz$ is the chemical potential gradient in the direction "z."

The membrane-phase concentration C_i^m is related to bulk-phase concentration C_i^b through the distribution coefficient k_i^{mb} as:

$$C_i^m = k_i^{mb} \cdot C_i^b \tag{5.13}$$

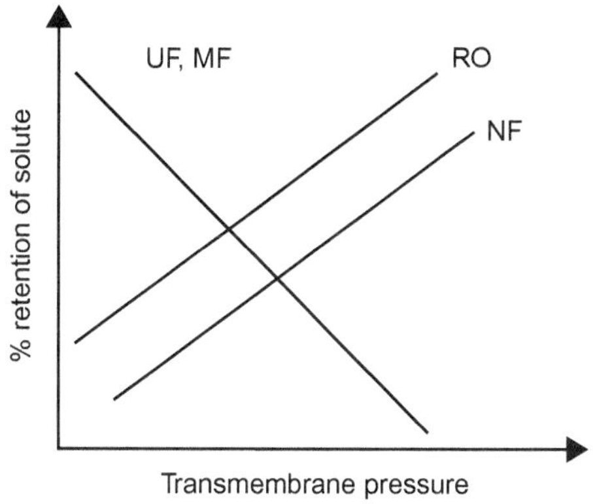

Figure 5.12 Solute-retention behavior of different membranes under varying pressure.

The mobility m_i of the component, which depends on diffusivity coefficient D, temperature (T), viscosity (μ), radius of the solute(r), can be computed from:

$$D_i^m = \frac{RT}{6\pi\eta_b r_i} \tag{5.14}$$

$$m_i = \frac{D_i^m}{RT} \tag{5.15}$$

In the solution-diffusion (SD) transport mechanism, the solute and solvent fluxes are uncoupled. Thus an increase in solvent flux will not necessarily lead to an increase in solute flux. In Fig. 5.12, such flux behavior under varying transmembrane pressure is shown. As seen with increase in transmembrane pressure, solute retention increases for RO and NF membranes, which are dense, and in SD transport when solvent flux increases the solute flux decreases resulting in higher solute retention. But UF and MF membranes show the reverse trend in solute retention as the size-exclusion mechanism, solvent, and solute fluxes are coupled. Under increased pressure, the solvent flux increases resulting in commensurate solute flux.

5.7.4 Donnan-Exclusion Mechanism

A membrane process is an open system where exchange of mass and heat with the surrounding environment takes place. Thus the Gibbs free energy that determines the driving force in transport for such an open system involving composition change may be expressed as:

$$dG = Vdp - SdT + \sum \left(\frac{\partial G}{\partial n_i}\right)_{p,T,n_j} dn_i(n_i\ n_1, n_2, \ldots, i \neq j) \tag{5.16}$$

where $\left(\frac{\partial g}{\partial n_i}\right)_{p,T,n_j}$ is the partial Gibbs free energy of the component i at constant temperature, pressure, and composition. This partial Gibbs free energy is defined as the chemical potential of a component. Thus the chemical potential (μ_i) of component "i" at constant temperature, pressure, and composition may be expressed as:

$$\mu_i = \left(\frac{\partial g}{\partial n_i}\right)_{p,T,n_j} \tag{5.17}$$

This chemical potential μ_i is a partial molar property representing the energy of component "i" in a system containing more than one component. In the same way, we can express the partial molar volume (V_i), partial molar pressure (p_i), and partial molar entropy (S_i) as:

$$V_i = \left(\frac{\partial V}{\partial n_i}\right)_{p,T,n_j}; \quad p_i = \left(\frac{\partial p}{\partial n_i}\right)_{p,T,n_j}; \quad S_i = \left(\frac{\partial S}{\partial n_i}\right)_{p,T,n_j} \tag{5.18}$$

In other words, the total Gibbs free energy, entropy, pressure, and volume, which are all state functions, can be derived from these partial state functions as:

$$G = \sum_i n\mu_i; \quad S = \sum_i n_i s_i; \quad P = \sum_i n_i p_i; \quad V = \sum_i n_i V_i. \tag{5.19}$$

As Gibbs free energy is a function of temperature, pressure, and composition, the chemical potential is also a function of these three variables and therefore may be expressed in terms of partial differentials as:

$$d\mu_i = \left(\frac{\partial \mu_i}{\partial p}\right)_{T,n_i} \mathrm{d}p + \left(\frac{\partial \mu_i}{\partial T}\right)_{p,n_i} \mathrm{d}T + \sum \left(\frac{\partial \mu_i}{\partial n_j}\right)_{P,T} \mathrm{d}n_j \tag{5.20}$$

where the term $\sum \left(\frac{\partial \mu_i}{\partial n_j}\right)_{P,T} \mathrm{d}n_j$ stands for the chemical potential of component i as function of the other components in the mixture at constant temperature and pressure.

In an ideal mixture, the chemical potential of component i at constant temperature and pressure is related to molar fraction X_i as:

$$(\mu_i)_{P,T} = \mu_i^0 + RT \ln X_i \tag{5.21}$$

where μ_i^0 is the chemical potential of the component at standard temperature and pressure.

For a real mixture, the ionic activity, i.e., the activity of a component "i" in a mixture, is expressed as $a_i = \gamma_i X_i$, where γ_i is the activity coefficient and X_i is the molar fraction.

Therefore the change in the chemical potential of component i in a mixture can be expressed as:

$$d\mu_i = V_i dp + RTd \ln a_i \tag{5.22}$$

When a mixture contains charged species, the Gibbs free energy should be expressed as a sum of chemical potential as well as electrical potential. Thus electrochemical potential may be written as:

$$d\overline{\mu}_i = d\mu_i + z_i F d\varphi \tag{5.23}$$

where z_i stands for the number of charges, F is the Faraday constant, and φ is the electrical potential. From Eqs. (5.20) and (5.21), the electrochemical potential gradient, which acts as the driving force in mass transport across the membrane in the direction "z" perpendicular to the surface of the membrane, can be expressed as [2,3]:

$$\frac{\partial \mu_i}{\partial z} = V_i \frac{dp}{dz} + RT \frac{d \ln a_i}{dz} \tag{5.24}$$

$$\frac{\partial \overline{\mu}_i}{\partial z} = V_i \frac{dp}{dz} + RT \frac{d \ln a_i}{dz} + z_i F \frac{d\varphi}{dz} \tag{5.25}$$

If a solution contains charged species or ion and the membrane is permeable to at least one ion, then the membrane will be in equilibrium with the adjacent solution provided the electrochemical potential of all ions in the solution equals the electrochemical potential of ions in the membrane phase where the electrochemical potential in the membrane phase is designated by $\overline{\mu}_i^m$ and that for the solution is designated as $\overline{\mu}_i^s$. μ_i^m stands for the chemical potential in the membrane phase.

$$\overline{\mu}_i^m = \overline{\mu}_i^s = \mu_i^m + z_i F \varphi^m \tag{5.26}$$

The Donnan potential (φ_{Donnan}) is the difference in electrical potential between the membrane phase and the solution phase expressed as [4]:

$$\begin{aligned} \varphi_{\text{Donnan}} &= \varphi^m - \varphi^s = \frac{1}{z_i F}\left(RT \ln \frac{a_i^s}{a_i^m} + V_i(p^s - p^m)\right) \\ &= \frac{1}{z_i F}\left(RT \ln \frac{a_i^s}{a_i^m} + V_i \Delta\pi\right) \end{aligned} \tag{5.27}$$

where V_i is the partial molar volume $\left(\frac{\partial V}{\partial n_i}\right)_{p,T,n_j}$ and $\Delta\pi$ is the osmotic pressure difference between the bulk solution (s) and membrane phase (m).

At Donnan equilibrium partitioning of the counterions takes place favorably while that of coions takes place unfavorably between a charged membrane and the adjacent solution. This leads to distribution of cation and anions between the solution and membrane phases and those ions with the same electrical charge as the fixed ions of the membrane get excluded from the membrane while those carrying opposite charge get adsorbed onto the membrane. The charge on a membrane surface may be due to dissociation of functional groups or due to adsorption of ions from the solution. In any practical membrane-separation system involving NF membranes, both steric and Donnan effects play active roles in partitioning of the components of the mixture or the solution and such Donnan-steric partitioning is expressed as:

$$\frac{c_i}{c_i^0} = \phi_i \exp\left(-\frac{z_i F}{RT}\Delta\psi_d\right) \tag{5.28}$$

where c_i is the concentration of solute i inside membrane pore, c_i^0 is the concentration of the same at the membrane surface, z_i is the stoichiometric coefficient or valence, F is the Faraday's constant, $\Delta\psi_d$ is the dimensionless Donnan potential, and ϕ_i is the steric partitioning coefficient. A steric partitioning effect arises from the sieve effect due to intrinsic membrane porosity.

Thus separation of ions takes place by charge repulsion in the Donnan-exclusion mechanism when electroneutrality both within the membrane as well as in the adjacent solutions are maintained. Such electroneutrality within the membrane is expressed as:

$$\sum_{i=1}^{n} z_i c_i = -X_{\mathrm{d}} \tag{5.29}$$

where X_{d} is the volumetric membrane-charge density (mol/m^3), z_i is the valence, charge number, or stoichiometric coefficient of ion i, and c_i is the concentration of the ion i. The electroneutrality within the bulk solution or permeate solution can be expressed as:

$$\sum_{i=1}^{n} z_i c_i = 0 \tag{5.30}$$

5.7.5 Dielectric Exclusion

In a partly porous, partly dense membrane like NF membrane, separation may be a result of a number of mechanisms such as sieving, solution diffusion, Donnan exclusion, and dielectric exclusion and transport may be by diffusion, convection, and electromigration. Dielectric exclusion is due to differences in dielectric constants of the solvent in the bulk and within the membrane pore, which results in rejection of solutes.

In a partly porous, partly dense membrane like NF membrane, almost all the mechanisms discussed so far barring Knudsen diffusion (which is mainly applied to gas separation) play a role during transport. Thus we describe the transport models of NF membrane separation in the next section to examine how different mechanisms contribute to the total mass transport.

5.8 TRANSPORT MODELING IN NANOFILTRATION

When drawing up specifications for the membrane, membrane module, and plant accessories and in determining optimum plant operational conditions for successful commissioning and running of any membrane-based plant, mathematical modeling of the relevant transport phenomena as well as that of the whole process is absolutely essential. Such models correlating design and operating parameters help system design, process optimization, and industrial scale-up. Modeling transport through NF membrane and hence the whole process has always been considered challenging because of the complexity of the interplay of multiple mechanisms. There have been several attempts to understand the involved transport phenomena from different angles discussed as follows.

5.8.1 Continuum Hydrodynamic Model

The continuum hydrodynamic model describes transport of uncharged solutes through NF membranes. In this modeling approach, NF membrane is assumed to be a porous membrane that behaves as a bundle of straight cylindrical pores. The appropriate correction factors are incorporated into the transport model to address the hindrance in convection and diffusion due to solute-membrane interactions. The major limitation of this model is its failure to describe ion transport [5].

5.8.2 Irreversible Thermodynamic Model

In the early years (1960s) of membrane-transport modeling, the principles of irreversible thermodynamics were adopted to describe transport through composite membranes [6–7]. Phenomenological equations defined through irreversible thermodynamics were used to describe ion transport through NF membranes, although originally this type of modeling approach was used to describe transport through RO membranes. In this model, the membrane was assumed to be a black box containing no descriptions of ion transport. Thus the inherent weakness of the model was evident in its failure to characterize structural and electrical properties of the membrane. However, in some studies involving dye-salt separation [8] and negative ion transport [9] the black-box model was successful in predicting membrane performance.

5.8.3 Electrokinetic Space-Charge Model

Electrokinetic space-charge models were developed in the 1960s and 1970s in the context of hyperfiltration and capillary filtration [10–11]. However, application of these models remained limited by the numerical complexity of calculations.

The basic features of the model are as follows:

1. Describe the creeping flow of charged solutes through charged capillaries.
2. Ions are treated as point charges.
3. Assumes radial distribution of charge across the pore.
4. Charge distribution together with potential is described by the Poisson–Boltzmann equation.
5. Ion transport along the pore is described by the extended Nernst–Planck equation.

A major limitation to this modeling approach is computational complexity due to consideration of radial variation of charge, concentration, and potential. Subsequent models developed on space-charge models assumed homogeneous distribution of concentration, charge, and potential [12,13], but were largely valid for small surface charge density and very narrow membrane pore only. NF membranes in many studies [11,14] have been found to have narrow pores with small charge density thereby validating the assumptions on homogeneity of charge, concentration, or potential.

5.8.4 Donnan–Steric–Pore Model

The most widely used NF model is the Donnan–steric–pore model (DSPM). Ion transport in this model is described by the extended Nernst–Planck approach [14–16]. As the name suggests, the model considers equilibrium partitioning of the solutes between the solution and the membrane at the membrane–solution interface under the combined effects of the Donnan (electrical) potential and steric hindrances (in sieving). The model is used to characterize membrane through analysis of rejection data as a function of volumetric flux where the flux and rejection depend on three major membrane characteristics: pore radius (r_p), effective ratio of membrane thickness to porosity ($\Delta x/A_p$), and volumetric membrane-charge density (X_d). The DSPM model largely succeeds in predicting the membrane performance of transport of univalent electrolytes and simple, small organic molecules. However, the reported success of the DSPM model is due to the fact that the characterization parameters (r_p, $\Delta x/A_k$, X_d) are in many ways adjustable and do not truly mirror the structural and electrical properties of the membrane. The DSPM model also fails to adequately predict transport involved in separating mixtures of electrolytes as well as multivalent cations (e.g., Mg^{+2}). For example, high rejection of bivalent ions (Mg^{+2}) cannot be predicted by the DSPM model. Thus to make the DSPM model more realistic research on other involved phenomena was done.

5.8.5 Steric, Electric, and Dielectric Exclusion Model

Dielectric exclusion in addition to Donnan–steric exclusion has been considered in modeling transport through NF membrane [17,18]. In the steric, electric, and dielectric exclusion (SEDE) models, it is assumed that water molecules are oriented in a layer in the pores thereby reducing the dielectric constant of the solvent within the membrane pore. This reduced dielectric constant of the solvent inside the membrane pore compared to that in the bulk solution suggests the presence of a solvation energy barrier for the ions in the pore. This causes further ion partitioning and contributes to additional ion rejection. Ion partitioning resulting from the presence of a solvation energy barrier in the membrane pores is expressed by Eq. (5.31) considering the originally proposed model of Born [19]:

$$\frac{c_i}{C_i} = \Phi_i \exp\left(\frac{-z_i F}{RT}\Delta\Psi_D\right)\exp\left(\frac{-\Delta W_i}{kT}\right) \tag{5.31}$$

where c_i and C_i are the ion concentration (of ion i) in the pore and bulk solute concentration (mol/m^3), respectively, $\Delta\psi_D$ is the Donnan potential (V), Φ_i is the steric partitioning coefficient $[=(1-\lambda)^2]$, λ is the ratio of ionic or uncharged solute radius to pore radius, k is the Boltzmann constant, and ΔW_i is the salvation energy barrier (J) and may be calculated from the Born model [19] as:

$$\Delta W_i = \frac{z_i^2 e^2}{8\pi\varepsilon_0 a_s}\left(\frac{1}{\varepsilon_p} - \frac{1}{\varepsilon_b}\right) \tag{5.32}$$

where ε_p and ε_b are pore dielectric constant and bulk dielectric constant (dimensionless), respectively, ε_0 is the permittivity of free space ($8.85419\times10^{-12}\,\mathrm{J}^{-1}$), and z_i is the valence of the ion.

An atomic force microscopic (AFM) study of the morphology of NF membranes showed that NF membranes can have porous structures and thus solvent velocities can be calculated based on the Hagen–Poiseuille equation as follows:

$$V = \frac{r_p^2 \Delta P_e}{8\eta\Delta x} \tag{5.33}$$

where$\Delta P_e = (\Delta p - \Delta\pi)$ is the effective transmembrane pressure, V is the solvent velocity, $\Delta\pi$ is the osmotic pressure difference, Δx is the thickness of membrane, r_p is the membrane pore radius, and η is the viscosity of the solvent within pores. The viscosity of solvent within the pores may be significantly higher than the bulk solvent viscosity. Osmotic pressure difference may be calculated based on concentrations of the solute at the membrane pore inlet and outlet. The fact that ΔP_e shows a linear relationship with solvent flux is an indicator of the insignificance of the electrokinetic effects in NF membranes because the pores are sufficiently narrow, the membrane charge is sufficiently small, and the pore counter-ion concentrations are sufficiently low to prevent development of electrical double layers. This is consistent with the assumption of homogeneous distribution of potential and concentration in the vast majority of NF models.

Flux Equations for Nanofiltration Membrane in SEDE Models

Charged-Solute Model

The latest flux equations for porous charged membrane such as NF membrane are primarily based on the extended Nernst–Planck (ENP) model where total ion transport is a result of transport contributions from concentration gradient-driven diffusion, electric potential-driven migration, and solvent flow-driven convention. The molar flux for ion "i" can be written as:

$$J_i = (K_{i,c}c_i V) + \left[\left\{-\left(K_{i,d}D_i\frac{dc_i}{dx}\right)\right\} + \left\{-\frac{z_i c_i D_i F}{RT}\left(\frac{d\psi}{dx}\right)\right\}\right] \tag{5.34}$$

where $K_{i,d}$, $K_{i,c}$ account for the hindrance factors for diffusion and convention, respectively, within the pores of the membrane. The local electric potential is ψ and V is the solvent velocity through the membrane pores.

The molar ionic flux (J_i) may be correlated to volumetric permeate flux (J_v) [16] as:

$$J_v = \frac{J_i A_k}{c_i(\Delta x^+)} \tag{5.35}$$

Rejection by the NF membrane may be calculated using:

$$R_i = 1 - \frac{c_i(\Delta x^+)}{c_i(0^-)} \tag{5.36}$$

where in volumetric flux computation, active membrane surface area is considered through porosity A_k of the membrane. In this case, $0^+ \text{I} 0^-$ represents the bulk solution membrane interface (0^+ is feed side, 0^- is membrane side) and $\Delta x^- \text{I} \Delta x^+$ represents the membrane pore permeate solution interface (Δx^- represents membrane pore side, Δx^+ represents permeate side) where Δx is the pore length or membrane thickness.

Nanofiltration Model for Uncharged Solutes

In the transport of uncharged solute, the electrical potential gradient has no role and the basic flux equation based on the hydrodynamic model modified for hindered diffusion may be expressed as [20]:

$$j_s = K_c c V - \frac{c D_p}{RT}\frac{d\mu}{dx} \tag{5.37}$$

where K_c = hindered diffusion factor, j_s = uncharged solute flux ($\text{mol/m}^2 \cdot \text{s}$).

Uncharged solute chemical potential μ is defined as:

$$\mu = RT \ln a + V_s P + \text{constant} \tag{5.38}$$

For dilute solution that can be assumed to behave ideally, differentiation of Eq. (5.38) and substitution into Eq. (5.37) results in

$$J_s = K_c cV - D_p \frac{dc}{dx} - \frac{cD_p}{RT} V_s \frac{dP}{dX} \tag{5.39}$$

Laminar flow conditions in nanopores permit the use of a Hagen–Poiseuille type relationship where the pressure gradient along the membrane pore may be assumed constant. We can find the effective pressure gradient as follows:

$$\frac{dP}{dX} = \frac{\Delta P_e}{\Delta x} = \frac{8\eta V}{r_p^2}$$

Now from the above equations, we may arrive at the concentration gradient equation as:

$$\frac{dc}{dx} = \frac{V}{D_p}\left[\left(K_c - \frac{D_p}{RT} V_p \frac{8\eta}{r_p^2}\right) - C_p\right] \tag{5.40}$$

Integrating Eq. (5.40) between the limits $x = 0$ to Δx (i.e., pore length) and incorporating the steric partition coefficient Φ (between membrane and solution) we get solute concentrations at the pore inlet and outlet as [21]:

$$C_{x=0} = \Phi C_f \text{ and } C_x = \Delta x = \Phi C_p$$

Rejection may be expressed as:

$$R = 1 - \frac{c_p}{c_f} \tag{5.41}$$

Concentration Polarization Effects on Nanomembrane Performance and Model

At high cross-flow rates, the CP effect may be insignificant. But at low cross-flow rates, CP should be included in NF flux and rejection models through an appropriate mass transfer coefficient term.

5.9 SELECTION OF MEMBRANE TECHNOLOGY IN WATER TREATMENT

The selection of membrane technology in the treatment of contaminated water or wastewater depends very much on the physicochemical characteristics of the water or wastewater stream, desired degree of purification, the target use of the treated water, and the cost of the treatment. For very rough preliminary treatment of water containing mostly suspended particles, the MF membrane process should be selected. UF may be selected for separating particulates much finer than those present in MF feedwater. In both cases separation—purification is based on a sieving mechanism so the relative sizes of the membrane pores and the impurities will determine the type of membrane process that needs to be applied. MF and UF membranes in general may be used in concentration of fine particles, for pretreatment for RO, NF feedwater, sterilization of various streams in pharmaceutical industry, production of ultrapure water, or for production of injection-grade water.

RO can be used in the recovery of freshwater from seawater or brackish water, for removal of ions from wastewater, for pretreatment of boiler feedwater, and for removal of color removal from wastewater. Pervaporation membranes can be used in the removal of small amounts of water from organic solutions (water from isopropanol) and for removal of small amounts of organics from water. Membrane distillation can be used for removal of trace contaminants from water with a high degree of purification. Electrodialysis membranes can be used for desalination of brackish water or deionization of boiler feedwater using the electrical potential gradient as the driving force in ionic transport. Ion-exchange membranes are used at large-scale in the removal of acid or alkali from water.

5.10 MICROFILTRATION TECHNOLOGY IN WATER TREATMENT

MF technology is generally considered suitable for removal of coarse particles or microbes in the size range of 0.025—10.0 μm. Bacteria, paint, pigment, yeast cells, and other suspended matters can be separated from water by MF technology using microporous membrane with large pore size (between 50 nm and 5 μm) are used. An operating pressure ranging from 1 to 5 bars is normally applied in this pressure-driven process. While there are other alternatives such as the use of fibrous material in removing such micron-sized particles, MF membrane with a precisely defined pore size is preferred for quantitative retention. MF membranes are often used as a prefiltration step prior to high-purity separation such as UF, NF, or RO.

Due to retention of particles of higher-molecular-weight MF membranes may quickly suffer from fouling by microbial cells, proteins, and other particulates, which may be very severe compared to that in NF/RO membranes. However, there are some membrane-based modules that

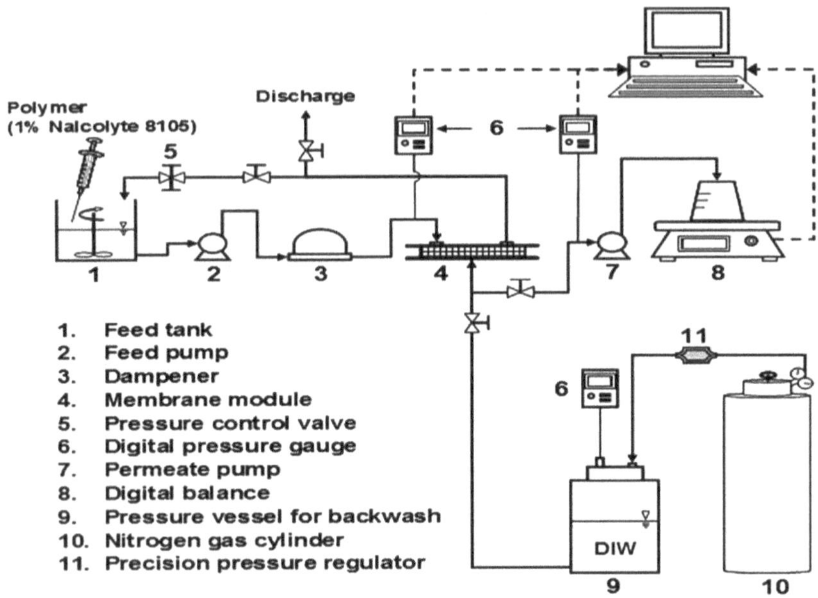

Figure 5.13 Schematic for constant-flux membrane filtration [22].

may be operated long term without much fouling like flat-sheet cross-flow types. Separation by MF membranes is based on the size-exclusion mechanism (or sieving effect) and depends on the molecular weight cut-off (MWCO) of the membrane and the molecular weight of feed particles.

Membrane material constituents are also key factors in minimizing fouling. PES membranes foul more quickly than polyamide membranes. Ceramic membranes suffer quick fouling, although complete disinfection is achievable using these membranes. Polyvinylidene fluoride membranes are also prone to fouling but the problem of CP and the effect of fouling can be effectively minimized by generating shear stress while controlling the tangential flow over the membrane. The mechanical strength of ceramic MF membranes allows much higher flux in a membrane module than a polymeric membrane. The life of a ceramic membrane can be 2–5 times the life of a polymeric membrane. Another great advantage of ceramic membranes over polymeric types is that ceramic membranes can be subjected to online chemical treatment, cleaning, and repeated autoclaving for sterilization.

MF alone is rarely used in water treatment. It is often used along with another process for complete treatment. For example, a computer-controlled coagulation and MF-integrated system as presented in Fig. 5.13 has been effective in decreasing fouling, increasing critical flux,

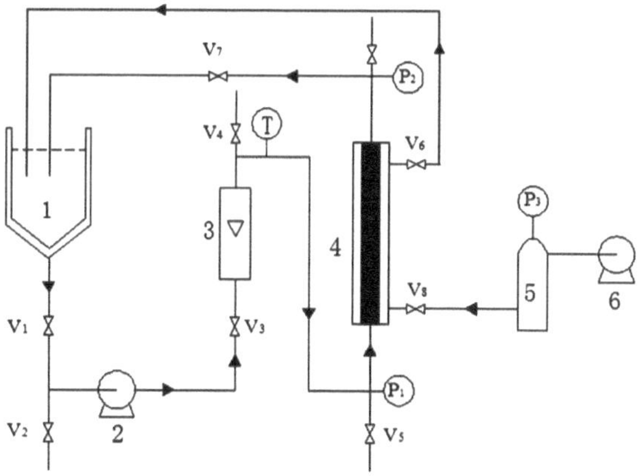

Figure 5.14 Schematic of cross-flow microfiltration (MF) apparatus. (1) Feed tank; (2) Centrifugal pump; (3) Flow meter; (4) Membrane module; (5) Buffer tank; (6) Air compressor; V1–V7: Valves. [23].

and increasing contaminant removals for laundry wastewater employing constant-flux PVDF flat-sheet MF membranes with 0.22 μm pores [22].

The use of a ceramic microfilter in pretreatment of industrial wastewater is common. For example, ZrO_2 ceramic MF membranes (Fig. 5.14) have been found effective in the pretreatment of dimethylformamide (DMF) wastewater from polyurethane (PU) synthetic leather factories [23].

5.11 ULTRAFILTRATION TECHNOLOGY IN WATER TREATMENT

UF-based water-treatment technology is applied for separation of contaminants from water that fall in the size range of 10–50 nm. All suspended solids and pathogens within this range can be very efficiently removed from water by applying a moderate operating pressure of 5–10 bars where the transport regime is mesoporous and the separation mechanism is sieving. Membranes for UF are commonly asymmetric and more porous.

UF membranes are specified by the MWCO, which is defined as the minimum molecular weight (in Dalton) of the molecules, 90% of which are retained or rejected by the membrane. The rejection may be expressed by Eq. (5.42).

5.11.1 Flux Equations and Concentration Polarization for Ultrafiltration

Solvent Flux

The flux equation for diffusion of solvent through the membrane is:

$$N_w = A_w(\Delta P - \Delta \pi) \tag{5.42}$$

where A_w = solvent permeability constant (a property of the membrane) = $\frac{P_w}{L_m}$ (kg solvent/s · m^2 · atm); ΔP = hydrostatic pressure difference = $P_1 - P_2$; $\Delta \pi$ = osmotic pressure difference between the two sides of the membrane; $\Delta P - \Delta \pi = \Delta P_e$ is effective transmembrane pressure that acts as driving force in transport; P_1 and P_2 are hydrostatic pressures on the feed side and permeate side, respectively; L_m = membrane thickness, m; P_w = solvent-membrane permeability

In UF, the membrane does not allow passage of the solute, which is generally a macromolecule. The solute concentration in moles per liter of the large solute molecules is usually small and the osmotic pressure is very low and can be neglected.

The above equation becomes:

$$N_w = A_w(\Delta P). \tag{5.43}$$

5.11.2 Concentration Polarization in Ultrafiltration

CP is considered a major hindrance in the long-term sustainable use of UF membrane. During solute transport and separation by membrane, the solute rejected or retained by the membrane starts accumulating on the surface of the membrane leading to a buildup of rejected solute and increased drop of pressure. This phenomenon, which is called concentration polarization, may occur both in porous UF membrane as well in dense RO membrane but is much more severe in UF than in RO. Fig. 5.15 depicts how such CP takes place in UF.

The increased operating pressure on UF membrane leads to increased solvent flux N_w, which in turn results in higher convective transport of the solute along with solvent to the membrane surface. An increased concentration of solute on the surface (C_s) results in larger back molecular diffusion of solute from the membrane to the bulk solution. At steady state, the convective flux of solute from bulk or FS to the membrane side equals the sum of the diffusive flux from the membrane to the feed side

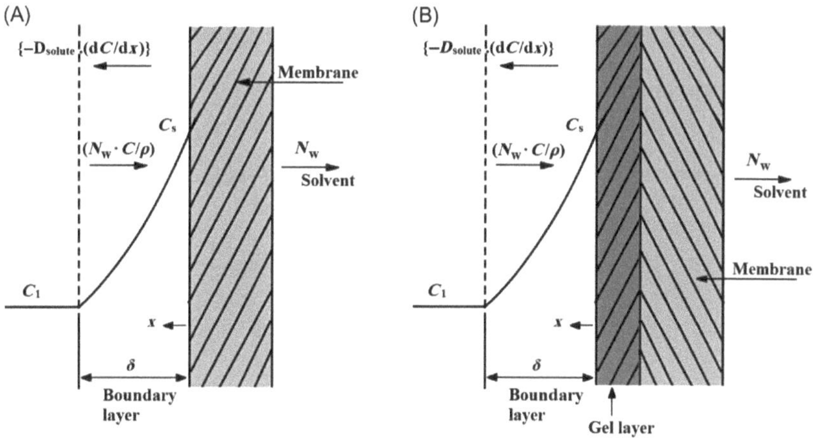

Figure 5.15 Concentration polarization (CP) in ultrafiltration (UF): (A) Concentration profile before gel formation and (B) Concentration profile with a gel layer formed at membrane surface.

and convective flux from the membrane to the permeate side. Thus at steady state, the mass flux of the solute may be expressed as:

$$\frac{N_w c}{\rho} = -D_{\text{solute}}\frac{dc}{dx} + \frac{N_w c_w}{\rho} \tag{5.44}$$

Integrating the above equation between the limits:

$$x = 0, c = c_s$$
$$x = \delta, c = c_1$$

$$\begin{aligned}\frac{N_w}{\rho} &= \left(\frac{D_{AB}}{\delta}\right)\ln\left(\frac{c_s - c_p}{c_1 - c_p}\right)\\ &= k_c \ln\left(\frac{c_s - c_p}{c_1 - c_p}\right)\end{aligned} \tag{5.45}$$

where k_c = mass transfer coefficient (m/s); c_1 = concentration of the solute in the bulk solution (kg solute/m^3); c_s = concentration of the solute at the surface of the membrane (kg solute/m^3); c_p = concentration of the solute in the permeate (kg solute/m^3).

As ΔP increases, c_s also increases to a limiting concentration, at which the accumulated solute forms a semisolid gel when c_s becomes equal to c_g.

Under this limiting condition, solute retention becomes almost complete implying:

$c_p = 0$ and that Eq. (5.45) turns into Eq. (5.46) as follows:

$$\frac{N_w}{\rho} = k_c \ln\left(\frac{c_s}{c_1}\right) \tag{5.46}$$

When further increase in ΔP does not change c_g, the membrane is said to be *gel polarized*. Under gel-polarized condition, Eq. (5.46) may be written as:

$$\frac{N_w}{\rho} = k_c \ln\left(\frac{c_g}{c_1}\right) \tag{5.47}$$

With increases in pressure drop, the gel layer increases in thickness, causing the solvent flux to decrease because of the added gel-layer resistance. Finally, the net flux equals the back diffusion of solute into the bulk solution due to the polarized concentration gradient.

The added gel-layer resistance next to the membrane multiplies the resistance to solvent flux. The solvent flux equation, therefore, is written as:

$$N_w = \frac{\Delta P}{\frac{1}{A_w} + R_g} \tag{5.48}$$

where $1/A_w$ is the membrane resistance and R_g is the gel-layer resistance $(s \cdot m^2 \cdot atm)/kg$ solvent.

The solvent flux in this gel-polarized regime is independent of pressure difference and is determined by:

$$\frac{N_w}{\rho} = k_c \ln\left(\frac{c_g}{c_1}\right) \tag{5.49}$$

There are ways to reduce fouling and to delay its onset by adopting a hydrodynamic and back-flushing technique or changing the membrane material and mode of operation. Most membrane manufacturers recommended a series of acid-alkali-cleaning cycles depending on the feed processed and membrane material. The degree of fouling relates directly to the membrane material. For example, ceramic membranes permit easy disinfection, but suffer from quick fouling.

Polymer-enhanced ultrafiltration (PEUF) membrane as the mercury-binding polymer has been found effective in the removal of mercury (II) from wastewater. Among the three metal-binding polymers, namely,

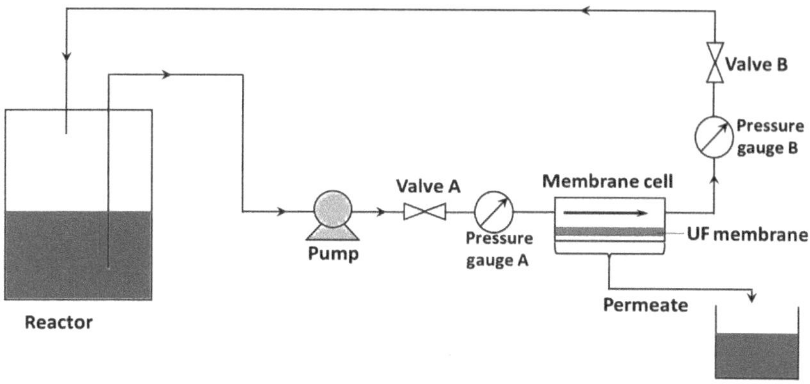

Figure 5.16 Schematic of the Polymer-enhanced ultrafiltration (PEUF) process for mercury separation [24].

polyethylenimine (PEI), polyvinylamine (PVAm), and polyacrylic acid (PAA), with identical dosage, the strongest interaction of mercury has been found with PVAm [24]. A schematic of the PEUF process using cross-flow membrane modules is shown in Fig. 5.16. Using chelating agent, a high level of efficiency in removing mercury from industrial wastewater has been achieved. By increasing the concentration of PVAm to more than 0.3 wt% and employing the pristine PES membrane for UF, mercury rejection of 99% has been achieved along with a flux of about 200 L/(m^2h) under a transmembrane pressure of 2 bar only.

A UF-coupled precipitation treatment system was recently developed for removal of fluoride from high fluoride-containing industrial wastewater such as that discharged by the fertilizer, aluminum, and semiconductor manufacturing industries. By the addition of calcium chloride at a molar ratio of calcium to fluoride ($[Ca^{2+}]/[F^-]$) of 0.7, and integrating a pre-flocculation step (for turbidity minimization) with PAA, the filtration by polyvinylidene fluoride (PVDF) membrane (MWCO of 30 kD and 100 kD) can be highly improved. The chemical reaction is as follows [25]:

$$Ca(aq)^{2+} + 2F(aq)^{-} \rightarrow CaF_2(s)$$

Steady-state flux in the system can be quickly achieved and maintained long term while using membrane with an MWCO close to 100 KDa. In separation—purification of oily wastewater, namely olive oil wastewater and petroleum refinery wastewater, PES UF membrane of MWCO 50 kDa has been effective. Photomodification of UF membranes

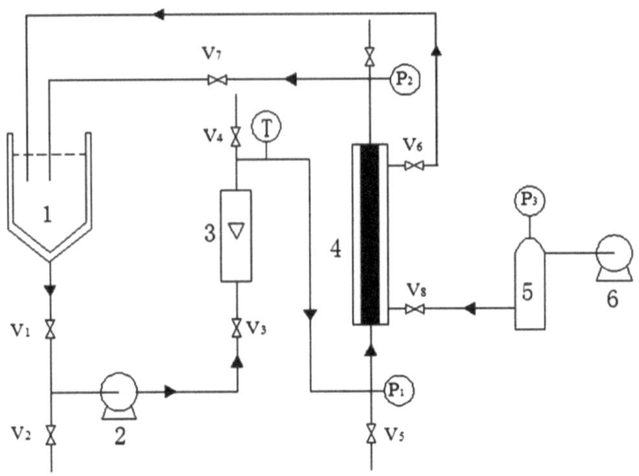

Figure 5.17 Schematic of a ceramic cross-flow ultrafiltration (UF) system. (1) Feed tank; (2) Centrifugal pump; (3) Flow meter; (4) Membrane module; (5) Buffer tank; (6) Air compressor; V1–V7: valves [22].

using UV rays can enhance the fouling-resistant membrane capability. Table-olive processing wastewaters have high salt (NaCl) concentration and total phenolic content that cannot be removed by simple UF. Surface modification using UV rays in the presence of two hydrophilic compounds (glycol and aluminum oxide) has thus been done on commercial PES membranes of 30 kDa. With much better fouling resistance these membranes could remove phenolic wastes from water. Combined amino-functionalized metal–organic frameworks with ceramic (zirconia) UF membrane have been reported for the adsorptive removal of zinc. Fig. 5.17 shows the schematic of a ceramic UF setup in recycling mode.

A cross-flow micellar-enhanced UF system developed [26] for abatement of zinc in hollow-fiber module with very low concentration of sodium dodecyl sulfate (SDS) is shown in Fig. 5.18.

A membrane UF technology developed for treating an oil refinery's end-of-pipe effluent to remove mercury has successfully reduced the concentration from as high as 22.7 ppt to 1.3 ppt. This technology known as GE's Zee Weed low-energy membrane technology (with 0.04 mm nominal pore size PVDF membrane) consists of outside-in, hollow-fiber modules (immersed directly in the wastewater feed source). The Zee Weed 500 system operates under a low-pressure vacuum that is induced within

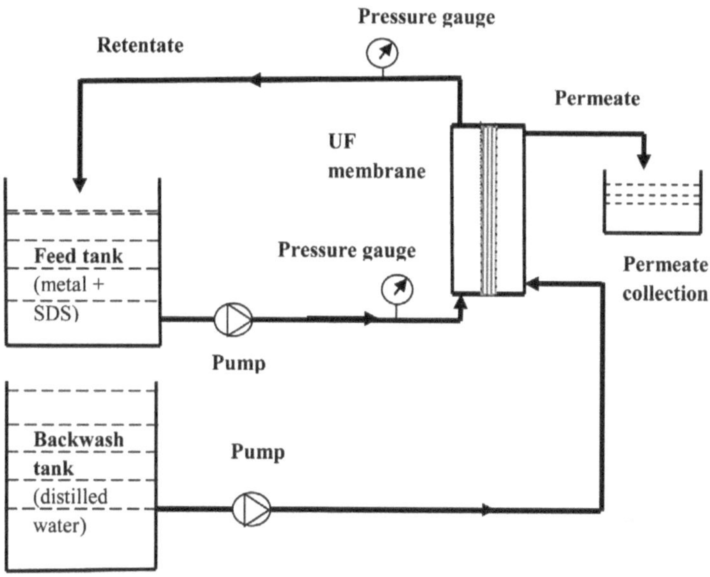

Figure 5.18 Schematic of experimental setup of micellar-enhanced ultrafiltration (MEUF) process for zinc (II) removal from synthetic wastewater [26].

the hollow-membrane fibers by a connection to the suction side of a permeate pump. Airflow is intermittently introduced at the bottom of the membrane module to create turbulence that scrubs and cleans the outside of the membrane fibers. This reduces the solids accumulation on the membrane surface. It also provides mixing within the process tank to maintain solids in suspension. The plant ensures excellent effluent quality by removing 97% of turbidity and consistent mercury concentration <0.5 ppt in effluent. Throughout the operation time, the total suspended solids concentration in the [27]. Nanocomposite membranes have been developed by incorporating nanocomposite materials (e.g., SiO_2, TiO_2, ZrO_2, ZnO, Al_2O_3) into polymer matrix toward modification of polymer membranes. Significant improvement in terms of hydrophilicity and permeability of the synthesized membranes has been achieved through structural modification in a phase-inversion technique. However, the proportion of nanomaterials to be dosed needs to be optimized because a relatively high proportion (1−4 wt%) of inorganic nanomaterials may enhance mechanical strength but also intensify the tendency for membrane clogging and fouling, which conversely weakens the output flux [28−30].

Carbon nanotubes have been investigated in recent years for possible application in water treatment. However, carbon nanotube materials are relatively costly, which increases the cost of the overall membrane modification. Graphene oxide (GO), which is also a carbon nanomaterial, contains a large number of hydrophilic oxygen functional groups on its basal planes and edges and provides excellent mechanical properties and strong hydrophilicity. GO is a low-cost material that is more competitive and accessible than carbon nanotubes for the design and development of advanced membranes [31–32]. A minute dosage of GO is enough to develop newly modified membranes, which also reduces the cost of process developing the hybrid membranes with improved performance.

5.12 NANOFILTRATION TECHNOLOGY IN WATER TREATMENT

The major advantage of an NF plant is that it can be operated at an operating pressure much lower than that required for an RO plant, which significantly lowers both capital investment as well operating costs [33] while ensuring very safe potable water from contaminated groundwater or from water with low concentration of dissolved solids. NF can remove viruses, bacteria, contamination metal, or metalloids, NOM (naturally occurring organic matter or SOM (synthetic organic matter) in water that act as precursors of disinfection byproducts known as carcinogens or suspected carcinogens at lower cost than RO plant technology. In groundwater treatment, the presence of NOM and iron often has to be considered. Considering these issues, McKim and Creed built an 8-mgd capacity NF water-treatment plant in Jacksonville North Carolina, United States, to treat groundwater from Castle Hayne wells. After treatment of the groundwater, residual concentrated water is discharged via a river diffuser line to a brackish water estuary. However, an NF-integrated water-treatment plant (Fig. 5.19) based on a patented technology for arsenic removal [33] addresses the concentrated arsenic rejects in a much better way than the Jacksonville, North Carolina NF plant. The arsenic rejects are totally stabilized in unleachable mineral matrix and disposed of for landfilling or road-making in a safer way. This NF-integrated water-treatment technology [33] targeting arsenic removal from contaminated groundwater can equally address other contaminants like fluoride, iron, chromium, or NOM.

This new treatment technology for removal of arsenic from contaminated water with simultaneous stabilization of the arsenic rejects offers a

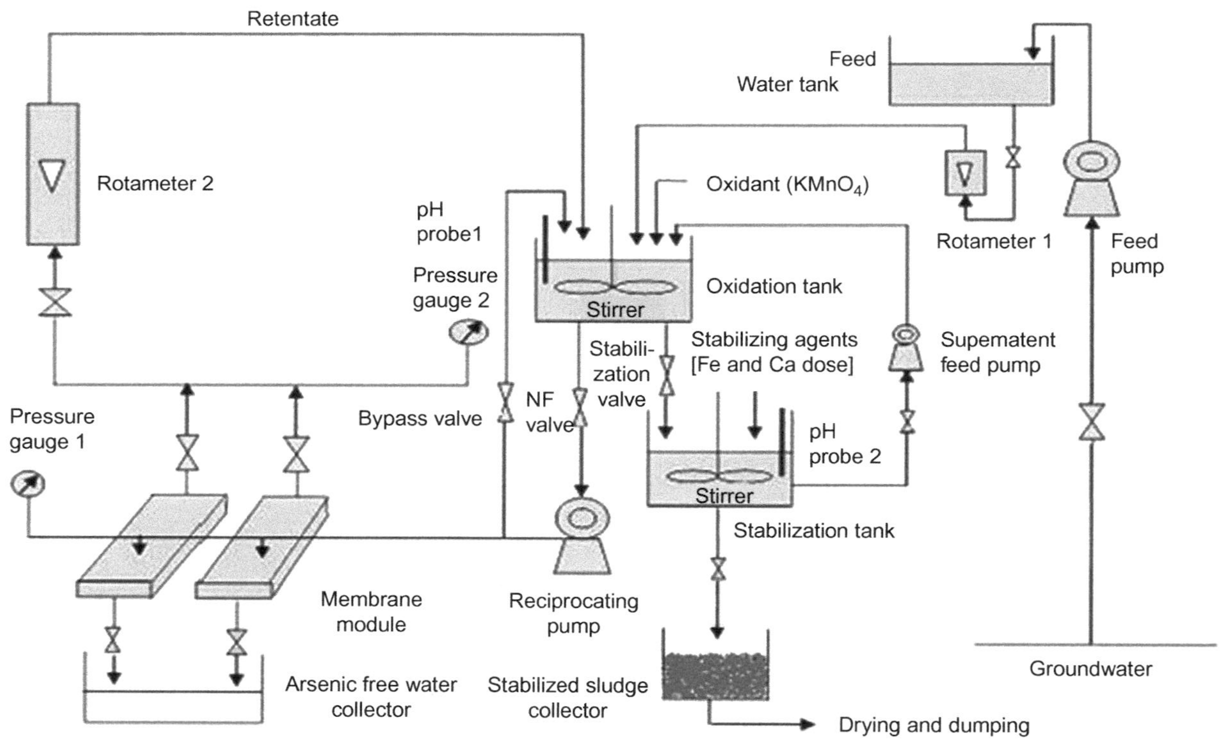

Figure 5.19 A nanofiltration (NF)-integrated water-treatment plant based on a patented technology for arsenic removal [33].

sustainable solution to the groundwater arsenic contamination problem across the world. While preoxidation in a continuously stirred tank reactor using $KMnO_4$ as oxidizing agent converts all arsenic to pentavalent form ensuring efficient retention by NF membrane the flat-sheet cross-flow NF membrane module ensures more than 98% removal of arsenic without any significant drop in pure water flux over prolonged hours of operation. The stabilization step offers a safe disposal route for arsenic rejects. The novelty and beauty of the membrane-integrated hybrid scheme thus lie in its efficient water purification with simultaneous stabilization of the concentrated arsenic rejects in a very simple treatment scheme that promises 1000 L of safe drinking water at a price of only $1.4. Fouling, which is often considered a major hindrance in the long-term operation of a membrane-based plant, is largely overcome here by the favorable hydraudynamics in the module. The sweeping action of the fluid flow itself on the membrane surface keeps the membrane free from CP. Achieving high flux with high target solute retention simultaneously is extremely difficult but is possible in this new module as evident in the high flux of around 140–150 LMH.

Considering that arsenic in trivalent form is not as efficiently removed as arsenic in pentavalent form from water by NF membrane, this plant introduces a chemical pretreatment unit (CSTR) prior to the membrane-filtration module for conversion of trivalent arsenic into pentavalent form through oxidation using $KMnO_4$ as oxidizing agent.

5.12.1 Nanofiltration Membrane Technology in Cyanide Removal From Industrial Wastewater

Cyanide or cyanide-bearing wastewaters are discharged from many industries such as those involved in coke manufacturing, petroleum refining, chemical manufacturing, metal plating and finishing, photography, and metal processing. In the gold and silver mining industry, extraction of these high-value metals is done using cyanide. While the largest use of cyanide is in leaching of gold and silver, cyanide has also been used for decades by a number of other industries for electroplating and metal processing. The high toxicity associated with cyanide has resulted in stringent discharge regulations for cyanide-bearing wastewater to the environment. The United States Environmental Protection Agency (US EPA) has set 0.01 mg/L as the guideline and 0.2 mg/L as the permissible limit for cyanide in effluent. It has also set disposal limits for total cyanide concentration in drinking water and aquatic-biota water at 200 ppb and 50 ppb, respectively. In India, the Central Pollution Control Board (CPCB) has set 0.2 mg/L as a minimum national standard (MINAS) limit for cyanide in effluent. The German and Swiss regulations have set a limit of 0.01 mg/L for cyanide in surface water and 0.5 mg/L for sewers. Mexico sets the limit at 0.2 mg/L. Because of the potential hazards associated with cyanide, it is imperative that wastewater containing cyanide be treated properly before discharge to the environment.

Treatment Plant Operation

In the NF-integrated treatment scheme as shown in Fig. 5.20, flat-sheet cross-flow NF membranes are employed for filtering prefiltered cyanide waste. Cyanide-bearing industrial wastewater, however, needs to be microfiltered prior to introduction to the NF unit to reduce the potential clogging materials thereby ensuring the long-term use of the NF membranes. A PVDF MF membrane with a pore size of 0.45 micron is used for this prefiltration. The NF module is operated at 12–15 bar pressure.

As rejection of cyanide by charged NF membrane depends on membrane-charge density and charge on cyanide species, significant variation in rejection is observed with variation of pH. pH can influence

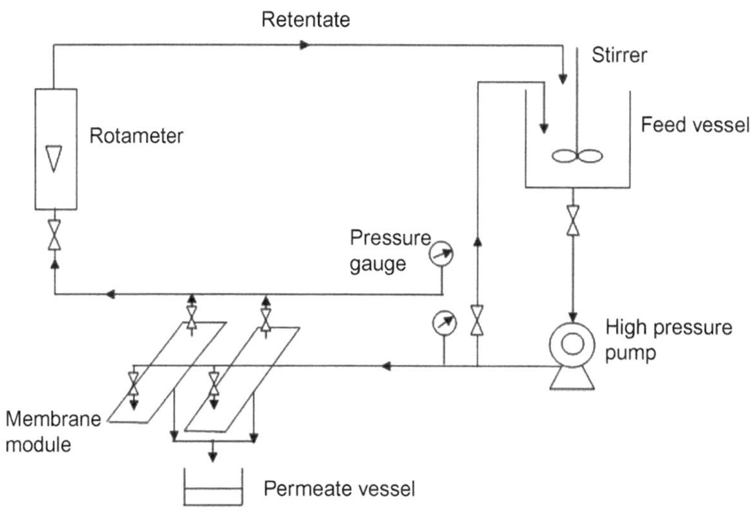

Figure 5.20 A nanofiltration (NF) scheme for cyanide removal from industrial wastewater [34].

cyanide speciation, and as has been shown cyanide rejection may increase from around 64% to 94% when changing pH from 4 to 10 [34]. Cyanide exists essentially in the form of HCN at pH below 9.3 (p*K*a) under non-oxidizing conditions where the level of free cyanide (CN^-) in equilibrium with HCN decreases further with increasing acidity. As we increase the pH, HCN gets converted to CN^- thereby enhancing retention of cyanide due to Donnan exclusion. The apparent pore size of polyamide NF membranes can also vary with solution pH. At the pore surface point of zero charge (isoelectric point) where the membrane functional groups are minimal in charge the pores open up more in the absence of repulsion forces that contribute to the widening of the membrane pores. At high or low pH values, the functional groups of membrane polymer can dissociate and take on positive or negative charge functions. Repulsion between these functions in the membrane polymer reduces or closes up membrane pores. At high ionic strength and high pH, pore size reduces remarkably.

5.12.2 Nanofiltration-Integrated Hybrid Technologies in Water and Wastewater Treatment

NF is often advisable as a final or polishing stage in industrial wastewater treatment as it will ensure long life of the membrane and reduce clogging. Thus many NF-integrated hybrid processes or technologies have evolved using NF membrane in the final polishing stage. Using NF coupled with chemical treatment or biological treatment, it has been possible to reuse the treated water. Such NF-integrated hybrid processes [35,36] have been discussed in Chapter 6, Industry-Specific Water Treatment: Case Studies, in the context of industry-specific wastewater treatment.

5.13 PERVAPORATION TECHNOLOGY IN WATER TREATMENT

Pervaporation has been used in the removal of volatile impurities such as hydrocarbon vapor (VOC) from water. This technique is particularly useful when due to azeotropism the traditional distillation process fails to separate the components beyond a certain level. For example, this process is successful at separating water from ethanol beyond 96%. For example, a mixture of benzene and cyclohexane can be separated using the treatment scheme shown Fig. 5.21.

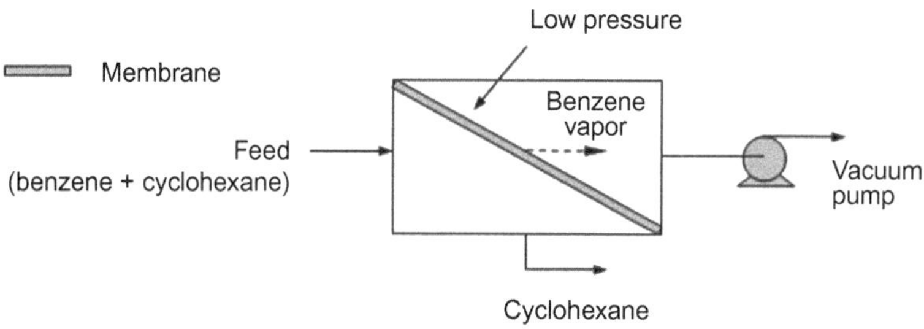

Figure 5.21 Schematic of pervaporation process.

5.14 REVERSE OSMOSIS TECHNOLOGY IN WATER TREATMENT

Water-treatment technology based on RO principles is used all over the world today in desalination of seawater to produce potable water. RO technology is possibly the only viable water-treatment technology in water-stressed regions of the world such as in countries like Saudi Arabia and Israel where even groundwater is salty. In the vast deserts, where power supply is another major problem, photovoltaic solar-driven RO plants are working very successfully. These large water-treatment plants have been commissioned in these areas for seawater desalination by RO. The largest water-treatment plant has been commissioned in Sorek 15 km south of Tel Aviv in Israel.

The basic principle in RO is reversing the natural osmotic flow of water by application of a hydrostatic pressure much in excess of the corresponding osmotic pressure. The osmotic pressure difference between the two compartments containing freshwater and saline water divided by a dense RO membrane is due to differences in concentration or chemical potential. The pure water side will have much higher chemical potential than the saline water side and thus pure water will flow from the pure water side to the saline water side due to osmotic pressure difference. However, as the requirement is to produce pure water from saline water, the direction of pure water flow must be reversed, which is done by applying hydrostatic much in excess of the osmotic pressure on the brine side thereby reversing the flow direction and facilitating production of freshwater from saline water as shown in Fig. 5.22.

The dense RO membrane retains almost everything such as dissolved solids, organic, pyrogens, submicron colloidal matter, color, nitrate, virus, and bacteria except water molecules. As the useful minerals like calcium

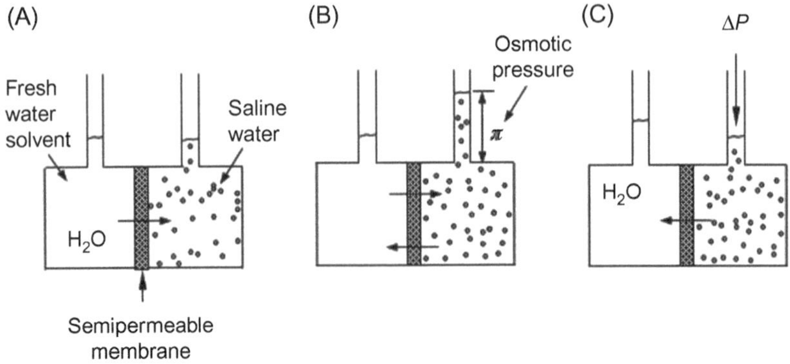

Figure 5.22 Schematic of reversing water flow across reverse osmosis (RO) membrane by applied pressure. (A) Osmosis, (B) Osmotic equilibrium, and (C) Reverse osmosis.

and magnesium are also retained, remineralization of water obtained from the RO unit is necessary to make it healthy potable water. RO plants are simple to design and operate and have low maintenance requirements. Modular design offers huge flexibility in scale of operation.

Osmotic pressure of solutions

The osmotic pressure π of a solution is proportional to the concentration of the solute and temperature T.

For dilute water solutions:

$$\pi = \frac{n}{V_{\mathrm{m}}} RT \tag{5.50}$$

where n = number of kmol of solute; V_{m} = volume of pure solvent water in m^3 associated with n kmol of solute; R = gas law constant = 82.057×10^{-3} ($\mathrm{m}^3 \cdot \mathrm{atm/kmol} \cdot \mathrm{K}$); T = temperature, K

If a solute exists as two or more ions in solution, n represents the total number of ions:

Osmotic pressure (Psi)

$$\pi = 1.12(T + 273) \sum m_i \tag{5.51}$$

where T = temperature, °C; $\sum m_{\mathrm{i}}$ = summation of molalities of all ionic and nonionic constituents in solution.

A solution with 300 mg/L TDS is found to have osmotic pressure of around 3 psi, whereas a solution with 1500 mg/L TDS is found to have osmotic pressure of around 15 psi indicating almost a linear relationship between TDS and osmotic pressure.

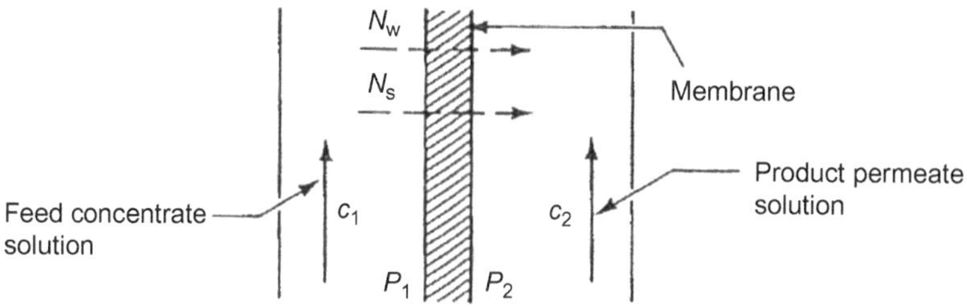

Figure 5.23 Concentration and fluxes in RO.

5.14.1 Flux Equations for Reverse Osmosis

There are two basic types of mass-transport mechanisms.

Diffusion-Type Transport

Both the solute and the solvent migrate by molecular diffusion in the polymer, driven by concentration gradients set up in the membrane by the applied pressure difference.

Membranes are capable of retaining solutes of about 10 Å in size or less.

Sieve-Type Mechanism

The solvent moves through the micropores in essentially viscous flow and the solute molecules small enough to pass through the pores are carried by convection with the solvent. Microporous membranes retain particles larger than 10 Å.

Diffusion-Type Model

Solvent Flux

Fig. 5.23 shows diffusion of solvent through an RO membrane.

Solvent flux is expressed as:

$$N_w = \frac{P_w}{L_m}(\Delta P - \Delta\pi) = A_w(\Delta P - \Delta\pi)$$
$$A_w = \frac{P_w}{L_m} \tag{5.52}$$

where N_w = solvent (water) flux, kg/m^2·s; P_w = solvent-membrane permeability, kg solvent/m·atm·s; L_m = membrane thickness, m; A_w = solvent permeability constant, kg solvent/m^2·atm·s; ΔP = hydrostatic pressure difference, atm; P_1 = pressure on feed side (subscript 1, feed or upstream side of the membrane), atm; P_2 = pressure on product side (subscript 2, product or downstream side of the membrane), atm; $\Delta\pi$ = osmotic pressure of FS−osmotic pressure of product solution, atm.

Solute Flux

For diffusion of the solute through the membrane, the solute flux is expressed as:

$$\begin{aligned} N_s &= \frac{D_s K_s}{L_m}(c_1 - c_2) = A_s(c_1 - c_2) \\ A_s &= \frac{D_s K_s}{L_m} \end{aligned} \tag{5.53}$$

where N_s = solute (salt) flux, kg solute/$m^2 \cdot s$; D_s = diffusivity of solute in membrane, m^2/s; K_s = the distribution coefficient = $\frac{c_m}{c}$ = $\frac{\text{concentration of solute in membrane}}{\text{concentration of solute in solution}}$; A_s = solute permeability constant, m/s; c_1 = solute concentration in upstream or feed (concentrate) solution, kg solute/m^3; c_2 = solute concentration in downstream or product (permeate) solution, kg solute/m^3.

The distribution coefficient K_s is approximately constant over the membrane.

Steady-State Material Balance for Solute

The solute diffusing through the membrane is the amount of solute leaving in the downstream or product (permeate) solution:

$$N_s = \frac{N_w c_2}{c_{w2}} \tag{5.54}$$

where c_{w2} = the concentration of solvent in stream 2 (permeate), kg solvent/m^3.

If stream 2 is dilute in solute, c_{w2} is approximately the density of the solvent.

Solute Rejection (R)

$$R = \frac{c_1 - c_2}{c_1} = 1 - \frac{c_2}{c_1}$$

This can be related to the flux equations as follows:

$$
\begin{aligned}
N_s &= \frac{N_w c_2}{c_{w2}} \\
&\Rightarrow \frac{D_s K_s}{L_m}(c_1 - c_2) = \frac{N_w c_2}{c_{w2}} \\
&\Rightarrow \frac{D_s K_s}{L_m}(c_1 - c_2) = \frac{P_w(\Delta P - \Delta\pi)c_2}{L_m c_{w2}} \\
&\Rightarrow c_1 - c_2 = \frac{P_w(\Delta P - \Delta\pi)c_2}{D_s K_s c_{w2}} \\
&\Rightarrow \frac{c_1}{c_2} - 1 = B(\Delta P - \Delta\pi) \quad \left[B = \frac{P_w}{D_s K_s c_{w2}} = \frac{A_w}{A_s c_{w2}}\right] \\
&\Rightarrow \frac{c_2}{c_1} = \frac{1}{1 + B(\Delta P - \Delta\pi)} \\
&\therefore R = 1 - \frac{c_2}{c_1} = 1 - \frac{1}{1 + B(\Delta P - \Delta\pi)} \\
R &= \frac{B(\Delta P - \Delta\pi)}{1 + B(\Delta P - \Delta\pi)}
\end{aligned}
\tag{5.55}
$$

where B is in atm^{-1}.

5.14.2 Water-Treatment Plants Using Reverse Osmosis Technology

Fig. 5.24 shows the treatment scheme followed in an RO-based water treatment plant working in Island. Seawater is pumped and stored first in an abandoned quarry pool from where water is passed through pretreatment units such as a dual-media filter or cartridge filter prior to RO [37].

Fig. 5.25 shows the world's largest desalination plant using RO technology. The Sorek plant is about 15 km south of Tel Aviv, Israel. The plant was commissioned in 2013 with a sea water-treatment capacity of 624,000 m^3/day making it the world's biggest seawater desalination plant. The plant was constructed at a cost of \$400 m. The major components of the plant include intake system facilities for adequate and consistent flow of feedwater, facilities for offshore seawater supply,

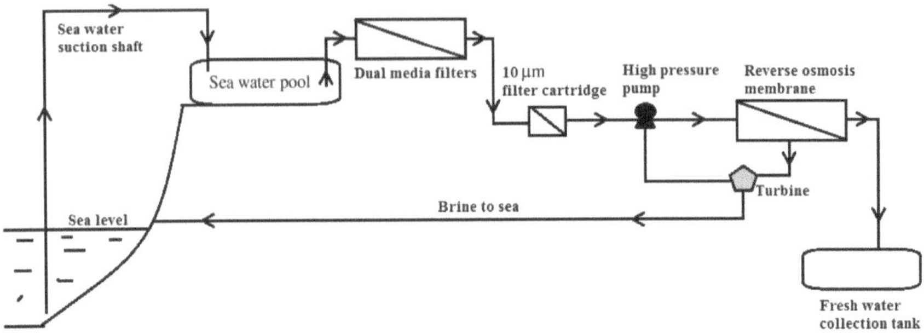

Figure 5.24 The scheme of treatment of seawater by reverse osmosis in La Rosiere desalination plant.

Figure 5.25 World's largest desalination plant (Sorek) 15 km south of Tel Aviv, Israel using RO technology [38]. *From IDE Technologies.*

and brine outfall pipelines. Two open sea intake heads located about 1.15 km offshore supply the saline water for desalination. The suction heads are provided with a slow suction velocity of 0.15 m/s only to minimize the effects of entrainment and impingement of marine organisms. As the piping and metal infrastructure are likely to undergo extreme corrosion in a seawater environment, special provisions for protection from corrosion are made. One such provision is a cathodic protection system. The pipe-jacking method is also an option in installing onshore pipelines. Feed pipelines are made of concrete. RO needs both pre- and posttreatment of the feedwater and treated water, respectively. Chemical coagulation–precipitation and MF cartridges are often installed at the pretreatment step and remineralization is necessary to make the water healthy since even the useful minerals like calcium and magnesium are totally removed by RO.

5.15 FORWARD OSMOSIS TECHNOLOGY IN WATER TREATMENT

Forward osmosis (FO) separates components based on osmotic pressure differentials as shown in Fig. 5.26. When it is difficult to separate the components from water, the option is to separate out water from the components. For example, if two compartments are separated by a semipermeable FO membrane containing wastewater and a concentrated aqueous medium with osmotic pressure, pure water will flow spontaneously from the feed side to the high osmotic pressure solution to eventually dilute the solution on the permeate side, which is called draw solution (DS). The draw solution can be prepared by dissolving a gas like ammonia or carbon dioxide in water or by dissolving glucose or sodium chloride. In this process, a concentrated draw solution is used to extract contaminants free water from a FS through the membrane. The magnitude of concentration or osmotic pressure difference between the two solutions (DS and FS) acts as the driving force for water permeation

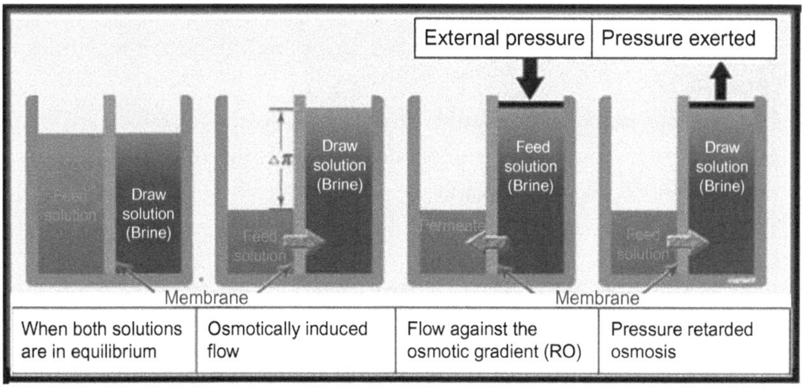

Figure 5.26 Forward-osmosis (FO) process.

through the membrane from the feed side to the draw solute side. The process does not directly produce pure water from wastewater, and another step involving RO or NF is required to get the recovered water back. FO offers a multitude of potential advantages.

Being mechanically strong and hydrophilic cellulose acetate has traditionally been used as FO membrane material. An ultrathin skin layer formed on polyether UF membrane can also lead to formation of a FO membrane. As FO typically requires zero or low hydraulic driving pressure, the possibility of fouling is negligible compared to that in a pressure-driven process. However, the problem of CP often stands in the way of FO application in water purification. Such CP, in particular internal concentration polarization (ICP), significantly reduces the effective osmotic pressure across the FO membrane, which is the driving force in the separation process. Efforts have been made to make FO membranes more hydrophilic while offering better fluxes through the use of innovative FO membrane materials like electrospun nanofibers. In new developments attempts have been made to prepare a membrane with low tortuosity, high porosity, and a thin yet highly selective hydrophilic skin layer to effect high rejection of solutes and low or zero back diffusion of the draw solute. Orientation of the FO membrane is an important consideration in FO membrane application where the active skin layer should face the FS to reduce fouling. By imparting higher hydrophilicity and charge negativity to the FO membrane during synthesis, performance of such membranes can be remarkably improved particularly during separation of NOMs.

The potential advantages of the FO process include:

1. Good quality of water comparable to conventional technology of desalination, such as RO
2. Suitability in extracting pure water from highly hazardous multiple components
3. No necessity for high hydraulic pressure for desalination
4. High osmotic driving force possible with appropriate draw solution
5. High water recovery attainable
6. Energy-saving since ΔP is practically zero

5.15.1 Transport Modeling of Forward Osmosis-Membrane Process

Mass transport models of FO for wastewater treatment are limited. Many of the existing models were developed by neglecting fouling due to the complexity and inconsistency in the nature of wastewaters. Internal-external concentration polarization phenomena, [38] solution-diffision phenomena and convection-diffusion phenomena are the four major phenomena used in FO model development. The theoretical water flux (J_v) in FO mode may be computed by [38,39]:

$$J_v = K_m \ln\left(\frac{A\pi_{DS} + B}{A\pi_{FS} + J_v + B}\right) \tag{5.56}$$

where K_m is the coefficient of mass transfer and A and B are the permeability constant of pure water and solute, respectively. Due to the phenomena of ICP, a nonlinear relationship between the combined effect of osmotic pressure difference of the feed (π_{FS}) and draw solutions (π_{DS}) with J_v is observed in Eq. (5.56).

5.15.2 Modeling Concentration Polarization

In the FO flux profile, CP plays a significant role. In FO systems, both dilutive and concentrative external concentration polarization (ECP) and ICP are encountered, although in most cases ICP effects predominate. The reverse-draw solute flux-accelerated osmotic pressure, which is enhanced by cake formation, also contributes to flux decline in FO. This happens when the porous support layer faces the draw solution and the back diffusion of the draw solute to the inside of the porous support layer takes place resulting in less osmotic pressure difference over the active layer. This is called ICP, which is considered a major hurdle in the application of a FO system (Fig. 5.27). In most FO studies, the dead-end flow module is employed where irreversible membrane fouling and hence flux decline is common, and is directly related to the type of membrane module. The point of osmotic equilibrium also influences the water-extraction capacity of the draw solutes. The FO modules operated in counter-current mode can yield more water for a unit mass of draw solute than under cocurrent mode.

Although ECP is not a significant impediment to the performance of FO it cannot be completely ignored, especially under low cross-flow velocity or high water flux conditions. Osmotic pressure on the draw-solution side decreases due to occurrence of ECP on both sides of the membrane surface, which in turn reduces water transportation through the membrane. The high cross-flow velocity reduces the ECP effects, which facilitates diffusion of the solute particle on the membrane surface. Moreover, water does not permeate well into the bulk draw solution under low cross-flow velocity of the draw solution as the reverse solute flux of the draw solute is concentrated on the surface of the active layer of

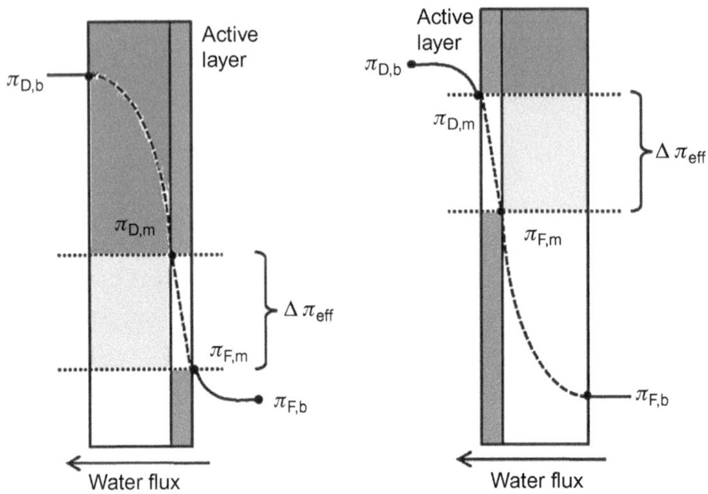

Figure 5.27 Internal concentration polarization (ICP) buildup with osmotic pressure in AL-facing feed solution (FS) and draw solute orientation.

the membrane. Likewise, simple ECP on the surface of the support layer results in high water flux. ECP becomes more significant under these conditions. Therefore it is necessary to develop a more precise FO model that can simulate the reverse-draw solute flux and ECP simultaneously. The dense cellulose membranes do not contain pores and operate by the transport mechanism of solution diffusion, which needs a partition of solute and solvent into the active layer phase before diffusing across it. The coupled diffusion effects for the dense membranes arise inside the membrane active layer, so the normal solute concentration is used in the active layer phase. If the contribution of the solute partition coefficient is considered in the total flux, the contribution from the convective term in the main flux equation goes down, which explains the reason even at higher draw solution concentrations solute–solvent coupling does not seem to influence the solute flux, but does impact the solvent flux [39,40].

5.15.3 Membrane-Structure Parameter

Recent efforts to improve transport modeling of FO membranes have focused on the membrane structure parameter (S), which is related to mass transfer through FO membranes as:

$$K_{m} = \left(\frac{D}{S}\right) \tag{5.57}$$

where D is the solute diffusion coefficient. S is directly proportional to the product of thickness (l) and tortuosity (τ), and is inversely proportional to the porosity (ε) as:

$$S = \left(\frac{\tau l}{\varepsilon}\right) \tag{5.58}$$

FO membranes with large structure parameter values, i.e., thicker and denser membrane supports, increase hindered diffusion and boundary layer thickness resulting in decreased membrane performance.

5.15.4 Forward Osmosis-Membrane Technology in Desalination

FO offers several benefits over RO in terms of economics and power consumption. One method of FO desalination includes the use of thermolytic compounds as draw solutes, which can be converted to volatile dioxide gases (like CO_2 or SO_2) by heating after osmotic dilution. Water can be recovered and the gases can be recycled during the thermal decomposition of the draw solution. After osmotic dilution, the dissolved SO_2 can be removed by standard means. Hydrophilic-based nanoparticles have also been used as the draw solutes for desalination nanoparticles are regenerated by UF. Optimization of the draw-solute composition and dosage is a key factor governing the FO process. For example, divalent salts (Na_2SO_4) have been used as draw solutes for brackish water desalination as the diluted draw solution can be recovered with NF [41]. Desalination has huge potential in the freshwater-starved Middle East countries. Most of the operational plants there use RO and thus to incorporate FO all the operational machinery would have to be changed, which is not an attractive proposition. Thus to reduce costs the best option is to propose a hybrid FO–RO process. In fact, there are a lot of operationally combined FO–NF or FO–RO processes. In such hybrid processes incorporation of FO has enhanced quality of drinking water, decreased the energy costs needed to make the solvent flow against the concentration gradient, and facilitated recovery of brine. There are also process design modifications that use FO as a pretreatment option and then use the RO or NF to recover pure water from the diluted draw solution. In the context of pharmaceutical wastewater treatment, an FO–NF-integrated system developed recently has proved very successful [42]. This is covered in Chapter 6, Industry-Specific Water Treatment: Case Studies.

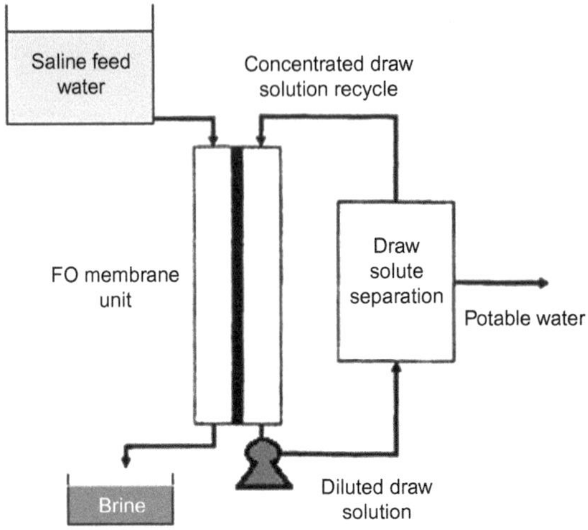

Figure 5.28 Scheme of an ammonia-carbon dioxide FO desalination process [43].

In FO system design, wide variations are observed in the selection of draw solution considering the potential of osmotic pressure buildup and ease of recovery. As ammonium bicarbonate is highly soluble in water and generates high osmotic pressure, it is often a good choice as draw solution. An ammonium bicarbonate draw solution was used to extract water from saline feedwater across a semipermeable polymeric membrane where the saline feedwater and draw solution was fed to the FO unit, which flows tangentially to the membrane in a cross-flow mode with cocurrent direction (Fig. 5.28). Highly soluble ammonium bicarbonate draw solution yields high osmotic pressure difference and high water fluxes as well as high feedwater recoveries. RO membranes are not suitable for the FO process because of inherently low associated water flux due to severe ICP in the porous support and fabric layers on the membrane surface.

The use of common chemical fertilizers in preparing draw solution is another economically viable option for desalination of saline water as the diluted fertilizer draw solution can be directly applied for agricultural purposes. Out of the very commonly used fertilizers, $NH_4H_2PO_4$ is found to have the lowest reverse slat flux (RSF), followed by $(NH_4)_2HPO_4$, $Ca(NO_3)_2$, and $(NH_4)_2SO_4$ whereas due to the presence of divalent anions, the ammonium compounds of sulfate and phosphate as well as $Ca(NO_3)_2$ are found have very low RSF, which is important for any FO process.

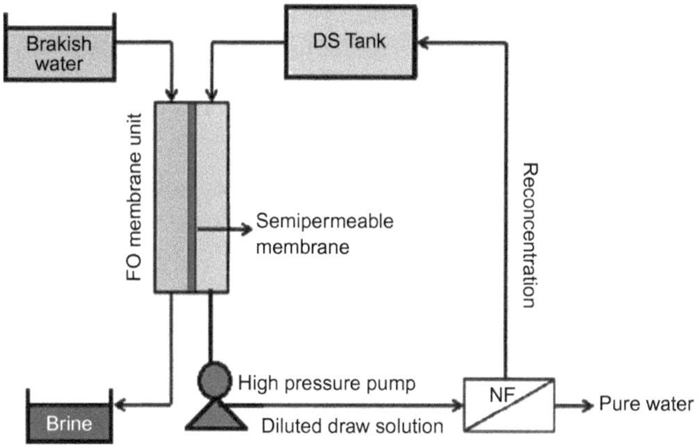

Figure 5.29 Schematic of hybrid forward Osmosis–nanofiltration (FO–NF) system [45].

A preliminary estimate shows that 1 kg of fertilizer can approximately extract 11–29 L freshwater from seawater [44].

An integrated FO–NF system can be effectively used in brackish water desalination, as shown in Fig. 5.29 [45], where Na_2SO_4 can be used as draw solute. Comparison of the hybrid FO–NF process with a standalone RO in brackish water desalination shows that the hybrid FO–NF process has many advantages over the standalone RO process such as lower hydraulic pressure, less flux decline resulting from membrane fouling, and higher flux recovery after cleaning. Moreover, the NF–FO system requires neither any pretreatment nor chemical cleaning during brackish water desalination.

Use of ammonium bicarbonate as draw solute in FO is common. As the draw solution gets diluted it can be decomposed into ammonia and carbon dioxide in a packed column again for subsequent absorption recovery reuse in ammonium bicarbonate solution [46].

An FO–RO integrated system has been found to be effective in removing sulfate, silica, di-sodium hydrogen phosphate (DHSP), sodium hexametaphosphate (SHMP), and sodium lignosulfonate (SLS) from coal mine wastewater [47]. An FO–RO integrated system is capable of concentrating salted coalmine wastewaters where almost 80% of the volume of mine water is reduced yielding dischargeable quality treated water. While the possibility of fouling is always there, the impact should be minimized by making frequent cleaning.

5.16 INTEGRATED MEMBRANE TECHNOLOGY IN GROUNDWATER AND WASTEWATER TREATMENT

A removal study of arsenic and boron from water by a laboratory-scale cross-flow FO system was performed by Jin et al. (2012) [48] who investigated the influence of membrane orientation and organic fouling on the performance of FO membrane. The results showed that inorganic contaminants such as boron, arsenic, and calcium were rejected 10%, 50%, and 96%, respectively, when the membrane active layer was facing the draw solution (AL–DS) and rejected 62%, 90%, and 100% when the active layer was facing the feedwater (AL–FW), as a result of the more severe ICP in the latter orientation. Due to higher permeability through the FO membrane of the boron, rejection between the two membrane orientations was greater. Due to the formation of an alginate fouling layer on the membrane surface (with active layer-facing feedwater) enhanced the sieving effect results high rejection of arsenious acid as it is relatively larger molecular size. However, in the AL–DS orientation of the membrane, alginate fouling in the membrane support layer had an adverse effect on boron rejection at a water flux below 15.3 L/m^2h, which was attributed to the fouling-enhanced concentrative ICP effect.

5.16.1 Hybrid Forward Osmosis Technology in Oily Wastewater Treatment

The recently developed hybrid FO-membrane distillation (FO–MD) has been found to successfully recover 90% of water from oily wastewater containing relatively high salinity, petroleum, surfactant, NaCl, and acetic acid employing cellulose triacetate thin-film composite (CTA–TFC) hollow-fiber membranes with relatively high water permeability and low salt permeability. The integrated FO–MD system yields high water flux and retains high NaCl and oil droplets (99.9%) and acetic acid while allowing only a small fraction of acetic acid to pass through. Acetic acid retained by CTA-based TFC membranes could be reused as a chemical additive [49].

5.16.2 Hybrid Forward Osmosis Technology in Treatment of Textile Wastewater

Textile wastewater containing multiple textile dyes, inorganic salts, and organic additives is an ideal case for application of FO technology to extract pure water from such a complex system instead of removing so many contaminants from water through several steps.

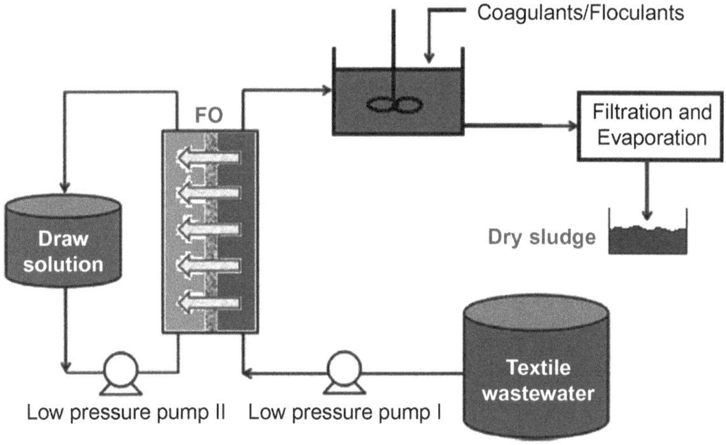

Figure 5.30 Schematic of the coagulation–flocculation-supported forward osmosis (FO) process for the treatment of textile wastewater [50].

Some FO-based treatment schemes have integrated coagulation–flotation as shown in Fig. 5.30 [50]. This treatment technology achieves high removal efficiency, high water flux, and high recovery rate while suffering from relatively low membrane fouling and adverse environmental impact. A lab-made TFC–FO membrane was used in an FO system that is basically a plate-and-frame-based membrane cell (surface area is 10 cm^2) that has a spacer-free rectangular waterway on both sides. Initially, the FO process reduces the volume of wastewater by recovery of water from wastewater through the osmosis process and intensely enhances dye concentration. The dye of the FO concentrated stream is then removed by the combined coagulation–flocculation techniques and results in high removal efficiency at a minimum chemical dosage. The system is capable of removing more than 99% of dye using 2 M NaCl as draw solution with initial water flux 36 LMH ($L/m^2 \cdot h$). The coagulation–flocculation process removes more than 95% of dye with a small dosage (500–1000 ppm) range of coagulants and flocculants. Significant reverse fouling is removed by freshwater flushing due to the great dye concentration in the FO-concentrated wastewater stream.

5.16.3 Hybrid Forward Osmosis Technology in Heavy Metal Removal

FO technology has also been effective in the removal of heavy metals from water. An FO system using TFC macrovoid-free polyimide support could remove up to 99% heavy metals using hydroacid complex $Na_4[Co(C_6H_4O_7)_2]$, $2H_2O$ (Na–Co–CA) as draw solute to reduce reverse solute flux. During FO of wastewater containing the heavy metals ($Na_2Cr_2O_7$, Na_2HAsO_4, $Pb(NO_3)_2$, $CdCl_2$, $CuSO_4$, $Hg(NO_3)_2$) in the concentration range of 2000–5000 ppm, the achieved water flux is around 11–12 LMH at 60°C temperature using 1–1.5 M draw solution (Na–Co–CA) [51].

5.16.4 Pharmaceutical Wastewater Treatment by Hybrid Forward Osmosis Technology

Treating complex pharmaceutical wastewater by FO–NF-integrated technology has been found to be very efficient [43,48,52]. FO systems using commercial membranes, namely CTA–HW and CTA–W (Hydration Technologies, Inc., Albany) and handcast TFC–FO membranes (TFC-1 and TFC-2), have been found successful. NaCl at 2 M solution is used as draw solution. Membrane performance in terms of rejection and water flux are found to depend on membrane interfacial properties, physicochemical characteristics of the pharmaceutical molecules, and feed-solution pH. TFC polyamide membranes show excellent overall performance in terms of water flux, rejection of all pharmaceuticals, and excellent pH stability while commercial CTA-based FO membranes exhibit strong influence on their rejection under acidic conditions due to hydrophobic interaction between the compounds and membranes. Under alkaline conditions, both electrostatic repulsion and size exclusion contribute to the removal of deprotonated molecules. TFC-2 membrane produces high water flux (8.15 ± 0.04 μm/s) whereas CTA–W membrane (3.29 ± 0.23 μm/s) shows the lowest water permeation capacity due to the higher water permeability values and lower structure parameter values of the TFC polyamide membranes over the CTA-based membranes. The pharmaceutical rejection by CTA–HW membrane at pH 8 is diclofenac (99%) > carbamazepine, (95%) > ibuprofen (93%), and naproxen (93%). Toxic pharmaceutically active compounds (PAC) such as carbamazepine and sulfamethoxazole from pharmaceutical wastewater have been effectively removed by FO [53]. It is observed that pH has pronounced effect on flux in both membrane orientations. The system offers higher flux (7–8 LMH) at higher pH (7–8) as with increase of pH the specific reverse salt flux and hydrogen ion flux decrease resulting in increased water flux. Water flux produced from normal FO mode is much lower than that in the pressure-retarded osmosis (PRO) mode as the osmotic-pressure differential is reduced due to the ICP phenomenon. In separation of neutral molecules, however, pH has no such effect. Due to the higher concentration gradient caused by concentrative ICP in porous supporting layer and the steric hindrance of the FO process, the rejection of carbamazepine in the PRO mode is lower than that in the FO mode. In the separation of charged species, however, electrostatic repulsion between the negatively charged FO-membrane surface and the charge of the species determines the extent of rejection.

5.17 FORWARD OSMOSIS TECHNOLOGY IN POWER GENERATION

FO technology opens the possibility of generating electricity exploiting natural salinity gradient. A fossil-fuel-based power plant generates an enormous amount of hazardous wastewater that in many cases pollutes surface water bodies such as rivers and lakes. Therefore a clean energy source (so-called blue energy, originally developed in the middle of the 20th century [54]) mixing freshwater with salt water to generate energy is ideal. Theoretically, 1.7–2.5 MJ (Mega Joule) of energy can be generated when 1 m^3 of river water is mixed with 1 m^3 of seawater or with a large surplus of seawater. It is estimated that the gross power potential of this unconventional energy source is massive, up to 2.4–2.6 TW.

5.18 MEMBRANE DISTILLATION TECHNOLOGY IN WATER TREATMENT

Membrane distillation (MD) or a slightly different version pervaporation is the only membrane-based water purification process that involves phase change. Therefore the energy requirement for a MD process is expected to be higher than in a pressure-driven membrane-separation process. However, in a thermally driven membrane-separation process, the heat-energy requirement is very low and the real driving force is the vapor-pressure difference created by the temperature difference across the membrane. Within a temperature range of 30°C–80°C, MD can be used for water purification. Similar to conventional distillation, in MD it is the vapor-liquid equilibrium that determines the separation of the components. However, unlike in conventional distillation, low-grade heat energy can cause the desired separation in MD. Unlike conventional distillation, MD is not effective in cases where the solution contains more than one volatile component. MD technology can be used in the purification of water from trace contaminants such as arsenic, boron, lead, fluoride, etc. When the required low-grade thermal energy can be extracted from sunlight such a process is likely to be more viable. In other words, a solar-driven MD to purify water from trace yet toxic or hazardous contaminants can be considered a sustainable technology. A technology developed recently [55] will be discussed later in this chapter.

5.18.1 Necessary Conditions for Membrane Distillation

For effective separation, the membrane should have micropores (0.1−1 μm) for the passage of water vapors. The temperature difference maintained across the membrane (between two sides) results in a water-vapor pressure difference across the membrane, which in turn acts as the driving force in diffusion of the water-vapor molecules through the pores, which must essentially remain dry. When reaching the exit point of the membrane pore the vapor molecules should drop in a relatively cold environment for instant condensation. One important specification of the membrane is its liquid entry pressure (LEP), which is the minimum hydrostatic pressure that must be required for the liquid-water molecules to enter the membrane pores. To avoid any liquid entry the synthesized membrane should have a high LEP. Membranes for MD may be prepared from a range of chemically resistant polymers such as polytetrafluoroethylene (PTFE), PP, and polyvinylidenedifluoride (PVDF) where moderate thermal stability over a temperature range of 30°C−90°C is desired. The material should also have low thermal conductivity to prevent heat loss. Overall, the MD process is a low-temperature and low-pressure process with relatively insignificant costs involved in operating the whole system and in designing the equipment. Compared to the pressure-driven membrane process, in this technology the issue of membrane fouling is almost absent. The molar flux (N) through the membrane pore very much depends on pore size (r_p), membrane thickness (Δx), tortuosity (τ), porosity (ε), and flow mechanism (Knudsen diffusion) or viscous flow through [56]:

$$N \propto \frac{r^a \varepsilon}{\tau \delta} \tag{5.59}$$

where the value of factor "a" is 1 for Knudsen diffusion and 2 for viscous flow.

Membrane Distillation Configurations

Depending on how the permeating vapor is recovered at the other side of the membrane there are four basic MD configurations: direct contact membrane distillation (DCMD), air-gap membrane distillation (AGMD), sweeping gas membrane distillation (SGMD), and vacuum membrane distillation (VMD).

In the DCMD configuration, the feed and the permeate liquid are in direct contact with the membrane in their respective compartments. The vapor is directly condensed and collected in a cold pure, liquid pool on the other side of the membrane. We examine such a configuration in the context of a flash vaporization DCMD module. This type of module offers higher flux than other modules despite the heat loss through membrane conduction.

In the AGMD configuration, the condensing surface is separated from the membrane by a stagnant air gap in the permeate side, which the permeated vapor molecules have to cross to eventually condense on a cold surface. As the air gap increases, the conductive heat loss first rapidly decreases and then slowly and eventually turns constant. The flux also decreases as the air gap increases.

In SGMD, condensation of vapor is done by a cold sweeping gas in the permeate compartment. In VMD, condensation on the permeate side takes place under a vacuum.

5.18.2 Membrane Distillation Technology for Purification of Water From Trace-Metalloid Contamination

A major drawback of industrial or large-scale applications of MD is its inherently low flux. Thus in this section we discuss development of a MD technology where low flux has been successfully overcome [56] by doubling the flux in this module design. Moreover, the system is self-sufficient in meeting the energy required for the involved phase change as it has been designed as a solar-driven module. This type of module is likely to be very successful in sunlight-parched South-East Asian countries where there are vast arsenic-affected regions and solar-driven pumps are being used increasingly in agriculture. In this new module, called a flash-vaporization module, a specially design provision causes flash vaporization of the warm contaminated feedwater leading to substantial improvement of pure water flux, almost double the normally reported values.

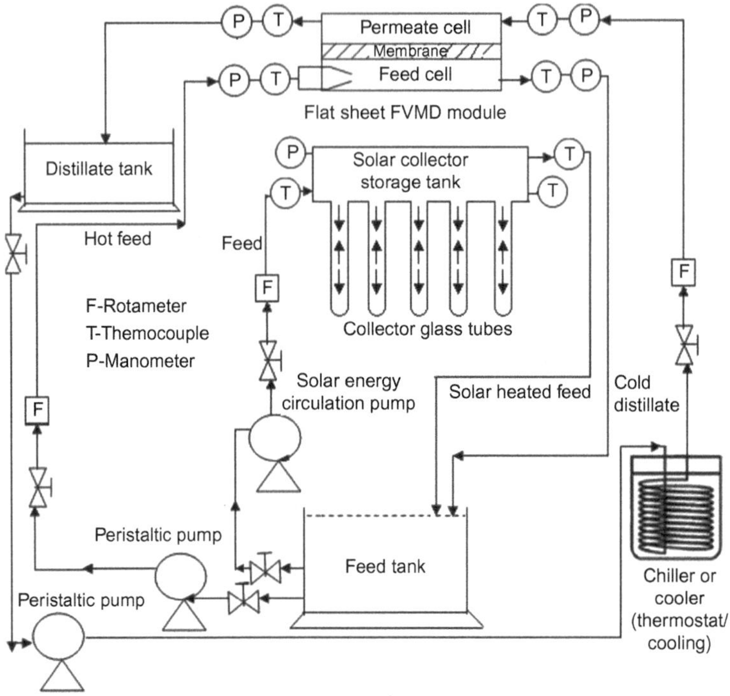

Figure 5.31 A flux-enhancing, flash-vaporization direct-contact membrane distillation (MD) module [56].

As shown in Fig. 5.31 the flux-enhancing module (FVMD) consists of a direct contact FVMD module and a solar-heating loop made of an evacuated glass panel. The flat-sheet cross-flow membrane module is made of polycarbonate-sandwiched flat membrane between the feed cell and the permeate cell. The polycarbonate material helps prevent thermal loss due to poor thermal conductivity. The module is designed in DCMD configuration where flash vaporization contributes to high flux. An evacuated glass tube-type solar energy collector is used to heat up the feed. In this DCMD hot-liquid feed and cold-liquid permeate are in direct contact with the membrane surfaces.

5.18.3 Mathematical Descriptions of Mass and Heat Transfer Phenomena in the Flash-Vaporization Module

Mass and heat transfer models of different MD modules relate the system performance in terms of water flux, evaporation efficiency (EE), and percentage rejection with the operating variables and membrane parameters

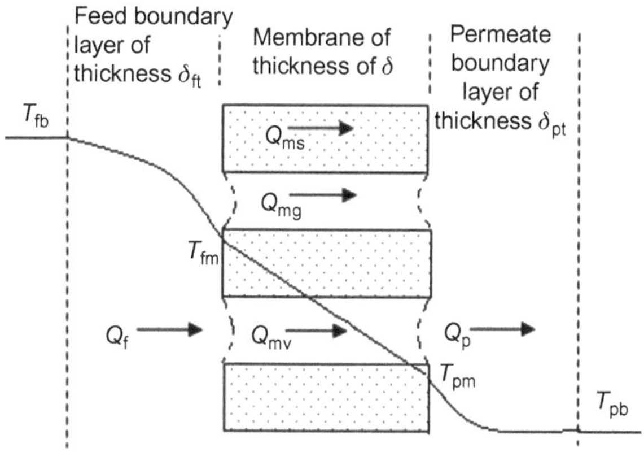

Figure 5.32 Heat transfer and temperature polarization in membrane distillation (MD) [56].

where the most significant phenomena impacting mass and heat transfer are temperature polarization and CP. We therefore first describe these phenomena.

5.18.4 Temperature-Polarization Effect

During MD, thermal boundary layers at the feed-membrane interface and membrane-permeate interface are formed that offer resistance to heat transfer as well as mass transfer resulting in a drop in the desired flux in MD. This is called temperature polarization. Such temperature polarization sometimes reduces the driving force by as high as 80%. This negative effect of temperature polarization on MD is expressed as the temperature polarization coefficient (TPC) and is defined as the ratio of the transmembrane temperature to the bulk temperature difference as:

$$\mathrm{TPC} = \frac{T_{\mathrm{fm}} - T_{\mathrm{pm}}}{T_{\mathrm{fb}} - T_{\mathrm{pb}}} \tag{5.60}$$

Fig. 5.32 shows that the presence of such a boundary layer on the feed side results in a drop in temperature at the interface from the bulk-level temperature, whereas on the permeate side, the temperature at the permeate–membrane interface rises over the bulk permeate temperature both being undesirable for good mass transfer. This reduces the available driving force.

This TPC actually represents the loss of thermal driving force due to thermal boundary layer resistance. In a DCMD system with good fluid dynamics where the process is controlled by mass transfer, the TPC approaches unity. For example, at high cross-flow velocity of around 1.85 m/s the TPC is around 0.8–0.9.

5.18.5 Concentration Polarization Effect

Another very significant phenomenon in MD is CP. We consider this in the context of nonvolatile solute(s) with one volatile component only. If we consider a liquid mixture consisting of a nonvolatile solute (B) and a volatile solvent (A), evaporation of the volatile component (A) at the membrane pore entrance will result in buildup of the nonvolatile component (B) near the membrane surface. This means that the concentration of the nonvolatile component at the membrane surface will be definitely higher than that at the bulk feed whereas the concentration of the volatile component at the membrane surface will be less than that at the bulk feed. This is what is called CP. The zone of the membrane on the feed side where the buildup of volatile and nonvolatile components takes place is called the CP layer. Fig. 5.33 illustrates this CP phenomenon [56].

The CP layer at the boundary represents resistance to mass transfer of volatile component from bulk feed to membrane surface and reduces the

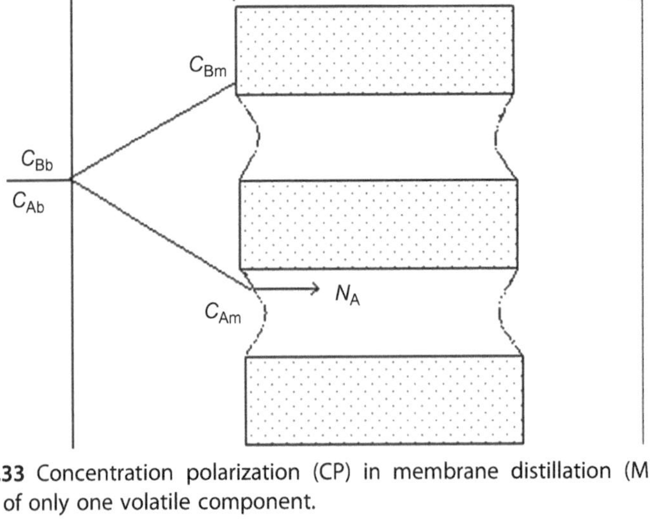

Figure 5.33 Concentration polarization (CP) in membrane distillation (MD) in the presence of only one volatile component.

transmembrane flux. As evaporation of the volatile component continues at the membrane surface, concentration of the nonvolatile solute continues increasing resulting in supersaturation of the solute. This affects the efficiency of the MD process. This supersaturation of the nonvolatile component also deposits a cake layer on the membrane surface thereby promoting fouling and scaling. All these developments eventually increase the chance of membrane wetting and also increase the heat transfer resistance.

CP is characterized by the concentration polarization coefficient (CPC), which is defined as the ratio of concentration of solute (B) at the membrane surface to that at the bulk feed. CPC is expressed as [44]:

$$\mathrm{CPC} = \frac{C_{\mathrm{Bm}}}{C_{\mathrm{Bb}}} \tag{5.61}$$

However, resistance due to CPC is less than that of TPC. While the effect of TPC is more pronounced than CPC, the effect of CPC will be significant for systems containing more than one volatile component. Both the effects of temperature polarization and CP can be minimized by promoting eddies and turbulence under high flow rates inside the flow channels, which in turn will enhance both heat and mass transfer. Following the theory of mass transfer in the boundary layer we may express the molar flux of the volatile component A through the feed-side concentration boundary layer from Fick's law as:

$$N_{\mathrm{A}} = Ck_{\mathrm{f}}^{\mathrm{m}} \ln \frac{C_{\mathrm{Bm}}}{C_{\mathrm{Bb}}} = Ck_{\mathrm{f}}^{\mathrm{m}} \ln CPC \tag{5.62}$$

where N_{A} is the mass flux of volatile component A, C is the summation of concentration of A and B (nonvolatile) in bulk feed, and $k_{\mathrm{f}}^{\mathrm{m}}$ is the mass transfer coefficient of the volatile component (A) through the concentration boundary layer.

5.18.6 Combined Impact of Temperature Polarization and Concentration Polarization

The combined effect of temperature and CPs is reduction of effective driving force that can be measured by the vapor-pressure polarization coefficient (VPC) as:

$$\mathrm{VPC} = \frac{p_{\mathrm{fm}} - p_{\mathrm{pm}}}{p_{\mathrm{fb}} - p_{\mathrm{pb}}} \tag{5.63}$$

The vapor-pressure difference is the real driving force in mass transport across the membrane pores, and the VPC measures the reduction of the imposed force ($p_{fb} - p_{pb}$) that results from the combined impacts of TPC and CPC. The VPC will coincide approximately with the TPC in the case of pure water as feed. However, for solutions of nonvolatile solutes, the VPC differs from the TPC as it depends on concentration of nonvolatile solutes and temperature. Thus we may say that VPC truly represents the driving force for mass flux where such mass flux depends on temperature, solution concentration, and recirculation flow rate as the VPC is a function of these factors.

For MD of water or dilute solutions when the bulk temperature difference ($T_{fb} - T_{pb}$) remains confined within 10°C, the vapor-pressure relations with respect to Fig. 5.34 can be expressed as:

$$\frac{(p_{fm} - p_{pm})}{(T_{fm} - T_{pm})} = \frac{(p_{fb} - p_{pb})}{(T_{fb} - T_{pb})} \tag{5.64}$$

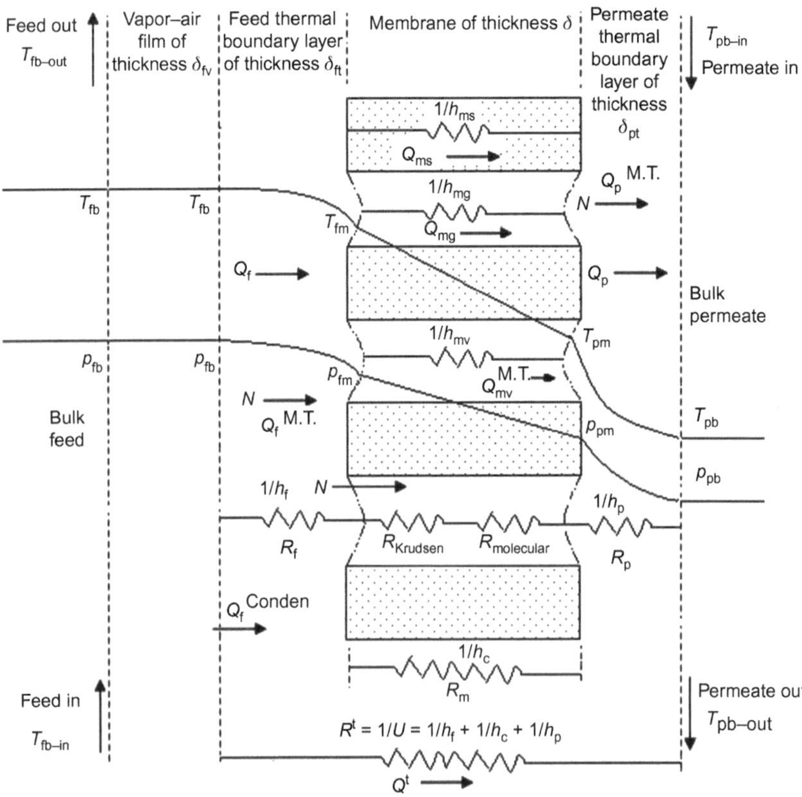

Figure 5.34 Mass and heat transfer resistance in direct-contact MD module [56].

The boundary layer mass transfer coefficient, $k_{\mathrm{f}}^{\mathrm{m}}$, can be obtained experimentally or estimated by empirical equations where we use the following dimensionless parameters:

$$Sh = Z\mathrm{Re}^{\alpha} Sc^{\beta} \tag{5.65}$$

$$Sh = \frac{k_{\mathrm{f}}^{\mathrm{m}} d_{\mathrm{h}}}{D_{\mathrm{AB}}}, \ \mathrm{Re} = \frac{d_{\mathrm{h}} \nu \rho}{\mu} \text{ and } Sc = \frac{\mu}{\rho D_{\mathrm{AB}}}, \ Nu = Z\mathrm{Re}^{\alpha}\mathrm{Pr}^{\beta} \tag{5.66}$$

where ρ is the liquid density and μ is the bulk liquid viscosity. The hydraulic diameter, which is the characteristic diameter of the flow channel, is designated as d_{h}, and the binary diffusion coefficient in the liquid is represented by D_{AB}. The liquid velocity is v, and Sh and Sc represent the Sherwood number and Schmidt number, respectively.

where Nusselt, Reynolds, and Prandtl numbers are designated by Nu, Re, and *Pr* respectively. Z, α, and β are characteristic constants of the module design and liquid flow regime. These constants are considered equal in both equations assuming heat-mass transfer analogy. The constant Z reflects the geometric characteristics and other conditions of the system where α and β represent the state of development of the velocity, temperature, and concentration profile along the flow channel in the MD module. The value of β is generally used as 0.33 for $Sc >> 1$ and when the length of the concentration profile entrance region is much larger than the length of the flow channel of the module. Flow regime determines the value of α.

5.18.7 Transport Phenomena in the Flash-Vaporization Membrane Distillation Module

During flash vaporization in the flux-enhancing module shown in Fig. 5.31, the hot feed is introduced to a relatively large feed membrane cell through narrow inlet tubing causing instant flash vaporization due to the sudden drop in pressure in the feed space. The rising vapor in the feed cell eventually diffuses through the membrane pores and directly comes in contact with the flowing cold pure water stream and gets condensed into liquid on the permeate side. The surface of the membrane on the feed side comes in contact with the rising air-vapor mixture while the permeate side of the membrane surface remains at a colder temperature. The temperature on the feed side of this membrane is lower than the dew point of the rising air—vapor mixture, which leads to formation of a film of condensate on the feed-side membrane surface along with an additional film of noncondensable air and water vapor. This layer of thickness δ_{fv} can be observed in Fig. 5.34. While operating at a feed temperature of 344 K, a distillate temperature of 278 K, and optimum distillate velocity of 0.12 m/s, this system achieves a vapor flux of 52 kg/($m^2 \cdot h$), which is considered remarkable in flash distillation literature. The corresponding mass and heat transfer model of this design is substantially different from existing DCMD models by incorporating the additional air-vapor thin film on the feed side of the membrane surface. The provision for flash vaporization as already discussed makes the new module design substantially more efficient than existing modules. The mass and heat transfer phenomena of this design are shown in Fig. 5.34. The best advantage of FVMD in the context of purification of water containing trace contaminants like metals or metalloids is that almost 100% pure water can be produced using low-grade thermal energy such solar energy.

In the schematic of the heat and mass transfer phenomena shown in Fig. 5.34, the associated terms and phenomena are defined and explained as follows:

1. Feed thermal boundary layer thickness δ_{ft}: In the permeate cell, formation of a permeate thermal boundary layer of thickness δ_{pt} takes place.
2. As water vapor generated from the bulk feed by flash vaporization continuously mixes in the vapor-air film, the temperature gradient across the vapor-air film is negligible.
3. Thermal boundary layers on both sides of the membrane offer resistance to heat and mass transfer.
4. Thermal energy required for phase change (evaporation of water) at the feed-membrane interface is derived from the interface of vapor-air film and feed thermal boundary layer.
5. Water vapor in vapor-air film condenses at the interface by releasing heat of condensation (Q_f^{conden}) and is explained by Nusselt's condensing mechanism related to film type condensation.
6. Vaporization at the feed-membrane surface is controlled by convective heat transfer Q_f to mass transfer ($Q_f^{M.T.}$) with and heat transfer due reference to Figure 5.34.
7. Water vapor condenses at the permeate-membrane surface after transport of vapor from the feed-membrane surface to the permeate-membrane surface through the membrane pores due to vapor-pressure difference. The associated heat of condensation is designated in the diagram as $Q_{mv}^{M.T.}$

8. The membrane conductive (Q_{mc}) that is transferred from the permeate-membrane interface to the bulk permeate across the permeate thermal boundary layer in the form of convective heat transfer (Q_p) and the heat transfer due to mass transfer ($Q_p^{M.T.}$).
9. The major difference between the modeling approach with respect to the flash vaporization module and earlier modules is the consideration of additional air-vapor film, which is developed in a relatively large feed vaporization cell. Continuous mixing of air and water vapor takes place in this zone where condensation of water vapor takes place following Nusselt's mechanism. The temperature gradient across the vapor-air film is, however, smoothed out as a large amount of water vapor produced in the bulk feed by flash vaporization continuously mixes in the film. This special film, therefore, is maintained at the same temperature as the bulk feed. This consideration is quite justified by flash-induced turbulence and mixing. Transport of water vapor through the porous membrane in the DCMD model takes place by both diffusive as well as convective modes where trapped air acts against the transport. But in the flash-vaporization model, vapor transport through the membrane takes place following combined Knudsen and molecular diffusion mechanisms of the dusty gas model (DGM) as justified by the value of the Knudsen number ($K_n = \lambda/d$), which lies here between 0.01 and 1.0 at a typical membrane surface temperature of 333 K. Further details of this modeling approach can be found in [55,56].

REFERENCES

[1] Pal P, Sikder J, Roy S, Giorno L. Process intensification in lactic acid production: a review of membrane-based processes. Chem Eng Process Process Intensif 2009;48:1549–59.
[2] Starzak ME. The physical chemistry of membranes. New York, USA: Academic Press; 1984.
[3] Strathmann H. Introduction to membrane science and technology. Wiley-VCH; 2011.
[4] Donnan FG. The theory of membrane equilibria. Chem Rev 1924;1:73–90.
[5] Dean WM. Hindered transport of large molecules in liquid-filled pores. AIChE J 1987;33:1409–25.
[6] Kedem O, Katchalsky A. Permeability of composite membranes: part 1. Electric current, volume flow and flow of solute through membranes. Trans Faraday Soc 1963;59:163.
[7] Perry M, Linder C. Intermediate RO-UF membranes for concentrating and desalting low molecular weight organic solutes. Desalination 1989;71:233–45.

[8] Schirg P, Widmer F. Characterisation of nanofiltration membranes for the separation of aqueous dye-salt solutions. Desalination 1992;89:89–107.
[9] Jacazio G, Probstein RF, Sonin AA, Young D. Electrokinetic salt rejection in hyperfiltration through porous materials, theory and experiment. J Phys Chem 1972;76:4015–23.
[10] Wang XL, Tsuru T, Nakao S, Kimura S. Electrolyte transport through nanofiltration membrane by space-charge model and comparison with Teorell-Meyer-Sievers model. J Membr Sci 1995;103:117–33.
[11] Wang XL, Tsuru T, Nakao S, Kimura S. The electrostatic and steric hindrance model for the transport of charged solutes through nanofiltration membrane. J Membr Sci 1997;135:19–32.
[12] Tsuru T, Nakao S, Kimura S. Calculation of ion rejection by extended Nernst-Planck equation with charged reverse osmosis membranes for single and mixed electrolyte solutions. J Chem Eng Japan 1991;24:511–17.
[13] Rios GM, Joulie R, Sarrade SJ, Carles M. Investigation of ion separation by microporous nanofiltration membranes. AIChE J 1996;42:2521–8.
[14] Bowen WR, Mohammad AW, Hilal N. Characterisation of nanofiltration membranes for predictive purposes—use of salts, uncharged solutes and atomic force microscopy. J Membr Sci 1997;126:91–105.
[15] Schaep J, Vandecasteele C, Mohammad AW, Bowen WR. Analysis of the salt retention of nanofiltration membranes using Donnan-steric portioning pore model. Sep Sci Tech 1999;34:3009–30.
[16] Vezzani D, Bandini S. Donnan equilibrium and dielectric exclusion for characterization of nanofiltration membranes. Desalination 2002;149:477–83.
[17] Szymxzyk A, Sbai M, Fievet P, Vidonne A. Langmuir, transport properties and electrokinetic characterization of an amphoteric nanofilter. Langmuir 2006;22:3910–19.
[18] Yaroschuk AE. Dielectric exclusion of ions from membranes. Adv Colloid Interface Sci 2000;85:193–230.
[19] Born M. Volumen and hydratationswarme der ionen. Z Physik Chem 1920;1:45–8.
[20] Deeen WM. Hindered transport of large molecules in liquid-filled pores. AIChE J 1987;33:1409–23.
[21] Schaefer AI, Fane AG, Waite TD, editors. Nanofiltration—principles and applications. Amsterdam: Elsevier; 2006.
[22] Zhang Q, Xu R, Xu P, Chen R, He Q, Zhong J, et al. Performance study of ZrO_2 ceramic micro-filtration membranes used in pretreatment of DMF wastewater. Desalination 2014;346:1–8.
[23] Huang Y, Du J, Zhang Y, Lawless D, Feng X. Batch process of polymer-enhanced ultrafiltration to recover mercury (II) from wastewater. J Memb Sci 2016;514:229–40.
[24] Huang Y, Du JR, Zhang Y, Lawless D, Feng X. Removal of mercury (II) from wastewater by polyvinylamine-enhanced ultrafiltration. Sep Purif Technol 2015;154:1–10.
[25] Liu C-C, Liu JC. Coupled precipitation-ultrafiltration for treatment of high fluoride-content wastewater. J Taiwan Inst Chem Eng 2016;58:259–63.
[26] Lee SH, Shrestha S. Application of micellar enhanced ultrafiltration (MEUF) process for zinc (II) removal in synthetic wastewater: kinetics and two-parameter isotherm models. Int Biodeterior Biodegrad 2014;95:241–50.
[27] Urgun-Demirtas M, Negri MC, Gillenwater PS, Agwu Nnanna AG, Yu J. Meeting world's most stringent Hg criterion: a pilot-study for the treatment of oil refinery wastewater using an ultrafiltration membrane process. J Environ Manage 2013;117:65–75.
[28] Yin N, Wang K, Wang L, Li Z. Amino-functionalized MOFs combining ceramic membrane ultrafiltration for Pb (II) removal. Chem Eng J 2016;306:619–28.

[29] Garcia-Ivarsa J, Iborra-Clara M-I, Alcaina-Miranda M-I, Mendoza-Roca J-A, Pastor-Alcañiz L. Treatment of table olive processing wastewaters using novel photo modified ultrafiltration membranes as first step for recovering phenolic compounds. J Hazard Mater 2015;290:51–9.

[30] Zhang X, Chen Y, Konsowa A, Zhu X, Crittenden JC. Evaluation of an innovative polyvinyl chloride (PVC) ultrafiltration membrane for wastewater treatment. Sep Purif Technol 2009;70:71–8.

[31] Kim J, Van der Bruggen B. The use of nanoparticles in polymeric and ceramic membrane structures: review of manufacturing procedures and performance improvement for water treatment. Environ Pollut 2010;158:2335–49.

[32] Safarpour M, Khataee A, Vatanpour V. Effect of reduced graphene oxide/TiO_2 nanocomposite with different molar ratios on the performance of PVDF ultrafiltration membranes. Sep Purif Technol 2015;140:32–42.

[33] Pal P. Membrane-integrated hybrid treatment system for arsenic removal. Indian Patent No.275244 2016.

[34] Pal P. Groundwater arsenic remediation—treatment technology and scale-up. 1st ed. Waltham, MA.: Elsevier Inc.; 2015.

[35] Kumar R, Bhakta P, Chakraborty S, Pal P. Separating cyanide from coke wastewater by cross flow nanofiltration. Sep Sci Technol 2011;46:2119–27.

[36] Kumar R, Pal P. Membrane-integrated hybrid bioremediation of industrial wastewater: a continuous treatment and recycling approach. Desalin Water Reuse 2013; 1(3):26–38.

[37] <https://www.jerseywater.je/about-us/learn-more/desalination-plant/>.

[38] <http://www.water-technology.net/projects/sorek-desalination-plant/>.

[39] Loeb S, Titelman L, Korngold E, Freiman J. Effect of porous support fabric on osmosis through a Loeb–Sourirajan type asymmetric membrane. J Membr Sci 1997;129:243–9.

[40] McCutcheon JR, McGinnis RL, Elimelech M. Desalination by ammonia–carbon dioxide forward osmosis: influence of draw and feed solution concentrations on process performance. J Membr Sci 2006;278:114–23.

[41] Gruber MF, Johnson CJ, Tang CY, Jensen MH, Yde L, Helix-Nielsen C. Computational fluid dynamics simulations of flow and concentration polarization in forward osmosis membrane systems. J Membr Sci 2011;379:488–95.

[42] Cath TY, Childress AE, Elimelech M. Forward osmosis: principles, applications, and recent developments. J Membr Sci 2006;281:70–87.

[43] Ling MM, Chung T-S. Desalination process using super hydrophilic nanoparticles via forward osmosis integrated with ultrafiltration regeneration. Desalination 2011;278:194–202.

[44] Thakura R, Chakrabortty S, Pal P. Treating complex industrial wastewater in a new membrane-integrated closed loop system for recovery and reuse. Clean Technol Environ Policy 2015;17(8):2299–310.

[45] Phuntsho S, Shon HK, Hong S, Lee S, Vigneswaran S. A novel low energy fertilizer driven forward osmosis desalination for direct fertigation: Evaluating the performance of fertilizer draw solutions. J Membr Sci 2011;375:172–81.

[46] Zhao S, Zou L, Mulcahy D. Brackish water desalination by a hybrid forward osmosis–nanofiltration system using divalent draw solute. Desalination 2012;284:175–81.

[47] Kim Y, Lee JH, Kim YC, Lee KH, Park IS, Park S-J. Operation and simulation of pilot-scale forward osmosis desalination with ammonium bicarbonate. Chem Eng Res Design 2014. Available from: http://dx.doi.org/10.1016/j.cherd.2014.08.015.

[48] Thiruvenkatachari R, Francis M, Cunnington M, Su S. Application of integrated forward and reverse osmosis for coal mine wastewater desalination. Sep Purif Technol 2016;163:181–8.

[49] Jin X, Shan J, Wang C, Wei J, Tang CY. Rejection of pharmaceuticals by forward osmosis membranes. J Hazard Mater 2012;227–228:55–61.
[50] Han G, Liang C, Chung T, Weber M, Staudt C, Maletzko C. Combination of forward osmosis (FO) process with coagulation/flocculation (CF) for potential treatment of textile wastewater. Water Res 2016;91:361–70.
[51] Cui Y, Ge Q, Liu X, Chung T. Novel forward osmosis process to effectively remove heavy metal ions. J Membr Sci 2014;467:188–94.
[52] Zhao P, Gao B, Yue Q, Shon HK. The performance of forward osmosis process in treating the surfactant wastewater: the rejection of surfactant, water flux and physical cleaning effectiveness. Chem Eng J 2015;281:688–95.
[53] Xie M, Price WE, Nghiem LD. Rejection of pharmaceutically active compounds by forward osmosis: role of solution pH and membrane orientation. Sep Purif Technol 2012;93:107–14.
[54] Pattle RE. Production of electric power by mixing fresh and salt water in the hydroelectric pile. Nature 1954;174:660.
[55] Lawson KW, Lloyd DR. Membrane distillation. J Membr Sci 1997;124:1–25.
[56] Pal P, Manna AK, Linnanen L. Arsenic removal by solar-driven membrane distillation: modeling and experimental investigation with a new flash vaporization module. Water Environ Res 2013;85:63–76.

CHAPTER 6

Industry-Specific Water Treatment: Case Studies

SUBCHAPTER 6.1

Treatment of Wastewater From Steel and Coke Industries

6.1.1 TREATMENT OF COKE-WASTEWATER: CONVENTIONAL TECHNOLOGIES

6.1.1.1 Introduction to Conventional Treatment Technology

In the history of human civilization, steel has played a crucial role in construction. Recent decades have witnessed rapid expansion of urban development all over the world where steel has played the most crucial role. In the steel-making process, metallurgical coke is the major ingredient second to iorn ore. Coal is also often burnt in steel industries to supply power to power to this highly energy-intensive industry. Coal is converted into metallurgical coke for use in the blast furnaces of steel-making industries across the globe. Millions of tons of coke are produced to meet the demand for metallurgical coke in the backdrop of the ever-growing demand for steel. Coke is also used in the foundries and for domestic purposes. Coal carbonization results not only in coke but also gas and other byproducts such as benzene, toluene, anthracene, naphthalene, and coal-tar products, which constitute the raw materials for the manufacture of many synthetic dyes, drugs, and high explosives [1,2]. More than 1000 m^3 of highly hazardous wastewater is generated to produce 1000 tons of coke and around 4000 m^3 of freshwater is consumed. The sources of this wastewater include quenching units of hot coke mass, ammonia washing still, condensing, and washing units of the coke-oven gases and units of processing and purification of the byproducts of coke

Industrial Water Treatment Process Technology.
DOI: http://dx.doi.org/10.1016/B978-0-12-810391-3.00006-0

industry. All these units consume enormous quantities of freshwater and eventually generate wastewater heavily laden with a range of hazardous and toxic contaminants like cyanides, thiocyanate, phenol, ammonia, oil, and grease in varied concentrations depending on the types of coal and the operating conditions in the coke ovens [3–7]. In many cases, to reduce the volume of wastewater, quenching of coke mass is done by wastewater resulting in emissions of carcinogenic organic compounds, cyanide, and ammonia into the air thus causing serious air pollution and literally shifting the pollutants from water to air. Wastewater discharged into the surface channels eventually contaminates both surface water and groundwater [8]. The phenolic compounds can particularly easily migrate within different aqueous environments and contaminate groundwater because of their high solubility in water [9].

6.1.1.2 Treatment Challenges and Possible Solutions

While biological treatment is often suggested for removing organic pollutants from wastewater due to its relatively low cost, the long residence time required for the slow biodegradation process is considered a drawback [10–14]. Moreover, the presence of toxic contaminants like phenol, cyanide, etc., often makes it difficult for the autotrophic bacteria to effectively oxidize ammonia during the biological nitrification process. The same difficulty is also encountered by the heterotrophic-oxidizing bacteria. In such cases, pretreatments like steam stripping, settling, aeration, chemical coagulation, sludge concentration, and sometimes dilution under different conditions (anaerobic, anoxic, and aerobic) can minimize the adverse effects of environmental conditions on biological treatments [15–18]. While the individual methods of biodegradation of phenol, cyanide (CN^-), thiocyanate (SCN^-), and ammonia–nitrogen (NH_4^+-N) are well known, their successful simultaneous treatment is difficult because of cross-interference. The presence of cyanide in coke and steel industry wastewater poses one of the biggest challenges in the treatment of coke wastewater. This challenge is evident by the development of many new processes/methods dedicated to cyanide removal such as the INCO process (process of International Nickel Company Ltd. by SO_2/air), Caro's acid method, ozonation, electrolytic oxidation, ion exchange, the AVR (acidification, volatilization, and reneutralization) process, reverse osmosis (RO), activated carbon adsorption, biological treatments, and photocatalytic and catalytic oxidation [19–25].

However, each of these methods suffers from one limitation or the other. For example, the main drawback of the most widely practiced alkaline chlorination is the formation of highly toxic cyanogen chloride, which along with the residual chlorine, creates additional environmental problems. Moreover, recovery and reuse of cyanide is quite difficult. The simultaneous presence of phenol and cyanide has been shown to exhibit a negative influence on removal. Phenol and cyanide also exhibit inhibitory effects on the degradation of thiocyanate as well as on the nitrification in an activated sludge process (ASP) [26–28]. In the presence of toxic contaminants in high concentrations, incineration, adsorption on activated carbon, chemical or enzymatic oxidation, and solvent extraction are found to be more effective in degrading phenolic compound [29,30].

Ammonia may be removed using a stripping process, but it is a very costly process that requires chemical feed, stripping tower, pump and liquid spray system, forced aeration, and carbonization systems [31]. Ion exchange may also be an option for removal of ammonia in which wastewater needs to be passed through a bed of clinoptilolite (a zeolite resin) that selectively removes the ammonium ions [32]. But the resinous material needs to be regenerated with a lime slurry containing sodium chloride. This process is also very tedious, labor intensive, and expensive.

Biological denitrification is one of the most environmentally acceptable technologies although it needs a carbon source (e.g., methanol) and the removal rate is highly influenced by the aqueous chemistry. The removal efficiency is also low at low concentration of pollutants like cyanide, phenol, and ammonia.

Membrane separation could be an alternative to the biological treatment of wastewater from coal-based power plants and steel industries since the separation process is fast. In recent years, the contaminants in the wastewater such as cyanide, phenol, ammonia, and suspended solids have been removed efficiently by nanofiltration (NF) membranes at moderate pressure instead of high-pressure RO membrane where membrane fouling has largely been overcome by better membrane modules [33,34].

6.1.1.3 Coke-Wastewater Composition and Hazardous Effects

Carbonization of coal during coke production generates a large volume of gas, which on subsequent treatment in the byproduct plant, produces a clean fuel gas while removing condensable, corrosive, or economically valuable components called coal chemicals generally with reasonably

Table 6.1.1 Typical compositions of industrial coke wastewater [35]

Pollutant	Coke wastewater				
	Australia [6]	Germany [36]	China [37]	Spain [7]	India [38]
BOD_5	450–720	1600–2600	200–380	500–1000	64–94
COD	1800–2200	4000–6500	630–860	800–1900	525–81
TSS	40–50	2–10	–	25–50	–
TKN	200–270	50–150	220–280	200–1100	336–562
Total P	<1	<1	–	<1	–
Phenols	60–330	400–1200	50–80	110–375	82–123
CN^-	70–95	4–15	–	15–40	8.2–21
SCN^-	180–200	200–500	–	130–375	–

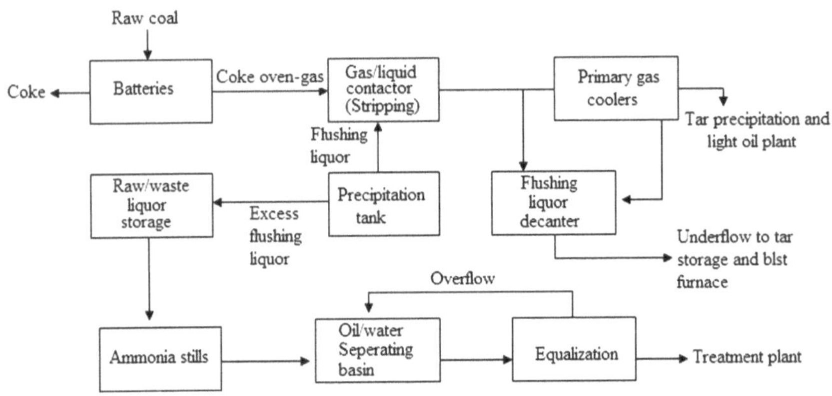

Figure 6.1.1 Simplified schematic of coke plant operation and wastewater generation [35].

good selling value. However, the coke wastewater generated during washing of coke gas contains major chemical and hazardous contaminants like ammonia, phenol, cyanide, and sulfide, and materials like oil, grease, and tar form an insoluble layer that hinders the access of oxygen from air. A detailed composition of coke wastewater is presented in Table 6.1.1.

Fig. 6.1.1 shows a simple schematic of a coke plant with the sources of process wastewater noted, and operation of a large coking plant is depicted in Fig. 6.1.2 indicating inflow of materials like coal, lime, nitric acid, and the outflow of products like coke, tar, sulfuric acid, benzol, fertilizer, and ammoniacal liquor.

The presence of both organic (phenols and other polynuclear aromatic hydrocarbons) and inorganic compounds (cyanide, thiocyanate, and

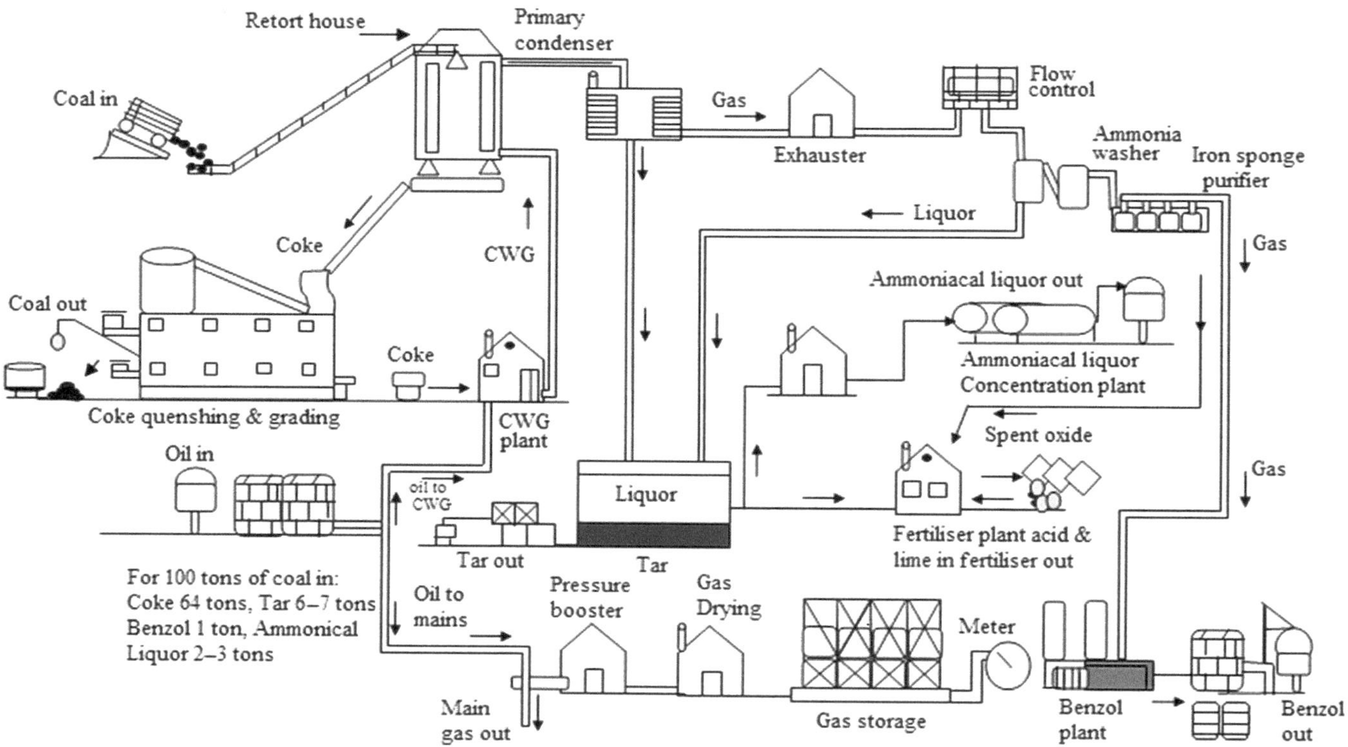

Figure 6.1.2 Diagram of a typical conventional coke manufacturing plant [35].

ammonia) makes the chemical composition of coke wastewater very complex. The heavily toxic wastewater may lead to sudden failure of the nitrification and denitrification process [39–41]. The pollutants discharged from the coke plant affect the biosphere through air, water, and soil. The toxic effects of coke wastewater have been investigated in plants like *Zea mays*, *Viciafaba,* and *Hordeumvulgare* through various parameters including growth, fresh biomass, mitotic index, micronucleus frequency, and antioxidant capacity [42,43].

6.1.1.4 Treatment Options

For removal of suspended solids, colloidal particles, floating matters, colors, and toxic compounds physicochemical techniques such as sedimentation, adsorption, steam stripping, wet-air oxidation (WAO), chemical oxidation, coagulation–precipitation, ozonation, electrochemical oxidation, and membrane-based techniques can be used. These techniques may collectively be designated as physicochemical treatment techniques. The other options include treatment using biological methods.

Sedimentation

The suspended matter present in coke wastewater is dominated by coke particles, oil, and grease. Settling tanks are used for the separation of total suspended solids (TSS). But often failure in the sedimentation process allows a large amount of coke breeze to escape with the effluent, creating trouble during the biological treatment of coke wastewater [44]. River water at the waste discharge point turns black due to deposited coke particles and oil and grease when effluent is not treated properly. Treatment of coke-plant effluent with the addition of lime slurry in a secondary sedimentation tank may be done more effectively while efficient ammonia removal might be achieved by synthetic zeolite columns [38].

Adsorption

Adsorption using different adsorbents (activated carbon being the dominant one) may be used to remove the hazardous components from coking wastewater. In batch adsorption 90%–95% removal of some contaminants of coke wastewater can be achieved. However, confidence in running continuous-mode treatment plants, in particular, the integrated types, is still lacking. Adsorption-based processes require frequent replacement/regeneration of the adsorption-bed material, which often makes the process cost-prohibitive [45–48].

Steam Stripping

Steam stripping of mainly ammonia from coke wastewater is very common. But there two major difficulties are encountered in this approach. One is imposed by pH limitation as at a pH lower than 7.0, ammonia cannot be removed by this method. While this method can be used in cases where ammonia is recovered as byproduct [49,50], it involves significant capital investment as lots of equipment including stripping tower, pump, liquid spray system, forced air, and carbonization system is needed for this type of separation. Moreover, removal of ammonia alone does not solve the problem of coke-wastewater treatment. The other major difficulty arises from the fact that in steam stripping, a significant amount of pollutants is shifted from water to air, which means integration with other operations enabling removal of major contaminants is required.

Wet-Air Oxidation

Wet-air oxidation has been found to be very effective in removal of organic contaminants in the form of chemical oxygen demand (COD) where an extent of 90% can be achieved. Supercritical WAO is an equally good technology. However, the main problem with this technology is salt precipitation, which often causes serious reactor plugging. A better design of salt separator can solve this problem [51–55]. Another critical problem of this process is the high possibility of inhibition of ammonia destruction by the presence of phenol and coexisting compounds, which often lead to failure of the complex reactor.

Chemical Oxidation

Chemical oxidation of the target pollutants is done using hydroxyl radicals, which react rapidly and nonselectively with nearly all organic compounds. During chemical oxidation by Fenton's reagent (H_2O_2 and $FeSO_4.7H_2O$), most of the persistent organic pollutants can be broken down in which complex cyanide compounds present in the coke wastewater facilitate biological treatment. Close control of the pH of the wastewater during oxidation by Fenton's reagent and also during post-treatment prior to discharge is required. Disposal of sludge is also considered a big problem in this approach [56–59].

Coagulation and Precipitation

Even adsorbable organic halides (AOX), total organic carbon (TOC), and coloring substances can be effectively removed by coagulation–precipitation, which in turn, makes downstream biological treatment more efficient in eliminating the residual concentration. The zero-valent iron process and coagulation–precipitation in treating coke wastewater are considered inexpensive and easy. Manganese (oxidation) and magnesium (precipitation) ore can also be used for treatment of coke wastewater, which can reduce more than 95% phenol, COD, and ammonia [43,60,61].

Ozonation

The high oxidation capability of Ozoe makes ozonation a potential technology in removal of coloring materials, to reduce COD, TOC, CN^-, SCN^-, and phenolic compounds generated not only in the steel and coke industries but also in a range of other industries such as petrochemical, textile, leather, paper and pulp industries. TOC, BOD (biological oxygen demand), SCN, and CN^- are also removed to by 70%–95% after 60 minutes of ozonation under neutral conditions in bubble column reactor [62–67]. However, the high cost of ozone generation and the expensive chemicals like potassium iodide required often make the process economically unattractive.

Electrochemical Oxidation

Electrochemical oxidation is an environmentally friendly and promising technology for treatment of biorefractory and toxic compounds to meet the discharge standards after conventional biological treatment. This energy-intensive oxidation is found to effectively treat refractory pollutants such as phenol, chlorophenols, nitrophenols, dyes, surfactants, and landfill leachate [68–75]. Since an electrochemical oxidation process strongly depends on anode material, many electrode materials have been examined and developed in recent years to improve the effectiveness of oxidation as well as current efficiency. These materials include graphite, platinum, IrO_2, RuO_2, and PbO_2. Boron-doped diamond (BDD) electrode has been successfully used for electrochemical oxidation to mineralize organic pollutants and to remove ammoniacal nitrogen (NH_4-N). However, the major challenge in using this technology is its high energy consumption [76–81].

Membrane Technology

Membrane-based technologies are emerging as replacements for many conventional treatment technologies due to their effectiveness at treating wastewater relatively inexpensively and using simple plant coonfigurations. Membrane technology is fast gaining importance in potable and industrial wastewater treatment and for reclamation of reusable water from various industrial effluent streams. A range of membranes from the microfiltration (MF) regime for rough separation—purification to the tight RO regime for extremely fine separation are employed today to treat diverse source waters depending on the desired quality of the treated water and the cost of treatment. Out of the range of membrane choices available, NF membrane appears to be the most popular due to its low operating pressure, high flux, high retention of multivalent anion salts and organic matter, and relatively low operational costs. NF membranes are in between ultrafiltration (UF) and RO membranes in terms of separation regime and operating pressure. NF membranes separate the solutes from the solution through both steric (sieving) and Donnan (electrostatic) mechanisms, the latter mechanism being the dominat one [82]. Rejection of ions by charged NF membranes can be explained by the combination of a few theorems such as degree of hydration, continuum hydrodynamic models, charge repulsion, Donnan exclusion, and the extended Nernst—Planck principles (ENP). At pH values greater than 7.0, most of the polyamide composite NF membranes possess negative zeta potential. Pore size also plays significant role in the retention of the neutral solutes by the size-exclusion mechanism, which depends on the parameters of the membrane such as the effective pore radius, thickness-porosity ratio, and the Stokes radius of solute. It is mainly the Donnan exclusion principle that is exploited in rejection of the charged solutes where the three parameters, namely effective pore radius, thickness-porosity ratio, and effective charge density of the membrane, determine the degree and efficiency of separation [82—87].

Fig. 6.1.3 shows a simple membrane-based plant configuration consisting of micro- and NF membrane modules. Wastewater from a stirred vessel is passed through a MF membrane module followed by a NF module, both in flat-sheet cross-flow mode. This two-stage membrane-integrated plant can remove hardness, dissolved fluoride, nitrate, and other metal ions. As the conventional chemical or biological treatments cannot effectively treat all the hazardous compounds of coke wastewater, conventional treatment has been integrated with membrane-based treatment [4,88—90].

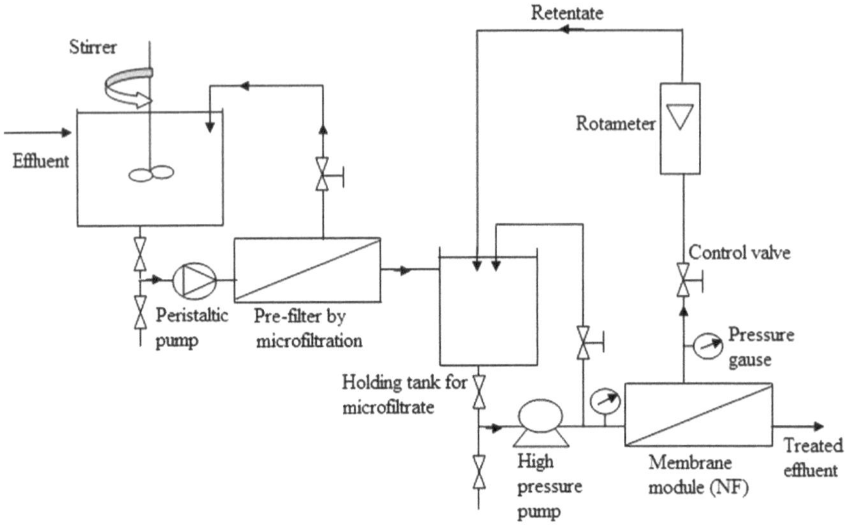

Figure 6.1.3 Schematic of treatment of coke wastewater by micro- and nanofiltration membranes [35].

Integration of biological treatment with RO has also been done in treatment plants but RO suffers from flux decline due to high osmotic-pressure build-up, making it inefficient. RO also involves expensive membranes and high operating pressure. To overcome these problems, there have been attempts [4,90] to integrate NF with steam stripping for the fractionation of pollutants as NF membranes demand much lower operating pressure and the associated costs are also than for RO membranes. Effective separation and concentration of ammonia from cyanide at 40% recovery rate with reasonable flux under optimized steam consumption has been achieved in an NF steam-stripping integrated plant with moderately high transmembrane pressure (TMP). The use of NF membrane in one stage has been successful in reducing the costs of separation of high concentration of bicomponent ammonium solution into monocomponent in removing the resources from the coke wastewater in the form of ammonium thiosulfate and ammonium thiocyanate. As membrane fouling has remained a problem in membrane-based plants, there have been various attempts at minimizing the membrane-solute negative interaction that leads to fouling under different pH regimes [91,92]. The other major approach is ensuring suitable hydrodynamics in the module that prevents concentration polarization (CP). For example, the use of a flat-sheet cross-flow module instead of a dead-end module can significantly reduce CP due to the sweeping fluid action on the membrane surface. Around 95% removal of cyanide from coke wastewater has been possible by such modification of the module hydrodynamics [93].

Biological Treatments

In today's steel and coke wastewater-treatment plants, biological treatment is widely practiced. Considering the enormous volume of wastewater generated in these plants, biological treatment seems to be one of the most economically viable options provided such biosensitive plants are operated by well-trained plant operators. Thermal and chemical shock often lead to sudden death of the entire microbial population, but this requires close monitoring of the operating conditions of the plant. In biological treatment, the microbes are the real reactors and for high rate of conversion of polluting substances into harmless components, hiatment unitsgh cell density needs to be maintained in the biological treatment units.

Classification of Biological Treatment Processes

The success of a plant very much depends on the growth of the microbial population and can subclassified as those populations with suspended growth mechanism and the attached growth mechanism.

Suspended growth mechanism: activated sludge process ASP is an efficient biological treatment process in which microbes within the aeration tank or bioreactor grow while remaining suspended in the aqueous medium. A part of the settled sludge from the downstream sedimentation unit is continuously recycled to the main aeration tank. Microbes that return to the main reactor through this sludge get reinvigorated or activated upon receiving fresh oxygen and food in the reactor. Thus the bioreactor always remains activated resulting in high conversion of pollution loads through microbial decomposition. Organic pollutants are quickly decomposed.

Fig. 6.1.4 shows a schematic of the ASP where the major provisions are for an aeration tank and a sedimentation tank. However, the complexity of coke wastewater due to the presence of a variety of pollutants, fluctuations in concentration of the pollutants, and the presence of high doses of certain inorganic contaminants like cyanide and thiocyanates often renders biological treatment ineffective. These problems have been overcome through several modifications to conventional single-step ASP.

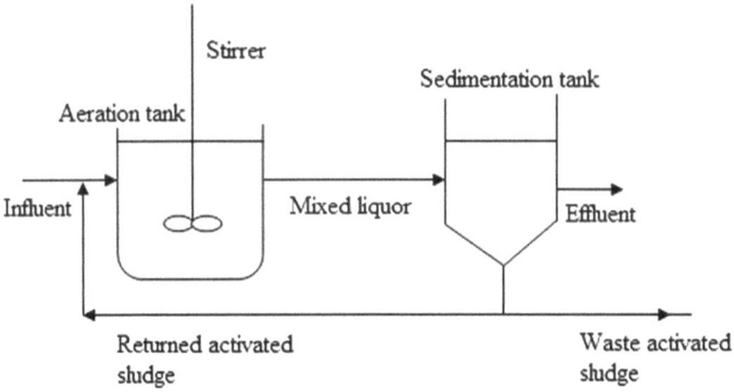

Figure 6.1.4 Schematic of the activated sludge process (suspended) [35].

Single-step treatment of coke wastewater In single-step treatment of coke wastewater, the activated sludge unit is developed into one aerated reactor followed by a downstream sedimentation tank. Removal of organic matter (substrate) in the ASP involves adsorption and agglomeration onto microbial flocs, subsequent assimilation and conversion into new microbial cell materials, and finally mineralization by complete oxidation. The removal of COD, phenols, SCN^-, and NH_4^+-N may reach 75, 98, 90, and 71%, respectively, in a single-step ASP [7]. The activated sludge from a landfill treatment plant is considered better acclimatized for biodegradation than the sludge from a municipal sewage treatment plant. External carbon sources like bicarbonate and methanol may be needed to stimulate the growth of microorganisms in order to remove the NH_4^+-N, organic matter, and SCN^- in a single step. High hydraulic residence time (HRT) and long solid-retention time are the major disadvantages of single-step treatment of coke wastewater due to the toxic effects of cyanide, thiocyanate, and phenol on nitrifying bacteria.

Multistep treatment of coke wastewater The difficulties in single-step activated sludge are largely overcome in multistep processes involving anoxic, aerobic, and anaerobic environmental conditions. In the anoxic condition, heterotrophic denitrifiers convert nitrite and nitrate into nitrogen gas using organic-carbon sources that provide the energy source and serve as electron donor for the denitrification reaction. In anaerobic reactor, some toxic compounds like cyanide are removed by anaerobes. In an aerobic condition, autotrophic nitrifiers convert ammonia into nitrite or nitrate while autotrophic thiocyanate-oxidizing bacteria convert thiocyanate into ammonia and sulfate. Low pH (6–6.5) favors the

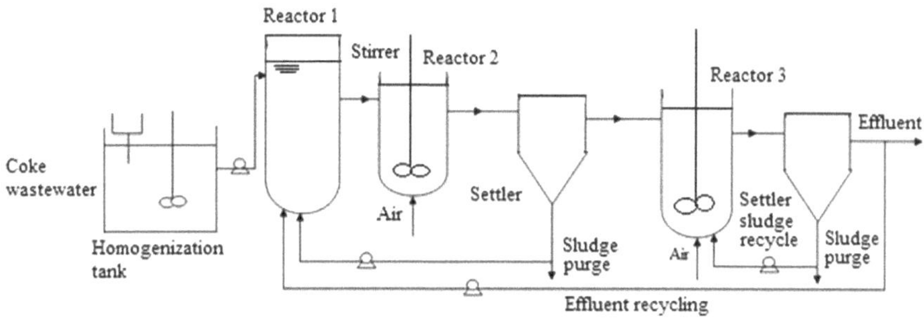

Figure 6.1.5 Scheme of multistep activated sludge plant for the biological treatment of coke wastewater [35].

biodegradation of thiocyanate whereas high pH (7.8–8.5) is required for nitrification and denitrification. In such a multistep process, the maximum removal efficiencies for COD, phenol, thiocyanate, and NH_4^+-N achieved are 90, 98, and 99%, respectively, at HRT of 98 hours in the first reactor and 86 hours in the second reactor with a recycling ratio of 2 at an operating temperature of 35°C. Different HRTs for different pollutants and different effluent recycling ratios help minimize the inhibition phenomena attributable to the high concentrations of the pollutants, albeit at the cost of speed of degradation of the organic and chemical compounds. For fast removal of nitrate, an additional reactor is operated to carry out the denitrification process in anaerobic conditions with the addition of methanol as an external carbon source. While this obviously adds to the overall cost of treatment, the cost may be brought down by putting the denitrification step at the beginning of the process to avoid the cost of an external carbon source like methanol for denitrifying bacteria. High removal efficiency of phenol, SCN, ammonia, COD, TOC, and TN (total nitrogen) can be achieved using this approach [94–100] with a two-staged reactor system. Despite all these modifications, the final removal efficiency of the total cyanides may still not be satisfactory due to slow biodegradation of ferric cyanide under anoxic as well as aerobic conditions. The removal efficiency of different types of contaminants present in coke wastewater is dependent on the concentration of cyanide. A schematic of a common multistep biological plant under different environmental conditions is shown in Fig. 6.1.5.

Sequential batch reactor The sequencing batch reactor (SBR) process is a sequential suspended growth (activated sludge) process designed to operate in nonsteady state. A SBR operates in true batch mode with

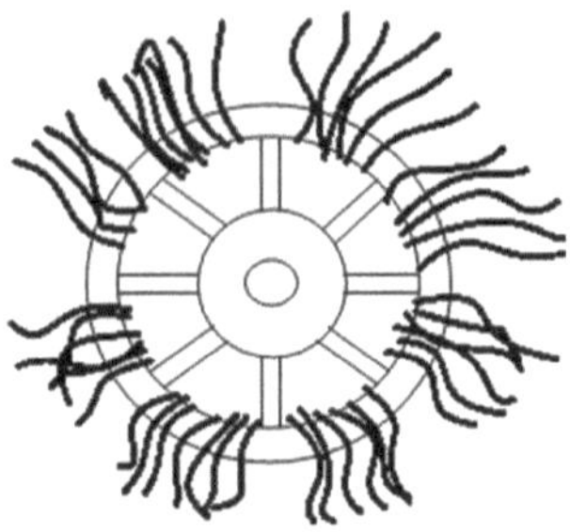

Figure 6.1.6 Configuration of semisoft media for attached growth of microbes [35].

aeration and sludge settlement both occurring in the same tank. SBR helps in providing kinetic data during the reaction period and provides important information on the rates of removal and also on the relative rate of removal of key compounds. Through evaluation of oxygen uptake rates, sulfur, and nitrogen balances, plant operators can determine which components (e.g., thiocyanates, thiocyanyanide, and phenol) will be degraded first and how thiocyanate degradation may be inhibited by cyanide concentration. As has been seen thiocyanate degradation may be totally inhibited at cyanide concentration >1 mg/L, but not due to phenol because phenol is degraded much quicker than thiocyanate [2]. In SBR, higher levels of sophistication are required for control through automatic valves and switches [50].

Attached growth mechanism Due to the attached growth mechanism, microbial mass grows like film and attaches to a solid or semisolid medium or support and acts as fixed film reactor. These fixed films may be made of plastic ring or synthetic fiber string forming a semisoft media that is not easily clogged and provides large surface area for the growth of microorganisms. Fig. 6.1.6 shows a semisoft medium for attached microbial growth. The choice of carrier or support material is a key factor in ensuring the success of a given biofilm process. Purification is achieved when the coke wastewater is brought into contact with this microbial film. As the active biomass is largely retained within the reactor, there is no need to recirculate any displaced biomass to maintain sufficient density of microorganisms, as in the case of complete mixed culture. However, the thickness of the film is critical, as the oxygen can only diffuse a certain distance through the film before being utilized, leaving the deeper areas of the film either anoxic or anaerobic.

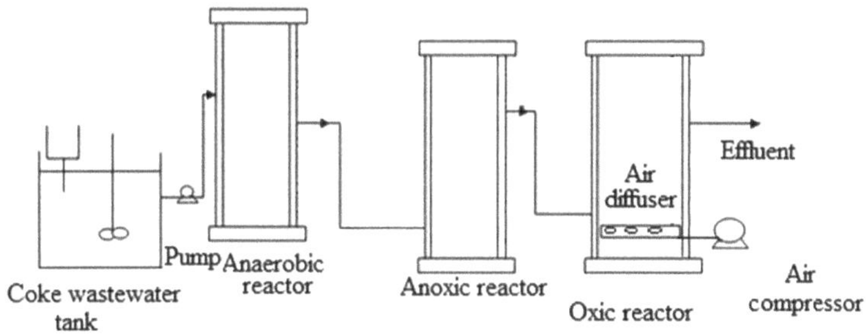

Figure 6.1.7 Schematic of anaerobic-anoxic-oxic system containing fixed biofilm for the treatment of coke wastewater [35].

Multistep treatment using a fixed biofilm system Biofilm process has proved to be reliable for organic carbon and nitrogen removal overcoming the problem of shock load effects of ASP [101]. Multistep biofilm system may be used in different combinations like anaerobic-anoxic-aerobic (A_1-A_2-O) or anoxic-aerobic (A-O) in treatment of coke wastewater and one such scheme of treatment is shown in Fig. 6.1.7 [40,43,102,103]. Total organic nitrogen (total-N) removal in an A_1-A_2-O system is found to be more comparable to A-O fixed biofilm during the treatment of coke wastewater due to an additional acidogenic stage [36]. High concentration of ammonia and organic compounds especially refractory and inhibitory organics have been treated successfully in a anaerobic-anoxic-oxic fixed biofilm system with more than 98% removal efficiency for NH_3-N and above 92% for COD [39].

Fig. 6.1.7 shows a typical multistep reactor operating under different environmental conditions packed with semisoft media. A biofilm system may be developed by combining A-O_1-O_2 (anaerobic-oxic1-oxic2) with special carriers like functional polyurethane foams (FPUFS) with a cubical structure with micro- and macropores. This type of system will have high water content capacity and porosity and can remove more than 90% COD. However, the long start-up time required for biofilm formation, difficulty in controlling biofilm thickness, and overgrowth of biofilm are some of the limitations of biofilm-mediated wastewater treatment [104,105].

Biological aerated filters In biological aerated filters (BAFs) as shown in Fig. 6.1.8, the active film grows over a support medium completely submerged in wastewater where oxygen is supplied by diffusers at the base of the reactor. High specific surface area of the medium

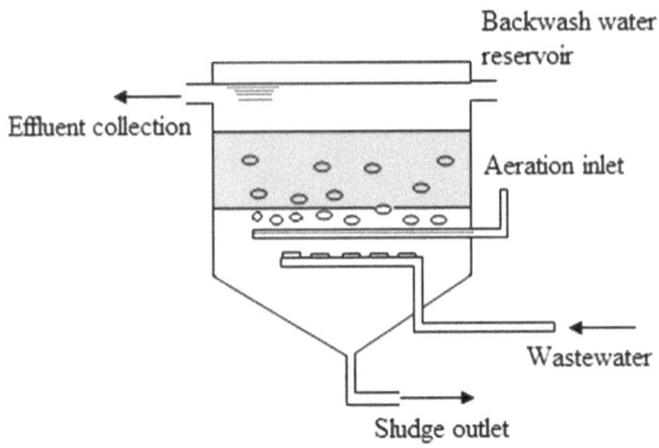

Figure 6.1.8 Submerged biological aerated filter [35].

promotes microbial growth of film with substantial thickness. These microbial films act as bioreactor as well as biofilter making the process quite efficient. In some modifications, biological activated carbon (BAC) is used in combination with Fenton's treatment. BAC functions as an adsorber, thus substantially reducing the need to replace activated carbon, making combined Fenton oxidation–BAC a cost-effective treatment process allowing recycle of the final effluent. However, the BAF requires long time for acclimatization of the microbes. High-capacity carbon is also required for effective removal of persistent compounds [57].

Aerobic fluidized bed reactor An aerobic fluidized bed (AFB) is made of small medium particles such as sand, glass or anthracite (0.2–2.0 mm diameter), with very high specific surface area compared to other media, allowing considerable biomass to develop (equivalent to mixed liquor suspended solid (MLSS) concentration up to 40,000 mg/L). Removal of phenol can be achieved at low cost in comparison to the depenalization plant by extraction-recovery system [106]. Fig. 6.1.9 shows a schematic of a typical AFB reactor. Coke wastewater is made to pass upward through a rectangular reactor containing a bed of sand or granular activated carbon at a velocity sufficient to expand the bed and thus produce a fluidized state. On fluidization the media particles provide a vast surface area for biological growth, in part leading to development of a biomass concentration approximately 5–10 times greater than that normally maintained in conventional bioreactors. The film-coated medium is periodically removed and cleaned by passing it through a hydrocyclone where high shear strips away any attached microbial growth. The cleaned

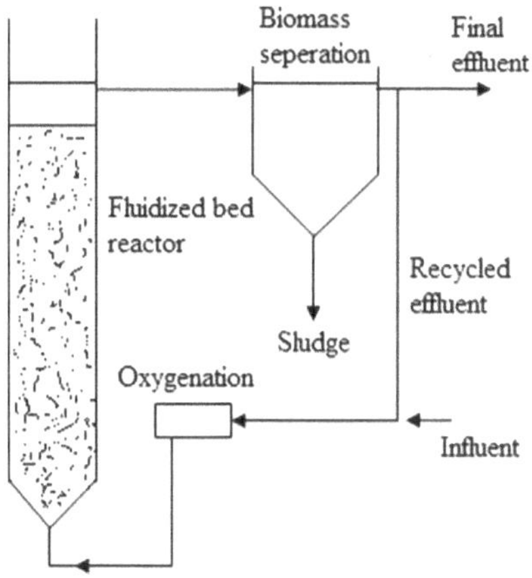

Figure 6.1.9 A typical fluidized bed reactor system [35].

medium is then returned but the sludge must be treated further before disposal. AFB is a highly compact system, although a bit expensive to run because of the need for pure oxygen and high pumping pressure.

Bioaugmented activated sludge process An activated sludge system based on bioaugmentation with specialized microorganisms can be a powerful tool to enhance the efficiency of a wastewater-treatment process through improved flocculation and settling of activated sludge. Coke wastewater has been treated with enhanced efficiency by bioaugmentation with specialized microorganisms identified as *Burkholderia picketti* in an anaerobic-anoxic-oxic (A_1-A_2-O) system. Specialized microorganisms added at different locations of the A_1-A_2-O system help optimize the total operating environment, which consists of the augmentation biomass and the different types of toxic compounds present in the coke wastewater. Bioaugmentation has been found to be effective for the treatment of metal complex cyanide by a laboratory-cultivated microorganism and cyanide-degrading yeast (*Cryptococcus humicolus*) in a fluidized-bed type process. Upgrading and special seeding of the wastewater-treatment plants by commercial-grade microorganisms may be necessary in such cases. By using the appropriate microbe cultures many toxic and recalcitrant waste can be treated successfully. Such microbes should be resistant to environmental stresses like sunlight, temperature, and predation by natural microbes. However, the use of commercially grown microbes is expensive and limited due to the problem of adaptability [107,108].

Hybrid Bioreactor Technology

Membrane-Integrated Bioreactor

Environmental regulations are likely to become more and more stringent in coming years, demanding better treatment of industrial effluent than what is required today. With the scarcity of freshwater in some regions of the world, reclamation of water for recycling is also urgently needed. Thus in addition to removal of all organic and inorganic contaminants, elimination of bacteria and viruses will have to be ensured in order to recycle effluent. Moreover, the disposal of sludge containing heavy metals is also becoming increasingly difficult and expensive. In recent years, specially developed membranes have made new membrane bioreactors (MBRs) promising alternatives to the well-known aerobic processes of wastewater treatment [109–113]. The use of submerged membranes has reduced the power consumption of MBRs significantly and hence increased the potential for the application of membranes in wastewater treatment. Moreover, UF and MF membranes with a pore size of 0.2 μm can retain bacteria as well viruses almost completely. Complete retention of sludge allows operation at much higher biomass concentrations. The higher the concentration of biomass the lower the food-to-microorganism ratio (F/M) and thus the microorganisms utilize a growing portion of the carbon content of the coke wastewater for maintenance purposes and consequently less for growth. High loading rate and significant fluctuation of influent raw wastewater can adversely affect the stability of treatment of a conventional activated sludge system, leading to unacceptable effluent quality. To comply with the increasingly stringent environmental regulations, an efficient and reliable treatment process is needed to handle this highly toxic wastewater. The MBR, regarded as an efficient wastewater-treatment process, is widely applied to treat both municipal and industrial wastewaters. The advantages of MBR over CAS are high-quality effluent even at high and fluctuant loading rates, improved treatment efficiency of refractory organic pollutants, and satisfactory pretreatment of wastewater for subsequent RO/NF stages for possible reuse. Membrane-integrated anaerobic-anoxic-oxic (A_1-A_2-O-MBR) systems are more efficient and reliable in pollutant and acute toxicity removal than controlled anaerobic-anoxic-oxic conventional activated sludge (A_1-A_2-O-CAS) systems especially at high and varying loading rates. The removal achieved in this case is 88, 100, and 100% for COD, turbidity, and NH_4^+-N, respectively [114–121].

Fig. 6.1.10 shows such a laboratory-scale anaerobic-anoxic-oxic reactor integrated with submerged MBR.

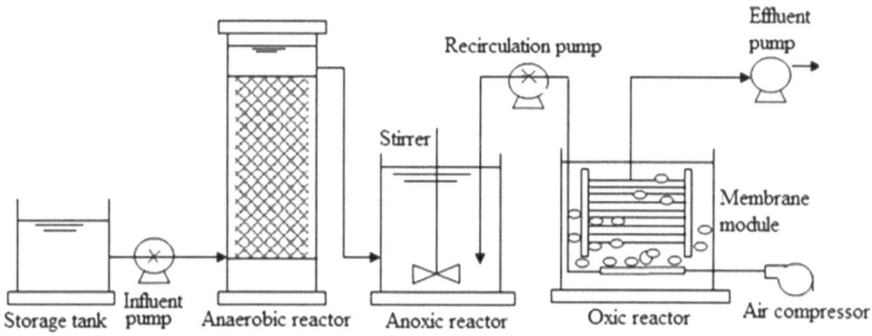

Figure 6.1.10 Schematic of anaerobic-anoxic-oxic (A_1-A_2-O) membrane bioreactor system [35].

Colloidal fractions of supernatant in suspension like polysaccharides and proteins fractions with molecular weight above 100 kDa largely determine operations of MBR. These materials are likely to accumulate in the supernatant and are responsible for flux deterioration during dead-end filtration tests. Although initially deposited particles can be removed easily by physical cleaning, subsequent cleaning is more difficult. The fouled membrane in these situations may be cleaned ex situ *by* physical cleaning (jetting the module with tap water under moderate pressure) and chemical cleaning (soaking the module in 0.05% NaClO solution for 24 hours). Improved hydrodynamics of may be considered for reducing such fouling on a significant scale [122,123].

Combined Method of Ultrasonic Irradiation, Catalytic Oxidation, and Activated Sludge

Ultrasonic energy can be effectively used to remove toxic and hazardous organic compounds from contaminated water by formation of acoustic cavitation [102,124,125]. The success of ultrasonic irradiation of the organic pollutants present in wastewater is largely determined by saturating gas, initial pollutants concentration, ultrasonic power density, and catalyst. Low COD concentration and high ultrasonic power density are the most favorable conditions for degradation of pollutants. At high-pressure and temperature, primary chemical reactions occur and result in the transient state formation during collapse of microbubles producing CO_2, H_2O, and radicals, such as hydroxyl and atomic hydrogen. Sometimes metallic ions act as catalysts that enhance degradation of the organic pollutants in wastewater by ultrasound. Compared with a single

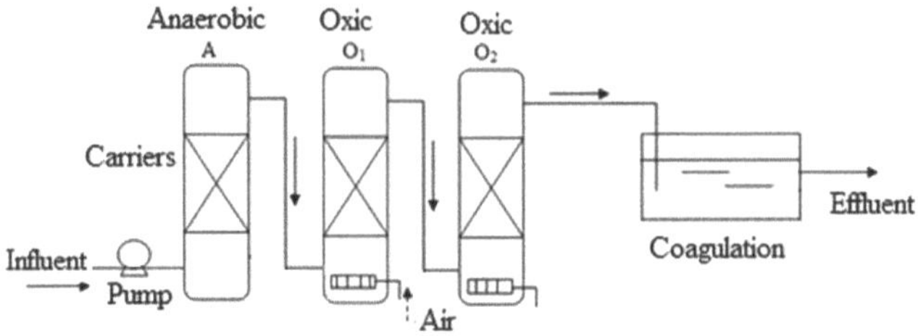

Figure 6.1.11 Schematic of anaerobic-anoxic-oxic reactors integrated with coagulation unit [35].

activated-sludge process, combining ultrasonic irradiation, catalytic oxidation, and activated sludge has the potential to greatly increase the efficiency of degradation of COD from 48 to above 95% [103,126].

Integrated Chemical-Biological Process

Normally a very high rate (up to 300%–400%) of effluent recirculation needs to be applied to ensure low toxic concentration (especially ammonium concentration) in biological reactors to eliminate the microbial inhibition caused by these agents (e.g., ammonium, cyanide, thiocyanates). However, this high recirculation ratio technically solves the problems of inhibition at the cost of volumetric loading of the reactors and results in high treatment costs as the treated volume is drastically reduced [17,61,97]. Moreover, requirement of space is also high for construction of these bioreactors. Chemical pretreatment can reduce the toxic concentration while avoiding the problem of too much recirculation. Coagulation may be adopted in solid–liquid separation and soluble-containment removal from aqueous solution. Different kinds of aluminum salts, iron salts, and inorganic/organic polymers are the mostly commonly used coagulants. Iron-containing materials of low in cost are usually adopted in ZVI (zero valent iron) processes. Oxidation–reduction coupled with a standard reduction potential of −0.440 V is formed in the ZVI reaction [43,127]. Fig. 6.1.11 shows a biological–chemical integrated scheme in which anaerobic anoxic-oxic reactors have been integrated with chemical coagulation–precipitation. The biological treatment units are multistep biofilm reactors. Due to the strong reductive capacity, ZVI can transform the structures and minimize the toxicity of hazardous contaminants. Fenton's oxidation and coagulation–flocculation–sedimentation are effective in removing many organic constit. The polyaromatic hydrocarbons present in the coke wastewater can be effectively mineralized using Fenton's reagent, but not by ozone. The high concentration of ammonium-nitrogen can be reduced by chemical precipitation of ammonium-N with equimolar concentration of magnesium and phosphate salts to form struvite, which is a slow-releasing fertilizer that facilitates the biotreatment. An integrated extraction–biodegradation technique can successfully remove COD from coking wastewater by more than 88%. Using cyclohexane and *n*-octane in the ratio of 1:1 for 5 minutes as extractants nitrogen-containing heterocyclic compounds such as pyridines and quinolones have been effectively removed from coke wastewater with enhanced biodegradability of wastewater [58,61,128,129]. Table 6.1.2 presents the performance of different coke wastewater-treatment schemes.

6.1.1.5 Conventional Treatment Technologies: Advantages and Disadvantages

Sedimentation is the most commonly used process by coke-producing industries to remove suspended solids. Coagulation and chemical precipitation is a preferred option for removing turbidity and color from the wastewater. These techniques are also capable of reducing COD and TOC considerably. Polyelectrolytes are better options than alum and produce less sludge and pose fewer problems in sludge dewatering.

Chemical oxidants such as ozone, photocatalysis, and ultraviolet (UV) are efficient in removing COD, TOC, and color. However, efficiency largely depends on the concentration of the toxic substances.

The ASP, UASB (upward flow anaerobic sludge blanket) system, and fluidized beds are widely practiced where ASP can remove 60%–90% of cyanide, 50%–70% phenol 80%–100% thiocyanate, and 70%–90% COD, respectively. High removal is achieved when two or more physicochemical are combined or with combined physicochemical and biological processes. Pilot scale data with economic evaluations significantly increase scale-up confidence.

Economic evaluation of various advanced oxidation processes to remove organic contaminants shows that some treatment processes such as ozonation, Fenton's reagent, and adsorption and membrane technology are efficient but are often expensive. While ozonation is expensive, it can remove 97% and 99% cyanide and thiocyanate, respectively in less treatment time (1 hour) than other processes [36,67–129,138–150].

Table 6.1.2 Coke wastewater treatment by different procedures [35]

No	Type[a]	Type of WW[b]	Influent[c]					Effluent (% removal)					HRT	References
			COD	CN	SCN	Phenol	NH_4^+	COD	CN	SCN	Phenol	$NH4^+$	(h)[d]	
1	Single-step ASP	synthetic	2041	–	277	230	1422	75	–	90	98	71	167	[7]
2	Two and three step ASP	real	1451	–	269	213	209	90.7	–	98.6	98.9	99.9	98	[97]
3	Ultrasonic irradiation + catalytic oxidation + + ASP	real	807	–	–	–	–	95.7	–	–	–	–	–	[124]
4	SBR	real	1900	–	315	240	300	85	–	98	99	96	115	[50]
5	Three-step ASP	synthetic	2500	27	400	1000	712	90	96	100	100	50	158	[16]
6	Two-step ASP	real	2050	15	490	217	240	85	100	100	100	97	16.7	[100]
7	Three-step ASP $A/O_1/O_2$	real	1335	26	312	160	481	63	99	97	99	98	98	[99]
8	A_1-A_2-O-MBR	real	2372	–	–	566	266	90	–	–	99.9	99.5	20	[121]
9	A_1-A_2-O-MBR	real	2372	–	–	566	266	88	–	–	–	99.7	20	[120]
10	Nanofiltration/Steam stripping	real	–	176	–	151	8700	–	88	–	90	99	–	[4]
11	Ozonation	real	1950	32	618	320	1118	–	97	99	–	–	1	[67]
12	Two-step Mn & Mg ores	real	3890	–	–	475	2213	70	–	–	99	94	1	[60]
13	A_1-A_2-O biofilm	real	860	–	–	–	232	85	–	–	–	98	19.7	[36]
14	Two-step UASB	real	2500	5	–	550	150	60	–	–	63	–	48	[130]
15	Bioaugmentation	real	300	14	–	–	–	73	–	–	–	–	–	[108]
16	Batch methanogenic fermentation	real	9100	–	–	30	800	72	–	–	–	–	–	[131]
17	Two-stage ASP	real (diluted)	22,000	18.8	175	12,375	5597	78	15	73	99	–	52	[29]
18	A_1-A_1-O fixed-bed biofilm system	real	1496	–	–	–	251	92.4	–	–	–	98.8	31	[39]

19	MBBR moving bed biofilm reactor	real	2026	–	120	414	220	81	–	94	89	93	48	[106]
20	Batch vacuum distillation	real	17,110	7.8	–	1230	408	99.7	–	–	100	100	3	[132]
21	Anaerobic and two aerobic biofilm	real	1802	–	–	496	263	91.0	–	–	100	96.8	60	[133]
22	Adsorption by activated coke	real	1600	0.1	–	1650	–	91.6	–	–	–	–	6	[48]
23	Supercritical water oxidation	real (HSCW)	24354	–	–	1490	1879	>99	–	–	>99	>99	–	[134]
24	A_1-A_2-O biofilm system	real	860	–	–	–	232	85	–	–	–	98	19.7	[135]
25	CPRT	real	200	–	–	–	520	–	–	–	–	89	1/2	[136]
26	Pulsed corona discharge	real	–	8	–	528	–	–	91.5	–	66.8	–	–	[137]
27	Biofilm reactors integrated by ZVI	real	1847	–	–	–	260	96.1	–	–	–	99.2	–	[104]
28	Membrane technology (NF)	real	2468	108	–	139	2620	94.2	94	–	85	75	–	[93]

[a]Reactor type: ASP, activated sludge process; SBR, sequential batch reactor; A_1-A_2-O, anaerobic-anoxic-oxic; A-O_1-O_2, anoxic, oxic1st, oxic 2nd; UASB, upflow anaerobic sludge bed; MBBR, moving bed biofilm reactor CPRT, chemical precipitation recycle.
[b]Type of wastewater used HSCW high strength coking wastewater.
[c]Influent: COD, chemical oxygen demand; CN, cyanide; SCN, thiocyanide; NH_4^+, ammoniacal-nitrogen; mg/L, milligram/liter.
[d]HRT, hydraulic retention time; h, hour.

Combination of methods such as Fenton's method followed by carbon adsorption and chemical oxidation followed by ammonia precipitation with addition of magnesium salt, ultrasonic irradiation, catalytic oxidation, and activated sludge are comparable to single activated-sludge processes, ultrasonic irradiation. Chemical oxidation followed by biological degradation of coke wastewater has also shown high removal efficiency. Recently used technology like two-continuous UASB system, MBBR, vacuum distillation, adsorption by activated carbon, and supercritical water oxidation have also shown high removal efficiency. Chemical precipitation with recycle technology for converting ammonium-N into valuable struvite byproduct appears technoeconomically sound. Mineral ores like manganese and magnesium ores are effective for the treatment of coke wastewater by oxidation and precipitation. Effluent from biological treatment plants still contains a high concentration of fatty acids and phenolic compounds, which can be effectively removed by methanogenic fermentation to produce gas [48,131].

Although biological treatment appears to be the least costly method for the treatment of enormous volumes of coke wastewater, pure biological treatment may require a long time to effectively treat contaminants and may not remove enough of the major contaminants to meet the permissible limits.

Biological treatment followed by chemical treatment for removal of toxic cyanide compounds has the potential to lead to greater success of the overall treatment plant. A judicious combination of chemical, biological, and membrane separation in a flat-sheet cross-flow module can help reach the goal of making hazardous coke wastewater reusable.

Precipitation of ammonia in the form of magnesium ammonium phosphate (MAP) can add substantially to the economy of the process. Recovery of a hazardous compound as a valuable fertilizer will not only make the process economically more attractive but will also help remove ammoniacal compounds very effectively without much difficulty.

Although few studies have been conducted on membrane-integrated hybrid treatment schemes, these schemes seem to be the most attractive option for fast removal of contaminants with a high degree of purification making water suitable for reuse and recycling. An integrated scheme such as this will not only save surface water from the toxic substances of the discharge of coke-making industries it will also save on consumption of freshwater and is likely to be a sustainable approach.

6.1.1.6 Introduction to Advanced Treatment Technology

A single treatment technique seems to be inadequate for effective treatment of steel and coke plant wastewater. Since environmental regulations are likely to become more and more stringent, new approaches toward achieving much higher degrees of separation—purification are needed. New approaches must attempt to recover as much resuable water as possible and eventually close the cycle and make the process/technology sustainable. In this section, we briefly touch on value addition of waste and recycling of treated water to close the loop.

Nutrient Recovery From Wastewater—Value Addition Approach

Treatment of wastewater alone is not an economically attractive option to process plants. Thus in order to make waste treatment attractive, new technologly attempts to recover some of the chemical constituents of wastewater as value-added products through chemical precipitation. For example, struvite or MAP ($MgNH_4PO4 \cdot 6H_2O$) from ammonia- or phosphate-rich wastewater (like coke wastewater, urine or animal waste, anaerobic waste, and landfill leachate) may be recovered and used as a fertilizer. In this chemical precipitation of ammonium nitrogen, magnesium salt and phosphate may be added to form MAP hexahydrate. Application of MAP in the agricultural sector is a profitable investment. 1 kg of MAP generated per day would be enough to fertilize 2.6 ha of arable land at a rate of 40 kg of phosphorus/hectare. Unlike other chemical fertilizers, application of MAP helps to create an ecofriendly environment by reducing the need for rock phosphate (rock-P). MAP application as fertilizer does not result in heavy-metal contamination of vegetables rather the higher levels of P and Mg are expected to stimulate growth of vegetables. MAP has been commercialized, and it is currently being sold to fertilizer companies in Japan. Highly soluble fertilizers are undesirable in grasslands and forests, where fertilizers are only applied once in several years. Thus a slow-releasing fertilizer would be effective in such environments. Struvite has been used commercially for ornamental plants and is known to be appropriate for use with turf, tree seedlings, and vegetable and flower boards as fertilizer. Ammonium thiosulfate and ammonium thiocyanate can also be recovered from coke wastewater by applying NF membrane. In addition to nutrients, energy in the form of methane gas has also been recovered from the high-coal wastewater during the degradation of phenol and m- and p-cresol by anaerobic methane-producing cultures in semicontinuous mode [91,138–149]. Advanced technology using this approach will be described in detail in the next chapter.

Advanced Membrane-Integrated Hybrid Treatment Approches—Closing the Loop

Membrane-integrated hybrid treatment approaches may integrate membrane separation with a biological process or chemical treatment, or combine chemical and biological treatments with membrane-based downstream purification. If such an advanced treatment option succeeds in treating wastewater up to the reusable criteria level a twofold benefit to the water resources would be gained. On the one hand, hazardous wastewater will be prevented from polluting otherwise clean surface water bodies where wastewater is normally discharged, and on the other, the process will save on freshwater consumption. It is essential to develop advanced treatment technology as conventional techniques fail to meet this reusable criteria. However, in this fast changing world, where there is "no time for waste" none of these treatment schemes is in tune with the "zero emission" standard. Although biological treatment plants are very sensitive to operational conditions, these aspects of operation normally do not receive adequate attention, often leading to insufficient management and hence failure. The innovative approaches needed to make waste-treatment plants economically attractive, technologically challenging and operationally interesting. Effective integration of slow biological remediation with fast chemical remediation and fast and highly selective membrane filtration has the potential to make coke wastewater reusable. Pressure-driven NF followed by MF can offer finishing or polishing treatment in an integrated advanced treatment scheme. Thus a sustainable treatment scheme for coke wastewater can be proposed that will integrate Fenton's treatment for cyanide removal, chemical precipitation for ammonia in the form struvite, microbial treatment for removal of phenolic compounds, MF for removal of major suspended solids, and NF for final polishing for recycling of treated water.

Wastewater is a misplaced resource and thus should be either reused or recycled through proper management. Thus a paradigm shift in wastewater management approach is absolutely essential. New approaches should be aimed at developing sustainable treatment technologies embracing the concepts of recycling, resource recovery, and reuse in addition to existing targets aimed at decreasing waste volume and reduction.

REFERENCES

[1] Bone WA. Coal and its scientific uses. FRS 1980.
[2] Ghose MK. Control of pollution by recycling and reuse of wastewater. Indian J Environ Prot 1994;12:884–7.
[3] Kim YM, Park D, Lee DS, Park JM. Instability of biological nitrogen removal in a cokes wastewater treatment facility during summer. J Hazard Mater 2007;141:27–32.
[4] Minhalma M, de Pinho MN. Development ofnanofiltration/steam stripping sequence for coke plant wastewater treatment. Desalination 2002;149:95–100.
[5] Papadimitriou CA, Dabou X, Samaras P, Sakellaropoulos GP. Coke oven wastewater treatment by two activated sludge systems. Global NEST J 2006; 8(1):16–22.
[6] Staib C, Lant P. Thiocyanate degradation during activated sludge treatment of coke-ovens wastewater. Biochem Eng J 2007;34:122–30.
[7] Vázquez I, Rodríguez J, Marañón E, Castrillón L, Fernandez Y. Simultaneous removal of phenol, ammonium and thiocyanate from coke wastewater by aerobic biodegradation. J Hazard Mater B 2006;137:1773–80.
[8] Ghose MK, Roy S. Status of water pollution due to coke oven effluent, a case study. J Indian Public Health Eng 1996;3:1–9.
[9] Chang YJ, Nishio N, Nagai S. Characteristics of granular methanogenic sludge grown on phenol synthetic medium and methanogenic fermentation of phenolic wastewater in a USAB reactor. J Ferment Bioeng 1995;79(4):348–53.
[10] González G, Herrera G, García MT, Peña M. Biodegradation of phenol in a continuous process: comparative study of stirred tank and fluidised-bed bioreactors. Bioresour Technol 2001;76:245–51.
[11] Ballinger SJ, Head IM, Curtis TP, Godley AR. Molecular microbial ecology of nitrification in an activated sludge process treating refinery wastewater. Water Sci Technol 1998;37(4–5):105–8.
[12] Fang HY, Chou MS, Huang CW. Nitrification of ammonia–nitrogen in refinery wastewater. Water Res 1993;27:1761–5.
[13] Grupta SK, Sharma R. Biological oxidation of high strength nitrogenous wastewater. Water Res 1996;30:593–600.
[14] Hirata A, Nakamura Y, Tsuneda S. Biological nitrogen removal from industrial wastewater discharged from metal recovery processes. Water Sci Technol 2001;44 (2–3):171–9.
[15] Chakraborty S, Veeramani H. Anaerobic–anoxic–oxicsequential degradation of synthetic wastewaters. Appl Biochem Biotechnol 2002;102–103:443–51.
[16] Chakraborty S, Veeramani H. Response of pulse phenol injection on an anaerobic–anoxic–aerobic system. Bioresour Technol 2005;96(7):761–7.
[17] Chakraborty S, Veeramani H. Effect of HRT and recycle ratio on removal of cyanide, phenol, thiocyanateand ammonia in an anaerobic–anoxic–aerobic continuous system. Process Biochem 2006;41:96–105.
[18] Amor L, Eiroa M, Kennes C, Veiga MC. Phenol biodegradation and its effect on the nitrification process. Water Res 2005;39:2915–20.
[19] Sexsena S, Prasad M, Amritphale SS, Chandra M. Adsorption of cyanide from aqueous solutions at pyrophyllite surface. Sep Purif Technol 2001;24:263–70.
[20] Desai JD, Ramakrishna C. Microbial degradation of cyanides and its commercial application. J Sci Ind Res 1998;57:441–53.
[21] Dash RR, Balomajumder C, Kumar A. Treatment of metal cyanide bearing wastewater by simultaneous adsorption and biodegradation (SAB). J Hazard Mater 2008;152:387–96.

[22] Gallerani P, Drake D. Wastewater management for the metal finishing industry in the 21st Century. Plat Surf Finish 1998;80:28–35.
[23] Young CA, Jordan TS. Cyanide remediation: current and past technologies. In: Proceedings of the 10th Annual Conference on Hazardous, Waste Res 1995; 104–29.
[24] Akcil A. Destruction of cyanide in gold mill effluent: biological versus chemical-treatments. Biotechnol Adv 2003;21:501–11.
[25] Pak D, Chang W. Oxidation of aqueous cyanide solution using hydrogen peroxide in the presence of heterogeneous catalyst. Environ Technol 1997;18:557.
[26] Banerjee G. Phenol and thiocyanate based wastewater treatment in RBC reactor. J Environ Eng ASCE, 122. 1996. p. 941–8.
[27] Wang RC, Kuo CC, Shyu CC. Adsorption of phenols onto granular activated carbon in a liquid–solid fluidized bed. J Chem Technol Biotech 1997;68 (2):187–94.
[28] Shivaraman N, Kumaran P, Pandey RT, Chatterjee SK, Chowdhury KR, Parhad NM. Microbial degradation of SCN^- Phenol and CN^- in a completely mixed aeration system. Environ Pollut 1985;39:141–50 (Series A).
[29] Pandey RA, Parhad NM, Kumaran P. Biological treatment of low temperature carbonization wastewater by activated sludge process—a case study. Water Res 1991;25:1555–64.
[30] Wilson GJ, Khodadoust AP, Suidan MT, Brenner RC, Acheson CM. Anaerobic/aerobic biodegradation of pentachlorophenol using GAC fluidized bed reactors: optimization of the empty bed contact time. Water Sci Technol 1998;38(7):9–17.
[31] Edgar TF. Coal processing and pollution control. Houston: Gulf Publishing Co; 1983.
[32] Qasim SR. Wastewater treatment plant: planning design and operation. Holt: Richart and Winston; 1985.
[33] Schhoeman JJ, Steyn A, Scurr PJ. Treatment using reverse osmosis of an effluent from stainless steel manufacture. Water Res 1996;30(9):1979–84.
[34] Wilf M, Alt S. Application of low fouling membrane elements for reclamation of municipal wastewater. Desalination 2000;132:11–19.
[35] Pal P, Kumar R. Treatment of coke wastewater: a critical review for development of sustainable management strategies. Sep Purif Rev 2014;43:89–123.
[36] Li YM, Gu GW, Zhao JF, Yu HQ, Qui YL, Peng YZ. Treatment of coke-plant wastewater by biofilm systems for removal of organic compounds and nitrogen. Chemosphere 2003;52:997–1005.
[37] Hoorn A. Biological wastewater treatment plant at an integrated coke oven plant. *Coke making international*. Band 4 Heft 2 1992; S: 87–90.
[38] Ghose MK. Complete physicochemical treatment for coke plant effluent. Water Res 2002;36:1127–34.
[39] Zhang M, Tay JH, Qian Y, Gu XS. Coke plant wastewater treatment by fixed biofilm system for COD and NH_3-N removal. Water Res 1998;32:591-27.
[40] Kim YM, Park D, Lee DS, Jung K, Park JM. Inhibitory effects of toxic compounds on nitrification process for cokes wastewater treatment. J Hazard Mater 2009;152:915–21.
[41] Kim YM, Park D, Lee DS, Jung K, Park JM. Sudden failure of biological nitrogen and carbon removal in the full-scale predenitrification process treating cokes wastewater. Bioresour Technol 2009;100:4340–7.
[42] Han M, Li G, Sang N, Dong Y. Investigating the biotoxicity of coking wastewater using*Zea mays* L. assay. Ecotoxicol Environ Saf 2011;74:1050–6.
[43] Lai P, Zhao HZ, Wang C, Ni J. Advanced treatment of coking wastewater by coagulation and zero-valent iron processes. J Hazard Mater 2007;147:232–9.

[44] Ghose MK, Kumar A. Impact on surface water quality due to the discharge of coal washery effluent and dispersion profile of pollutant in Damodar river. Asian Environ 1993;15(1):32–40.
[45] Vázquez I, Rodríguez J, Marañón E, Castrillón L, Alvarez M. Removal of residual phenols from coke wastewater by adsorption. J Hazard Mater 2007;147:395–400.
[46] Moussavi G, Khosravi R. Removal of cyanide from wastewater by adsorption onto pistachio hull wastes: parametric experiments, kinetics and equilibrium analysis. J Hazard Mater 2010;183:724–30.
[47] Ahmaruzzaman M, Sharma DK. Adsorption of phenols from wastewater. J Colloid Interface Sci 2005;287:14–24.
[48] Zhanga M, Zhaob Q, Baib X, Ye. Adsorption of organic pollutants from coking wastewater by activated coke. Colloids Surf., A 2010;362:140–6.
[49] Marr R, Koucar M. Recovery of ammonia from industrial wastewater. Int Chem Eng 1993;33(3):416.
[50] Maranon E, Vazquez I, Rodriguez J, Castrillón L, Fernandez Y. Treatment of coke wastewater in a sequential batch reactor (SBR) at pilot plant scale. Bioresour Technol 2008;99:4192–8.
[51] Yang M, Sun Y, Xu AH, Lu XY, Du HZ, Sun CL, Li C. Catalytic wet air oxidation of coke-plant wastewater on ruthenium-based eggshell catalysts in abubbling bed reactor. Bull Environ Contam Toxicol 2007;79(1):66–70.
[52] Kritzer P, Dinjus E. An assessment of supercritical water oxidation (SCWO): Existing problems, possible solutions and new reactor concepts. Chem Eng J 2001;83:207–14.
[53] Marrone PA, Hodes M, Smith KA, Tester JW. Salt precipitation and scale control in supercritical water oxidation-part B: commercial/full-scale applications. J Supercrit Fluids 2004;29:289–312.
[54] Du X, Zhang R, Gan Z, Bi J. Treatment of high strength coking wastewater by supercritical water oxidation. Fuel 2010; article in press.
[55] Oshima Y, Inaba K, Koda S. Catalytic supercritical water oxidation of coke waste with manganese oxide. SekiyuGakkaishi 2001;44(6):343–50.
[56] Kang YW, Hwang KY. Effects of reaction conditions on the oxidation efficiency in the Fenton process. Water Res 2000;34:2786–90.
[57] Jiang W, Zhang W, Li B, Duan J, Lv Y, Liu W, Ying W. Combined fenton oxidation and biological activated carbon process for recycling of coking plant effluent. J Hazard Mater 2011;189:308–14.
[58] Park D, Kim YM, Lee DS, Park JM. Chemical treatment for treating cyanides-containing effluent from biological cokes wastewater treatment process. Chem Eng J 2008;143:141–6.
[59] Robinson T, McMullan G, Marchant R, Nigam PR. Remediation of dyes in textile effluent: a critical review on current treatment technologies with a proposed alternative. Bioresour Technol 2001;77:247–55.
[60] Chen T, Huang X, Pan M, Jin S, Peng S, Fallgren PH. Treatment of coking wastewater by using manganese and magnesium ores. J Hazard Mater 2009;168:843–7.
[61] Li XZ, Zhao QL. Efficiency of biological treatment affected by high strength of ammonium-nitrogen in leachate and chemical precipitation of ammonium-nitrogen as pretreatment. Chemosphere 2001;44:37–43.
[62] Arslan I, Balcioglu IA. Advanced oxidation of raw and bio treated textile industry wastewater with O_3, H_2O_2/UV-C and their sequential application. J Chem Technol Biotechnol 2001;76:53–60.
[63] Sevimli MF, Sarikaya HZ. Ozone treatment of textile effluent and dyes: effect of applied ozone dose, pH and dye concentration. J Chem Technol Biotechnol 2002;77:842–50.

[64] Wang C, Yediler A, Lienert D, Wang Z, Kettrup A. Ozonation of an Azo Dye C.I. Remazol Black 5 and toxicological assessment of its oxidation products. Chemosphere 2003;52:1225–32.
[65] Soto H, Nava F, Leal J, Jara J. Regeneration of cyanide by ozone oxidation of thiocyanate in cyanidation tailings. Miner Eng 1995;8:273–81.
[66] Karageorgos P, Coz A, Charalabaki M, Kalogerakis N, Xekoulotakis NP, Mantzavinos D. Ozonation of weathered olive millwastewaters. J Chem Technol Biotechnol 2006;81:1570–6.
[67] Chang EE, Hsing HJ, Chiang CP, Chen MY, Shyng JY. The chemical and biological characteristics of coke-oven wastewater by ozonation. J Hazard Mater 2008;156:560–7.
[68] Mollah MYA, Schennach R, Parga JR, Cocke DL. Electro coagulation (EC)—science and applications. J Hazard Mater 2001;84(1):29–41.
[69] Mollah MYA, Morkovsky P, Gomes JAG, Kesmez M, Parga J, Cocke DL. Fundamentals, present and future perspectives of electro coagulation. J Hazard Mater 2004;114(1–3):199–210.
[70] Feng YJ, Li XY. Electrocatalytic oxidation of phenol on several metal-oxide electrodes in aqueous solution. Water Res 2003;37:2399–407.
[71] Rodgers JD, Jedral W, Bunce NI. Electrochemical oxidation of chlorinated phenols. Environ Sci Technol 1999;33(9):1453–7.
[72] Cañizares P, Sáez C, Lobato J, Rodrigo MA. Electrochemical treatment of 4-nitrophenol-containing aqueous wastes using boron-doped diamond anodes. Ind Eng Chem Res 2004;43(9):1944–51.
[73] Chen XM, Chen GH, Yue PL. Anodic oxidation of dyes at novel Ti/B-diamond electrodes. Chem Eng Sci 2003;58(3–6):995–1001.
[74] Panizza M, Cerisola G. Application of diamond electrodes to electrochemical processes. Electrochim Acta 2005;51(2):191–9.
[75] Cabeza A, Urtiaga AM, Ortiz I. Electrochemical treatment of landfill leachate using a boron-doped diamond anode. Ind Eng Chem Res 2007;46(5):1439–46.
[76] Awad YM, Abuzaid NS. Electrochemical oxidation of phenol using graphite anodes. Sep Sci Technol 1999;34(4):699–708.
[77] Torres RA, Torres W, Peringer P, Pulgarin C. Electrochemical degradation of p-substituted phenols of industrial interest on Ptelectrodes. Attempt of a structure reactivity relationship assessment. Chemosphere 2003;50:97–104.
[78] Fóti G, Gandini D, Comninellis C, Perret A, Haenni W. Oxidation of organics by intermediates of water discharge on IrO_2 and synthetic diamond anodes. Electrochem Solid State Lett 1999;2(5):228–30.
[79] Li XY, Cui YH, Feng YJ, Xie ZM, Gu JD. Reaction pathways and mechanisms of the electrochemical degradation of phenol on different electrodes. Water Res 2005;39(10):1972–81.
[80] Zhou MH, Dai QZ, Lei LC, Ma C, Wang DH. Long life modified lead dioxide anode for organic wastewater treatment: electrochemical characteristics and degradation mechanism. Environ Sci Technol 2005;39:363–70.
[81] Zhu X, Ni J, Lai P. Advanced treatment of biologically retreated coking wastewater by electrochemical oxidation using boron-doped diamond electrodes. Water Res 2009;43:4347–55.
[82] Strathmann H, Ho WS, Sirkar KK, editors. Membrane handbook. Van Nostrand Reinhold; 1992.
[83] Ferry JD. Ultrafilter membranes and ultrafiltration. Chem Rev 1936;18:373–455.
[84] Bowen WR, Mukhtar H. Characterisation and prediction of separation performance of nanofiltration membranes. J Membr Sci 1996;112:263–74.

[85] Peeters JMM, Noom JP, Mulder MHV, Strathmann H. Retention measurements of nanofiltration membranes with electrolyte solutions. J Membr Sci 1998;145:199–209.
[86] Schaep J, Bruggen BV, Vandecasteele C, Wilms D. Influence of ion size and charge in nanofiltration. Sep Purif Technol 1998;14:155–62.
[87] Schafer AI, Nghiem DI, Waite TD. Removal of natural hormone estrone from aqueous solutions using nanofiltration and reverse osmosis. Environ Sci Technol 2003;37(1):182–8.
[88] Diawara CK. Nanofiltration process efficiency in water desalination. Sep Purif Rev 2008;37:302–24.
[89] Hippen A, Rosenwinkel KH, Baumgarten G, Seyfried CF. Aerobic deammonification: a new experience in the treatment of wastewater. Water Sci Technol 1997;35 (10):111.
[90] Minhalma M, de Pinho MN. Integration of nanofiltration/steam stripping for the treatment of coke plant ammoniacal wastewaters. J Membr Sci 2004;242:87–95.
[91] Yin N, Yang G, Zhong Z, Xing W. Separation of ammonium salts from coking wastewater with nanofiltration combined with diafiltration. Desalination 2011;268:233–7.
[92] Korzenowski C, Minhalma M, Bernardesb AM, Ferreirab JZ, dePinho MN. Nanofiltration for the treatment of coke plant ammoniacal wastewaters. Sep Purif Technol 2011;76:303–7.
[93] Kumar R, Bhakta P, Chakraborty S, Pal P. Separating cyanide from coke wastewater by cross-flow nanofiltration. Sep Sci Technol 2011;46:2119–27.
[94] Melcer H, Nutt SG. The application of predenitrification nitrification technology for trace contaminant control. Water Sci Technol 1985;17:399–408.
[95] Richards DJ, Shieh WK. Anoxic oxic activated sludge treatment of cyanidesand-phenols. Biotechnol Bioeng 1989;33:32–8.
[96] Jeong Y-S, Chung JS. Biodegradation of thiocyanate in biofilm reactor using fluidized-carriers. Process Biochem 2006;41:701–7.
[97] Ramalho RS. Tratamiento de aguasresiduales, Editorial Reverte͂, S.A. 1996.
[98] Vázquez I, Rodríguez J, Marañón E, Castrillón L, Fernández Y. Study of aerobic biodegradation of coke wastewater in a two and three-step activated sludge process. J Hazard Mater 2006;137:1681–8.
[99] Maranon E, Vazquez I, Rodriguez J, Castrillon L, Fernandez Y. Coke wastewater treatment by a three-step activated sludge system. Water Air Soil Pollut 2008;192:155–64.
[100] Kim YM, Park D, Jeon CO, Lee DS, Park JM. Effect of HRT on the biological predenitrification process for the simultaneous removal of toxic pollutants from cokes wastewater. Bioresour Technol 2008;99:8824–32.
[101] Yang S, Yang FL, Fu ZM, Lei RB. Comparison between a moving bed membrane bioreactor and a conventional membrane bioreactor on organic carbon and nitrogen removal. Bioresour Technol 2009;100(8):2369–74.
[102] Cheung HM, Bhatnagar A, Jansin G. Sonochemical destruction of chlorinated hydrocarbons in dilute aqueous solution. Environ Sci Technol 1991;25:1510–23.
[103] Okouchi S, Nojima O, Arai T. Cavitation induced degradation of phenol by ultrasound. Water Sci Technol 1992;26(9–11):2053–6.
[104] Lai P, Zhao H, Zeng M, Ni J. Study on treatment of coking wastewater by biofilm reactors combined with zero-valent iron process. J Hazard Mater 2009;162:1423–9.
[105] Li H, Han H, Du M, Wang W. Removal of phenols, thiocyanate and ammonium from coal gasification wastewater using moving bed biofilm reactor. Bioresour Technol 2011;102:4667–73.

[106] Marvan IJ, Craig F, Sutton PM. Treatability Evaluation of coking plant effluent. Int Biodeterior Biodegradation 1992;30:313–29.
[107] Jianlong W, Xiangchun Q, Libo W, Yi Q, Hegemann W. Bioaugmentation as a tool to enhance the removal of refractory compound in coke plant wastewater. Process Biochem 2002;38:777–81.
[108] Park D, Lee DS, Kim YM, Park JM. Bioaugmentation of cyanidedegrading microorganisms in a full-scale cokes wastewater treatment facility. Bioresour Technol 2008;99:2092–6.
[109] Lubello C, Caffaz S, Mangini L, Santianni D, Caretti C. MBR pilot plant for textile wastewater treatment and reuse. Water Sci Technol 2007;55:115–24.
[110] Xing CH, Wu WZ, Qian Y, Tardieu E. Excess sludge production in membrane bioreactors: a theoretical investigation. J Environ Eng ASCE 2003;129:291–7.
[111] Liu R, Huang X, Xi JY, Qian Y. Microbial behavior in a membrane bioreactor with complete sludge retention. Process Biochem 2005;40:3165–70.
[112] Pollice A, Laera G, Saturno D, Giordano C. Effects of sludge retention time on the performance of a membrane bioreactor treating municipal sewage. J Membr Sci 2008;317:65–70.
[113] Marrot B, Barrios-Martinez A, Moulin P, Roche N. Industrial wastewater treatment in a membrane bioreactor: a review. Environ Prog 2004;23:59–68.
[114] Chiemchaisri C, Wong YK, Urase T. Yamamoto K. Organic stabilization and nitrogen removal in a membrane separation bioreactor for domestic wastewater treatment. Water Sci Technol 1992;25:231–40.
[115] Yang WB, Cicek N, Ilg J. State of the art of membrane bioreactors: worldwide research and commercial applications in north America. J Membr Sci 2006;270:201–11.
[116] Glen TD, Bruce ER, Samer A, Gianni A. Are membrane bioreactors ready for widespread application? Environ Sci Technol 2005;39:385A–408A.
[117] Qin JJ, Wai MN, Tao GH, Kekre KA, Seah H. Membrane bioreactor study for reclamation of mixed sewage mostly from industrial sources. Sep Purif Technol 2007;53:296–300.
[118] Reemtsma T, Zywicki B, Stueber M, Kloepfer A, Jekel M. Removal of sulfur organic polar micropollutants in a membrane bioreactor treating industrial wastewater. Environ Sci Technol 2002;36:1102–6.
[119] Visvanathan C, Ben Aim R, Parameshwaran K. Membrane separation bioreactors for wastewater treatment. Crit Rev Environ Sci Technol 2000;30:1–48.
[120] Zhao WT, Huang X, Lee D, Wang XH, Shen YX. Use of submerged anaerobic–anoxic–oxic membrane bioreactor to treat highly toxic coke wastewater with complete sludge retention. J Memb Sci 2009;330:57–64.
[121] Zhao WT, Huang X, Lee D. Enhanced treatment of coke plant wastewater using an anaerobic–anoxic–oxic membrane bioreactor system. Sep Purif Technol 2009;66:279–86.
[122] Zhao WT, Shen YX, Xiao K, Huang X. Fouling characteristics in a membrane bioreactor coupled with anaerobic–anoxic–oxic process for coke wastewater treatment. Bioresour Technol 2010;101:3876–83.
[123] Rahman MM, Al-Malack MH. Performance of a cross-flow membrane bioreactor (CF-MBR) when treating refinery wastewater. Desalination 2006;191:16–26.
[124] Agramomil YR, Karpukhin VF, Faymgold ZL. Ntibiotikiihimioteraplya (Russia) 1986;35:48–51.
[125] Mason TJ. Chemistry with ultrasound. UK: Published for the Society of Chemical Industry, Elsevier *Science Publishers Ltd*; 1990.

[126] Ning N, Bart HJ, Jiang Y, de Haand A, Tien C. Treatment of organic pollutants in coke plant wastewater by the method of ultrasonic irradiation, catalytic oxidation and activated sludge. Sep Purif Technol 2005;41:133–9.
[127] Cheng SF, Wu SC. The enhancement methods for the degradation of TCE by zero-valent metals. Chemosphere 2000;41:1263–70.
[128] Jainae K, Sanuwong K, Nuangjamnong J, Sukpirom N, Unob F. Extraction and recovery of precious metal ions in wastewater by polyster-coated magnetic particles functionalized with 2-(3-(2-aminoethylthio) propylthio) ethanamine. Chem Eng J 2010;160(2):586–93.
[129] Yuan X, Sun H, Guo D. The removal of COD from coking wastewater using extraction replacement–biodegradation coupling. Desalination 2012;289:45–50.
[130] Wang W, Han H, Yuan M, Li H, Fang F, Wang K. Treatment of coal gasification wastewater by a two-continuous UASB system with step-feed for COD and phenols removal. Bioresour Technol 2011;102(9):5454–60.
[131] Kuschk P, Stottmeister U, Liu YJ, Wiessner1 A, Kästner M, Müller RA. Batch methanogenic fermentation experiments of wastewater from a brown coal low-temperature coke plant. J Environ Sci 2010;22(2):192–7.
[132] Mao W, Ma H, Wang B. Performance of batch vacuum distillation process with promoters on coke-plant wastewater treatment. Chem Eng J 2010;160:232–8.
[133] Lai P, Zhao H, Ye Z, Ni J. Assessing the effectiveness of treating coking effluent using anaerobic and aerobic biofilms. Process Biochem 2008;43:229–37.
[134] Veriansyah B, Park TJ, Lim JS, Lee YW. Supercritical water oxidation of wastewater from LCD manufacturing process: kinetic and formation of chromium oxide nanoparticles. Superscript Fluids 2005;34:51–61.
[135] Lim BR, Hu HY, Fujie K. Biological degradation and chemical oxidation characteristics of coke-ovenwastewater. Water Air Soil Pollut 2003;146:23–33.
[136] Zhang T, Ding L, Ren H, Xiong X. Ammonium nitrogen removal from coking wastewater by chemical precipitation recycle technology. Water Res 2009;43:5209–15.
[137] Shao G, Li J, Wang W, He Z, Li S. Desulfurization and simultaneous treatment of coke-oven wastewater by pulsedcorona discharge. J Electrost 2004;62:1–13.
[138] Taylor A, Frazier A, Gurney E. Solubility products of magnesium ammonium phosphate. Trans Faraday Soc 1963;59:1580–4.
[139] Abbona F, Boistelle R, Lundager H. Crystallization of two magnesium phosphates, struvite and newberyite: effect of pH and concentration. J Cryst Growth 1982;57:6–14.
[140] Ohlinger K, Young T, Schroeder E. Predicting struvite formation in digestion. Water Res 1998;26:2229–32.
[141] Hao XD, van Loosdrecht MCM. Model-based evaluation of struvite recovery from P-released supernatant in a BNR process. Water Sci Technol 2006;53(3):191–8.
[142] Ronteltap M, Maurer M, Gujer W. Struvite precipitation thermodynamics in source-separated urine. Water Res 2007;41:977–84.
[143] Wilsenach J, Schuurbiers C, van, Loosdrecht M. Phosphate and potassium recovery from source separated urine through struvite precipitation. Water Res 2007;41:458–66.
[144] Pastor L, Mangin D, Barat R, Seco A. A pilot-scale study of struvite precipitation in a stirred tank reactor: conditions influencing the process. Bioresour Technol 2008;99:6285–91.
[145] European Fertilizer Manufacturers Association. Phosphorus Essential Element for Food Production 2000;9–10.

[146] Zheng F, Huang CH, Norton LD. Effects of near-surface hydraulic gradients on nitrate and phosphorus losses in surface runoff. J Environ Qual 2004;33 (6):2174–82.
[147] Ueno Y, Fujii M. Three years' experience of operating and selling recovered struvite from full-scale plant. Environ Technol 2001;22:1373–81.
[148] Munch EV, Barr K. Controlled struvite crystallisation for removing phosphorus from anaerobic digester side streams. Water Res 2001;35:151–9.
[149] Fedorak PM, Hrudey SE. Anaerobic treatment of phenolic coal conversion wastewater in semicontinuouscultures. Water Res 1986;20:113–22.
[150] Perez M, Torrades F, Domenech X, Peral J. Removal of organic contaminants in pulp effluent by AOPs: an economic study. J Chem Technol Biotechnol 2002;77 525–32.

SUBCHAPTER 6.2

Advances in Steel and Coke Wastewater-Treatment Technology

6.2.1 TURNING HAZARDOUS WASTE INTO A VALUE-ADDED BYPRODUCT

6.2.1.1 Introduction

When thousands of gallons of wastewater laden with toxic substances like ammonium nitrogen, phenolic compounds, cyanide, thiocyanate, and polynuclear aromatic hydrocarbons, are discharged into surface water bodies, they heavily degrade the water. Moreover, hydrolytic and fermentative processes involved in biological nutrient removal from such wastewater lead to gradual enrichment of water with nitrogen and phosphorus content. This in turn leads to increase in eutrophication in freshwater bodies. Eutrophication by nutrient discharge to natural water bodies results in grave consequences for aquatic life as well as for water-supply systems for industrial and domestic uses. Ammonium nitrogen is one such nutrient that can lead to water eutrophication, and can also pose a threat to the environment, aquatic life, and human health if discharged to water bodies without proper treatment. Nutrient removal from wastewater is increasingly a challenge to plant operators due to increased discharge standards by regulatory bodies and pollution control boards. Conventional biological treatment processes, although economical for treatment of enormous quantities of wastewater, often fail due to high concentration of ammonium nitrogen and the presence of toxic substances like cyanide and phenol in high concentration. Thus a new technology to recover nitrogen is needed. NH_4-N in high concentration may be separated and recovered by adding magnesium and phosphate salt, which on reaction with ammonium nitrogen, precipitate out as MAP hexahydrate, a useful byproduct. MAP, commonly known as struvite, is a white crystalline substance consisting of equal molar concentrations of magnesium, ammoniacal-N, and phosphorus. The ratio of Mg^{2+}:NH_4^+-N:PO_4^{3-} and the pH of the solution are the two main factors that determine the struvite precipitation (struvite is a premium-grade slow-releasing fertilizer because it is sparingly soluble in water and is used as a base material in the phosphate industry for making fire-resistant panels, and as a binding material in cements; efficient recovery of struvite demands proper optimization of the process). In conventional optimization, researchers usually focus on the one-factor-at-a-time approach to find the effect of one parameter on the output while keeping other parameters constant. However, this approach does not take into consideration cross-effects from other factors and leads to poor optimization that necessitates higher consumption of reagents and also results in inefficient chemical conversion. In the response surface methodology (RSM), the mutual interactions of the parameters are smoothed out through a central composite design (CCD) of experiments. RSM is very useful for developing, improving, and optimizing processes. In classical factorial design, a range of a parameters (upper limits and lower limits) cannot be determined, but in RSM with the help of CCD, a fixed range may be obtained. It is very difficult in classical factorial design to develop a mathematical model for interaction of parameters, but it is possible in RSM. Moreover, RSM is faster and less expensive than classical methods. In recent years, membrane-based technologies have gained popularity and are promising for the recycling of wastewater. Among the membrane processes, NF and RO are capable of achieving a high standard of purification due to their high selectivity for the small organic molecules and ions present in wastewater. However, treatment of ammoniacal wastewater has failed to gain momentum largely due to the fact that it relatively time-consuming and unprofitable. Thus new approaches need to be adopted that make such waste-treatment schemes attractive to plant operators. This section deals with the development of a new membrane-integrated process for struvite recovery from coke wastewater in terms of effectiveness of precipitation of NH_4^+-N as struvite from coke wastewater, the associated kinetics, optimization aspects, the effects of operating parameters on struvite precipitation, and the physicochemical properties of the produced struvite. The development an advanced technology developed by Kumar and Pal (2014) [1] combining chemical precipitation with membrane separation will be discussed here.

6.2.1.2 Chemistry of Struvite Precipitation: Theoretical background

The basic chemical reaction to form MAP hexahydrate can be expressed by [2]:

$$Mg^{2+} + NH_4^+ + PO_4^{3-} + 6H_2O \Leftrightarrow MgNH_4PO_4 \cdot 6H_2O\downarrow; pK_s = 12.6\ (25^\circ C) \tag{6.2.1}$$

As the precipitation of struvite reduces the pH of the solution, HPO_4^{2-} will take part in the reaction rather than PO_4^{3-} as follows:

$$Mg^{2+} + NH_4^+ + HPO_4^{2-} + 6H_2O \Leftrightarrow MgNH_4PO_4 \cdot 6H_2O\downarrow + H^+ \tag{6.2.2}$$

Coke wastewater generally contains a low concentration of both magnesium and phosphorus (Table 6.2.1) and thus an external source of these

Table 6.2.1 Characteristics of raw industrial wastewater (coke-oven) [1]

Major characteristics	Concentration (mg/L)
NH_4^+-N	3500
COD	3548
Phenol	85
Thiocyanate	152
Cyanide	38
Fluoride	112
Chloride	5927
Sodium	1856
Sulfate	38.8
Phosphate	56
Magnesium	31
Bicarbonate	4854
Oil & Grease	56
TDS	24,120
Conductivity (mS/cm)	12.3
Salinity	7.4
pH	9.3

salts must be added. The combination $MgCl_2 \cdot 6H_2O$ and $Na_2HPO_4 \cdot 12H_2O$ reacts with NH_4^+-N as follows:

$$MgCl_2 \cdot 6H_2O + NH_4^+ + Na_2HPO_4 \cdot 12H_2O \Leftrightarrow MgNH_4PO_4 \cdot 6H_2O \downarrow + 2NaCl \tag{6.2.3}$$

This reaction basically follows a first-order kinetic model as verified by experimental data. A slightly modified first-order kinetic model was used to calculate the kinetic constants from the experimental data obtained. The interrelationships of disappearance of a reactant ($-dC/dt$), the rate constant (K) and the reactant concentration (C) at time t minus the reactant concentration at equilibrium (C_e) may be expressed through Eq. (6.2.3) as:

$$-dC/dt = K(C - C_e) \tag{6.2.4}$$

The following linear form of the first-order rate equation was obtained by integrating Eq. (6.2.4):

$$-\ln[(C - C_e)/(C_0 - C_e)] = Kt \tag{6.2.5}$$

where C_0 is the initial concentration of the reactant. Assuming first-order kinetics, a plot of $-\ln[(C - C_e)/(C_0 - C_e)]$ against t should give a straight line with slope K and intercept equal to zero. The value of equilibrium concentration (C_e) can be obtained from the intercept with the abscissa at $-dC/dt$ equals zero when this parameter is plotted versus the NH_4-N concentration remaining in the supernatant. The possible production of struvite as a function of reaction time can be obtained by integrating Eq. (6.2.5) as follows:

$$(C_0 - C_e) - (C - C_e) = (C_0 - C_e) - (C_0 - C_e)e^{-Kt} \tag{6.2.6}$$

When terms are grouped, the following equation can be obtained:

$$C_0 - C = (C_0 - C_e)(1 - e^{-Kt}) \tag{6.2.7}$$

In Eq. (6.2.7), $(C_0 - C_e)$ represents the amount of product of reaction per liter of wastewater.

6.2.1.3 Membrane-Integrated Treatment Plant

Matrials Requirement

Standard solutions (1000 mg/L) of cyanide, NH_4^+-N, fluoride, chloride, sodium, magnesium chloride, magnesium oxide, magnesium sulfate, sodium biphosphate, ortho-phosphoric acid, calcium hydrogen phosphate, hydrochloric acid, sodium hydroxide, potassium bromide, poly vinylidene fluoride (PVDF) MF membranes, and thin-film composite polyamide NF membrane. Typical characteristics of the wastewater are presented in Table 6.2.1.

Detailed characteristics of both types of membranes as provided by manufacturers are presented in Table 6.2.2.

Continuous-Process Plant Configuration

A schematic of this type of plant is shown in Fig. 6.2.1, which consists of three continuous stirred tank reactors (CSTRs) for pretreatment, a crystallization reactor, and a struvite-separation unit consisting of flat-sheet,

Table 6.2.2 Characteristics of microfiltration and nanofiltration membrane (Sepro, USA) [1]

Characteristics	Material	Geometry	Thickness (μm)	Solute rejection (%)		Maximum Temp (°C)	Maximum pressure (bar)	Pore size
				$MgSO_4$	NaCl			
MF membrane	PVDF	Flat-sheet	100	–	–	80	–	0.45 (μm)
NF1 membrane	Polyamide	Flat-sheet	165	99.5	90.0	50	83	0.53 (nm)

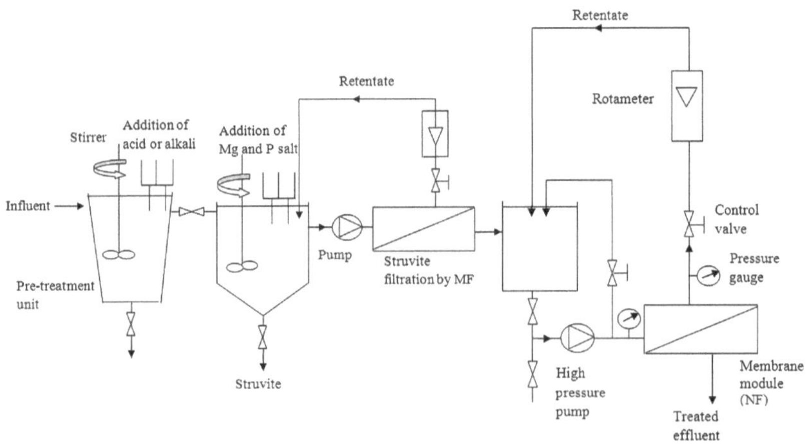

Figure 6.2.1 A detailed schematic of the membrane-integrated plant [1].

cross-flow membrane. Each CSTR has provisions for measurement of pH (using pH probe), concentration of ammonia (using ammonia electrodes), and temperature (using temperature probe). A circulation bath with a temperature controller is attached to the feed tank of the MF unit to maintain constant temperature. The pH is maintained by adding either NaOH (2M) or HCl (2N) as required. Flat-sheet, cross flow microfiltration(CFMF) and NF membrane modules with circulation pumps along with flow meters and pressure gauges are used for separation of residual chemical contaminants during continuous treatment.

Process Optimization: Response-Surface Optimization and Statistical Analysis

The chemical process of conversion of NH_4-N into struvite with the addition of magnesium and phosphate salt is optimized through batch experiments using the RSM of the Design Expert software and optimum values of the concentrations of reagents (magnesium and phosphate salt) and the pH, determined during continuous treatment study. RSM utilizes its statistical tools for analysis of experimental data obtained from definite experimental design to model and optimize any process in which several variables influence the desired response. Statistical design of experiments, estimation of the coefficients of the mathematical models, prediction of responses, and finally examination of the adequacy of the model are the major steps involved in response-surface optimization [3]. RSM helps to compute the relationships between input variables, which are called factors (X_is), and output variables, which are called responses (Y) [4]. CCD is used in the RSM as it is suitable for fitting a quadratic surface while involving a minimum number of experiments to optimize the process and to analyze interactions among the parameters [5]. Generally, the CCD consists of 2^k factorial runs with $2k$ axial runs and k_c central runs. Each variable in CCD is investigated at two levels and as the number of variables or factors (i.e., k) increases, the number of runs for a complete replicate of the design increases rapidly. The central points are used to evaluate the experimental errors and reproducibility of the data. For three factors ($k = 3$) the rotatable designs are most effective and suggested. The total number of experiments with different combinations of the three variables (ratio of Mg: NH_4^+ (0.75−2.5) and PO_4^{3-}: NH_4^+ (0.5−1.5)) and pH (7.0−10.0) was 20 ($2^k + 2k + k_c$), where k denotes the number of variables and $k_c = 6$, the number of central points for a three-variable system.

The required number of experiments was calculated from Eq. (6.2.8) as:

$$N = 2^k + 2k + k_c = 2^3 + 2 \times 3 + 6 = 20 \quad (6.2.8)$$

The coded form of five different levels for each experiment are $+\alpha$, -1, 0, $+1$, $-\alpha$. In this way the coded variables are within $-\alpha$ (minimum value) and $+\alpha$ (maximum value). The number of factors decides the value of α in the factorial portion of the design.

Eq. (6.2.9) shows the relationship between the coded and uncoded form of the variables:

$$x_i = (X_i - \overline{X_i})/\Delta x \quad (6.2.9)$$

where x_i is the coded value, X_i is actual value (uncoded form) of the *i*th factor in the uncoded form, $\overline{X_i}$ is the average of minimum and maximum values for the *i*th factor, and Δx represents the step change. An empirical model can be built to find the true relationship between the dependent variable and set of independent variables, i.e., single-response model using the RSM correspond to independent variables. The following quadratic equation may explain the behavior of the system:

$$Y = b_0 + \sum_{i=1}^{n} b_i x_i + \sum_{i=1}^{n} b_{ii} x_{ii}^2 + \sum b_{ij} x_i x_j \quad (6.2.10)$$

where Y is the predicted response, b_0, b_i, $b_{ii,}$ and b_{ij} are the offset terms, the linear effect, the squared effect, and the interaction effect, respectively, and x_i and x_j represent the coded independent variables. In this work, a second-order polynomial equation is obtained using the uncoded independent variables as follows:

$$Y = b_0 + b_1 X_1 + b_2 X_2 + b_3 X_3 + b_{11} X_{11}^2 + b_{22} X_{22}^2 + b_{33} X_{33}^2 + b_{12} X_1 X_2 + b_{13} X_1 X_3 + b_{23} X_2 X_3 \quad (6.2.11)$$

To evaluate the coefficients of the model, multiple regression analysis is used.

The necessary experiments as suggested by Eq. (6.2.8) for optimization are carried out in batch mode in 1 L glass beakers with a working volume of 500 mL. A magnetic stirrer at 125 rpm is used to mix the sample and keep the precipitate in suspension during the precipitation–crystallization process at ambient temperature for a duration of 200 min. Another set of batch experiments was conducted for the selection of magnesium and phosphate salt corresponding to the ammonium ion present in the wastewater and to arrive at the reaction rate constant. The pH of the treatment medium is adjusted by addition of hydrochloric acid (HCl, 2N) or sodium hydroxide (NaOH, 2M) depending on the pH of the medium. The samples obtained during the experiment are filtered through a 0.45 μm filter to check the residual NH_4^+-N concentration.

Plant Operation

Precipitation of NH_4-N as struvite takes place continuously in the plant (as shown in Fig. 6.2.1) under the optimized values of molar ratios of Mg^{2+}:NH_4^+, PO_4^{3-}:NH_4^+, and pH as obtained from optimization studies. After initial adjustment of pH in the pretreatment unit, settling of the suspended particles present in the raw wastewater takes place. After this the wastewater is passed to the next CSTR where magnesium and phosphate compounds are added at optimized doses and again pH is maintained during the reaction. The struvite resulting from the reaction of ammonium ion of wastewater and the added magnesium and phosphate compounds is separated by flat-sheet PVDF MF membrane under low pressure (around 2.5 bar) in a cross-flow module, which settles in the reactor. The settled struvite is collected from this reactor. Microfiltered water is subsequently passed to a second NF membrane module where almost all the charged as well as noncharged particles get separated from the stream following the Donnan-steric mechanisms of NF [6]. Finally treated water is collected as permeate from the NF membrane module under TMP of 15–16 bar.

Chemical Analysis of the Process Performance

Analysis of cyanide, NH_4-N, fluoride, sodium, chloride, and pH are done using a pH-ion meter. The electrodes are first calibrated using the previously prepared standard solutions of the respective compounds. The COD is measured using a COD Vario tube test MR (0–1500 mg/L) COD analyzer. The phenol content is determined by High Performance Liquid Chromatography (HPLC) with Zorbax SB-Phenyl column with mobile phase methanol: water (70:30) at flow rate 1 mL/minute, residence time of 3.5 minutes, and injection volume of 5 μL. Total dissolved solids (TDS), conductivity, and salinity are measured by conductivity-meter. Sulfate, bicarbonate ion, magnesium, phosphate, oil, and grease are determined following the procedures described in the standard methods (APHA, 1995). During NF percentage removal of pollutants (R%) like TDS, salinity, conductivity, cyanide, phenols, remaining NH_4^+-N, and COD are calculated using the initial concentration (C_i) of the pollutants in the feed sample and the residual concentration (C_f) in the permeate side, respectively, using:

$$R(\%) = (1 - C_f/C_i) \times 100 \quad (6.2.12)$$

Product Analysis: FT-IR Analysis for Struvite

Fourier transform infrared (FT-IR) spectroscopy is done for struvite precipitates. The collected precipitates are washed with pure water three times, dried in an oven at 45°C for 48 hours, and then analyzed by FT-IR spectroscopy. For FT-IR study, 40 mg of KBr is mixed well with 2 mg of finely ground sample for the preparation of transparent pellets and for the determination of functional groups.

SEM Analysis of Struvite

Scanning electron microscopy (SEM) with energy-dispersive X-ray analysis of struvite is done to find its crystalline structure, size of crystal, and surface composition. SEM images are taken at the required magnification at room temperature. A working distance of 25 mm is maintained and the acceleration voltage used is 15 keV, with the secondary electron image as a detector.

X-Ray Diffraction (XRD)

The collected precipitates are washed with pure water three times, dried in an oven at 45°C for 48 hours, and then analyzed by X-ray diffraction (XRD) with a Cu-target operated at 40 kV, 100 mA at a scan rate of 0.1 degree per seconds.

Thermogravimetric Analysis (TGA)

TGA of struvite is done to get the heating profile at 10°C/minute under nitrogen gas environment. The conversion of the TGA curve to its derivative mode (DTGA) is determined from the rate of mass loss curve as a function of temperature.

Monitoring Plant Performance

Selection of chemicals and effect of molar ratio of Mg: NH_4^+-N: PO_4^{3-} and pH on ammonium-ion precipitation

In the precipitation of NH_4^+-N as struvite, the molar ratios of Mg^{2+} and PO_4^{3-} with respect to NH_4^+-N concentration (1:1:1) as well as pH values have profound effects. Fig. 6.2.2A shows that the combination of

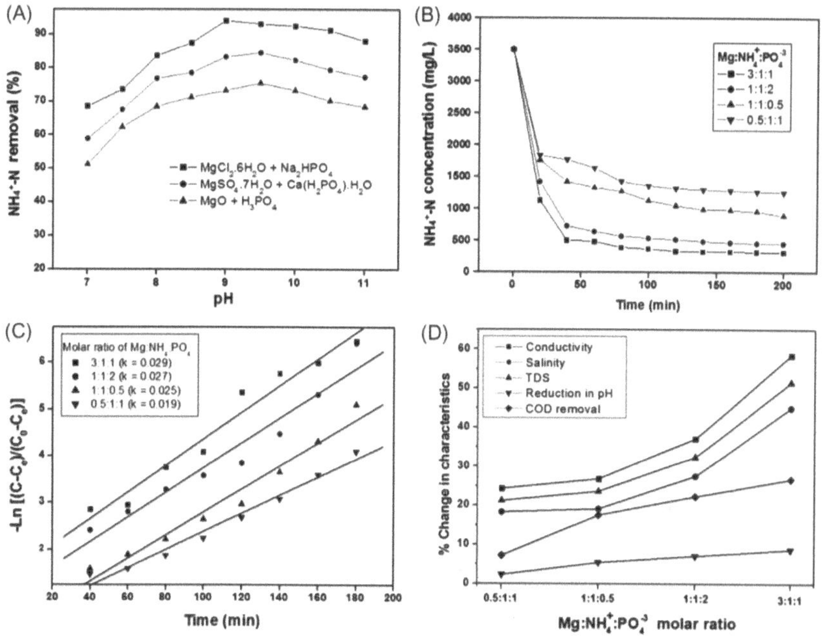

Figure 6.2.2 (A) Selection of chemicals and effect of pH on NH_4^+-N removal; (B) Variation of NH_4-N concentration with time at different molar ratio at pH 9.0; (C) Linearization of the first-order kinetic equation for determining the kinetic constants at different molar ratio at pH 9.0; (D) Effect on TDS, conductivity, salinity, COD, and pH of wastewater by using different molar ratio of Mg^{2+}:NH_4^+:PO_4^{3-} [1].

$MgCl_2 \cdot 6H_2O$, $Na_2HPO_4 \cdot 12H_2O$ is the most efficient in terms of NH_4^+-N removal from the wastewater, while the addition of $MgO + H_3PO_4$ results in the poorest efficiency. The limited solubility of MgO in the wastewater sample results in low efficiency of the NH_4^+-N precipitation. The combination of $MgCl_2 \cdot 6H_2O$ and $Na_2HPO_4 \cdot 12H_2O$ is the most efficient for NH_4^+-N removal, but this also leads to high salt concentration in the effluent compared to the other two combinations of chemicals as evident from Eq. (6.2.3). The salts produced at this stage are subsequently removed by NF membrane. Fig. 6.2.2A shows that pH plays an important role in the ammonium-ion precipitation as struvite. Under a fixed molar ratio of Mg:NH_4:PO_4^{3-}, the NH_4^+-N removal shows reflects that if the pH values are maintained at <9.0, the removal efficiency is increased with increased pH, but this decreases for pH above 9.5 in all cases of chemical combinations. In terms of thermodynamic equilibrium, hydrogen ion is released into solution during the struvite precipitation, resulting in a decrease in pH (from Eq. 6.2.2). Thus further crystallization and precipitation of struvite is inhibited due to low pH of the solution (≤7) because the ionic activity of HPO_4^{2-} decreases, whereas high pH values (≥11) encourages other precipitates like $Mg_3(PO_4)_2$ and $Mg(OH)_2$ as shown in the following:

$$3Mg^{2+} + 2PO_4^{3-} \rightarrow Mg_3(PO_4)_2 \tag{6.2.13}$$

$$Mg^{2+} + 2H_2O \leftrightarrow Mg(OH)_2\downarrow + 2H^+ \tag{6.2.14}$$

A high pH value lowers the chance of struvite precipitation. The molar ratio of $Mg{:}NH_3{:}PO_4^{3-}$ also strongly influences the NH_4^+-N precipitation among other various factors in industrial wastewater. Fig. 6.2.2B shows the variation of NH_4-N concentrates in the filtrate over time for four different types of stoichiometric ratios of magnesium and phosphate salt with NH_4-N concentration. A rapid loss of NH_4-N concentration is achieved during the first 40 minutes, which becomes more pronounced for high magnesium concentration. Moreover, the increase in molar ratio contributes to the increase in ammonium-ion removal and causes a decrease in equilibrium concentration (C_e), indicating that the reaction is more complete. As can be seen in Fig. 6.2.2C a plot of $-\ln[(C - C_e)/(C_0 - C_e)]$ versus time produces a straight line with different slopes and intercept equal to zero in all four different combinations of chemicals. The straight lines suggest that the proposed first-order kinetic model corroborates well with the experimental data. The values of the linear regression coefficients are greater than 0.94 in all cases. The values of the slopes are equivalent to the reaction constants of NH_4^+-N removal according to Eq. (6.2.5) and the values of the kinetic constants of final product formation (struvite), according to Eq. (6.2.7). Fig. 6.2.2D shows the effect of dosing of magnesium and phosphate salt concentration (with respect to NH_4-N concentration) on the percentage increase of conductivity, salinity, TDS, and percentage decrease in pH and COD. Eqs. (6.2.2) and (6.2.3) indicate that the produced hydrogen ion (H^+) and NaCl led to lowering of pH and enhancement of TDS, salinity, and conductivity. The molar ratio of $Mg^{2+}{:}NH_4^+\text{-N}{:}PO_4^{3-}$ has a similar positive effect on reduction of COD as on removal of NH_4-N. The maximum efficiency of COD reduction achieved was 27% at a molar ratio of 3:1:1 for $Mg^{+2}{:}NH_4\text{-N}{:}PO_4^{3-}$. The total COD removal efficiency depends on the amount of Mg as it flocculates particulate organic matter in the wastewater. Total reduction in COD is not as high after the chemical precipitation reaction compared to the corresponding NH_4^+-N removal in the experiments. This implies that struvite precipitation has significant selectivity to remove NH_4^+-N from the wastewater.

Response-Surface Optimization of Struvite Precipitation Using Design-Expert Software

Table 6.2.3 presents the results of struvite precipitation using CCD for different molar ratios (minimum and maximum levels) of Mg:NH_4^+ (0.75 to 2.50), PO_4^{3-}:NH_4^+ (0.5–1.5), and pH (7.0–10.0) for a fixed NH_4-N concentration of 3500 mg/L. These ranges of variables parameters are fixed based on the data reported in literature in various studies of NH_4^+-N removal through struvite precipitation. In determining the interrelationships of those variables, a second-order polynomial equation was fitted to the experimental data of NH_4-N removal (%). From fit summary section in design, the model *F*-values (108.58) imply the model is significant. Value of *P* (0.0001) in this case being less than 0.05 also indicates

Table 6.2.3 Experimental response matrix under suggested operating conditions by central composite design [1]

Std	run	Mg:NH_4^+	PO:NH_4^+	p	NH_4-N (%removal)
12	1	1.63	1.84	8.50	94.7
10	2	3.10	1.00	8.50	98.7
4	3	2.50	1.50	7.00	80.8
14	4	1.63	1.00	11.02	78.4
1	5	0.75	0.50	7.00	31.3
8	6	2.50	1.50	10.00	98.5
19	7	1.63	1.00	8.50	93.4
13	8	1.63	1.00	5.98	41.5
2	9	2.50	0.50	7.00	65.5
17	10	1.63	1.00	8.50	93.5
3	11	0.75	1.50	7.00	38.4
20	12	1.63	1.00	8.50	93.5
16	13	1.63	1.00	8.50	93.6
7	14	0.75	1.50	10.00	66.1
6	15	2.50	0.50	10.00	73.5
9	16	0.15	1.00	8.50	33.1
15	17	1.63	1.00	8.50	93.4
18	18	1.63	1.00	8.50	93.5
5	19	0.75	0.50	10.00	54.2
11	20	1.63	0.16	8.50	57.9

Table 6.2.4 Statistical parameters obtained from ANOVA for the regression models [1]

Response	R^2	Adjusted R^2	C.V. (%)	S.D.	A.P.
NH_4-N removal	0.99	0.96	4.44	3.27	31.5

that the model term is significant. The final regression equation developed through analysis of variance (ANOVA) shows the empirical relationship among the target variable (NH_4^+-N removal) and the three operating conditions or variables.

The equation in terms of coded factors is represented by Eq. (6.2.15) and the statistical parameters obtained from the ANOVA for this regression model are listed in Table 6.2.4.

Final equation in terms of coded factors:

$$\begin{aligned}\text{Ammonium-N removal} = {} & 93.56 + 17.47X_1 + 8.87X_2 + 10.13X_3 \\ & + 2.66X_1X_2 - 3.11X_1X_3 + 1.81X_2X_3 \\ & - 10.23\,X_1^2 - 6.55X_2^2 - 12.33X_3^2\end{aligned} \tag{6.2.15}$$

where X_1, X_2, and X_3 represent the three variables Mg:NH_4^+-N, PO_4^{3-}:NH_4^+, and pH, respectively. The quality of the model is evaluated based on the correlation coefficient R^2 and standard deviation value. The closer the value of R^2 to unity, the smaller the standard deviation and the more accurate the response. The R^2 value for NH_4-N removal is found to be 98.99% in this case. The value of the adjusted determination coefficient is in reasonable agreement with the predicted R^2 in this response, which implies the model is significant. The effects of molar ratio Mg:NH_4^+, PO_4^{3-}:NH_4^+, and pH are highly significant as the P values <0.0001 in this cases. The high significance of the model is also established in the plot of calculated values against the experimental values of NH_4-N removal % (Fig. 6.2.3).

Clustering of points around the diagonal line show the capability of the model to predict the experiment. The coefficient of variance (CV) should not be more than 10%, and is defined as the percent ratio of the standard error of the estimate and the mean value of the observed responses. This is a measure of the reproducibility of the model [6]. Thus the low CV values in this model (4.44%) indicate good reproducibility.

Adequate precision is a measure of the range in the predicted response relative to its associated error or, in other words, a signal-to-noise ratio,

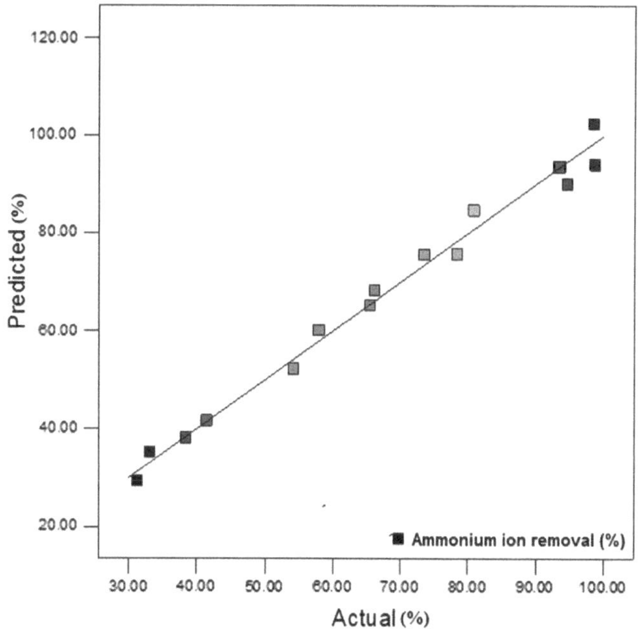

Figure 6.2.3 Distribution of experimentally determined values versus statistically predicted values of NH_4^+-N removal (%) [1].

and should be 4 or greater [7]. The present system shows an adequate precision value of 31.5.

Fig. 6.2.4 shows a 3D representation of the response-surface modeling reflecting the effects of molar ratio Mg:NH_4^+, PO_4^{3-}:NH_4^+, and pH on NH_4-N removal after 2 hours of reaction. As a general trend, it is observed that the effects of a molar ratio of Mg: NH_4^+ and PO_4^{3-}:NH_4^+ on removal of NH_4-N is pH-dependent. Precipitation of struvite was difficult at very low pH ($\leq$7) or at very high pH ($\geq$10.5). At high pH, Mg is converted into $Mg(OH)_2$ or $Mg_3(PO_4)_2$ salt. Maximum removal of NH_4-N is obtained at an optimum molar ratio of 1.6 for Mg: NH_4^+ and 1.0 for PO_4^{3-}:NH_4^+. Some optimized solutions with different criteria are suggested by the software. From those solutions, one solution is taken as follows: at pH 9.1, molar ratio Mg: NH_4^+, and PO_4^{3-}:NH_4^+ (1.6 and 1.0, respectively), while the expected NH_4-N removal is 95%. As the selected optimum criteria for variables are not among the 20 experiments previously designed by CCD and are assumptions, experiments with selected criteria are performed at a shaker flask level where 95% removal of NH_4-N is achieved along with production of 10 g/L of struvite. Thus the

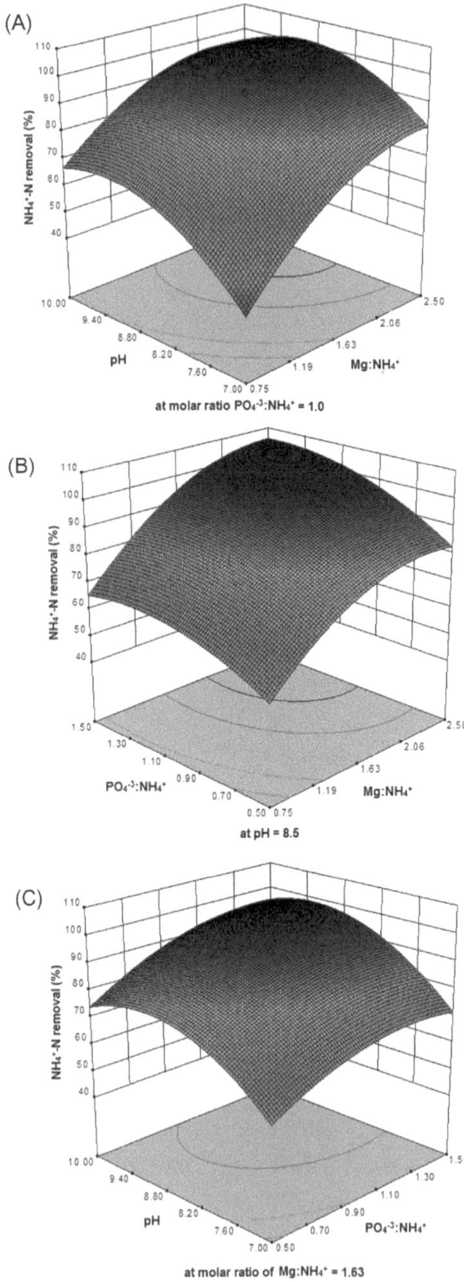

Figure 6.2.4 Response-surface plot showing the removal % of NH_4^+-N (A, B, C) with variable parameters: molar ratio of Mg:NH_4 and PO_4^{3-}:NH_4 with pH [1].

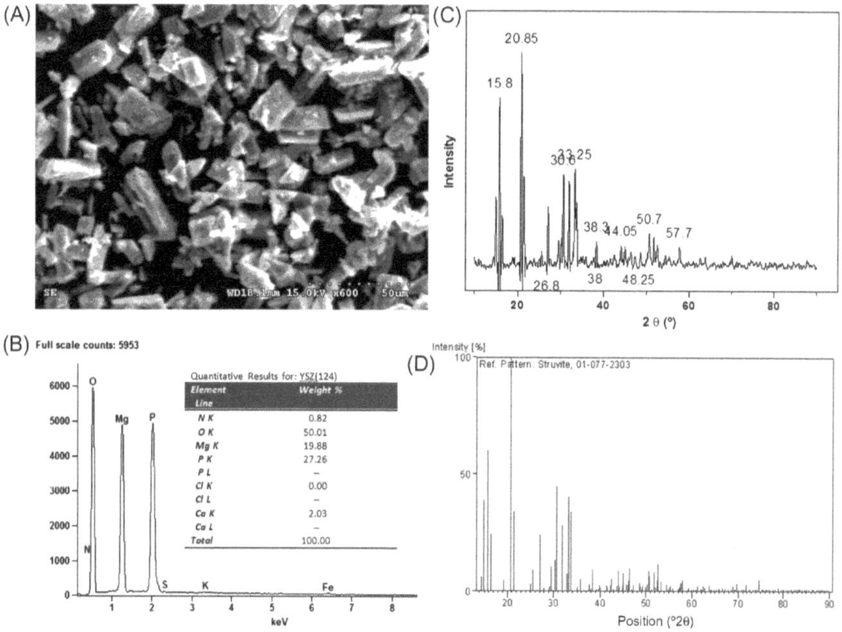

Figure 6.2.5 Surface analysis of struvite crystal during the chemical precipitation of NH_4^+-N: (A) Scanning electron microscopy analysis of struvite at two magnification; (B) Elemental composition analysis by SEM–EDS of struvite; (C) XRD pattern of the struvite crystal; and (D) Reference XRD pattern of struvite (JCPDS 01-077-2303) [1].

predicted response is in close agreement with the actual experimental observation. The optimum values thus obtained are used in continuous mode operation to control the doses of reagents.

Monitoring the Product Quality by SEM-EDS, XRD, TGA, and FT-IR

Struvite crystals are collected during the continuous treatment of coke wastewater when precipitation of ammonium ion with the optimized values (by RSM) of molar ratio of Mg^{2+}: NH_4^+ (1.6), PO_4^{3-}:NH_4^+ (1.0), and pH (9.1). The content of struvite in precipitates is confirmed through SEM image, SEM-EDS (Scanning electron microscopy-Energy-dispersive spectrometry), FT-IR, XRD, and TGA.

Microphotography and elementary chemical composition of precipitates (struvite) are determined using SEM–EDS. SEM image analysis shows that the size of the crystal is nonuniform (40–60 μm length) as illustrated in Fig. 6.2.5A. EDS analysis shows that the surface composition of the precipitates contains high amounts of O, P, and Mg as shown in

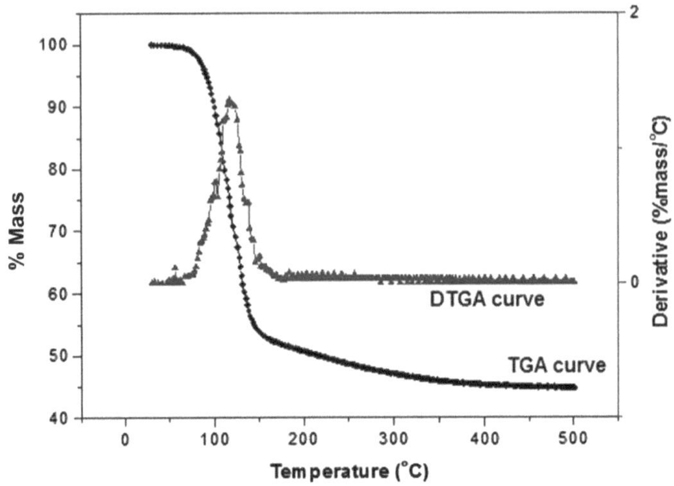

Figure 6.2.6 TGA and DTA curves of the harvested struvite precipitates at heating rate 10°C/minute [1].

Fig. 6.2.5B. EDS analysis indicates that the phosphorus atomic concentration is equal to the magnesium concentration. The EDS results of the struvite match the standard struvite pattern very well [8]. The morphology of the precipitated crystals is characterized by XRD. As shown in Fig. 6.2.5C and D, the XRD analysis indicates that the prominent characteristic peaks of the precipitate are similar to that of the pattern of the reference struvite (JCPDS PDF# 01-077-2303).

The curves of Fig. 6.2.6 were obtained from TGA and DTG analysis for struvite at heating rate 10°C/minute. As indicated by the data mass loss begins at around 55°C and is really complete when the temperature exceeds 250°C. At this temperature, about 52.2% of the original mass loss occurs. The following decomposition reaction for struvite shows the mass loss:

$$MgNH_4PO_4 \cdot 6H_2O \rightarrow MgHPO_4 + NH_3 \uparrow + 6H_2O \uparrow \qquad (6.2.16)$$

Fig. 6.2.6 shows a single peak attained at 102°C for the DTGA curve for struvite. The thermal decomposition occurs at 102°C, and the observed weight loss of 52.2% matches the literature values [9].

Fig. 6.2.7A and B illustrates the FT-IR spectrum (500–4000 cm^{-1}) of the precipitate obtained during ammonium-ion precipitation before and after TGA, respectively. The FT-IR analysis (Fig. 6.2.7A) shows that the infrared spectrum of the precipitates is close to that of the struvite as elucidated elsewhere [9]. The water-stretching broadband is observed at

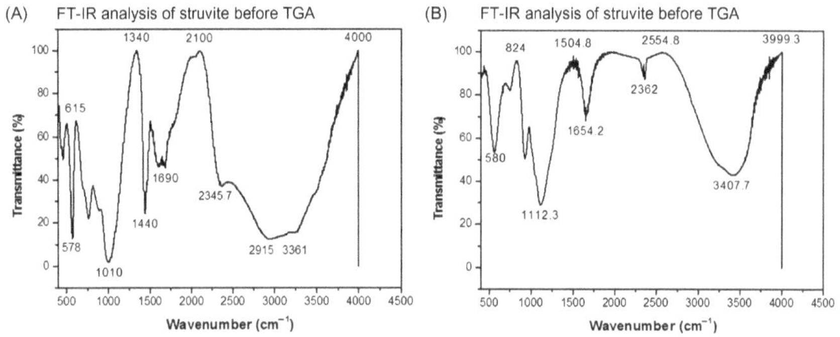

Figure 6.2.7 FT-IR spectral analysis of the harvested struvite precipitates (A) Before TGA and (B) After TGA [1].

3361–2915 cm^{-1}, which indicates the presence of crystalline hydrate. After TGA high-intensity broad band at 2915–3361 cm^{-1} is shifted to low intensity band at 3407–3650 cm^{-1}, and is assigned to adsorbed water [9]. The water-phosphate hydrogen bonding is attributed to the 2345 cm^{-1} band. The band at 1690 cm^{-1} is assigned to the Hydrogen Oxygen Hydrogen (HOH) deformation of water; other two bands 1440 and 1340 cm^{-1} correspond to the Hydrogen Nitrogen Hydrogen (HNH) deformation modes of the NH_4 units. The band at 1010 cm^{-1} is ascribed to the water NH_4 rocking modes. The two bands at 615 and 578 cm^{-1} are observed in the infrared spectrum and are assigned to the υ_4 bending modes of the PO_4 units. These bands are either not present or shifted in the thermally treated struvite as shown in Fig. 6.2.7B.

Effect of Pressure and Struvite Concentration on Microfiltrate Flux Performance

Fig. 6.2.8A illustrates the different TMP profiles for permeate fluxes with variation of cross-flow velocities (CFVs) as observed during MF with the PVDF membrane. The CFV is the ratio of volumetric flow rate of fluid through the membrane module (m^3/s) to the cross-sectional area of the inlet pipe of that module (m^2). The maximum flux 825 L/(m^2 hour) is observed at 3 bar TMP at a CFV of 1.25 m/second. The relationship between permeate flux and TMP is observed to be sigmoid. Generally, an increase in CFV reduces membrane fouling by higher sweeping action and thus flux also increases up to a certain level; further increase in cross-flow rate (CFR) induces higher membrane fouling due to higher hydrostatic pressure when adsorption on membrane surface and pore blocking

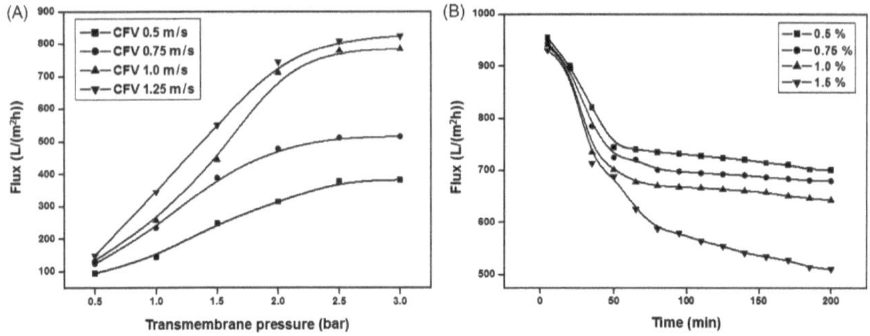

Figure 6.2.8 (A) The effect of increasing TMP on microfiltrate permeate fluxes of struvite solution at different cross-flow velocity; (B) Influence of struvite concentrations on membrane (MF) performance at TMP 2.5 bar and CFV 2.5 m/second [1].

become dominant. The effect of struvite concentration of synthetic feed on membrane performance is shown in Fig. 6.2.8B.

Four different concentrations of struvite (0.5, 0.75, 1.0, and 1.5 wt%) solution are used at TMP 2.5 bar and CFV 2.5 m/second. Initially, the membrane flux rapidly decreases due to precipitation of struvite particles and formation of a cake layer on the membrane surface, which block the pore of the membrane. The flux approximately reaches a steady-state value after 65 minutes except for the feed with 1.5 wt% struvite concentration. Increased struvite concentration marginally reduces flux. The maximum flux loss (45%) is observed after 200 minutes of separation runs with feed of 1.5 wt% struvite concentration. Thus the optimized values of TMP, CFV, and struvite concentration of the feed are 2.5 bar, 2.5 m/second, and 0.75–1.0 wt%, respectively.

Nanofiltration of the Major Chemical Contaminants Present in Wastewater: The Final Polishing Step

NH_4-N removal through struvite precipitation results in increased values of TDS, salinity, and conductivity due to the introduction of new ionic species and salts in the solution as can be seen in Eq. (6.2.3). The huge amount of salt (chloride and sodium ions) enhances the concentration of TDS, salinity, and conductivity. Cyanide, phenol, remaining NH_4-N, and COD are also present in the wastewater after ammonium-ion precipitation as indicated in Table 6.2.5.

MF can only remove particles bigger than 0.45 μm like struvite, but not the ionic species that should be removed prior to its disposal to any water body. NF plays a significant role in the removal of these phenolic

Table 6.2.5 Characteristics of wastewater after chemical precipitation of NH_4^+-N and final treatment (membrane separation) [1]

Major characteristics (mg/L)	Influent after NH_4^+-N precipitation	Final effluent after NF1 membrane	Permissible limit
Cyanide	35	1.7	<0.2
Phenol	80	3.2	<5.0
Ammonium-N	150	37	<30
COD	2750	206	<250
Fluoride	110	0.23	<1.5
Chloride	16,458	558	<1000
Sodium	3541	35	<60
Bicarbonate	4148	567	–
TDS	42,800	2140	<2100
Oil & Grease	31.5	N.D	<10
Conductivity (mS/cm)	39.56	0.93	–
Salinity	26.9	0.86	–
pH	9.5	7.5–8.0	6.5–9.0

intermediate compounds (COD), ionic species and salts (TDS, salinity, and conductivity), and cyanide in the final stage of treatment of coke wastewater. Polyamide composite nanofiltration membranes (NF1) fitted with a largely fouling-free cross-flow module do the filtration job at an optimized pressure of 15 bar and CFR of 800 L/hour. At pH values >7.0, most of the polyamide composite NF membranes have negative zeta potential [10]. The pH of feed solution is maintained at 10.0, so that all pollutants such as cyanide, phenol, etc., have a positive-negative charge. Due to the charged nature of the NF membrane, solutes with an opposite charge than the membrane (counterions) are attracted, while solutes with a similar charge (coions) are repelled. Apparent pore size of polyamide NF membranes can also vary with solution pH. At higher pH values, the addition of sodium hydroxide leads to an increase in osmotic pressure and ionic strength, thus reducing the membrane permeability and increasing rejection. Moreover, functional groups such as carboxyl and hydroxyl groups present on the surface of the membrane become deprotonated at high pH. High pH favors increased thickness of diffuse double layer of charged functional groups over the surface of membrane and thus reduces the apparent pore size and results in greater rejection of charged solutes (Braghetta et al., 1997). Struvite precipitation results in an effluent with 42,800 mg/L TDS, 26.9 salinity, and 39.56 mS/cm conductivity reflecting an increase of 77%,

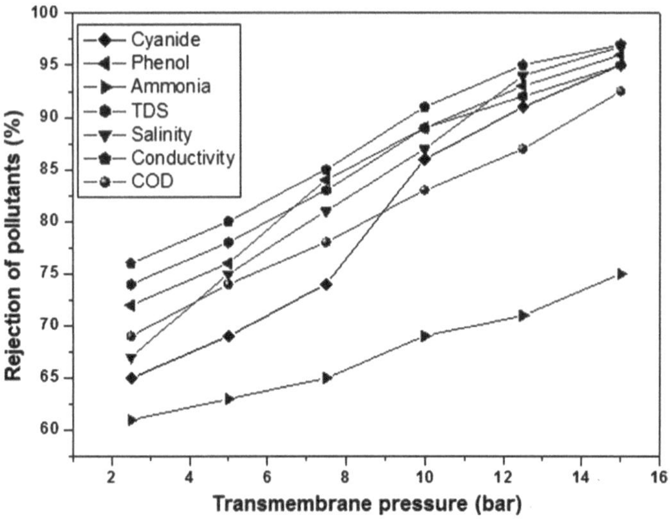

Figure 6.2.9 The effect of applied pressure on other pollutant rejection percentage for NF-1 membrane. Operating conditions: TMP 15 bar, cross flow rate 800 L/hour, pH 9.5, and temperature 35°C [1].

263%, and 220% of these three parameters compared to those in initial raw wastewater. Fig. 6.2.9 shows the pattern of cyanide, phenol, NH_4-N, COD, TDS, conductivity, and salinity rejection with increase in applied pressure. More than 90% of these pollutants can be reduced by NF1 membrane at a pressure of 15 bar and a CFR of 800 L/hour.

The solution-diffusion mechanism and Donnan-exclusion mechanism that apply to NF play major roles in the rejection of charged pollutants. The high TDS, salinity, and conductivity are due to charged inorganic ions such as bicarbonate ions, Cl^-, SO_4^{2-}, etc., which are rejected in contact when they come into contact with the negatively charged NF-1 membrane. The solution–diffusion mechanism refers to the diffusion of solution through NF membrane. In solution–diffusion mechanism T, solute flux and solvent flux are uncoupled and hence an increase of solvent flux with an increase in TMP does not result in an increase of solute flux. Rather, an increase of solvent flux obstructs the transport of solute through the membrane. An increase in TMP leads to an increase in solvent flux. This means that solute rejection will increase with an increase in TMP. In the case of removal of residual NH_4-N by membrane, the pore size of the NF-1 membrane plays a significant role. Table 6.2.5 shows the major characteristics of water after membrane separation against the discharge limits with respect to such characteristics.

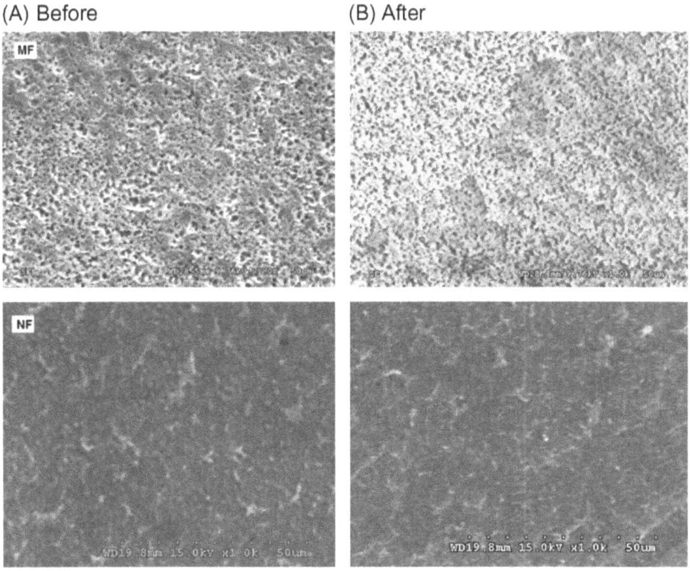

Figure 6.2.10 Surface SEM images of PVDF MF and NF membranes: (A) before and (B) after cleaning with NaOCl [1].

Chemical Cleaning of Fouled PVDF MF and NF Membrane

During separation of struvite membrane fouling by the minor components or contaminants present in the feed side may change the physical or chemical properties of the membrane. If the membrane module is operated in dead-end mode, CPs build up rapidly, resulting in rapid decrease in flux. However, build of CP may be substantially reduced due to operation of the membrane module in cross-flow mode as the fluid travels parallel to the surface of the membrane imparting a sweeping action on the membrane surface and thus leaving very little room for the formation of the CP layer. However, this fouling problem cannot be completely ignored. Chemical cleaning with HCl, NaOH, and NaOCl as acidic, alkaline, and alkali oxidizing agent, respectively, is the most commonly used method to obtain flux recovery. Out of the three chemical agents sodium hypochlorite solution (0.01 M) shows the highest flux recovery [11]. The used membrane is not regenerated, but is only rinsed with hypochlorite solution (0.01 M) when the flux falls drastically. Normally when rinsing with water or chemical cleaning fails to revive the flux, the membrane is discarded. Fig. 6.2.10 shows the surface characterization of fouled membrane of MF and NF before use and after cleaning with sodium hypochlorite.

LIST OF ABBREVIATIONS

ANOVA	Analysis of variance
C	Concentration of ammonium-N at time t
CCD	Central composite design
C_e	Equilibrium concentration of ammonium-N
C_f	Final concentration of pollutant
CFV	Cross-flow velocity
C_i	Initial concentration of pollutant
COD	Chemical oxygen demand
CSTR	Continuous stirred tank reactor
CV	Coefficient of variance
DTGA	Derivative thermogravimetric analysis
EDS	Energy-dispersive spectrometry
FT-IR	Fourier transforms infrared
k	Number of variables
K	Reaction rate constant
k_c	Number of central points
MAP	Magnesium ammonium phosphate hexahydrate
NF	Nanofiltration
PVDF	Polyvinylidene fluoride
RSM	Response surface methodology
SEM	Scanning electron microscopy
t	Time
TDS	Total dissolved solids
TGA	Thermogravimetric analysis
TMP	Transmembrane pressure
$\overline{X_i}$	Average of minimum and maximum value
X_i	Actual value of the ith factor
$X_{i,s}$	Factors
XRD	X-ray diffraction
Δx	Step change
Y	Response

6.2.2 TRANSPORT MODELING AND ECONOMIC EVALUATION OF AN ADVANCED MEMBRANE-INTEGRATED HYBRID TREATMENT TECHNOLOGY

6.2.2.1 Introduction

Effective treatment of coke wastewater has for long remained a challenge to the scientific community. Since the early 1970s wastewater-treatment processes have been modeled in an effort to address water scarcity and related issues. For example, the International Water Association has developed models such as the ASP. However, these models have been used primarily for carbon oxidation, nitrification, denitrification, and biological phosphorus removal in domestic wastewater treatment and have limited use in industrial wastewater treatment. In the area of separation and purification, membrane-integrated novel processes are now emerging as better quality technologies for treated water with highly selective membrane. However, membrane-integrated treatment models are not as common. In most highly water-intensive industries such as in coke-making steel plants, coal-based power plants, ceramic plants, and metallurgical industries aqueous effluent containing high concentration of ammonium-N, cyanide, and phenolic compounds is produced [12–14]. Effective treatment of enormous quantities of complex wastewater is still lacking today in the absence of scale-up confidence.

Among the membrane-based processes, NF appears to be an adequate solution due to the decrease of osmotic pressure associated with solutes partial retentions, which is useful for the confinement of multicomponent solutions, like coke-wastewater containing cyanide, phenol, and NH_4^+-N (Kumar and Pal, 2013a). NF membranes, with operational properties of both UF and RO membranes, have the potential to separate pollutants at lower operating pressure than RO and with better rejection than UF. Separation of impurities by NF membrane may also exploit both steric (sieving) and Donnan (electrical) mechanisms depending on the characteristics of the impurities.

This section describes the development of a dynamic model using NF as a fractionation technique for coke wastewater for generating two streams: a cyanide- and phenol-enriched stream on the retentate side and NH_4^+-N on the permeate side. The NF membrane module is operated in flat-sheet, cross-flow mode, and treatment of the cyanide- and phenol-enriched stream generally involves a chemical technique while the NH_4^+-N-enriched stream may be recovered as magnesium ammonium phosphate (MAP, $MgNH_4PO_4$ $6H_2O$).

6.2.2.2 Theoretical Background of the Model

NF membranes separate solutes from the solution using both steric (sieving) and Donnan (electrostatic) mechanisms depending on whether the solutes are charged or uncharged, which can be explained by continuum hydrodynamic models such as the one originally proposed by Ferry (1936) [15] and ENP model, respectively. The modified ENP equation was developed by Bowen and Welfoot [16] as a linearized model for the determination of ionic flux (Na^+) from NaCl solution, which is passed through an NF membrane. The ENP and CP models can adequately describe the fouling behavior of NF membrane. The model discussed here has been developed based on the following assumptions for describing transport through NF membrane in the context of separation–purification of industrial wastewater rather than synthetic solutions [17].

Model Assumptions

1. The effective membrane-charge density (X_{DC}) is constant throughout the membrane and is mainly controlled by the concentration of the solute and the pH of the feed solution.
2. The membrane consists of a bundle of identical straight cylindrical pores with uniform radius (r_p) and length Δx (with $\Delta x >> r_p$).
3. At the interface of the membrane, the steric effect in the Donnan portioning is negligible.
4. Inside the membrane, the solute concentration and electric potential are all defined in terms of radially averaged quantities.
5. The solvation barrier energy (ΔW_i) and osmotic-pressure difference ($\Delta\pi$) may be assumed to be insignificant.

Model Development [17]

The mathematical model is developed based on these assumptions. The flux for charged particle through the NF membrane can be measured using ENP equation. That modified ENP equation can describe transport of negatively charged ions for cyanide and phenol rejection through NF membrane may be expressed as:

$$J_{s,i} = K_{H,i} C_{m,i} V_s - (D_{s,i}) \frac{dc_m}{dx_m} - \frac{z_i C_{m,i} D_{s,i} F}{RT} \frac{d\psi_s}{dx_m} \tag{6.2.17}$$

The flux (J) of ion i is the sum of the fluxes due to convection, diffusion, and electro migration. Solute flux during NF may also be expressed as:

$$J_{s,i} = V_s \times C_{p,i} \tag{6.2.18}$$

where V_s is the solvent velocity and can be calculated using the Hagen–Poiseuille equation:

$$V_s = \left(\frac{r_p^2 \Delta P_T}{8\eta_{ws} \Delta x_m} \right) \tag{6.2.19}$$

where ΔP is the effective pressure driving force and is expressed as $\Delta P = dp = (\Delta p - \Delta\pi)$. In this case, $\Delta\pi$ is the osmotic-pressure difference.

The membrane pore-inlet concentration gradient may be derived from the modified ENP equation obtained by combining Eq. (6.2.18) with Eq. (6.2.19):

$$\frac{dc_{\mathrm{m,i}}}{dx_{\mathrm{m}}} = \left[\left(\frac{V_{\mathrm{s}}K_{\mathrm{c,i}}C_{\mathrm{m,i}} - V_{\mathrm{s}}C_{\mathrm{p,i}}}{D_{\mathrm{c,i}}}\right)\right] - \left[\left(\frac{F\, z_{\mathrm{i}}\, C_{\mathrm{m,i}}}{RT}\right) \times \left(\frac{d\psi_{\mathrm{s}}}{dx_{\mathrm{m}}}\right)\right] \tag{6.2.20}$$

The electromigration gradient through the membrane pores may be derived from the Eq. (6.2.17) with the help of Eqs. (6.2.18) and (6.2.20):

$$\frac{d\psi_{\mathrm{s}}}{dx_{\mathrm{m}}} = \left(\frac{\left[\dfrac{z_{\mathrm{p}}(K_{\mathrm{c,p}}C_{\mathrm{m,p}} - C_{\mathrm{p,p}})}{D_{\mathrm{c,p}}}\right] + \left[\dfrac{z_{\mathrm{n}}(K_{\mathrm{c,n}}C_{\mathrm{m,n}} - C_{\mathrm{p,n}})}{D_{\mathrm{c,n}}}\right]}{(z_{\mathrm{p}}^2 C_{\mathrm{m,p}} + z_{\mathrm{n}}^2 C_{\mathrm{m,n}})}\right) \times \left(\frac{V_{\mathrm{s}}RT}{F}\right) \tag{6.2.21}$$

The electroneutrality conditions within the pore and the permeate solutions are:

$$z_{\mathrm{p}}C_{\mathrm{m,p}} + z_{\mathrm{n}}C_{\mathrm{m,n}} = -X_{\mathrm{d}} \tag{6.2.22}$$

$$z_{\mathrm{p}}C_{\mathrm{p,p}} = -z_{\mathrm{n}}C_{\mathrm{p,n}} \tag{6.2.23}$$

where X_{d} = membrane-charge density, mol/m^3

From Eqs. (6.2.22) and (6.2.23) $C_{\mathrm{p,m}}$ and $C_{\mathrm{p,p}}$ can be calculated and substituted into Eq. (6.2.21) as:

$$\frac{F}{RT}\frac{d\psi_{\mathrm{s}}}{dx_{\mathrm{m}}} = \frac{\left(\dfrac{K_{\mathrm{c,p}}V_{\mathrm{s}}}{D_{\mathrm{p,p}}} - \dfrac{K_{\mathrm{c,n}}V_{\mathrm{s}}}{D_{\mathrm{p,n}}}\right)C_{\mathrm{n,m}} - \left(\dfrac{V_{\mathrm{s}}}{D_{\mathrm{p,p}}} - \dfrac{V_{\mathrm{s}}}{D_{\mathrm{p,n}}}\right)C_{\mathrm{p,n}} - \left(\dfrac{K_{\mathrm{c,p}}V_{\mathrm{s}}X_{\mathrm{d}}}{D_{\mathrm{p,p}}}\right)}{2c_{\mathrm{m,n}} - X_{\mathrm{d}}} \tag{6.2.24}$$

Membrane-surface concentrations of both ions were computed using the principle of electroneutrality. Neglecting the salvation energy barrier, the Donnan equilibrium can be expressed as:

$$\frac{C_{\mathrm{m,i}}}{C_{\mathrm{p,i}}} = \varphi_{\mathrm{i}} \exp\left(\frac{-z_{\mathrm{i}}F\Delta\psi_{\mathrm{s}}}{RT}\right) \tag{6.2.25}$$

Substituting Eq. (6.2.24) into Eq. (6.2.20) yields an equation with a numerator higher than the denominator. The concentration gradient will be effectively constant, which shows that the effect of the c_2 term is relatively small. Under these conditions, the concentration gradient may be approximated as:

$$\frac{\Delta C_{m,n}}{\Delta x_m} = \left[\frac{\left(\frac{K_{c,p}V_s}{D_{p,p}}+\frac{K_{c,n}V_s}{D_{p,n}}\right)C_{m,n_{av}}\left[C_{m,n_{av}}-X_d\right]-\left[C_{m,n_{av}}C_{p,n}\left(\frac{V_s}{D_{p,p}}+\frac{V_s}{D_{p,p}}\right)\right]+\left(\frac{V_sC_{p,n}X_d}{D_{p,n}}\right)}{2C_{m,n_{av}}-X_d}\right] \tag{6.2.26}$$

The Donnan potential at the pore inlet ($x=0$) is the same for both ions and may be obtained from Eq. (6.2.25) as:

$$\Delta\psi_s(0) = -\frac{RT}{F}\left[\ln\left(\frac{C_{m,p}(0)}{\varphi_p C_{f,p}}\right)\right] = \frac{RT}{F}\left[\ln\left(\frac{C_{m,n}(0)}{\varphi_n C_{f,n}}\right)\right] \tag{6.2.27}$$

Numerical manipulation of Eq. (6.2.27) with Eq. (6.2.22) produces:

$$(C_{m,n}-X_d)^2 + X_d(C_{m,n}-X_d) - \varphi_p\cdot\varphi_n\cdot C_{f,n}^2 = 0 \tag{6.2.28}$$

Eq. (6.2.28) has two roots representing the membrane-wall concentration of charge ions at the membrane-pore inlet and outlet. The concentration at the pore inlet may be expressed as:

$$C_{m,n}(0) = \left(\frac{X_d + \sqrt{(X_d^2 + 4\varphi_p\cdot\varphi_n\cdot C_{f,n}^2)}}{2}\right) \tag{6.2.29}$$

Correspondingly, an equivalent quadratic expression at the pore outlet provides an expression where $x=\Delta x$, $C_{F,i}$ is substituted by $C_{p,i}$ and is expressed as follows:

$$C_{m,n}(\Delta x_m) = \left(\frac{X_d + \sqrt{(X_d^2 + 4\varphi_p\cdot\varphi_n\cdot C_{p,n}^2)}}{2}\right) \tag{6.2.30}$$

$C_{m,n}(\Delta x)$ is calculated with a predicted value of $C_{p,n}$, which is cross-checked with a new $C_{p,n}$ value, evaluated by Eq. (6.2.33).

$$\Delta C_{m,n} = [C_{m,n}(\Delta x_m) - C_{m,n}(0)] \tag{6.2.31}$$

$$C_{m,n} = \left[\frac{C_{m,n}(0) + C_{m,n}(\Delta x_m)}{2}\right] \tag{6.2.32}$$

The following explicit expression for $C_{p,n}$ can be obtained by rearranging Eq. (6.2.26):

$$C_{p,n} = \left[\frac{\left[(\mathrm{Pe_p}+\mathrm{Pe_n})\cdot\left(X_d C_{m,n_{av}}\right)\right] - \left[(\mathrm{Pe_p}+\mathrm{Pe_n})\cdot C^2_{m,n_{av}}\right] + \left[\left(2C_{m,n_{av}} - X_d\right)\cdot \Delta C_{m,n}\right]}{\left(\frac{\mathrm{Pe_n} X_d}{K_{c,n}}\right) - \left(\frac{\mathrm{Pe_p} C_{m,n_{av}}}{K_{c,p}} + \frac{\mathrm{Pe_n} C_{m,n_{av}}}{K_{c,n}}\right)}\right] \tag{6.2.33}$$

where $\mathrm{Pe_i}$ stands for the Peclet number and is expressed as:

$$\mathrm{Pe_i} = \left(\frac{K_{c,i} V_s \Delta x_m}{D_{c,i}}\right) \tag{6.2.34}$$

Rejection is calculated by:

$$R_{j,n} = \left(1 - \frac{C_{p,n}}{C_{f,n}}\right) \tag{6.2.35}$$

where volumetric water flux can be calculated by:

$$J_{v,w} = \left(\frac{J_{s,i}}{C_{p,n}}\right) \tag{6.2.36}$$

The membrane-wall concentration (C_w) of the solute can be measured by the "concentration polarization" model equation, which is changed with the operating time. Thus the equation may be described as a function time as:

$$C_{m,n} = \left[(1 - R_{j,n}) + \left\{R_{j,n}\cdot \exp\left(\frac{J_{s,i}}{k_{mass}}\right)\right\}\right] \times C_{f,n} \tag{6.2.37}$$

Due to cross-flow system, CFV is expected to be an important parameter for the present model and particularly the flux calculation during long-term operation. Thus the CFV along with the operation time of the system plays a crucial role in the membrane-wall concentration. The modified equation for the present cross-flow system may be described as:

$$C_{m,n} = \left[(1 - R_{j,n}) + \left\{R_{j,n}\cdot \exp\left(\frac{J_{s,i}}{\mathrm{k}_{mass}}\right)\right\}\right] \times \frac{C_{f,n}}{\alpha} \tag{6.2.38}$$

where $\alpha = t^{0.5} \times \nu^{0.2}$.

At pH values greater than 7.0 most of the polyamide composite NF membranes have negative zeta potential (Schafer et al., 2003). Due to the charged nature of the NF membrane, solutes with an opposite charge to the membrane (counterions) are attracted, while solutes with a similar charge (coions) are repelled.

Determination of Physicochemical Parameters of the Model

The physicochemical parameters of the model equations were computed using empirical relations as follows:

Determination of Peclet number (Pe_i)
The Peclet number is defined by:

$$\mathrm{Pe_i} = \left(\frac{H_{c,i} \cdot V \cdot \Delta x}{D_{s,i}}\right)$$

Determination of hindered diffusivity ($D_{s,i}$)
The hindered diffusivity ($D_{s,i}$) is the product of the bulk-diffusion coefficient (D_{BC}) and hindered diffusivity ($H_{D,i}$), which is expressed as:

$$D_{s,i} = D_B \times K_{D,i}$$

where the hindered diffusivity is expressed as:

$$K_{D,i} = (1.0 - 2.3\lambda_i + 1.15\lambda_i^3) \text{ and } \lambda_i = \left(\frac{r_{s,i}}{r_p}\right)$$

Computation of pore radius and effective membrane thickness
The membrane pore radius (r_p) and effective membrane thickness (Δx) are calculated by validating the rejection and flux data with experimental values by separating the uncharged solutes (sucrose).

Computation procedure
The membrane-charge density as predicted by this model is validated by plant data. The typical model parameters used in this computation are listed in Table 6.2.6. To predict membrane fouling, the CP model is used and validated by experimental data. This linearized mathematical model is developed as follows:

1. Determine the membrane pore-inlet concentration of different charge ions by using Eq. (6.2.25).
2. Eqs. (6.2.29) and (6.2.30) are then used to calculate $C_{n,m}(0)$ and $C_{n,m}(x)$ with a known feed concentration ($C_{f,n}$) and an assumed permeate concentration ($C_{p,n}$) value, respectively.
3. Eqs. (6.2.31) and (6.2.32) are used to compute $\Delta C_{n,m}$ and $C_{n,m_{av}}$ by guessing the value of permeate concentration of charged ions and checking the value against Eq. (6.2.33).

Table 6.2.6 Typical set of experimental conditions and model parameters [17]

Pore radius of NF-1	0.53×10^{-9} m
Pore radius of NF-2	0.57×10^{-9} m
Pore radius of NF-3	0.55×10^{-9} m
Pore radius of NF-20	0.54×10^{-9} m
Membrane thickness of NF-1	0.0000012 m
Membrane thickness of NF-2	0.00000048 m
Membrane thickness of NF-3	0.0000009 m
Membrane thickness of NF-20	0.0000061 m
Cyanide concentration in CSTR	76 mg/L
Phenol concentration in CSTR	110 mg/L
Ammonium concentration in CSTR	2650 mg/L
Temperature maintained in units (T)	305 k
Feedwater flow rate	800 L/h
Cross-flow velocity	1.28 m/s
Operating pressure	15 kgf/cm^2
Solution velocity	0.001 kg/m · s
pH in the oxidant unit	10
Solute radius of H^+ ion ($r_{s,H}$)	0.025×10^{-9} m
Solute radius of CN^- ion ($r_{s,CN}$)	0.192×10^{-9} m
Solute radius of phenolic ions ($r_{s,ph}$)	0.17×10^{-9} m
Solute radius of ammonium ions ($r_{s,NH3}{}^+$)	0.151×10^{-9} m
Bulk diffusivity of H^+ ion (D_H)	9.3×10^{9} m^2/s
Bulk diffusivity of CN^- ion (D_{CN})	0.41×10^{-9} m^2/s
Bulk diffusivity of phenolic ions (D_{PH})	0.83×10^{-9} m^2/s
Bulk diffusivity of ammonium ions (D_{NH3})	0.98×10^{-9} m^2/s
Boltzmann constant (k)	1.38066×10^{-25} J/K
Faradays constant	96,500

4. Charge-particle separation is then calculated using Eq. (6.2.35) where the volumetric flux of the solvent is calculated using Eq. (6.2.36).
5. An iterative method is used to compute the rejection and flux using an assumed value of surface charge density (X_D) until the assumed value converges with the experimental value.
6. From the CP modulus (Eq. 6.2.37), the membrane-wall concentration of the charge particle is determined using the calculated flux and rejection value.
7. This membrane-wall concentration depends on the operation time and CFV. Thus with change in operation time and CFV, the membrane-wall concentration will also change. This altered value is then used as the membrane-surface concentration of solute for further model calculation, which is computed before with the help of Eq. (6.2.25). The typical model parameter values and plant operating conditions are listed in Table 6.2.6.

Table 6.2.7 Membrane characteristics and its performance at 15 $\times 10^2$ kPa operating pressure at (10 pH) [17]

Membranes Characteristics	NF1	NF2	NF3	NF20
Membrane geometry	Flat-sheet	Flat-sheet	Flat-sheet	Flat-sheet
Solute Rejection,%				
$MgSO_4$:	99.5	97	98	98
NaCl:	90.0	50	60	35
pH	2–11	2–11	2–11	2–11
Maxi^{m} temp (K)	323	323	323	323
Maxi^{m} pressure (kPa)	83×10^2	83×10^2	83×10^2	83×10^2
Thickness (μm)	165	165	165	165
Material	Polyamide	Polyamide	Polyamide	Polyamide
Membrane-surface area used (cm^2)	100	100	100	100
Molecular weight cut-off (g/mol)	150–250	250–300	250–300	200–300
Pore size (nm)	0.53	0.57	0.55	0.54
Water flux (L/m^2 per hour)	126	206	140	132
Investigated flux (L/m^2 per hour)	120	295	133	125
Cyanide rejection (%)	95	73	78	85
Phenol rejection (%)	96	74	80	87

6.2.2.3 Membrane-Integrated Chemical Treatment Technology

Plant-Operation Materials

The materials necessary for a treatment plant include methanol for calibrating phenyl column and phenol standard, cyanide and ammonia standard solutions (1000 mg/L), phenol, sodium hydroxide pellets, HCl, ferrous sulfate, hydrogen peroxide, thin-film composite polyamide NF (NF-1, NF-2, NF-3, and NF-20) as well as PVDF MF membrane (Sepromembranes Inc. of United States). The membrane-surface area of each module is 0.1 m^2. Some of the major characteristics of the membranes used are listed in Table 6.2.7. The typical wastewater characteristics are listed in Table 6.2.8.

Plant Configuration

The operating plant should be made of high-grade stainless steel such as SS-316 to avoid rusting. The basic plant components include stirred

Table 6.2.8 Typical physicochemical characterization of coke wastewater [17]

Parameter (mg/L)	Influent	Final effluent after chemical treatment	Tolerance limit
Cyanide	76	BDL[a]	0.2
Phenol	110	BDL[a]	0.5
Ammonium-N	2650	53	50
COD	3850	160	250
TDS	10,420	2500	2100
Total carbon	1150	4.2	–
Total organic carbon	250	1.1	–
Chloride ion	5930	181	<1000
Oil & grease	50	N.D	–
Conductivity (mS/cm)	9.12	5.5	–
pH	9–9.5	6.5	5.0–9.0

[a]Below detection limit.

stainless-steel feed tanks, and cross-flow membrane modules in any desired number may be arranged in parallel combinations along with accessories for monitoring flow, pH, temperature, and pressure. The microfiltrate is circulated through the NF membrane by a reciprocating pump capable of more than 15 bar pressure. The permeate is collected in separate tanks containing ammoniacal wastewater, and the retentate with enriched cyanide and phenol concentration is collected in separate tanks. A schematic of the plant is shown in Fig. 6.2.11.

Plant Operation

The plant runs in continuous flow mode. Effective membrane filtration surface of each membrane module is 100 cm^2. Permeate and retentate samples are analyzed at different time intervals to monitor filtration performance. The total plant operation consists of two basic functions—one is the fractionation of pollutants into two streams and the other is the chemical treatment of these pollutants. The operating pH is maintained at 10.0, which enables confinement of cyanide and phenol concentration on the retented side, which is treated by Fenton's reagents (effluent 1). The permeate side contains ammoniacal water collected in the MAP byproduct unit and is treated by magnesium and phosphate salt to produce struvite (effluent 2).

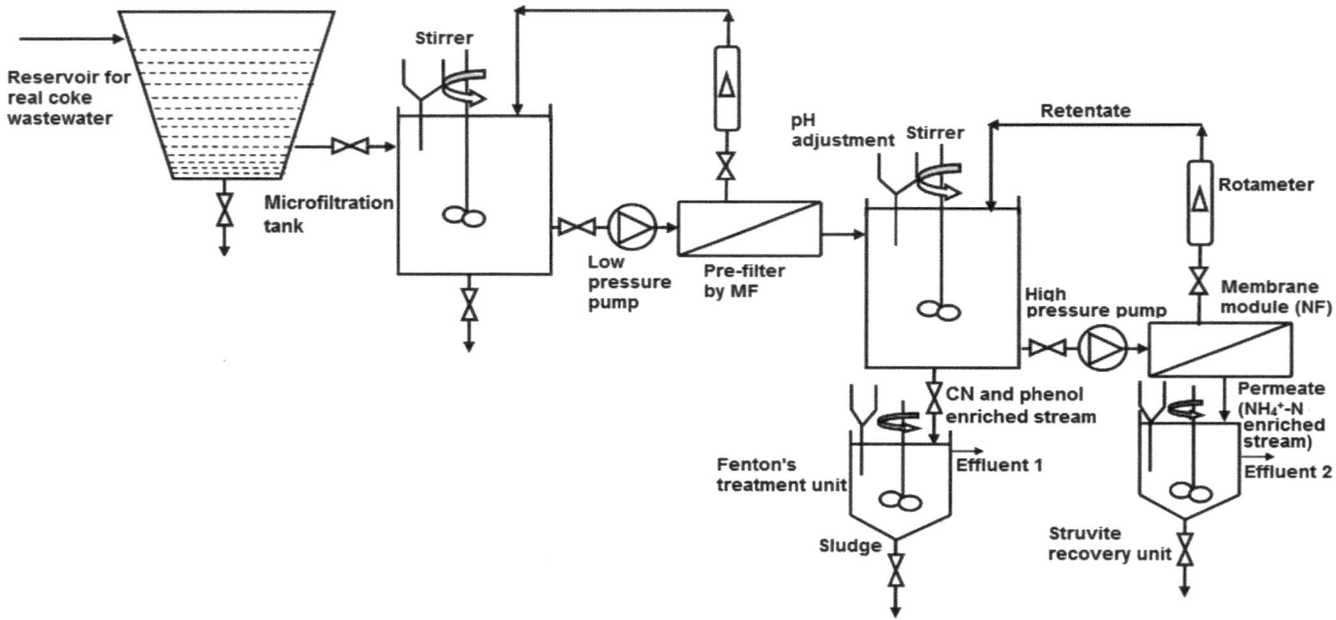

Figure 6.2.11 Schematic of the membrane-integrated hybrid treatment plant [17].

Table 6.2.9 RE and R^2 value for volumetric flux, rejection estimated by the four NF membranes under different conditions [17]

	NF1		NF2		NF3		NF20	
	RE	R^2	RE	R^2	RE	R^2	RE	$R^2 <$
Fig. 6.2.12	0.010	0.99	0.025	0.99	0.011	0.99	0.013	0.98
Fig. 6.2.13A	0.020	0.97	0.136	0.96	0.108	0.95	0.148	0.96
Fig. 6.2.13B	0.131	0.98	0.089	0.96	0.110	0.98	0.089	0.98
Fig. 6.2.13C	0.091	0.97	0.100	0.98	0.121	0.97	0.097	0.98
Fig. 6.2.14A	0.121	0.97	0.1155	0.95	0.127	0.97	0.098	0.96
Fig. 6.2.14B	0.121	0.97	0.119	0.96	0.114	0.97	0.129	0.97
Fig. 6.2.14C	0.105	0.96	0.130	0.96	0.127	0.97	0.137	0.97
Fig. 6.2.15	0.120	0.96	0.143	0.95	0.126	0.96	0.139	0.96

Analysis of Model Performance

Error analysis can be done using a standard statistical method that consists of computation of the following parameters.

The root mean square error (RMSE) is calculated as:

$$\text{RMSE} = \sqrt{\frac{\sum_{i=1}^{n} (P_i - E_i)^2}{n}}$$

where E_i = experimental value, P_i = predicted (model) value, and n = no of points analysis. The relative error (RE) is then calculated as:

$$\text{RE} = \frac{\text{RMSE}}{\overline{E}},$$

where $\overline{E}$ = mean of experimental data. Good model performance is indicated if $R^2 > 0.98$ and RE < 0.10. Table 6.2.9 shows typical model performance.

Plant Performance Analysis Under Different Operating Conditions

Flux Behavior During NF Under Varying Operating Pressure

Fig. 6.2.12 shows that the permeate flux increases with the increase of TMP from 5×10^2 to 15×10^2 kPa for all four membranes for using microfiltered coke wastewater. Flux varies linearly with applied pressure. The data suggest that NF2 is the loosest as it produces the most flux compared to the other three membranes, whereas NF1 is the tightest membrane yielding the lowest flux at the same operating conditions. The flux

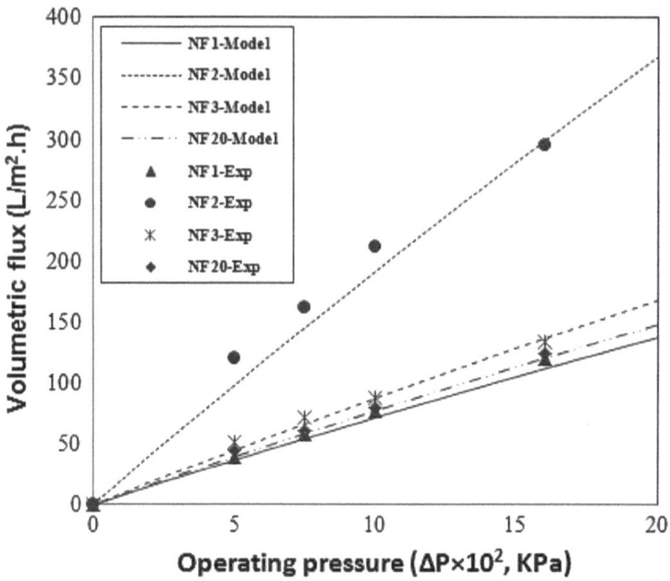

Figure 6.2.12 The effect of transmembrane pressure on flux NF1, NF2, NF3, and NF20 membranes. Operating conditions: Feed microfiltrate real coke wastewater, pressure range 5×10^2–15×10^2 kPa, CFR 800 L/hour, CFV 1.28 m/second, and pH = 10, at ambient temperature [17].

of the NF membrane largely depends on the molecular weight cut-off (MWCO) and pore size. The NF2 membrane has the maximum MWCO (300 g/mol) and pore size (0.57 nm) and exhibited the highest flux 295 L/m^2 per hour, whereas NF1 membrane has the lowest MWCO (200 g/mol) and pore size (0.53 nm) and thus exhibited the lowest flux 120 L/m^2 per hour at TMP 15×10^2 kPa, which is reasonably high for industrial use. The model predictions are in good agreement with the plant operation data.

Rejection Performance of Pollutants (Cyanide, Phenol, and NH_4^+-N) Under Varying Pressure

A steady increase in the rejection of pollutants present in the coke wastewater is observed for all types of membranes with increase of TMP. However, beyond 15×10^2 kPa, further improvement in pollutant rejection is negligible. Thus an operating pressure of 15×10^2 kPa may be considered for this configuration as the optimum pressure.

Fig. 6.2.13 shows that the NF1 membrane is the best performing membrane in terms of cyanide and phenol removal followed by NF20,

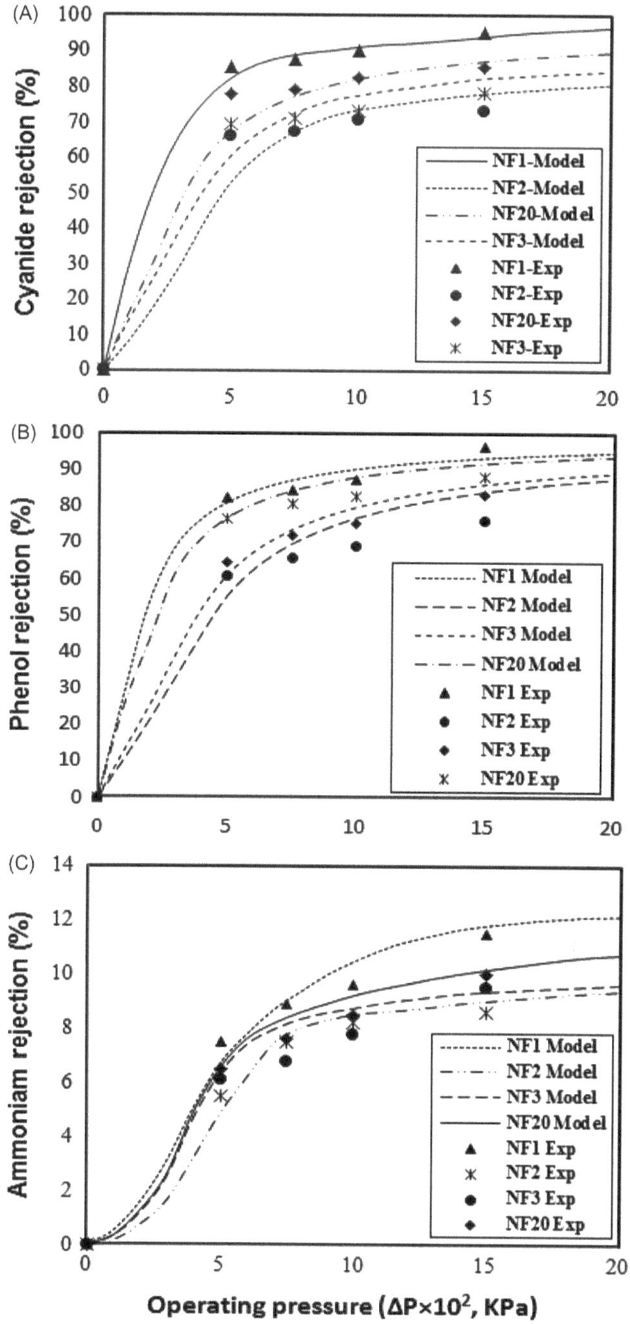

Figure 6.2.13 The effect of transmembrane pressure on the rejection of cyanide (A), phenol (B), and ammonium-N (C) in recirculation mode by using four NF membranes. Operating conditions: Feed microfiltrate real coke wastewater, pressure range $5 \times 10^2 - 15 \times 10^2$ kPa, CFR 800 L/hour, CFV 1.28 m/second, and pH = 10, at ambient temperature [17].

Table 6.2.10 Predicted membrane-charge density (X_d, mol/m^3) of the four nanofiltration membranes [17]

Membranes	NF1	NF2	NF3	NF20
X_d value (mol/m^3)	−49.60	−23.15	−27.68	−38.16

NF3, and NF2. More than 95% of cyanide and 96% phenol were rejected by the NF1 membrane, whereas NF2 shows the least rejection capability (73% and 74%, respectively). There was very little effect of pressure on NH_4^+-N rejection. The plant performance data may be fitted with the linearized model in order to obtain the effective membrane-charge density (X_{DC}), which is shown in the Table 6.2.10.

The membrane-charge density can also be measured directly using empirical relations along with streaming potential data. According to Table 6.2.10, the NF1 membrane has the highest negative charge, which explains the highest rejection for charged pollutants like cyanide and phenol in comparison to the other three NF membranes. In the electrostatic-charge repulsion (Donnan exclusion) between the negatively charged membrane surface and the negative solute ions enhances the rejection of the major contaminants. When negatively charged cyanate ions and phenolate ions at pH 10 come into contact with the negatively charged membrane surface, charge repulsion occurs, which results in rejection of cyanide and phenol. In addition to the Donnan-exclusion principle, the sieving mechanism (size exclusion) also causes (albeit to a small extent) rejection of some small species where the relative sizes of the membrane pores and solute species determine rejection. The four NF membranes in increasing order of pore size are NF1 > NF3 > NF20 > NF2 with an average pore size of 0.53, 0.57, 0.55, and 0.54 nm, respectively. Since the role of size exclusion is limited, NF membranes are not specified by MWCO. However, the MWCO is noted here to indicate the capability of the membranes to separate some small species such as residual components of oil and greasy substances and suspended solids escaping MF. The solution–diffusion mechanism is one of the transport mechanisms of an NF membrane where solute flux and solvent flux are uncoupled and as a result, with an increase in applied pressure, the solvent flux increases without a corresponding increase in solute flux [18]. Thus with increasing pressure, pure water flux increases, while the solute flux (pollutants) remains constant. This shows that rejection of pollutants of interest increases with increase in TMP. Model predictions on rejection satisfactorily corroborate with plant data.

Permeate Flux and Solute Rejection: Influence of Cross-flow rate

A significant effect of CFR on permeate flux was observed during the NF of cyanide, phenol, and ammonium-N containing wastewater. As shown in Fig. 6.2.14 the flux increases with increasing CFR from 350 to 800 L/hour at a constant pressure 15×10^2 kPa for all four NF membranes. Out of all four membranes, the NF2 membrane produces a high permeate flux of 365 L/m^2 per hour at a CFR of 800 L/hour while the NF1 membrane produces far less flux of 120 L/m^2 per hour under the same operating conditions. The high flux of NF2 may be due to its large pore radius, which helped to provide much more flux than the other three membranes. The lowest permeate flux of NF1 indicates its tight nature. The CFR plays a significant role in reducing CP due to its sweeping action on the membrane surface and thus reduces fouling. As shown in Fig. 6.2.14, rejection of cyanide and phenol increases with increasing CFR from 350 to 800 L/hour for all NF membranes. For the NF1 membrane, rejection increases from 88% to 95% for cyanide and 89% to 96% phenol with an increase in CFR from 350 to 800 L/hour.

The uncoupling nature of the solute and solvent fluxes results in higher retention of these pollutants in this case. The sweeping action on the active membrane-surface area increases with increasing CFR, which reduces the effect of CP and leads to a fouling-free nature. Achieving the maximum available effective membrane-surface area results in a higher degree of separation. Again, reduction of CP enhances convective force, which in turn enhances solvent flux.

Effect of Fouling on Permeate Flux With Respect to Operation Time

Membrane is normally fouled due to the build-up of CP layer, which causes a decline in permeate flux during long-term operation. However, with the proper selection module such as a flat-sheet, cross-flow module, fouling can be largely eliminated by the sweeping action of the fluid on the membrane surface, which reduces the possibility of CP. Fig. 6.2.15 shows the decline in flux due to fouling over time.

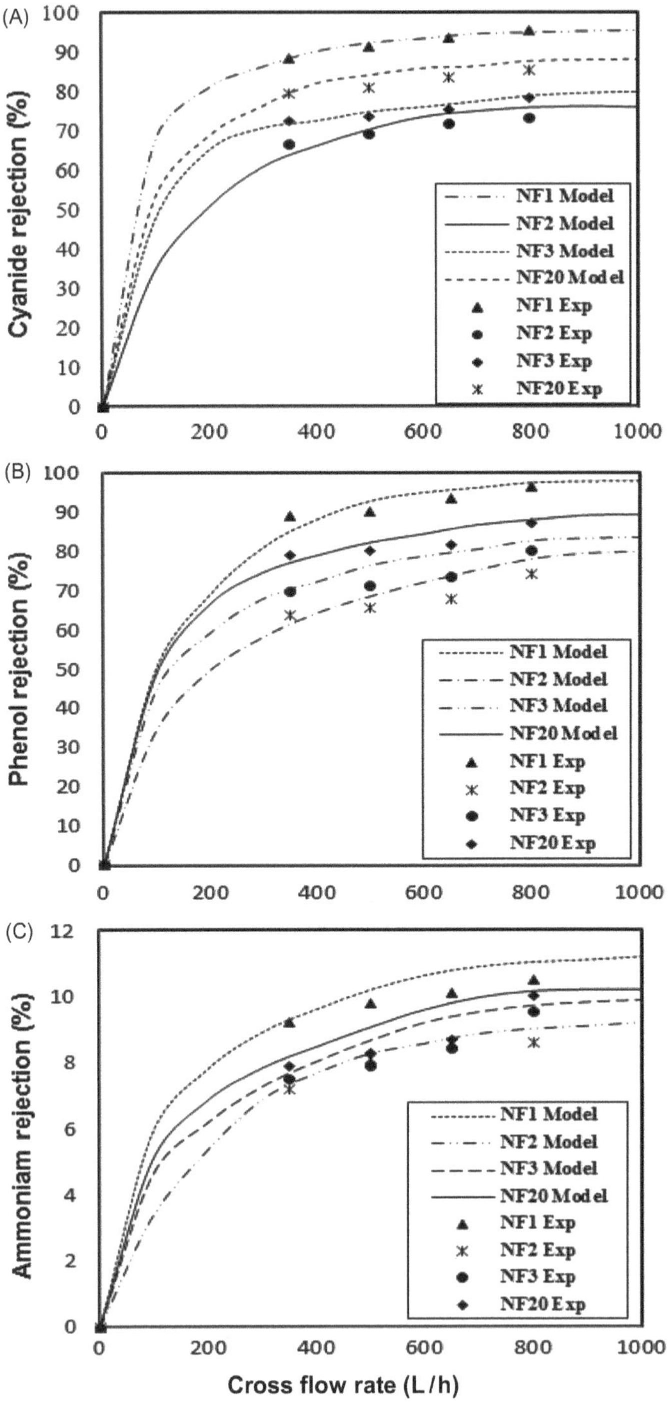

Figure 6.2.14 The effect of cross flow rate on the rejection of cyanide, phenol, and ammonium-N in recirculation mode using four NF membranes. Operating conditions: Feed microfiltrate real coke wastewater, pressure range 15×10^2 kPa, CFR range 350–800 L/hour, cross flow velocity 1.28 m/second, and pH = 10, at ambient temperature [17].

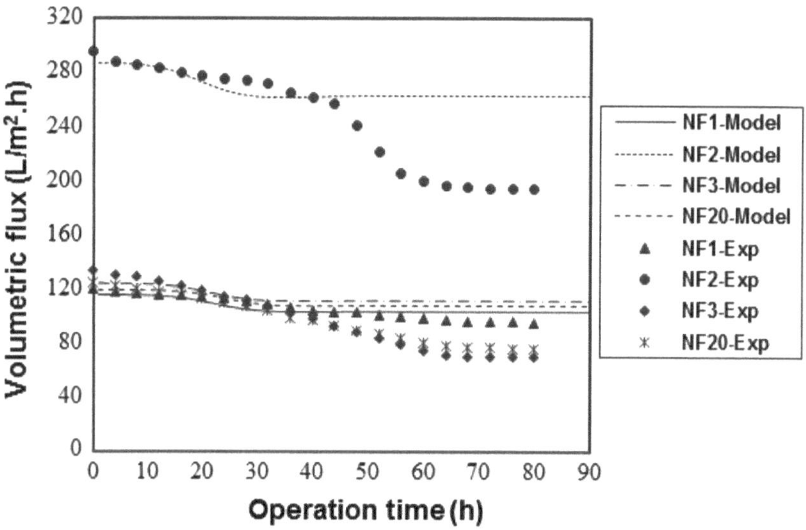

Figure 6.2.15 The effect of fouling on permeates flux with respect to operation time through NF1, NF2, NF3, and NF20 membranes. Operating conditions: Operation time range 0–80 hours, feed microfiltrate real coke wastewater, pH 10, cross flow rate 800 L/hour, pressure 15×10^2 kPa, and cross flow velocity 1.28 m/second, at ambient temperature [17].

The NF1 membrane shows a minimum decline of flux (19%), while NF3 membrane shows the maximum (47%) with respect to time. However, rejection performance was better for the NF1 membrane compared to the other membranes and the volumetric flux is industrially acceptable. Considering fractionation of pollutants, the NF1 membrane is the best. SEM images of the NF1 membrane before and after the filtration run are shown in Fig. 6.2.16.

As shown in the figure, the membranes do not undergo significant morphological changes possibly due to the very flow pattern of the cross-flow module. Reuse of these membranes after thorough rinsing with 0.1(N) NaOH and 10^{-2} molar HNO_3 is possible without significant flux decline, indicating the reversible nature of the minor fouling.

Chemical Treatment of Concentrated Pollutants

After NF the retented side contained high concentration of cyanide (375 mg/L) and phenol (525 mg/L) up to 80% recovery by NF1

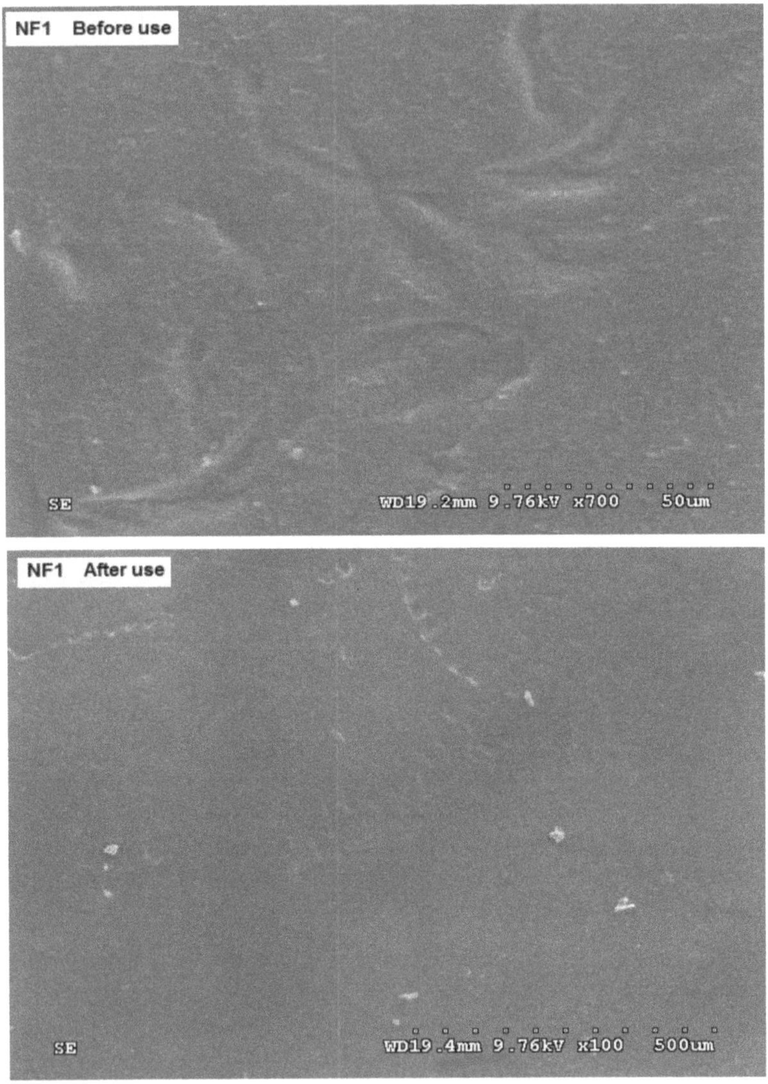

Figure 6.2.16 Surface SEM images of NF membrane before and after experiment, cleaning with NaOCl [17].

membrane of 1000 L microfiltered coke wastewater. To degrade the pollutants, Fenton's reagent is used at varying concentration H_2O_2 and $FeSO_4 \cdot 7H_2O$ from 1.5 to 9.0 g/L and 0.75 to 4.5 g/L, respectively. The H_2O_2 concentration is chosen with a concentration of COD of ferrous

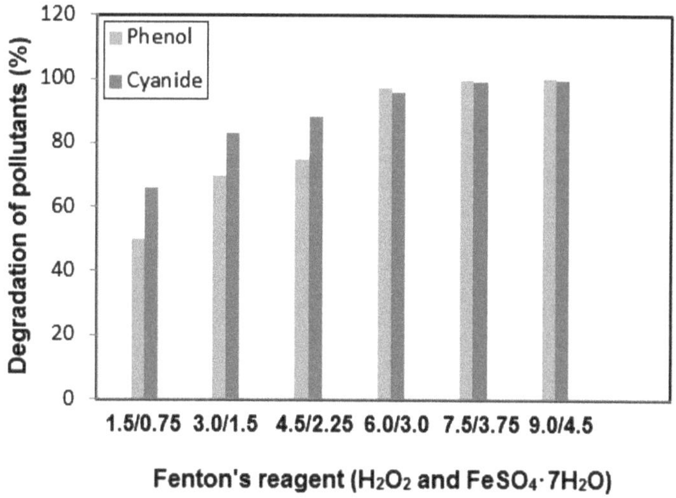

Figure 6.2.17 Degradation of cyanide and phenol by different concentrations of Fenton's reagent during treatment by nanofiltration of coke wastewater at pH 6.0 [17].

ion to avoid an additional step to remove the iron contamination. The results are shown in Fig. 6.2.17 [17].

As can be seen, Fenton's reagent has tremendous potential to degrade mixed-pollutant waste. Although the ratio is fixed, with increasing H_2O_2, $FeSO_4 \cdot 7H_2O$ the cyanide and phenol is reduced and the maximum reduction reaches around 99 in both cases when the amount of H_2O_2 7.5 g/L and $FeSO_4 \cdot 7H_2O$ 3.75 g/L is used for treatment time of 2 hours and stirring speed of 200 rpm at ambient temperature. The results show that with increases in H_2O_2 and $FeSO_4 \cdot 7H_2O$ in Fenton's reagents, the generation of $OH\cdot$ (hydroxyl) ion increases, which effectively decreases the pollutants in the wastewater. The phenol oxidation produces hydroquinone and catechol and a strong brown color is obtained at H_2O_2 dosages perhaps be due to quinine condensation. The cyclic intermediates can be further oxidized to the short-chain acids (acetic, formic, maleic, and oxalic acid) and to CO_2 [19]. As observed, high pH is better for degradation of cyanide but not for phenol. At pH 10, cyanide is degraded up to 99.9%, but phenol is degraded only 12.5% at time 2 hours. At pH 3.0 the case is reversed: 100% of phenol and 98% COD is degraded but only 7.8% of cyanide is degraded at the same time.

Cyanide oxidation by Fenton's reagent is highly pH-dependent and at high pH cyanide is present as CN^- ion so it reacts easily with H_2O_2 and Fe^{2+} ions but in acidic environment cyanide is present as HCN gas, which is very difficult to oxidize. At high pH generation of hydroxyl radical ($OH^•$) is reduced because of the formation of the ferric hydroxo complexes, which subsequently form $\{Fe\ (OH)_4\}$ at higher pH [20]. The critical pH is found to be 6.0 at which more than 99% phenol and cyanide are degraded.

After 80% recovery rate for 1000 L coke wastewater by NF, the permeate side contains high concentration of NH_4^+-N (2385 mg/L) in uncontaminated form, as impurities are rejected. More than 98% of NH_4^+-N is recovered as struvite (5 g/L) through chemical precipitation by adding an external source of magnesium and phosphate salt to keep the molar ratio of $Mg^{2+}:NH_4^+:PO_4^{3-}$ fixed at 1:1:1 at pH 9.0. The combination of $MgCl_2 \cdot 6H_2O + Na_2HPO_4 \cdot 12H_2O$ is the most efficient for ammonium-N removal:

$$M_gCl_2 \cdot 6H_2O + NH_4^+ + Na_2HPO_4 \cdot 12H_2O \rightarrow MgNH_4PO_4 \cdot 6H_2O \downarrow + 2NaCl \tag{6.2.39}$$

pH is an important factor in struvite formation. When the pH is lower than optimum (i.e., pH 9.0) hydrogen ion in the reaction solution inhibit struvite formation. As a result, the removal of NH_4^+-N is lower. When the pH is higher than optimum, $Mg_3\ (PO_4)_2$ is formed instead of struvite, which results in a decrease in NH_4^+-N removal. The precipitate is formed rapidly and settles quickly at the bottom when agitation is turned off. The struvite precipitate content is confirmed through characterization by SEM with energy-dispersive X-ray analysis, X-ray diffraction, thermogravimetric analysis, and FT-IR analysis. The FT-IR analysis, SEM − EDS, and XRD patterns show that the infrared spectrum of the precipitate and elemental profile are close to that of the MAP as reported elsewhere (Kumar & Pal, 2013b). The thermogravimetric analysis and differential thermogravimetric analysis curves for struvite at 10°C/minute indicate that mass loss begins at a temperature around 55°C and is essentially complete when the temperature reaches above 250°C and ~51% corresponds to:

$$MgNH_4^+PO_4 \cdot 6H_2O \rightarrow MgHPO_4 + NH_3 \uparrow + 6H_2O \uparrow \tag{6.2.40}$$

The DTGA curve of the struvite for heating rate 10°C/minute shows a single peak that is attained at 103°C due to simultaneous loss of both ammonia and water molecules, which indicates that the precipitate material is struvite.

Economic Evolution

A coke-oven wastewater (COWW) plant with a capacity of 70,000 L per day at a flux of 2880 L/m^2 per day by NF1 membrane at 15×10^2 kPa pressure is considered for the cost estimation. Assuming an average of 16 working hours in a day and one module with membrane-surface area of 0.5 m^2 the membrane area required is 34 m^2 and the number of modules (n) required will be 68 using:

$$\text{Number of modules (n)} = \frac{\text{Plant capacity } (L \text{ per day})}{\text{Flux obtained } (L/m^2 \text{ per day}) \times \text{Membrane area per module } (m^2)} \quad (6.2.41)$$

The operating cost involves the consumption cost of chemicals, electricity, membrane, and labor while the capital cost involves mechanical engineering cost, membrane module cost, civil investment for the installations, and electrotechnical costs. The equations for the calculation of capital costs are given as follows:

$$\text{Mechanical Engineering (\$)} = 1092\ Q_F^{0.19} + 204\ n \quad (6.2.42)$$

$$\text{Membrane module cost (\$)} = 500\ n \quad (6.2.43)$$

$$\text{Civil Investment (\$)} = 85\ Q_F^{0.38} + 235\ n \quad (6.2.44)$$

$$\text{Electrotechnical Investment (\$)} = 2.71 \times 10^4 + 62\ PQ_F^{0.27} \quad (6.2.45)$$

The estimated operating and capital costs are given in Tables 6.2.11 and 6.2.12.

Table 6.2.11 Operating costs for coke-oven wastewater-treatment plant with a capacity of 70,000 L/day [17]

Cost	Item name/character	Total cost ($/year)
Chemicals	Cost of H_2O_2, $FeSO_4$, and $MgSO_4$ with reduction of cost of struvite	447.125
Electricity	Electricity consumption 44,000 units/year	4000
Membrane cost	34 m^2 NF and 34 m^2 MF membranes (life of 6 months)	5100
Labor	1 Labor with salary of $100/month	1200

Table 6.2.12 Capital costs for a coke-oven wastewater-treatment plant with a capacity 70,000 L/day [17]

Cost	Item name/character	Total cost ($)
Mechanical engineering	(i) Valves (3000)	3000
	(ii) 5 tanks (70 m^3)	14,000
	(iii) Pipe (300 m)	3000
Membrane module	136 membrane modules	68,000
Civil investment	Treatment room	20,000
Electrotechnical investment	(i) 2 pumps	40,000
	(ii) 2 pressure gauges	160
	(iii) 3 pH probes	150
	(iv) 2 rotameters	1200

The annualized operating cost, i.e., the operating cost involved for treatment of 1 m^3 of wastewater, is calculated using Eq. (6.2.41):

$$\text{Annualized operating cost }(\$\ m^{3-}) = \frac{(\text{Total operating cost})}{Q_F} \qquad (6.2.46)$$

The economic evaluation is carried out on the basis of annualized capital, which is calculated for Q m^3 per year using Eq. (6.2.47):

$$\text{Annualized capital cost} = \frac{(\text{Total capital cost} \times \text{CRF})}{Q_F} \qquad (6.2.47)$$

where water production rate (Q_F) is calculated as:

$$Q_F = \left(\frac{70{,}000 \times 365}{1000}\right) = 25,550 \text{ m}^3 \text{ per year} \quad (6.2.48)$$

and capital recovery factor (CRF) is calculated as:

$$\text{CRF} = \left(\frac{i(1+i)^n}{(1+i)^{n+1} - 1}\right)$$

where n is the project life and i is the interest rate. Assuming a high-quality stainless-steel membrane module a project life of 15 years with an annual interest rate of 5% is a realistic assumption. The CRF value calculated is 0.0878:

$$\text{Annualized operating cost } (\$ \text{ m}^{3-}) = \frac{10{,}747.125}{25{,}550} = 0.421$$

$$\text{Annualized capital cost} = \frac{149{,}510 \times 0.0878}{25{,}550} = 0.513$$

In order to calculate the total annualized cost of treatment of 1 m^3 of COWW with operating and capital cost, i.e., $(0.421 + 0.513) = \$0.934$, this estimation covers the main factors for the treatment of COWW; but more detailed analysis needs to be carried out before installation of a full-scale treatment plant.

NOMENCLATURE

$C_{m,i}$	Concentration in membrane of ion i (mol/m^3)
$C_{m,i\ av}$	Average concentration of ion i (mol/m^3)
$C_{p,i}$	Concentration in permeate of ion i (mol/m^3)
$C_{f,i}$	Feed concentration of ion i (mol/m^3)
$D_{s,i}$	Hindered diffusivity of ion i (m^2/second) ($D_{s,i} = D_{B,i} \times K_{d,i}$)
$D_{B,i}$	Bulk diffusivity of ion i (m^2/second)
d	Thickness of oriented solvent layer (m)
F	Faraday constant
$J_{s,i}$	Ion flux (mol/m^2 per second)
J_v	Volumetric flux (m^3/m^2 per second)
$K_{c,i}$	Challenge factor for convection of ion i
$K_{d,i}$	Challenge factor for diffusion of ion i
k	Boltzmann constant, 1.38066×10^{-23} J/K
ΔP	Applied pressure difference (kPa)
ΔP_T	Effective pressure difference (kPa)

$\mathbf{Pe_i}$	Peclet number of ion i, dimensionless
$R_{j,i}$	Rejection (%) of ion i
r_p	Effective pore radius (nm)
$r_{s,i}$	Solute radius of ion i (nm)
R	Universal gas constant $(J/mol \cdot K)$
T	Absolute temperature (K)
Δx_m	Effective membrane thickness (m)
X_d	Effective charge membrane density (mol/m^3)
z_i	Valence of ion i

GREEK LETTERS

η_{ws}	Dynamic viscosity of the wastewater (kg/m per second)
Φ_i	Steric coefficient of ion i
λ_i	Ratio of solute radius to pore radius of ion i
$\Delta\Psi_s$	Donnan potential difference (V)

6.2.3 FORWARD OSMOSIS-NANOFILTRATION TECHNOLOGY FOR COKE-OVEN WASTEWATER RECLAMATION

6.2.3.1 Introduction

Thousands of tons of metallurgical coke are produced due to carbonization of coal in coke ovens due to the huge demands of the steel industry. One of the harmful byproducts of the coke-production process is COWW, which contains highly hazardous contaminants including carcinogenic compounds [21]. Combating the environmental consequences of COWW discharge is a major challenge. Characterized as a typical refractory industrial wastewater, coking wastewater contains cyanide, thiocyanide, high-strength ammonia, phenolic compounds, heterocyclic nitrogenous compounds, and polynuclear aromatics compounds. A heavily loaded industrial wastewater like this discharged without proper treatment to any surface waterbody will have severe long-term environmental and ecological impacts.

However, because of the tremendous fluctuations of the influent as well as the inherent instability of systems, conventional treatments rarely meet the stringent wastewater-discharge standards [22]. Methods such as adsorption, coagulation, wet oxidation, and advanced oxidation have been investigated to treat coking wastewater. However, these methods are either technically complicated or economically unfavorable, which often make them difficult to be used in practice. Biological treatment plants are highly sensitive and slow and often fail because of the presence of toxic contaminants. The adsorption bed is quickly saturated and is unable to treat complex wastewater on a long-term basis without frequent replacement of the adsorbent, which is expensive. In WAO, salt precipitation and inhibition of ammonia destruction in the presence of phenol and toxic contaminants often lead to process failure.

In countries with huge populations, surface water contamination from industrial or municipal sewer lines is so serious that immediate intervention from policymakers for reclamation of water from the waste stream for reuse instead of discharging into the river bodies is needed. In this context, membrane-based technology has been gaining importance because of its potential to produce reusable water from wastewater streams. Among the pressure-driven membrane-based processes, NF and RO are capable of achieving a high standard of purification. However, RO involves much higher TMP and hence high energy. Furthermore, RO membranes are expensive and require high-pressure vessels for housing. Fouling is another challenge in almost all pressure-driven membrane-separation processes. Today forward osmosis (FO) is an emerging technology that promises a high degree of separation of contaminants from water with low energy consumption, low fouling, and high recovery in an easy and simple design. In FO water is transported through a semipermeable membrane by a natural osmotic process using a highly concentrated solution (called draw solution, or DS) that draws water from the feed solution. As an emerging technology, FO has attracted significant interest in the treatment of seawater/brackish water desalination, liquid food processing, complex industrial streams such as those from textile industries, oil and gas well fracturing, landfill leachate, nutrient-rich liquid streams, wastewater from activated sludge plants, wastewater from municipal sources, and even nuclear wastewater. However, removal of pollutants from COWW through FO has received little attention, even though an FO scheme could be a complete and sustainable option when used with feedwater and DS of constant composition. This can be ensured only by periodic discharge of concentrated cyanide, phenol, and NH_4^+-N rejects for subsequent treatment and addition of fresh feed COWW to feeding tank and economical separation of draw solute for recycle/reuse of permeate water in the same industry.

Researchers have attempted to address the issues of CP through better design of membranes and draw solutes. However, CP and fouling must also be addressed. The use of RO in the recovery of draw solute offsets the gain of low-energy use in FO. Thus the major problem of flux decline over time following CP or reverse salt diffusion (RSD) (diffusion of the draw solute to the feed solution) limits the application of FO in wastewater purification. Diffusion of draw solute is normally accompanied with back diffusion of feed solute from the active membrane surface to the bulk-feed solution. Thus RSD progressively increases the osmotic pressure of the feed solution thereby seriously limiting separation of the desired solute.

A design was developed [23] to overcome the major hurdles of FO in the context of removal of contaminants from coking wastewater. The proposed scheme is a complete system comprising an FO loop upstream where water is separated from the wastestream and a downstream loop of flat-sheet, cross-flow NF membrane for recovery of draw solute for recycling and producing reusable water. This new design succeeds in reducing CP and RSD and in enhancing pure water flux while ensuring efficient draw solute recovery involving low energy.

6.2.3.2 Theory of Transport

When two aqueous solutions of different concentrations are separated by a membrane permeable to the solvent but impermeable to the solute, transport of water takes place across the membrane from the dilute solution to the concentrated solution even at the same temperature, electrical potential, and hydrostatic pressure. This water transport takes place due to the difference in the chemical potential of the water on the two sides of the membrane and is known as osmosis. The minimum pressure required to stop this water transport from the dilute-solution side to the concentrated-solution side is called osmotic pressure, denoted by $\Delta\pi$. This osmotic-pressure differential ($\Delta\pi$) is the driving force in water transport across the membrane in FO, while the hydraulic pressure differential acts as the driving force in RO. In FO, applied pressure ΔP is equal to zero, in RO $\Delta P >> \Delta\pi$, and in pressure-retarded osmosis (PRO), $\Delta P < \Delta\pi$. The direction of water flow is the same in FO and PRO. In RO the direction of water flow is reversed by applied ΔP where it is in excess of the osmotic-pressure differential. Water flux in all these processes can be in general described as:

$$J_{\text{water}} = A(\alpha\Delta\pi - \Delta P) \tag{6.2.49}$$

where A is the water permeability constant of the membrane, α is the reflection coefficient, and J_{water} is water flux.

6.2.3.3 Flux Calculation of the FO Process

Considering the osmotic pressures and mass transfer coefficient on the feed side and DS side, the volumetric flux for an FO system may be computed using:

$$J_{v,cw} = L_w\left[\pi_{DS} \times \exp\left(-\frac{J_{v,cw}}{k_{m,DS}}\right) - \pi_{FS} \times \exp\left(\frac{J_{v,cw}}{k_{m,FS}}\right)\right] \quad (6.2.50)$$

where $J_{v,cw}$ is the volumetric flux of water, L_w is the water permeability, and π_{DS} and π_{FS} are the osmotic pressure on the draw and feed side, respectively, where $k_{m,DS}$ and $k_{m,FS}$ are denoted as the mass transfer coefficient for external CP in the draw side and feed side, respectively. Based on the mass transfer correlation $k_{m,DS}$ and $k_{m,FS}$ can be described by:

$$k_{m,DS} = \left(\frac{D \times \varepsilon}{\tau \times x}\right) \text{ and } k_{m,FS} = \left[\frac{1.85 \times D}{d_h^{0.67} \times L^{0.33}}\right] \times (\mathrm{Re} \times \mathrm{Sc})^{0.33} \quad (6.2.51)$$

where D, ε, τ, x, dh, and L are the diffusivity coefficient, porosity of the support layer, tortuosity of the support layer, thickness of the support layer, and hydraulic diameter of the module and channel length, respectively, where Re and Sc are denoted as the Reynolds number and Schmidt number, respectively.

6.2.3.4 Flux and Rejection Calculation of NF Process

The charge particles can be separated by NF membranes by the Donnan-exclusion mechanism. The modified ENP equation adequately describes separation of major contaminants from COWW by such membranes and can be expressed as:

$$J_{cw} = (K_i C_{m,i} V) - \left(D_i \frac{dc_{m,i}}{dx}\right) - \left(\frac{z_i^{C_{m,i}} D_i^F}{RT} \frac{d\psi_m}{dx}\right) \quad (6.2.52)$$

The flux (J_{cw}) is the sum of the fluxes due to convection, diffusion, and electromigration. D_i is the diffusion coefficient of i through the membrane pores, which accounts for the friction of the components with the pore walls where K_i is the challenge factor for convection and $C_{m,i}$ is the ionic concentration in feed. F, R, and T denote the Faraday constant, universal gas constant, and temperature, respectively, and z_i is the valence of the respective ions.

V_{Sol} is the solvent velocity and may be expressed using the Hagen–Poiseuille type equation as:

$$V_{Sol} = \left(\frac{r_p^2 \, \Delta P_T}{8\eta_w \Delta x_m}\right) \tag{6.2.53}$$

where r_p is the membrane pore radius, η_w is the viscosity of the water, Δx_m is the membrane thickness, and ΔP_T is termed as effective TMP and is expressed as:

$$a\Delta P_T = dp = (\Delta P - \Delta\pi) = \left[\Delta P - \left\{RT\sum(C_{i,F} - C_{i,P})\right\}\right].$$

$\Delta\pi$ is the osmotic-pressure difference and can be computed using the bulk-feed concentration ($C_{i,F}$) and permeate concentration ($C_{i,P}$) of the solute ions. The membrane-surface concentration of solute ions can be measured by the Donnan equilibrium condition, which is a function of the Donnan potential of the membrane (ΔD_{Pot}) expressed as:

$$C_i = (C_{i,F}\ \varphi_i)\exp\left(\frac{-z_i F \ \Delta D_{Pot}}{RT}\right) \tag{6.2.54}$$

This membrane-surface concentration of the solute ions is used to determine the permeate concentration, which is also used to compute the rejection of the charge particles (R_j), which can be described by:

$$R_{i,j} = \left(1 - \frac{C_{i,P}}{C_{i,F}}\right)$$

$C_{i,P}$ and $C_{i,F}$ are the concentration of solute in the permeate and feed side, respectively.

6.2.3.5 Advanced Treatment Plant

Plant Configuration

The experimental setup used in the investigation made of stainless steel (SS 316) as shown in Fig. 6.2.18 consists of two loops: one for FO and the other for NF.

The FO loop consists of two continuous stirred tanks for feed and DS, respectively, connected to the FO-membrane system via circulating pumps, flow meters, and pressure gauges. The feed and DS are circulated through the FO cross-flow module with the help of diaphragm pumps. The upstream and downstream pressure transducers indicate TMP.

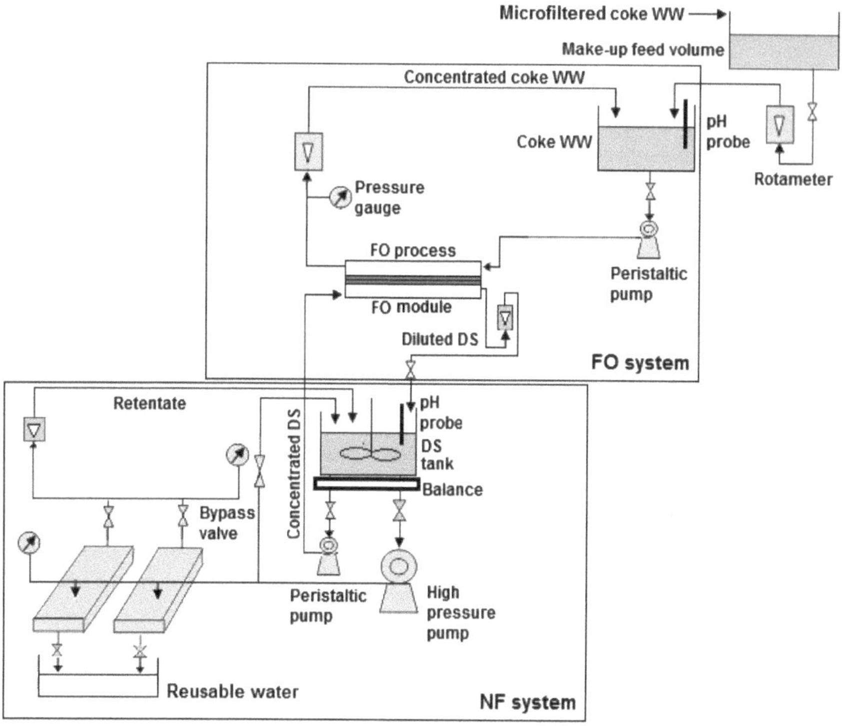

Figure 6.2.18 Schematic of the FO–NF system for treatment of complex wastewater [23].

The CFR through the system is monitored and controlled using a rotameter and control valve by increasing/decreasing the rotational speed of the pump. The module is designed so that the flat-sheet membrane remains on a horizontal plane on a perforated stainless-steel support while the feedwater and DS flow tangentially countercurrently along the top and bottom surfaces of the membrane, respectively. The lower chamber of the cross-flow module is positioned so that the DS enters one of its inlets at one end and on crossing tangentially along the membrane surface leaves the chamber through an outlet maintained at a much higher level than the level of the membrane. In this way the bottom surface of the membrane is completely immersed in the DS while allowing a continuous sweeping fluid action along its bottom surface. The feed wastewater flows tangentially along the top surface of the membrane ensuring sweeping fluid action on the top layer. This novel design minimizes CP, membrane

Table 6.2.13 Membranes physical and chemical properties [23]

	Unit	NF1	NF2	NF3	NF20
Pure water permeability (1.5 bar) L/m^2hour (FO)		–	49	–	–
Pure water permeability (10 bar) L/m^2hour (NF)		94	296	112	102
Pore size (nm)		0.53	0.57	0.55	0.54
Solute rejection (5 bar TMP)					
$MgSO_4$	%	99.5	97	98	98
NaCl	%	90	50	60	35
pH		2–11	2–11	2–11	2–11
Maximum temperature	K	323	323	323	323
Maximum pressure	bar	83	83	83	83

fouling, and also back diffusion of the DS thereby ensuring a high unidirectional volumetric flux, i.e., water transport from feed to DS only.

As continuous FO results in dilution of the DS, the DS concentration needs to be maintained to sustain the FO process. Thus a downstream NF membrane module in flat-sheet, cross-flow mode was simultaneously operated for recovery of pure water from the DS while recycling the draw solute to the system. A diaphragm pump (Milton Roy India Pvt. Ltd.) operated at 8–10 bar TMP maintains circulation of the DS through the flat-sheet, cross-flow NF module. Operating pressure and CFR are maintained by the bypass control valve, rotameter, and control valves.

Materials Needed to Run the Plant

Thin-film composite polyamide NF and MF membrane (0.45 μm) are used (Sepro Membranes USA and membrane solution USA). Table 6.2.13 shows the characteristics of the membranes used in the FO–NF system.

Plant Operation

The treatment plant can be run in continuous-flow mode using cross-flow, flat-sheet membrane modules with COWW as a feed. MF of COWW is done prior to FO to remove any suspended particles in the cross-flow membrane module. Then COWW containing cyanide and phenol ammonium-N with high COD is pumped through the FO

Table 6.2.14 Operating conditions for FO–NF system [23]

Parameters	
Feed tank volume	15 L
DS tank volume	15 L
Configuration	Submerged
Module type	Cross-flow
Membrane type	Flat-sheet, polyamide
Membrane area (cm^2)	150
Active layer facing direction	Feed
Draw solution (M/kg Milli-Q water)	1.5
Circulation flow rate of DS (L/h)	10
Circulation flow rate of feed (L/h)	30–200
Transmembrane pressure during FO process (bar)	1.5
Transmembrane pressure during NF process (bar)	10

module from where water is extracted in a DS of $NaCl/MgSO_4/CaCl_2$ leaving behind the major contaminants. The diluted DS is then passed on to another NF membrane module to recover pure water as the filtrate and recycle the draw solute in the DS loop. Filtration modules are run in continuous mode where feed COWW and DS compositions are maintained at the same level through continuous withdrawal of major contaminants and corresponding addition of fresh microfiltered COWW in the FO loop and recovery of pure water from the downstream NF module at the same rate of dilution of the DS. The draw solute can be very effectively separated out from aqueous medium by the downstream NF module operated in flat-sheet, cross-flow mode and recycled. The system is closed loop, providing for continuous recovery of draw solute and its reuse. The typical plant operating conditions are listed in Table 6.2.14.

The active layers of the membrane always face the feed solution and the support layer faces the DS in order to afford higher rejection with higher water fluxes due to the reduced internal CP. The DS and feed solution (COWW) flow countercurrently in their respective channels on two sides of the flat membrane. The CFRs for both draw and feed solution are varied between 21 and 42 cm/second whereas the pressure of the FO module is maintained in the range of 0–2 bars. The NF modules are run at 8–10 bar operating pressure. Both the feed and the DS are maintained at the same temperature. Cyanide, COD, phenol, and NH_4^+-N rejections and reverse salt flux (RSF) were analyzed by taking a sample from the feed and DS tank, respectively, while the water flux was calculated using the change of weight of COWW in the feed tank positioned on the analytical balance. The NF module may have to be run initially at different TMPs and CFRs to arrive at the critical operating pressure and CFR.

Monitoring Plant Performance Under Vaying Conditions

FO Process

Selection and effect of DS concentration on rejection and flux The effects of composition and concentration of DS can always be checked by running the plant using DS combinations such as NaCl, $MgSO_4$, and $CaCl_2 \cdot H_2O$ at different concentrations. Fig. 6.2.19 illustrates how a change in DS concentration affects water flux and rejection of major contaminants of COWW like cyanide, phenol, NH_4^+-N, and COD. Fig. 6.2.19A shows that the DS concentration has a positive correlation with volumetric flux. A higher osmotic potential than the COWW is essential to induce high water flux but low reverse leakage. A concentration of 1.5 M of NaCl using the NF-2 membrane yields a volumetric water flux of 46 L/(m^2hour) while water flux using $MgSO_4$ and $CaCl_2 \cdot H_2O$ as draw solutes is 42 L/(m^2hour) and 43 L/(m^2hour), respectively. Fig. 6.2.19B also shows the effects of DS concentration on cyanide, phenol, NH_4^+-N, and COD rejection using the NF-2 membrane.

Rejection of these major contaminants increases linearly with increasing DS concentration and >97% overall rejections are observed at a draw solute concentration of 1.5 M NaCl. A higher concentration of DS produces a higher osmotic pressure or driving force for water transport through the membrane. A higher concentration of DS and hence higher osmotic pressure to force a larger amount of water as water flux through the membrane eventually reduces solute flux across the membrane in an uncoupled transport process. The DS and the corresponding osmotic pressure it develops are important factors influencing mass transport and overall process performance in FO. This explains the higher major contaminant rejection present in COWW following higher water flux. Due to its high solubility, NaCl produces higher osmotic pressure than $MgSO_4$ and $CaCl_2 \cdot H_2O$ for the same concentration in a. In addition to the low diffusivity of $MgSO_4$ and $CaCl_2 \cdot H_2O$ compared to that of NaCl is another reason behind its significantly high osmotic pressure which in turn high flux.

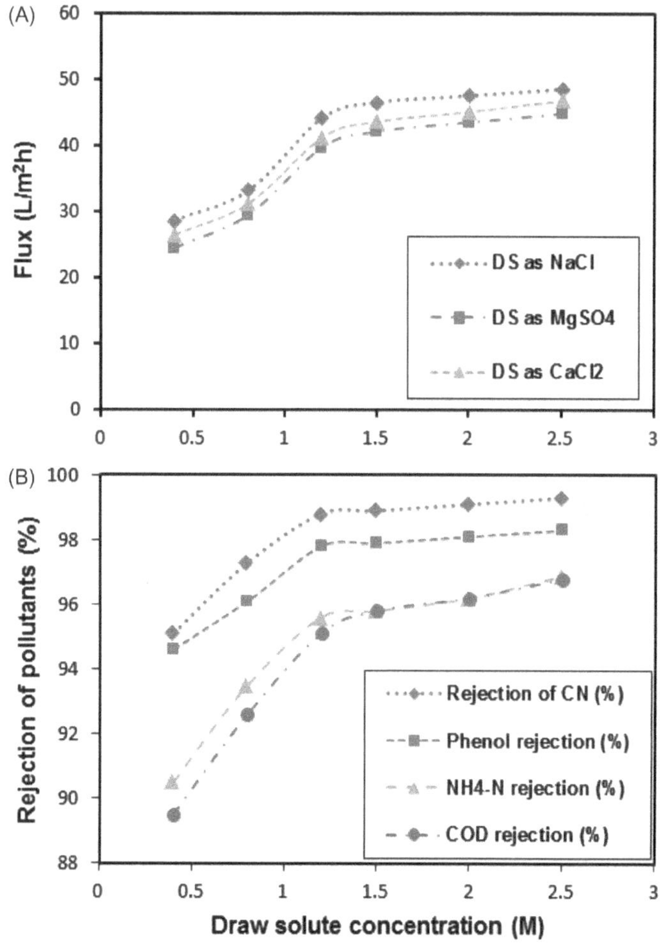

Figure 6.2.19 The effects of three different types of draw solute concentration (0.4–2.5 M) on the (A) flux and (B) on the rejection of pollutants present in the COWW (operating conditions: TMP = 1.5 bar, cross flow rate = 150 L/hour, and feed = microfiltered COWW) [23].

Pure Water Flux: Influence of TMP and CFR

Water flux is calculated after each FO experiment to identify the effects of applied hydraulic TMP and CFR on water permeability through membrane. Water flux as a function of TMP and CFR for each experiment was conducted with fresh COWW feed volume, temperature, and DS concentration to eliminate any effects of operating conditions on membrane performance. In this way a constant osmotic driving force

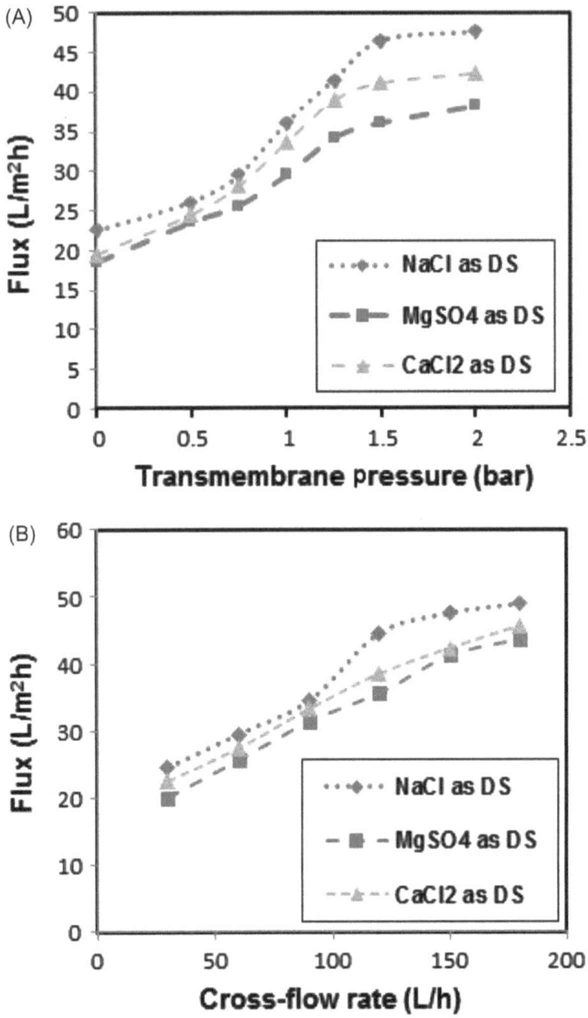

Figure 6.2.20 The effects of (A) Transmembrane pressure and (B) Cross-flow rate on flux in case of FO (operating conditions: pressure range = 0–2 bar, cross flow rate = 30–200 L/hour, and feed = microfiltered COWW) [23].

could be achieved, allowing for an unbiased comparison of water flux between experiments with different hydraulic feed pressures.

Fig. 6.2.20A and B shows the effect of TMP and CFR on volumetric water flux at hydraulic TMP varied from 0 to 2 bar and feed flow rate of 30–180 L/hour with fixed DS flow rate of 10 L/hour by using different draw solutes like NaCl, $MgSO_4$, and $CaCl_2 \cdot H_2O$ (1.5 M) in the FO system. The water flux shows a positive correlation with the TMP and CFR and the highest flux of 46.4 L/(m^2hour) was achieved with 1.5 M NaCl DS and 40 L/(m^2hour) and 42 L/(m^2hour) with $MgSO_4$ and $CaCl_2 \cdot H_2O$ draw solute. The higher water flux due to the increase of TMP and CFR was attributed to the decrease of fouling potential due to the greater turbulence and sweeping action of the fluid on the membrane surface over time. Thus it was concluded a TMP of 1.5 bar and a CFR of 150 L/hour are the best operating parameters for maximum flux.

Rejection of Major Contaminants Present in COWW: Influences of TMP and CFR

Fig. 6.2.21A and B illustrates the effect of hydraulic TMP and CFR on major contaminants like cyanide, phenol, NH_4^+-N, and COD rejection. Although water transport in FO is due to osmotic-pressure difference only, hydraulic TMP often plays a significant role in spiral wound or capillary type modules where it helps overcome the hydraulic resistance of the flow channels. The effect of up to 2 bar hydraulic TMP has been investigated in the present study in cross-flow module with flat-sheet NF-2 membrane in removing the major contaminants present in the COWW close to 100% except NH_4^+-N, which was 96% at a feed flow rate of 150 L/hour and a DS flow rate of 10 L/hour as shown in Fig. 6.2.21B. A lower degree of NH_4^+-N rejection compared to other contaminants such as cyanide and phenol can be attributed to its occurrence in neutral molecules in natural COWW against the anionic form of cyanate and phenolate, which gets repelled (by Donnan exclusion) to some extent by the composite polyamide membrane (due to negative charge developed on membrane surface) used here. Since the applied pressure adds to the already high osmotic pressure, which forces water through the membranes, transport of other small species can't be completely stopped, thus explaining the less than 100% rejection of major contaminants present in COWW in the permeated water.

While the addition to the solution–diffusion mechanism of transport through composite polyamide membranes increases the rejection, because the solute flux and solvent flux are uncoupled and under enhanced pressure, the solvent flux increases with commensurate decrease of solute flux or rather increase of solute rejection [24].

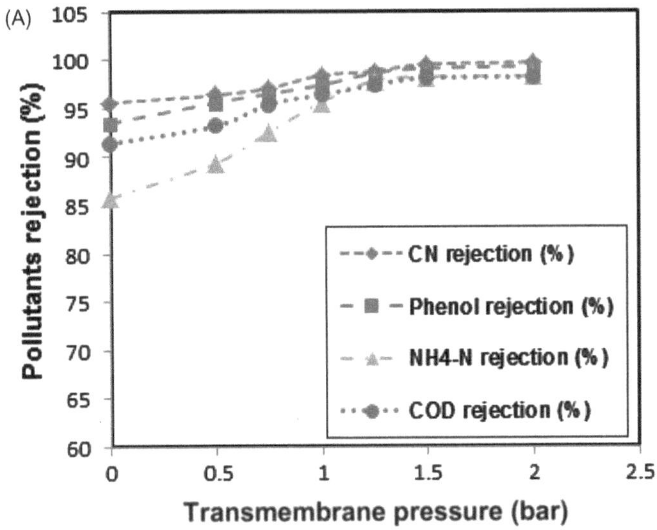

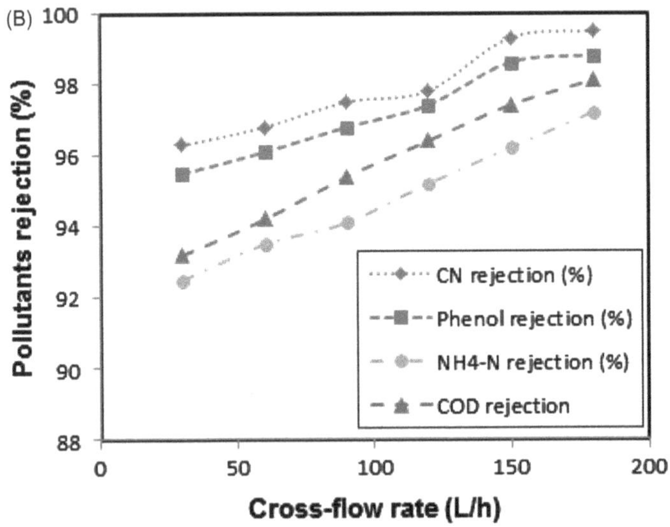

Figure 6.2.21 The effects of (A) transmembrane pressure and (B) cross-flow rate of feed on removal of pollutants (operating conditions: pressure range = 0–2 bar, cross-flow rate = 30–200 L/hour, cyanide = 120 mg/L, phenol = 165 mg/L, ammonium-N = 2500 mg/L, and COD = 2548 mg/L) [23].

Effect of pH of feed solution on flux and rejection of major contaminants present in COWW Fig. 6.2.22A and B illustrates the effect of pH of feed solution on flux and rejection of major contaminants present in the COWW using different draw solutes like NaCl, $MgSO_4$, and

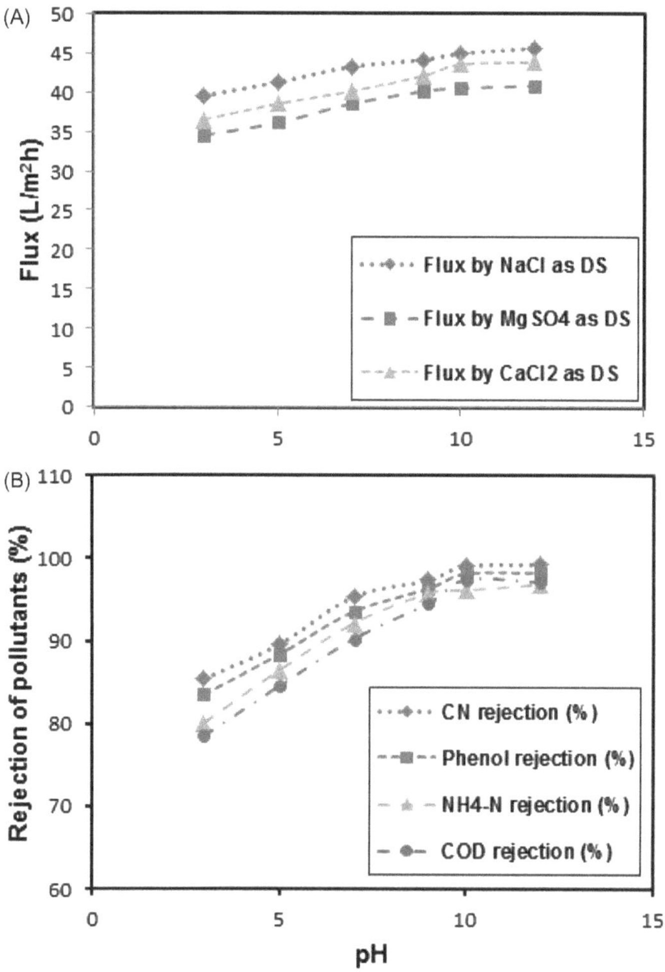

Figure 6.2.22 The effect of feed solution pH on (A) permeate flux and (B) removal of pollutants by NaCl as a DS for FO cross-flow membrane module (operating conditions: pH range = 4.0–10.0, TMP = 1.5 bar, cross flow rate = 150 L/hour, cyanide = 120 mg/L, phenol = 165 mg/L, ammonium-N = 2500 mg/L, and COD = 2548 mg/L).

$CaCl_2 \cdot H_2O$ (1.5 M) in the FO system. The pH of the feed solution was varied from 3 to 12 at fixed TMP and CFR 1.5 bar and 150 L/hour, respectively, and the feed sample temperature was maintained at 25°C. The water flux shows minimum variation (7%–9%) with changes in pH for all three draw solutes. The maximum flux of 45.6 L/(m^2hour) was achieved with 1.5 M NaCl DS in comparison to 40 L/(m^2hour) and 43 L/(m^2hour) with $MgSO_4$ and $CaCl_2 \cdot H_2O$ draw solutes at pH range 9–10 as shown in Fig. 6.2.22A. The minimum change in the flux due to pH variation might be due to the filtration mechanism as osmotic pressure is the driving force in FO. The osmotic potential between the COWW and DS should have the greatest influence on membrane flux. Thus a change of pH in the feed solution can not significantly influence the membrane flux. However, high pH (12) might affect the structure of the FO membrane.

The effect of the pH of the feed solution on cyanide, phenol, NH_4^+-N, and COD removal in cross-flow, flat-sheet polyamide membrane is shown in Fig. 6.2.22B. The rejection of contaminants from COWW increased as the pH of the feed solution was increased using NaCl as draw solute (1.5 M) at fixed TMP and CFR 1.5 bar and 150 L/hour, respectively. Rejection of cyanide, phenol, NH_4^+-N and COD was achieved at 99%, 98%, 96%, and 97%, respectively, and was found to have a positive correlation with pH. This can be explained by considering the dissociation equilibrium of cyanide and phenol, which was dissociated cyanate and phenolate ions along with other species take negative charge at pH 10. Moreover, at high pH, the functional groups of polymers (NF2 membrane) are dissociated and take on negative charge and thus charged species are repelled more quickly (by Donnan exclusion). In addition to the functional groups such as carboxyl and hydroxyl groups present at the surface of the membrane that become deprotonated at high pH and thus reduce the apparent pore size and resulting in greater rejection of the charge [25].

Effect of TMP and Operation Time on RSF

The effects of TMP on RSD in the form of conductivity and Cl^- ion concentration in the permeation are shown in Fig. 6.2.23A where NaCl was used in preparing the DS.

An inversely proportional relationship between hydraulic TMP and RSF is observed. High RSF for NaCl is attributed to its monovalent ions possessing high diffusivity. The decline in RSF with applied TMP may be attributed to physical changes in the membrane active layer as a function of the applied pressure. Compression of the interface between the thin active layer and support layer of the membrane by increasing pressure at the membrane surface reduces the possibility of RSF and salt permeability of the membrane, while marginally affecting water flux.

The most common phenomenon observed during long-term operation of the membrane is the build-up of CP, which is the main reason for the flux decline. Fig. 6.2.23B shows the RSF decline, which is measured

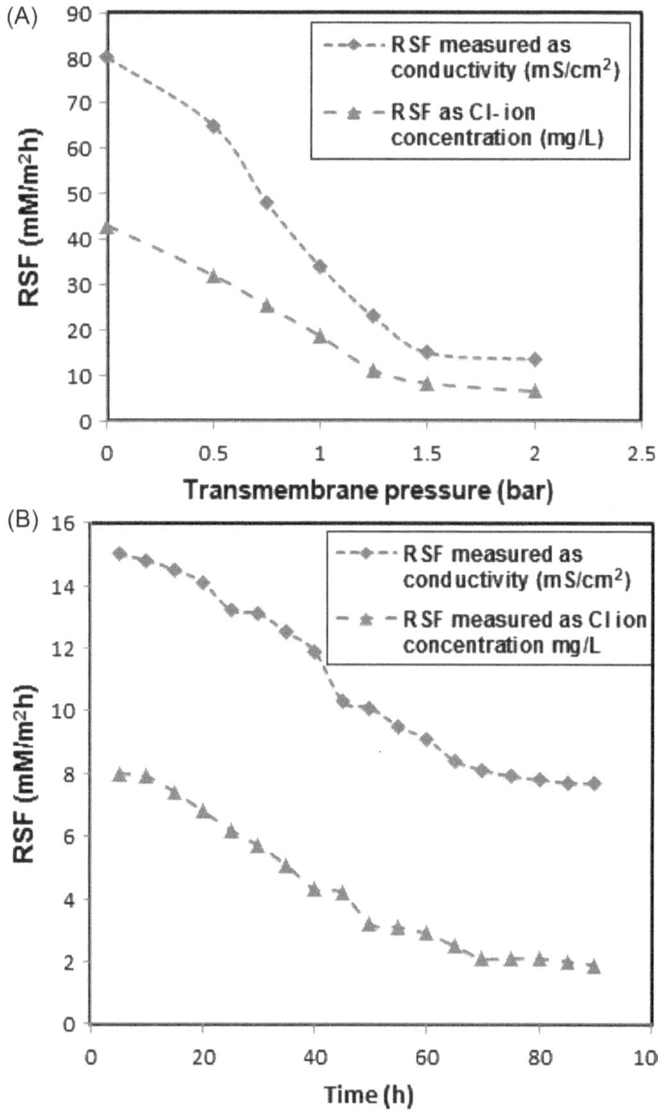

Figure 6.2.23 The RSF of DS (NaCl) measured in the feed solution as a function of (A) transmembrane pressure and (B) time (operating conditions: NaCl concentration = 1.5 M, TMP range = 0.2–2.0, and time = 0–94 hours) [23].

in the form of conductivity and chloride-ion concentration in the permeate side during long 94 hours of operation. Over time the RSF decrease may be due to formation of the layer due to slow CFR of DS in the system.

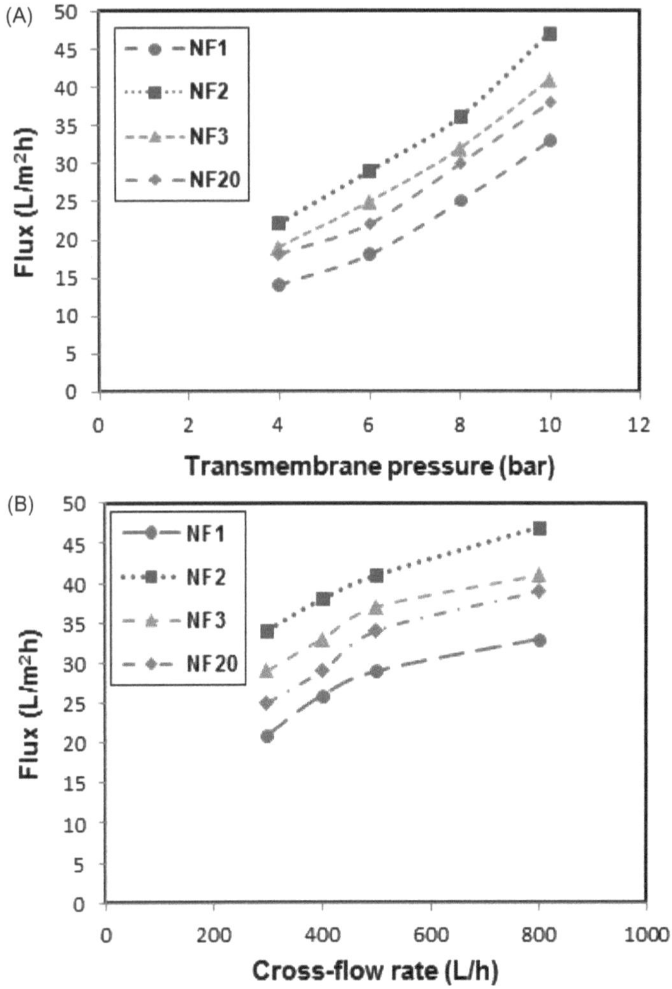

Figure 6.2.24 The effect of (A) transmembrane pressure (b) cross-flow rate on flux through NF-1, NF-2, NF-3, and NF-20 membrane (operating conditions: pressure range 4–10 bar and cross flow rate 300–800 L/hour at ambient temperature) [23].

6.2.3.6 Downstream Purification of DS by NF

Effect of TMP and CFR on Flux

Fig. 6.2.24A and B shows that hydraulic TMP and CFR have a positive correlation with the water flux of DS (NaCl) with NF membrane. As hydraulic TMP is increased from 4 to 10 bar for the NF-1 membrane, the salt-free pure water flux increases from 24 to 45 L/(m^2hour) while NF-2 shows the maximum water flux at 65 L/(m^2hour) (Fig. 6.2.24A). The flux pattern of the four polyamide flat-sheet NF membranes based on pore size were in increasing order (NF-2 > NF-20 > NF-3 > NF-1). Similarly as the CFR is increased from 300 to 800 L/hour, NF-2 shows the maximum water flux 65 L/(m^2hour) while the minimum water flux was obtained by NF-1 (44 L/(m^2hour)) at fixed TMP of 10 bar due to different pore size (Fig. 6.2.24B).

Through rigorous experiments it was found that 10 bar TMP and 800 L/hour CFR are the optimum parameters for maximum flux for NF for the recovery of draw solute. CFR also plays a crucial role in reducing membrane fouling and minimizing the membrane-area requirement in a cross-flow module. With an increase in the CFR, the sweeping action on the active membrane-surface area also increases thereby reducing CP. In other flow modes, an increase in solvent flux normally accompanies an increase in CP. However, in flat-sheet, cross-flow module such enhancement of flux following increased cross-flow does not really lead to any increase in CP. Again, reduction of CP results in convective force, which in turn enhances the solvent flux.

Effect of TMP and CFR on Rejection of Salt From DS

Fig. 6.2.25A and B indicates that hydraulic TMP and CFR have a positive correlation with retention of the draw solute (NaCl) by NF membrane. Retention of NaCl is maximum for NF-1 membrane, which increases from 92% to 98% on increase of CFR from 300 to 800 L/hour at fixed TMP 10 bar (Fig. 6.2.25A). Similarly, retention of NaCl by the NF-1 membrane increases from 88% to 98% on increase of TMP from 4 to 10 bar at a fixed CFR of 800 L/hour (Fig. 6.2.25B). This solute retention behavior is attributed to the solution-diffusion mechanism of mass transport through the NF membrane, in which the solute and solvent fluxes are uncoupled and thus an increase in TMP causes an increase in solvent flux and a decrease in solute flux results in higher retention or recovery for recycling. The uncoupled nature of the solute and solvent fluxes results in higher retention of ionic species like Na^+ and Cl^- as flux increases with increasing CFR and TMP. Being negatively charged the polyamide composite membrane mainly retains the draw solute due to the Donnan exclusion mechanism.

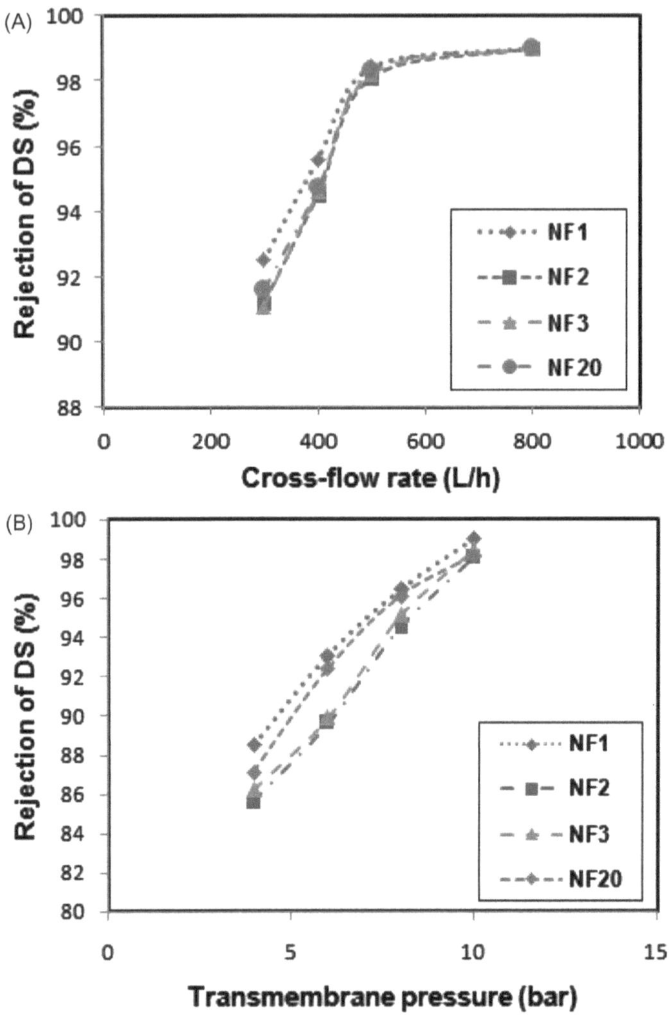

Figure 6.2.25 The effect of applied (A) cross flow rate (CFR) and (B) transmembrane pressure (TMP) on concentrated NaCl rejection percentage for NF-1, NF-2, NF-3, and NF-20 membranes (operating conditions: TMP = 4–10 bar, CFR 300–800 L/hour, pH 10.0, and at ambient temperature) [23].

Apart from Donnan exclusion, the sieving mechanism also plays a significant role in the separation of species with large hydrated radii. In the size exclusion or sieving mechanism, where the relative sizes of the membrane pores and the solute dimension assume important roles in determining the degree of separation.

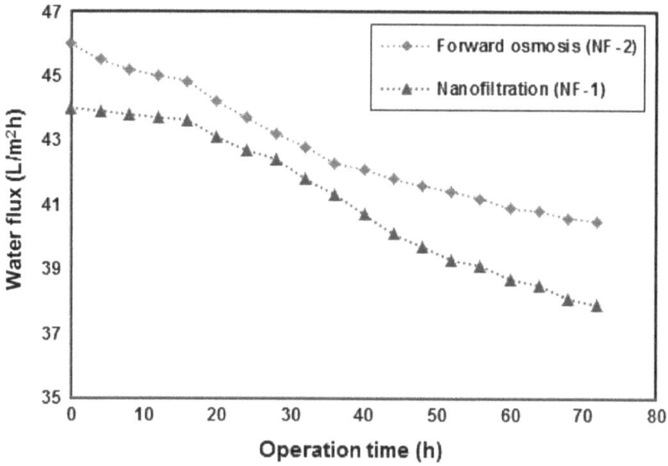

Figure 6.2.26 The effect of fouling on flux with respect to time during FO process and NF process during 72 hours of operation [23].

6.2.3.7 Time Profile of Water Flux During FO and NF

The most common phenomenon observed during long-term operation of a membrane-filtration system is the build-up of CP, the main reason for flux decline. However, build-up of CP is very dependent on the mode of operation and type of membrane modules as well as the prevailing hydrodynamics. If the membrane module is operated in dead-end mode, CP builds up rapidly. Fouling can be avoided to a large extent by choosing a proper module such as a flat-sheet, cross-flow module where the sweeping action of the fluid on the membrane surface reduces the possibility of CP significantly. This study uses a flat-sheet, cross-flow membrane module that largely eliminates fouling. Fig. 6.2.26 shows that during FO the NF-2 membrane suffered only 12% flux decline due to fouling in comparison to a 14% flux decline in the NF-1 membrane during NF over 72 hours of operation. This significant improvement of this design over existing ones has the potential to control CP and membrane fouling and ensure long-term operation at steady flux.

Fig. 6.2.27 shows the SEM images of the NF-2 and NF-1 membranes before and after the FO and NF investigation, which show that the membranes do not undergo major morphological changes perhaps due to the flow design of the cross-flow module. Whatever small fouling occurs can be removed by thorough rinsing with 0.1(N) NaOH, NaOCl (0.01N), and 0.01 (M) HNO_3.

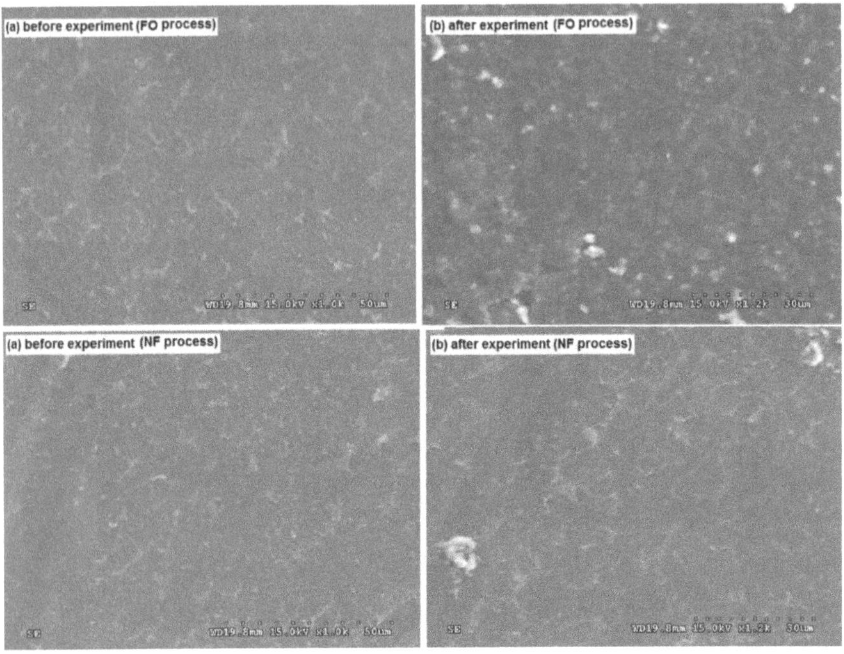

Figure 6.2.27 Surface SEM images of NF-2 and NF-1 membranes (A) before and (B) after experiments [23].

Chemical cleaning with HCl, NaOH, and NaOCl as acidic, alkaline, and alkalioxidizing agents to obtain flux recovery is the most widely used method for reducing fouling. Out of these three chemical reagents, sodium hypochlorite solution (0.01 M) shows the highest flux recovery [26]. The treatment time may vary from 30 to 60 minutes. This system was flushed with deionized water (with no recycling of permeate) using the same flow condition and temperature for 30 minutes. In this experiment, only cleaned membranes were repeatedly employed as long as the difference between pure water flux clean and fresh membrane was within 4%–5%. During downstream NF for recovery of draw solute and pure water, fouling is insignificant and reversible as the feed here comprises only water and sodium chloride and appropriate hydrodynamics are maintained to ensure minimum fouling.

6.2.3.8 Safe Disposal Route of Concentrated Rejects

In the FO–NF system continuous feeding and withdrawal of pure water take place. The system was made to run continuously for 7 days (12 hours per day) to get 25 L of permeate water with a flux of 45 L/(m^2hour). With this aim five modules each with FO and NF membrane of 0.01 m^2 membrane-surface area were used. After seven days of operation, the concentrations of the major contaminants are high (cyanide 950 mg/L; phenols 1300 mg/L; NH_4^+-N 19,200 mg/L; COD 19,700 mg/L). Proper treatment is required prior to disposal of these pollutants. Fenton's reagent ($FeSO_4 \cdot 7H_2O$ and H_2O_2) has tremendous potential to degrade mixed-pollutant waste. For example, Fenton's reagent converts cyanide into cyanate and then bicarbonate ion and phenol is converted into CO_2 and H_2O. The high concentration of ammonium-N may be recovered as struvite through chemical precipitation by adding magnesium and phosphate salt. Struvite recovered after chemical precipitation can be used as a slow-releasing fertilizer for ornamental and agricultural plants.

6.2.3.9 Technoeconomic Analysis

A coke wastewater-treatment plant with a capacity of 70,000 L/day (25,550 m^3 per annum) is considered here for cost assessment. The FO setup yielded 46 L/(m^2hour) (considering NF-2 membrane) water flux by separating more than 95% pollutants from the COWW at 1.5 bar applied pressure. Assuming an average of 24 working hours in a day, the water flux produced by the plant is 1080(45 × 24) L/m^2 day and $\frac{70{,}000}{1080 \times 0.25} = 260$, the number of membrane modules (0.25 m^2 area) required. As flux is the same during NF of DS to recover reusable water and concentrated salt, again 260 cross-flow modules were required during NF. Using the standard equation, the cost of the equipment and process is calculated for scale-up. Using the sixth-tenth power law [27], the scale-up cost is defined as:

$$\text{Scale-up cost} = \text{Equipment cost at lab scale}\left[\frac{\text{Capacity at industrial scale}}{\text{Capacity at lab scale}}\right]^{0.6}$$

The calculated cost (capital and operational cost) is shown in Table 6.2.15. The operating costs include the cost of chemicals, electricity, membrane, and labor while the capital costs include mechanical engineering, membrane module, and civil investment cost for the installation and electrotechnical considerations.

The cost assessment is based on the annualized investment and annualized operational costs. The annualized capital cost is computed by:

$$\text{Annualized capital cost} = \left(\frac{\text{Total capital(\$)} \times \text{Cost recovery factor}}{\text{Water flux per year (m}^3\text{/year)}}\right)$$

Table 6.2.15 Capital and operating cost of a 70,000 L/day capacity COWW treatment plant [23]

Cost parameters	No of equipment with specification	Cost value ($)
Capital cost		**Cost ($)**
i. Civil investment	80 m^2 (20 m × 4 m) space	3500
ii. Membrane module cost for FO	260 no of module perspex sheet (0.25 m^2 area)	30,000
iii. Membrane module cost for NF	260 no of module SS 316 (0.25 m^2 area)	130,000
iv. Membrane module cost for MF	30 no of module perspex sheet (0.25 m^2 area)	3500
v. Large volume tank cost	2 (Fiber tank, 80,000 L capacity)	10,600
vi. High flow pump	1 (Submersible pump)	300
vii. High-pressure pump cost	2 (Diaphragm pump, Max. pr. 50 bar)	6600
viii. Low-pressure pump cost	2 (Peristaltic pump)	800
ix. Cost for main feed pipe	65 m long and 0.075 m dia	800
x. Others pipe fittings and electrotechnical cost	Rotameter (4), Pr. Gauge (3), pH probe (2)	500
Total cost		186,600
Operating cost		**Cost ($/Year)**
i. Electricity cost	Power consumption-4000 KWh/month	4000
ii. Membrane cost for NF	Membrane needed-65 m^2, Cost-50 $/$m^2$ Membrane life-6 months	6500
iii. Membrane cost for FO	Membrane needed-65 m^2, Cost-50 $/$m^2$ Membrane life-6 months	6500
iv. Membrane cost for FO	Membrane needed-7.5 m^2, cost-50 $/$m^2$ Membrane life-6 months	750
v. Labor cost	No of labor 4	2880
vi. Chemical cost	NaCl, NaOH, HCl	370
Total cost		21,000

The cost recovery factor was dependent on the plant project life ($n = 15$ years) and interest ($i = 8\%$) and can be calculated by:

$$\text{Cost recovery factor} = [i(1+i)^n/((1+i)^{n-1} - 1)]$$

Again, the annualized operational cost can be computed by:

$$\text{Annualized operational cost} = \left(\frac{\text{Total operational cost (\$/year)}}{\text{Water flux per year (m}^3\text{/year)}}\right)$$

Thus the annualized cost for production of 1000 L reusable water is the sum of the annualized capital and annualized operating costs and is $(0.73 + 0.82) = \$1.5$. This estimation covers the main factors for the treatment of COWW; however, more detailed analysis needs to be carried out before the installation of a full-scale treatment plant.

The effective treatment of a large amount of complex wastewater from the coke-making industries is difficult and can result in serious environmental pollution. This study focuses on development of a new integrated FO–NF system that succeeds in separating more than 98% of the major contaminants present in the COWW thereby turning it reusable. This design reduces CP significantly and results in long-term operation at a reasonably high flux of 46 L/(m^2hour). Out of the three different DSs, 1.5 M NaCl is the best for FO. Treatment plants based on this type of scheme are expected to be operationally fast and environmentally friendly. The findings are expected to be very useful in the design and operation of industrial-scale coke wastewater-treatment plant. Preliminary economic evaluations indicate economic viability as the cost of treatment of 1000 L wastewater at only \$ 1.5. The technoeconomic evaluation is expected to increase to the level of confidence in installation of a new FO–NF plant for treatment of not only COWW but also other similar wastewater-treatment plants.

6.2.4 ADVANCED MEMBRANE-INTEGRATED BIOCHEMICAL TREATMENT TECHNOLOGY

6.2.4.1 Introduction

With the availability of highly selective and durable membranes in the area of separation and purification, membrane-integrated novel processes are now emerging with the promise of better-quality treated water. While there are several approaches to developing membrane-integrated hybrid plants, in earlier sections, development of membrane-integrated chemical treatment technology and NF–RO integrated with chemical treatment technology was described. In this section, technology combining chemical, biological, and membrane separation will be discussed with emphasis on the development of mathematical models as they are essential for industrial scale-up. The technology developed by Kumar and Pal [28] along with its mathematical model and economic analysis paves the way for industrial scale-up. The microbial degradation of phenol by *Pseudomonas putida* is modeled with the Haldane–Andrew approach. Degradation of residual NH_4^+ by nitrification and denitrification has been modeled using modified Monod kinetics. The model successfully predicts the plant performance with reasonably low RE (0.03–0.18) and high Willmott d-index (>0.98).

6.2.4.2 Theory and Model Development

Chemical and Biological Treatment Unit

Toxic cyanide compounds as found in COWW can be degraded into less harmful compounds using Fenton's reagent. Cyanide removal in this type of chemical treatment follows the first-order kinetics. Similarly, NH_4^+ present in high concentration may also be precipitated out in the form of magnesium ammonium phosphate, or MAP (also known as struvite, $MgNH_4PO_4 \cdot 6H_2O$), by adding magnesium and phosphate salts to wastewater.

Modeling of phenol, NH_4^+, and nitrate degradation involves mass balance of these pollutants, and requires knowledge of the flow pattern of the liquid phase and process kinetics. The following assumptions are made here: (a) microorganisms are preacclimated to the substrate and can metabolize substrate in the aqueous phase only; (b) during biodegradation, the pH of the system remains constant; (c) no degradation of substrate takes place in the settling tanks; (d) dissolved oxygen (DO) content is sufficient enough to sustain the aerobic reactions in the nitrification process; (e) food-to-microorganism ratio (F/M) is maintained in the ranges of 0.06–0.1 kg COD kg^{-1} VSS d^{-1} in the nitrification unit; (f) DO concentration is maintained at less than 0.5 mg/L in the denitrification unit; and (g) substrate-specific treatment occurs in each reactor and other substrates remain inert if present simultaneously.

To simulate the continuous operation of the treatment plant resulting in degradation of phenol, NH_4^+, and nitrate in the reactor, the treatment scheme shown in Fig. 6.2.28 is modeled. The mass balance for the

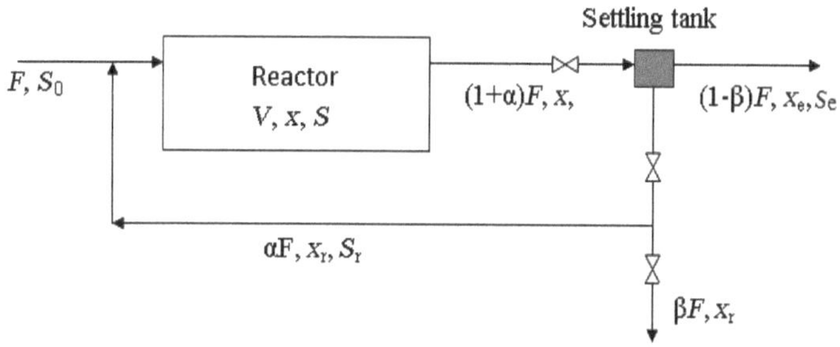

Figure 6.2.28 Flow pattern of substrate and biomass wastewater in all three reactors with settling tank used for the material balance during biological treatment [28].

reactors is done with variable parameters like substrate, biomass, and specific growth rate of microorganism. The reflux ratio for all three cycles is taken to be 10%. The intentional wasting coefficient (β) is a parameter used to maintain the sludge retention time (SRT). When effluent substrate concentration (S_e) drops below the detection limit β can be optimized to stabilize the system.

Chemical and Biological Treatment Unit

Following the material flow scheme the mass-balance equations may be written as:

Biomass

$$V_r \frac{dx_i}{dt} = \alpha_i F_i x_{r,i} - (1 + \alpha_i) F_i x_i + V_r \mu_i x_i \tag{6.2.55}$$

Or

$$\frac{dx_i}{dt} = \alpha_i D_i x_{r,i} - (1 + \alpha_i) D_i x_i + \mu_i x_i \tag{6.2.56}$$

where V_r is the volume of reactor (L), x_i is the biomass in each reactor (mg/L), α_i is the reflux ratio, μ_i is the specific growth rate of microorganism (h^{-1}), and D_i is the dilution rate (F_i/V_r):

Substrate

$$V_r \frac{dS_i}{dt} = F_i S_{0,i} + \alpha_i F_i S_{r,i} - (1 + \alpha_i) F_i S_i - \frac{V_r \mu_i x_i}{Y_i} \tag{6.2.57}$$

Or

$$\frac{dS_i}{dt} = D_i(S_{0,i} - S_i) + \alpha_i D_i(S_{r,i} - S_i) - \frac{\mu_i x_i}{Y_i} \tag{6.2.58}$$

where S_o is the concentration of substrate in the reactor (mg/L), S_r is the concentration of substrate in the recycle stream (mg/L), and Y_i is the yield coefficient for substrate consumption.

The material balance for the clarifier (assuming no accumulation and negligible volume of the clarifier biomass) is:

$$(1 + \alpha_i)F_i x_i = (\alpha_i + \beta_i)F_i x_{r,i} + (1 - \beta_i)F_i x_{e,i} \tag{6.2.59}$$

where β_i is the intentional wasting rate, F_i is the flow rate (mg/L), and x_i, x_r, and x_e are the biomass concentration at reactor, recycle stream, and effluent, respectively:

$$(1 + \alpha_i) = (\alpha_i + \beta_i)\frac{x_{r,i}}{x_i} + (1 - \beta_i)\frac{x_{e,i}}{x_i} \tag{6.2.60}$$

Or

$$1 + \left(1 - \frac{x_{r,i}}{x_i}\right)\alpha_i = \beta_i \frac{x_{r,i}}{x_i} + (1 - \beta_i)\frac{x_{e,i}}{x_i} \tag{6.2.61}$$

The volume of reactor (V) may be expressed as (Bailey and Ollis, 1986):

$$V_r = \mathrm{SRT}\left(1 + \alpha_i - \alpha_i \frac{x_{r,i}}{x_i}\right)F_i \tag{6.2.62}$$

where the SRT is the sludge retention time (d).

Thus from Eqs. (6.2.61) and (6.2.62):

$$1 + \left(1 - \frac{x_{r,i}}{x_i}\right)\alpha_i = \beta_i \frac{x_{r,i}}{x_i} + (1 - \beta_i)\frac{x_{e,i}}{x_i} = \frac{1}{D_i \cdot \mathrm{SRT}_i} \tag{6.2.63}$$

From Eq. (6.2.63) the recycle ratio and β value may be obtained:

$$\frac{x_{r,i}}{x_i} = \frac{(1 + \alpha_i) - (1 - \beta_i)\dfrac{x_{e,i}}{x_i}}{(\alpha_i + \beta_i)} = 1 + \left(\frac{1 - \dfrac{1}{D_i \cdot \mathrm{SRT}_i}}{\alpha_i}\right) \tag{6.2.64}$$

and

$$\beta_i = \frac{\left(\frac{x_i}{D_i \cdot SRT_i} - x_{e,i}\right)}{(x_{r,i} - x_{e,i})} \tag{6.2.65}$$

x_e may be assumed to be zero here as the effluent must not contain any cell mass that can contaminate other downstream cultures.

Thus:

$$\beta_i = \frac{1}{D_i\left(\frac{x_{r,i}}{x_i}\right)SRT_i} \tag{6.2.66}$$

substrate:

$$(1 + \alpha_i)F_i S_i = (\alpha_i + \beta_i)F_i S_{r,i} + (1 - \beta_i)F_i S_{e,i} \tag{6.2.67}$$

Or:

$$S_{r,i} = \left(\frac{1 + \alpha_i}{\alpha_i + \beta_i}\right)S_i - \left(\frac{1 - \beta_i}{\alpha_i + \beta_i}\right)S_{e,i} \tag{6.2.68}$$

The material balance for the combined reactor and clarifier system may be written as:

Biomass

$$\frac{dx_i}{dt} = x_{r,i}(\alpha_i - \beta_i)D_i + \mu_i x_i \tag{6.2.69}$$

Substrate

$$\frac{dS_i}{dt} = D_i S_{0,i} + D_i(\alpha_i - \beta_i)S_{r,i} - D_i(1 - \beta_i)S_{e,i} - \frac{\mu_i x_i}{Y_i} \tag{6.2.70}$$

where S_o, S_r, and S_e are the concentration of the substrate in the reactor, recycle stream, and effluent (mg/L). To solve the differential equation the iterative Euler method is used and the algorithm is implemented in a MATLAB program.

NF Membrane Separation

The modified ENP equation was developed by Bowen and Welfoot (2002) [29] as a linearized model for the determination of ionic flux (Na^+) from NaCl solution that is passed through a NF membrane. In the present investigation, transport of chloride and bicarbonate ions through the NF membrane may be expressed as:

$$J_i = K_{i,c} C_i V - (D_{i,p}) \frac{dc}{dx} - \frac{z_i C_i D_{i,p} F}{RT} \frac{d\psi}{dx} \tag{6.2.71}$$

The flux (J) of ion i is the sum of the fluxes due to convection, diffusion, and electromigration where $K_{i,c}$ is the challenge factor for convection, C_i is the concentration in the membrane of ion i (mol/m), $D_{i,p}$ is the hindered diffusivity of ion i (m^2/second), F is the Faraday constant, z_i is the valence of ion i, R is the universal gas constant (J/mol · K), T is the absolute temperature (K), and $\Delta\Psi_d$ is the Donnan potential difference (V).

The solvent velocity through NF membrane may be expressed using the Hagen–Poiseuille type equation as:

$$V = \frac{r_p^2 \, \Delta P_e}{8\eta\Delta x} \tag{6.2.72}$$

where ΔP_e is the effective pressure driving force and is expressed as $\Delta P_e = dp = (\Delta p - \Delta\pi)$. In this case, $\Delta\pi$ is the osmotic-pressure difference, r_p is the effective pore radius (nm), Δx is the effective membrane thickness (m), and η is the dynamic viscosity of the solution (kg/m per second). The potential gradient through the membrane as derived from the ENP equation (Eq. 6.2.73) can be expressed as:

$$\frac{d\psi}{dx} = \frac{\frac{z_1 V}{D_{1,p}}(K_{1,p} C_1 - C_{1,p}) + \frac{z_2 V}{D_{2,p}}(K_{2,c} C_2 - C_{2,p})}{\frac{F}{RT}\left(z_1^1 C_1 + z_2^2 C_2\right)} \tag{6.2.73}$$

The electroneutrality conditions within the pore and the permeate solutions are [29]:

$$z_1 C_1 + z_2 C_2 = -X_d \tag{6.2.74}$$

$$z_1 C_{1,p} = -z_2 C_{2,p} \tag{6.2.75}$$

where X_d = membrane-charge density (mol/m^3).

From Eqs. (6.2.74) and (6.2.75) C_1 and $C_{1,p}$ can be calculated and substituted into Eq. (6.2.73) yielding:

$$\frac{F}{RT}\frac{d\psi}{dx} = \frac{\left(\frac{K_{1,c} V}{D_{1,p}} - \frac{K_{2,c} V}{D_{2,p}}\right) C_2 - \left(\frac{V}{D_{1,p}} - \frac{V}{D_{2,p}}\right) C_{2,p} - \left(\frac{K_{1,c} V X_d}{D_{1,p}}\right)}{2C_2 - X_d} \tag{6.2.76}$$

Membrane-surface concentrations of both ions were computed using the principle of electroneutrality. Neglecting the salvation energy barrier, the Donnan equilibrium may be expressed as:

$$\frac{C_i}{C_i} = \Phi_i \exp\left(\frac{-z_i F \Delta\psi_d}{RT}\right) \tag{6.2.77}$$

where Φ is the steric coefficient.

The concentration gradient for ion 2 may be derived from the ENP (Eq. 6.2.73) and expressed as:

$$\frac{dC_2}{dx} = \frac{V}{D_{2,p}}(K_{2,c}C_2 - C_{2,p}) - z_2 C_2 \frac{F}{RT}\frac{d\psi}{dx} \tag{6.2.78}$$

Substitution of Eq. (6.2.76) into Eq. (6.2.78) yields an equation with a numerator higher than the denominator. The concentration gradient will be effectively constant showing that the effect of C_2 term is relatively small. Under these conditions, the concentration gradient may be approximated as:

$$\frac{\Delta C_2}{\Delta x} = \frac{\left(\frac{K_{1,c}V}{D_{1,p}} + \frac{K_{2,c}V}{D_{2,p}}\right) C_{2,av}\left[C_{2,av} - X_d\right] - \left[C_{2,av}C_{2,p}\left(\frac{V}{D_{1,p}} + \frac{V}{D_{2,p}}\right)\right] + \left(\frac{VC_{2,p}X_d}{D_{2,p}}\right)}{2C_{2,av} - X_d} \tag{6.2.79}$$

The Donnan potential at the pore inlet ($x = 0$) is the same for both ions and may be obtained from Eq. (6.2.77):

$$\Delta\psi_d(0) = -\frac{RT}{F}\left[\ln\left(\frac{C_1(0)}{\varphi_1 C_f}\right)\right] = \frac{RT}{F}\left[\ln\left(\frac{C_2(0)}{\varphi_2 C_f}\right)\right] \tag{6.2.80}$$

Algebraic manipulation of Eq. (6.2.80) with Eq. (6.2.73) yields:

$$C_2(0) = \frac{X_d + \sqrt{\left(X_d^2 + 4\varphi_1\varphi_2 C_f^2\right)}}{2} \tag{6.2.81}$$

Similarly, an equivalent quadratic expression at the pore outlet ($x = \Delta x$) gives:

$$C_2(\Delta x) = \frac{X_d + \sqrt{\left(X_d^2 + 4\varphi_1\varphi_2 C_{2,p}^2\right)}}{2} \tag{6.2.82}$$

Here C_2 (Δx) is calculated with an estimated value of $C_{2,p}$ that is checked with a new $C_{2,p}$ value from Eq. (6.2.81):

$$\Delta C_2 = C_2(\Delta x) - C_2(0) \tag{6.2.83}$$

$$C_{2,av} = \frac{C_2(0) + C_2(\Delta x)}{2} \tag{6.2.84}$$

Rearrangement of Eq. (6.2.79) yields the following explicit expression for $C_{2,p}$:

$$C_{2,p} = \frac{(Pe_1 + Pe_2)X_d C_{2,av} - (Pe_1 + Pe_2)C_{2,av}^2 + (2C_{2,av} - X_d)\Delta C_2}{\left(\frac{Pe_2}{K_{2,d}} X_d\right) - \left(\frac{Pe_1}{K_{1,c}} + \frac{Pe_2}{K_{2,c}}\right) C_{2,av}} \tag{6.2.85}$$

where Pe is the Peclet number.

With the help of Eq. (6.2.71) the solvent flux may be calculated as:

$$J_{2,v} = \frac{J_{2,s}}{C_2} \tag{6.2.86}$$

Rejection can be calculated by:

$$R_j = 1 - \frac{C_{2,p}}{C_f} \tag{6.2.87}$$

where C_f and $C_{2,p}$ are the feed and permeate concentration, respectively.

Determination of Physicochemical Parameters Involved in Nanofiltration

The physicochemical parameters of the model equations were computed by empirical relations as follows.

Computation of Pore Radius (r_p) and Effective Membrane Thickness (Δx)

Membrane pore radius (r_p) and effective membrane thickness (Δx) were calculated by validating the rejection and flux data against the experimental values by separation of uncharged solutes (sucrose).

Determination of Hindered Diffusivity ($D_{i,p}$)

Diffusivity of the solutes ($D_{i,p}$) affects the solution viscosity. The hindered diffusivity ($D_{i,p}$) is the product of the bulk-diffusion coefficient ($D_{i,a}$) and hindered diffusivity ($K_{i,d}$), which is expressed as:

$$D_{i,p} = D_{i,a} \times K_{i,d}$$

where $K_{i,d}$ is expressed as:

$$K_{i,d} = (1.0 - 2.3\lambda_i + 1.154\lambda_i^2 + 0.224\lambda_i^3) \text{ and } \lambda_i = \left(\frac{r_{i,s}}{r_p}\right)$$

Here, λ represents the ratio of the uncharged solute radius to pore radius (r_s/r_p). Similarly, the challenge factor for the diffusion of ionic components can also be determined. With an assumed value of r_p for NF membrane and using standard values of r_s and r_i the diffusion challenge factors for uncharged and charged anions are estimated under different conditions and are used to solve the model equations.

Determination of Peclet number (Pe_i)

The Peclet number is defined by the help of:

$$Pe_i = \left(\frac{K_{i,c} \cdot V \cdot \Delta x}{D_{i,p}}\right)$$

where $\varphi_i = (1 - \lambda_i)^2$.

6.2.4.3 Computational Procedure

The membrane-charge density is predicted by this model calculation by validating the experimental data with the model-predicted data. The CP model, another new approach, is introduced here to calculate the fouling model data. The model is developed as follows:

1. The first step for the calculation of the model predictive value is membrane-surface concentration calculation, computed using Eq. (6.2.77) for both ions.
2. Eq. (6.2.77) is then used to calculate C_2 (0) with a known feed concentration (C_f) and Eq. (6.2.82) is used to find C_2 (x) with an assumed permeate concentration (C_p) value.
3. ΔC_2 and $C_{2,ave}$ are computed using Eqs. (6.2.83) and (6.2.84) and checking the assumed of C_p using Eq. (6.2.85).
4. Chloride and bicarbonate ion separations are then calculated using Eq. (6.2.33) where the volumetric flux of the solvent is calculated using Eq. (6.2.86).
5. An iterative method is adopted to compute rejection and flux using some assumed value of surface-charge density (X_d) until the assumed value converges with the experimental value.

Table 6.2.16 Typical set of model parameters used in computation [28]

Parameters	Values
Solute radius of H^+ ion (r_{a1})	0.025×10^{-9} m
Solute radius of HCO_3^- ion (r_{a2})	0.142×10^{-9} m
Solute radius of Na^+ ion (r_{b1})	0.167×10^{-9} m
Solute radius of Cl^- ion (r_{b2})	0.116×10^{-9} m
Bulk diffusivity of H^+ ion ($D_{1,}a$)	9.3×10^{9} m^2/s
Bulk diffusivity of HCO_3^- ion ($D_{2,}a$)	1×10^{-10} m^2/s
Bulk diffusivity of Na^+ ion ($D_{1,}b$)	3×10^{-11} m^2/s
Bulk diffusivity of Cl^- ion ($D_{2,}b$)	5.7×10^{-12} m^2/s
Faraday constant (F)	96,500
Boltzmann constant (k)	1.38066×10^{-25} J/K
pH of the solution	10
Solution viscosity	0.01 kg/m · s
Operating pressure	1.5×10^{3} kPa
Cross-flow velocity	1.25 m/s
Feedwater flow rate	800 L/h
Temperature	30°C

Membrane-charge density (X_{DC}, mol/m^3) obtained from the linearized

	Chloride	Bicarbonate
NF1	−402	−385.6
NF2	−210	−260.8
NF3	−260	−345.2
NF20	−485	−545.3

Table 6.2.16 shows the values of the different parameters taken during modeling.

NF-Membrane Separation

Based on the present scheme of separation of ionic contaminants, specifically chloride and bicarbonate ions from chemically and biologically treated effluent using cross-flow NF membrane module, a mathematical model is developed using the ENP approach. Model equations are formulated with the assumptions that (a) the membrane consists of a bundle of identical straight cylindrical pores of radius r_p and length Δx (with $\Delta x >> r_p$); (b) the effective membrane volume charge (X_d) is constant throughout the membrane and is mainly controlled by the feed concentration; and (c) concentration of component i inside the membrane and electric potential are defined in terms of radially averaged quantities.

6.2.4.4 Treatment Plant Configuration and operation

The integrated wastewater-treatment plant consists of a series of CSTR. All the CSTRs (made of stainless steel, SS316) are positioned in vertically lowering order allowing free flow of the liquid to the next stage reactors by gravity eliminating the need for use of pumps in between the reactors as shown in Fig. 6.2.29.

Two membrane modules are connected in series for MF and NF. While membrane modules like hollow-fiber and spiral-bound types are prone to rapid fouling the flat-sheet, cross-flow membrane module used in the present scheme is largely fouling-free. This permits extended use of the same membrane without much decline in flux and reuse of the membrane with simple rinsing and cleaning.

Running the Plant

The setup starts with a cyanide-treatment reactor of volume 30 L where real coke wastewater is first treated with Fenton's reagent ($FeSO_4 \cdot 7H_2O$ and H_2O_2) for removal of cyanide. A second reactor is used for conversion of NH_4^+-N into struvite through addition of magnesium and phosphate salts. The clear supernatant solution passes down to the series of three CSTRs via a settling tank and rotameter. The sequentially arranged CSTRs of volume 15 L each have, respectively, the microbial cultures of *Psuedomonas putida* (NCIM 2152), nitrifying mixed culture (*Nitrosomonas* 5076 *and Nitrobacter* 5078), and *Pseudomonas aeruginosa* (NCIM 5032) in minimal salt medium where all the microbial strains are collected from a national collection of industrial microorganism NCIM from Pune, India. In the last step, biologically treated effluent is microfiltered under low pressure (around 0.2 MPa) and then subsequently passed to a second NF membrane cross-flow membrane module (with effective filtration surface of 0.1 m^2 each module) via a settling tank where almost all the charged and noncharged particles are separated out.

6.2.4.5 Plant-Performance Analysis

Conversion in Chemical Treatment Unit and Precipitation of Struvite

Fenton's reagent ($FeSO_4 \cdot 7H_2O$ and H_2O_2) is used for degradation of the cyanide. The ratios of H_2O_2:COD and H_2O_2:$FeSO_4 \cdot 7H_2O$ are 5:1 w/w and 1:50 w/v, respectively which are found to be appropriate for the complete removal of cyanide in 10 minutes at pH 10.0 as indicated in Fig. 6.2.30A.

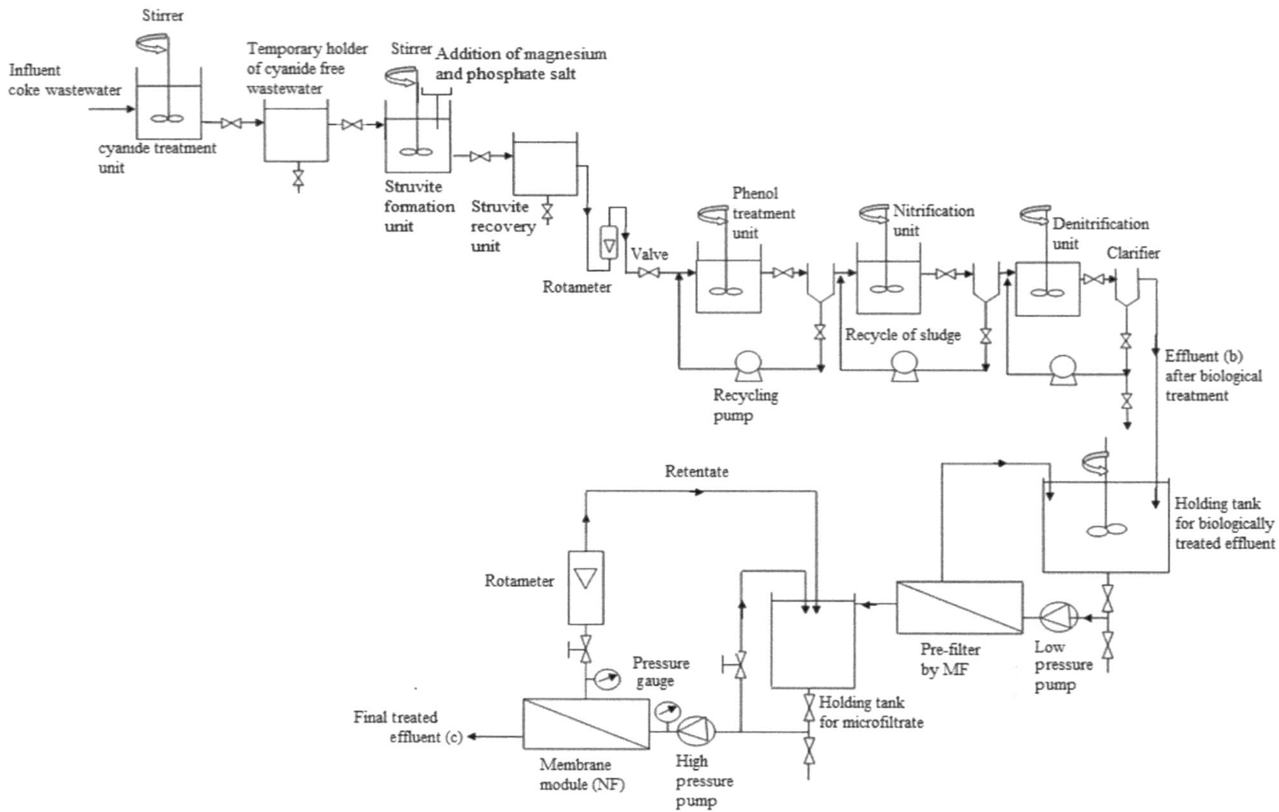

Figure 6.2.29 Membrane-integrated biochemical treatment plant [28].

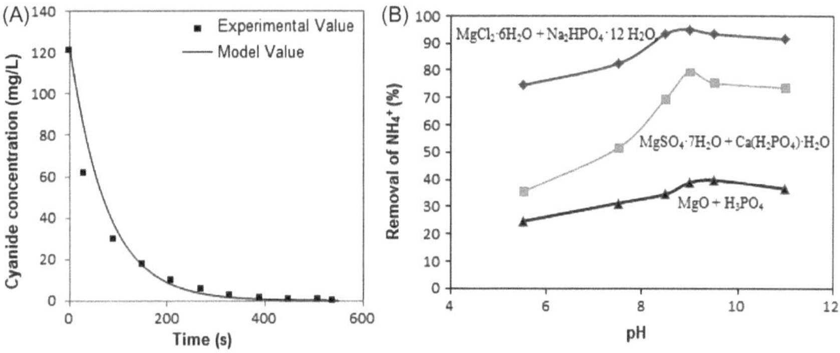

Figure 6.2.30 (A) Concentration profile of cyanide when treated with H_2O_2:COD (5:1) w/w and H_2O_2:$FeSO_4 \cdot 7H_2O$ (1:50) molar ratio; (B) efficiency of NH_4^+-N removal profile from coke wastewater by selection of different combinations of chemicals added at different pH [28].

The plot $-\ln(C/C_o)$ versus t is a straight line ($R^2 = 0.99$) confirms the first-order reaction kinetics of cyanide degradation, where C_o and C are the initial concentration and reactant concentration at time t of cyanide, respectively. Te kinetics constant was obtained as $0.012\ s^{-1}$ where H_2O_2/Fe^{2+} is a critical parameter for improving the efficiency of the Fenton's process. Cyanide can be mineralized to bicarbonate and NH_4^+ by using Fenton's reagent as follows:

$$CN^- + 2H_2O \xrightarrow{H_2O_2} NH_4^+ + HCOO^- \qquad (6.2.88)$$

Fenton's oxidation for cyanide degradation is known as a highly pH-dependent process, and alkaline pH has shown to be better for degradation of cyanide. At high pH, cyanide is present as CN^- ions so it reacts easily with H_2O_2 and Fe^{2+} ions. But at high pH, degradation of aromatic compounds like phenols is very low by Fenton's reagent. In acidic pH, cyanide is present mainly as HCN gas, which is very difficult to oxidize. Ferrous doses lead to the generation of more $OH^{\bullet}$ and act as a catalyst. Hydroxyl radicals are devoid of any charge, have high affinity for electrons, and can quickly strip any chemical of electrons including cyanide causin g their oxidation.

The effluent from Fenton's treatment unit contains enhanced concentration of NH_4^+ (4300 mg/L) due to conversion of cyanide to NH_4^+ and bicarbonate ion. NH_4^+-N is precipitated out as MAP from the Fenton's treated effluent in the struvite byproduct section. In the first step of

MAP precipitation, different combinations of the chemicals $MgCl_2 \cdot 6H_2O$, $Na_2HPO_4 \cdot 12H_2O$, MgO, H_3PO_4, $Ca(H_2PO_4)_2 \cdot H_2O$, and $MgSO_4 \cdot 7H_2O$ at different molar ratios of Mg^{2+} and PO_4^{3-} with respect to NH_4^+ concentration are investigated to find the best combinations for maximum precipitation of NH_4^+-N as struvite.

Fig. 6.2.30B shows that the combination of $MgCl_2 \cdot 6H_2O$, $Na_2HPO_4 \cdot 12H_2O$ and NH_4^+ is the best when used at the molar ratio of 1:1:1 at pH 9.0. Struvite precipitation follows (Eq. 6.2.89):

$$MgCl_2 \cdot 6H_2O + NH_4^+ + Na_2HPO_4 \cdot 12H_2O \rightarrow MgNH_4PO_4 \cdot 6H_2O \downarrow + 2NaCl + 12\,H_2O + H^+ \quad (6.2.89)$$

The combination of $MgCl_2 \cdot 6H_2O$ and $Na_2HPO_4 \cdot 12H_2O$ is the most efficient for NH_4^+ removal, but this combination also leads to high salt concentration in the effluent compared to the other two combinations of chemicals as evident from Eq. 6.2.89. The salts produced at this stage are subsequently removed by NF membrane. A first-order kinetic model is applied to the plant data obtained during NH_4^+ precipitation. By integrating and ordering the terms, the following linear form of the first-order rate equation may be obtained:

$$-\ln\frac{C - C_e}{C_0 - C_e} = kt \quad (6.2.90)$$

where C_o is the initial concentration of the reactant (NH_4^+). Assuming first-order reaction kinetics of NH_4^+ precipitation, a plot of $-\ln[(C - C_e)/(C_0 - C_e)]$ against the reaction time generates a straight line ($R^2 = 0.99$) with slope (k) equal to 0.12 per minute and intercept equal to zero. The equilibrium concentration of NH_4^+ (C_e) can be obtained from the intercept with abscissa at $-dc/dt$ equal to zero when this parameter is plotted against the concentration of NH_4^+ remaining in the filtrate. It is found that the equilibrium C_e is 201 mg/L NH_4^+-N.

The MAP precipitates form rapidly and settle quickly at the bottom of the reactor in the absence of stirring. The content of struvite in the precipitates can be confirmed through SEM with energy dispersive X-ray analysis, FT-IR analysis, and X-ray diffraction as shown in Fig. 6.2.31A–D. The FT-IR analysis, SEM–EDS, and XRD patterns show that the infrared spectrum of the precipitate and elemental profile are close to that of the MAP. The thermogravimetric analysis and differential thermogravimetric analysis curve for struvite at 10°C per minute

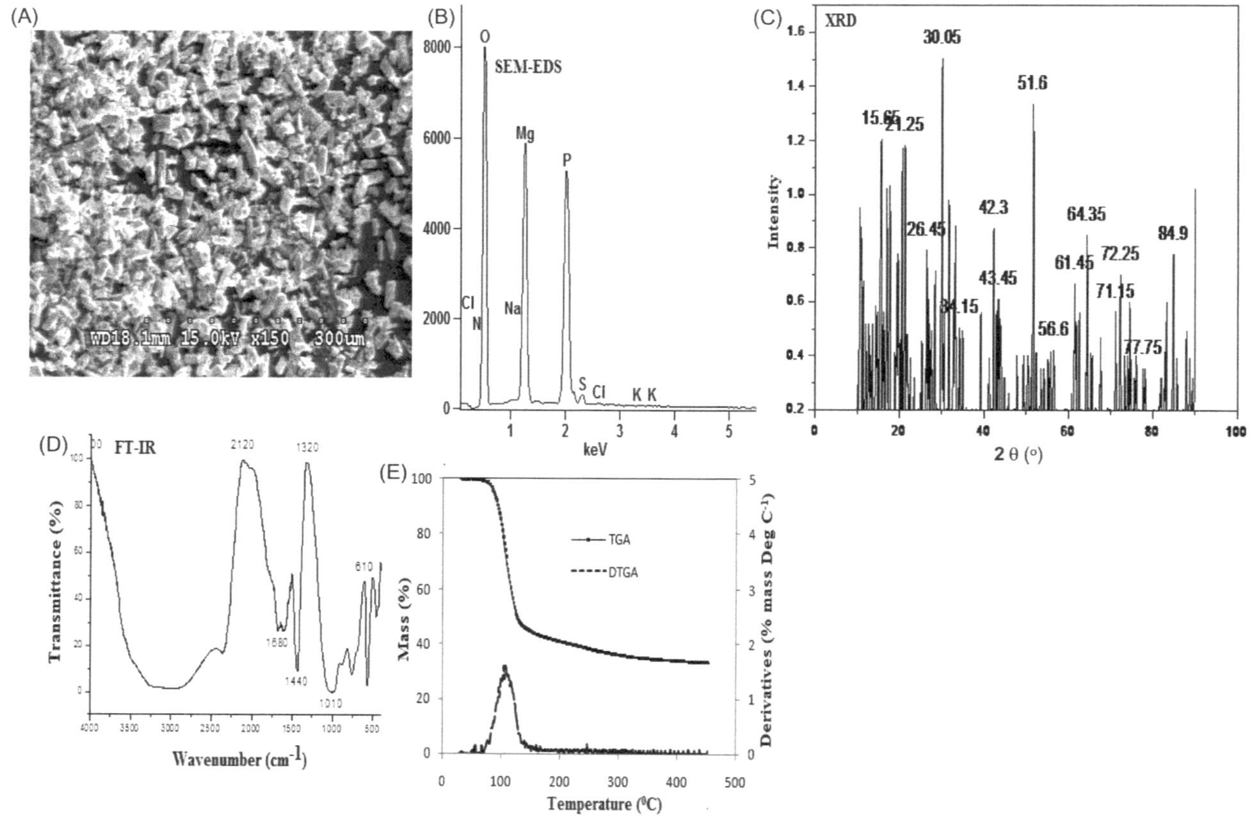

Figure 6.2.31 Surface characterization analysis of struvite (MAP precipitate) obtained during the chemical treatment of NH_4^+. (A) Scanning microscopy analysis of struvite; (B) Energy-dispersive X-ray analysis of struvite; (C) FT-IR spectroscopy analysis of struvite; (D) XRD pattern of the struvite crystel; and (E) TGA and DTGA curves for struvite for heating rate 10°C per minute [28].

are shown in Fig. 6.2.31E. These data indicate that mass loss begins at a temperature around 55°C and is essentially complete when the temperature exceeds 250°C and ~51% corresponds to:

$$MgNH_4PO_4 \cdot 6H_2O \rightarrow MgHPO_4 + NH_3\uparrow + 6H_2O\uparrow \quad (6.2.91)$$

The DTGA curve of the struvite for heating rate 10°C per minute shows a single peak, which is attained at a temperature of 103°C due to simultaneous loss of both ammonia and water molecules, indicating the precipitate material is struvite.

Biodegradation

Experimentally the effects of phenol, NH_4^+, and nitrate concentration on the growth kinetics of *P. putida,* nitrifying bacteria, and *P. aeruginosa* for different initial substrate concentrations can be found. The kinetic constants (k_s) thus obtained are 14.0, 10.9, and 26.9 mg/L, respectively. The specific growth rate (μ) is determined in the exponential growth phase for each respective microorganism by using different substrate concentrations of phenol, NH_4^+, and carbon source (methanol during denitrification). The variation of experimental specific growth rate (*P. putida*) as a function of phenol concentration fits the Haldane–Andrews model curve with a specific growth rate 0.45 per hour. The substrate inhibition constant and decay coefficient of *P. putida* during phenol biodegradation are found to be 500.8 mg/L and 3×10^{-5} per hour, respectively. The experimental data indicate that nitrification by nitrifying bacteria followed by Monod kinetics with μ value of 0.24 per hour. Degradation of nitrate is found to be of zero order where the growth pattern of *P. aeruginosa* on methanol follows the Monod kinetics with a μ value of 0.29 per hour.

Fig. 6.2.32 shows that experimental and model predicted profiles of degradation of phenol, NH_4^+-N, and nitrate by *P. putida,* nitrifying bacteria, and *P. aeruginosa*, respectively, with initial phenol, NH_4^+-N, and nitrate concentration of 159, 200, and 191 mg/L, respectively. Degradation times are 13, 23, and 14 hours, respectively, for phenol, NH_4^+-N, and nitrate. The kinetic parameters obtained from the batch study are used here to model a continuous system in an integrated plant.

After precipitation of NH_4^+, the supernatant fluid (effluent "a") was transferred for the biological treatment under controlled flow rate. Fig. 6.2.33 shows the biodegradation potential of different microorganisms in degrading their respective substrates and their biomass yields during continuous process. As continuous flow of wastewater containing

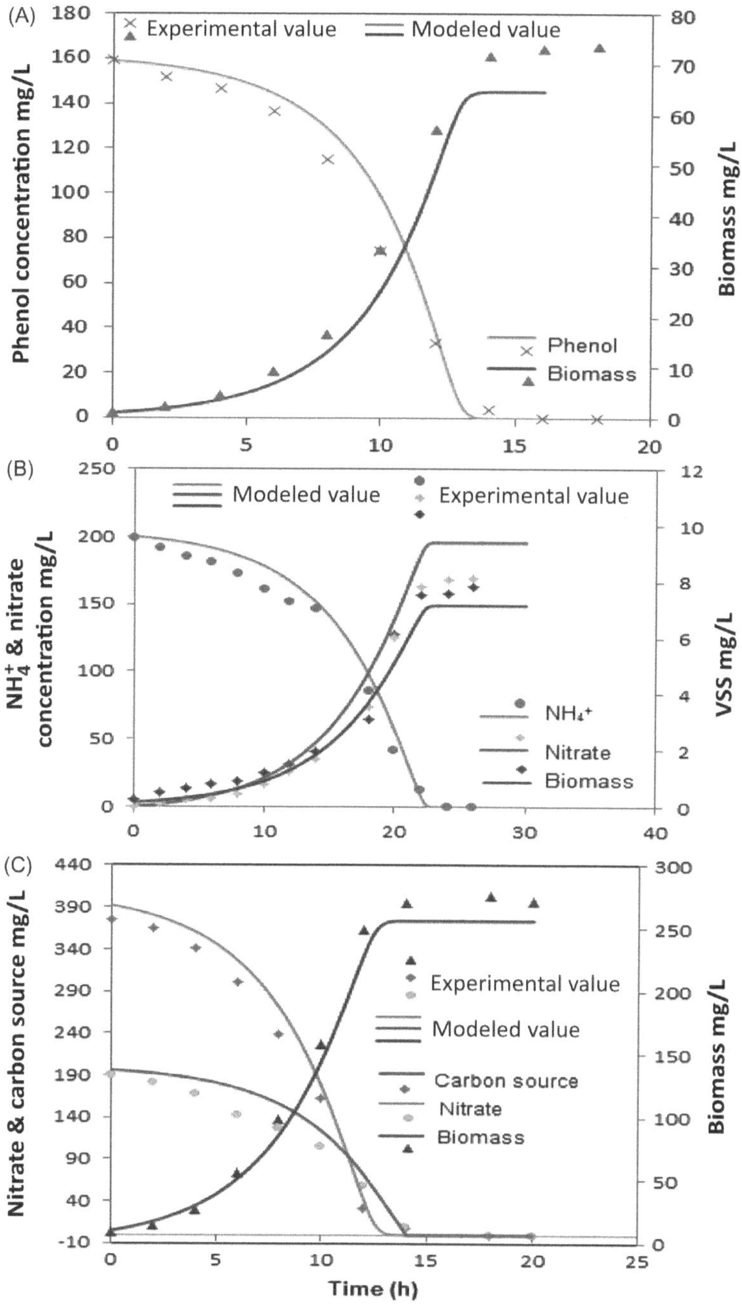

Figure 6.2.32 Experimental and model profiles during batch studies. (A) *Pseudomonas putida* during phenol degradation; (B) Mixed nitrifying bacteria (*Nitrosomonas and Nitrobacter*) during conversion of NH_4^+ to nitrate; (C) *Pseudomonas aeruginosa* during nitrate and carbon source (methanol) degradation [28].

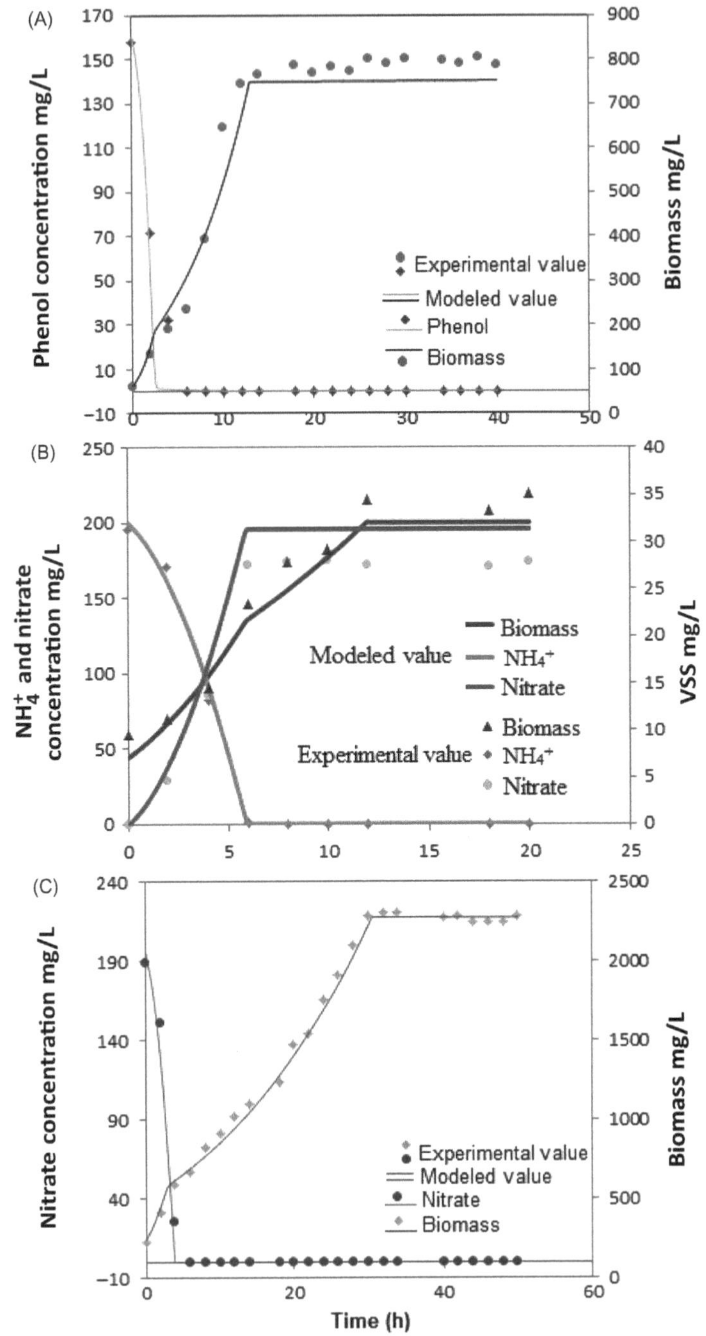

Figure 6.2.33 Experimental and model predicted profiles during continuous treatment of (A) phenol, (B) NH_4^+, and (C) nitrate [28].

phenol, NH_4^+, and nitrate was allowed into the respective reactor system, another exponential growth phase of the microbe population is observed. The exponential growth phase ends as the microbial population becomes high enough to consume influent substrate instantly. At this point, the system is allowed to bleed a part of the biomass to maintain its concentration within a desired level. The HRT and SRT are optimized at 19 hours and 20 d, 23 hours and 16 d, and 15 hours and 30 d, respectively, for the successful treatment of phenol, NH_4^+, and nitrate, respectively. Phenol is almost completely removed regardless of the loading variation. The removal efficiency is always higher than 99% even at very low HRT. The nitrification reactions are generally coupled and proceed rapidly to form nitrate from NH_4^+. Therefore nitrite levels at any given time are usually low as detected in the analysis. Nitrification requires a long retention time of 25–30 hours, a low food-to-microorganism ratio (F/M) 0.08 g COD g^{-1} VSS d^{-1}, a high mean cell residence time (sludge age) 15–25 d, adequate buffering (alkalinity) 3.0 g $NaHCO_3$ L^{-1}, and DO concentration of 4.5 mg/L. Concentration of nitrate in the reactor increases slowly at the beginning of the continuous process. This is due to the continuous removal of reactor contents to maintain flow. As the NH_4^+ concentration depletes nitrate concentration increases. *P. aeruginosa* is used as denitrification heterotrophic bacteria and methanol is used as the carbon source for the growth of the bacteria. The optimum dose of methanol that would result in maximum removal of nitrate is determined experimentally. The methanol:nitrate ratio is 2:1. The facultative bacteria draw DO that results from nitrate molecules. The process is performed under anoxic conditions, when the DO concentration is less than 0.2 mg/L. Methanol was selected as a suitable external carbon source because of its low cost and effective role in denitrification in an environment of relatively low carbon concentration. The rate of denitrification is reported to be zero order for NO_3 concentrations below 400 mg/L (Foglar et al., 2005). The nitrification reaction is described as follows:

$$6NO_3^- + 5CH_3OH \rightarrow 3N_2\uparrow + 5CO_2 + 7H_2O + 6OH^- \qquad (6.2.92)$$

The recycle ratio is maintained at 0.1 in each pair of reactor and settling tank.

Membrane Separation

After chemical treatment and then biological treatment, the effluent is enriched in terms of TDS, salinity, and conductivity due to the

Table 6.2.17 Characteristics of wastewater after final treatment (membrane separation) [28]

Parameter	influent concentration (mg/L)	After chemical & biological treatment (mg/L)	Final effluent (mg/L)	Permissible limit (mg/L)
Cyanide	120	ND[a]	ND	<0.1
Phenol	159	ND	ND	<0.5
NH_4^+-N	4195	ND	ND	<30
Chloride ion	5930	18,100	181	<1000
COD	2470	1570	102	<250
Total carbon	1150	192	4.2	–
Total organic carbon	246	15.5	1.1	–
TDS	11,600	37,800	1360	–
Fluoride	112	110	0.23	<1.5
Sodium	1850	3540	35	<60
Oil & grease	51.5	3.5	N.D	<10
Conductivity (mS/cm)	12.02	38.56	0.92	–
Alkalinity-$CaCO_3$	4250	29,100	582	–
Salinity	8.4	25.9	0.85	
pH	8.5–9.5	8.5–9.0	7.5–8.5	

[a]ND = Nondetectable

introduction of new ionic species such as bicarbonate and chloride ions as reflected in Table 6.2.17.

NF plays a significant role in the removal of these ionic species and the salts in the final stage of treatment of coke wastewater. Fig. 6.2.33A shows the permeate flux data. It shows that the permeate flux increases with increase of TMP for all the NF membranes and varies linearly with applied pressure. The NF-2 membrane yields the highest flux (at 1.5 MPa, 294 L/m^2 per hour flux) due to its large pore radius (0.57 nm) whereas the NF-1 membrane (0.53 nm) yields the lowest flux (at 1.5 MPa, 116 L/m^2 per hour flux) as permitted by its smallest pore radius. However, at 1.5 MPa pressure, the flux of NF-1 of around 115–116 L/m^2 per hour is reasonably good and can be accepted for scale-up. The NF-20 and NF-3 membranes yield a flux intermediate between the two former types as their pore radius (0.54 and 0.55 nm) is also in between. The developed model predicts this increase of flux following increase in operating pressure and the plant data agrees well with these predictions.

Fig. 6.2.34B and C shows the model versus experimental data of removal of bicarbonate and chloride ion by different NF membranes

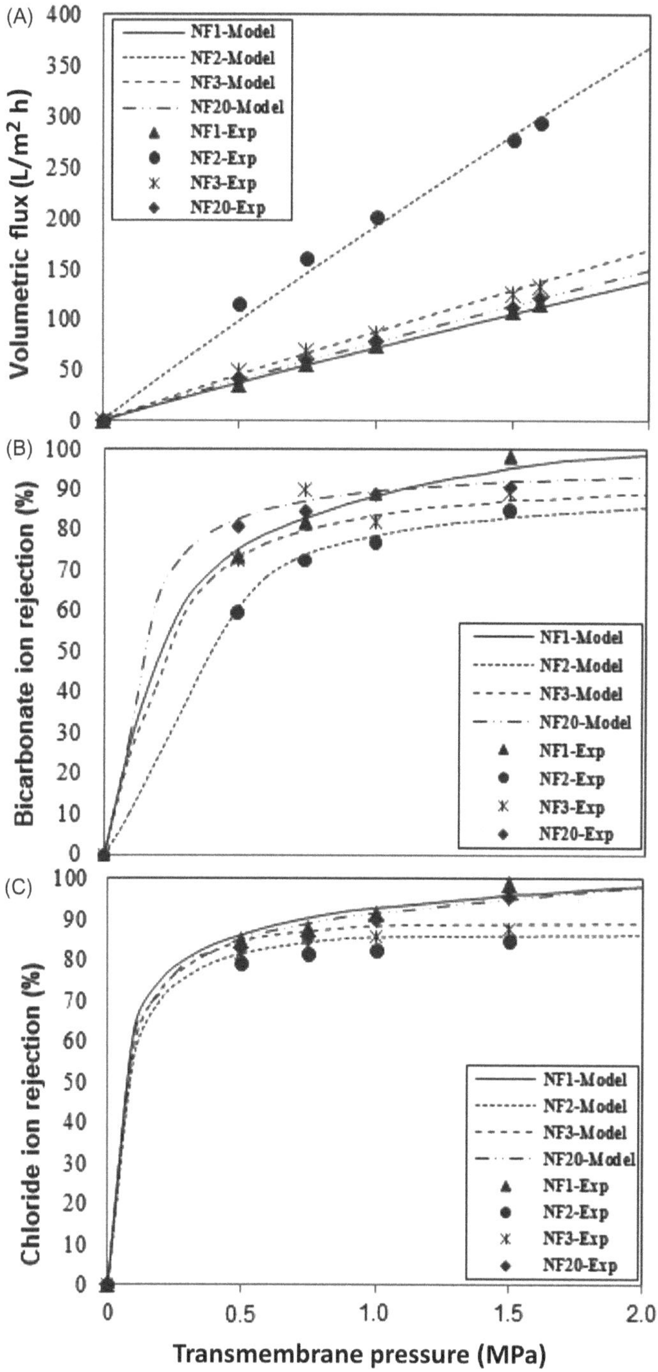

Figure 6.2.34 Comparative (A) flux behavior and (B) rejection of bicarbonate ions and (C) chloride ions performances of the membranes under varying pressure at cross-flow rate of 800 L/hour and pH 9.5 [28].

under varying pressures (from 0.5 to 1.6 MPa) at a pH of around 9.5. However, beyond the operating pressure of 1.5–1.6 MPa, no further improvement in rejection of bicarbonate and chloride ions were observed except a very marginal increase in flux under the investigated maximum pressure range. It is observed that the rejection order followed NF-1 > NF-3 > NF-20 > NF-2 at any particular applied pressure. The data set as shown in Fig. 6.2.33B and C is fitted to the linearized model in order to obtain the effective membrane-charge density X_d. Two transport mechanisms may work for NF membranes. In solution diffusion, solute flux and solvent flux are uncoupled and as a result increase of solvent flux occurs with increase of TMP without increasing solute flux. Thus with increasing pressure, pure water flux will increase, while the solute flux (bicarbonate and chloride ion) is constant and due to the low concentration of the solutes in the permeate side, the overall solute passage decreases. With size exclusion, the relative sizes of the membrane pore and solute dimension play important roles i n determining the degree of separation. Higher pH values lead to an increase in osmotic pressure and ionic strength thus reducing membrane permeability and increasing rejection. High pH favors increased thickness of diffuse double layer of charged functional groups due to deprotonation over the surface of membrane and thus reduces the apparent pore size and results in greater rejection of charged solutes [30]. Donnan exclusion or charge repulsion plays the dominant role as it remains as negatively charged ion. Thus after NF, the final effluent may be recycled in the same plant.

Error Analysis

During investigation, random error is kept at a minimum by averaging. The mean of three values is always considered as the final value. To validate the model, the experimental data are compared with the model-predicted values. The deviations of the experimental findings from the model predictions are calculated in terms of the model errors by calculating the RE and Willmott index (d). The values of d should take from 0 to 1.0, with an index of 1.0 indicating perfect agreement [31] and expressed as:

$$d = 1 - \left[\frac{\sum_{i=1}^{n} (P_i - E_i)^2}{\sum_{i=1}^{n} \left\{ \left|(P_i - \overline{E})\right| + \left|(E_i - \overline{E})\right| \right\}^2} \right] \quad 0 \le d \le 1 \qquad (6.2.93)$$

where E_i = experimental value, P_i = predicted (model) value, and n = no of points analyzed.

where RE and d were found to be in the range of 0.03–0.18 and 0.98–0.99, respectively, indicating that the model can successfully predict system performance.

CONCLUSIONS

The advanced treatment technology described here is a major departure from existing systems of wastewater-treatment systems. Statistical analysis indicates the model (with overall correlation coefficient being of the order of 0.99) can successfully predict the performance of the hybrid treatment plant. The first of its kind this model deals with a novel, continuous treatment process for hazardous industrial wastewater that integrates conventional chemical and biological treatments with membrane-based treatment following a well-investigated, logical sequencing of operations and leading to production of valuable struvite byproduct and reusable water. The treatment scheme and model depart from traditional, conventional activated sludge-based processes and appear to be sustainable due to the economicl and fast treatment of industrial wastewater that is not confined to the coke-making industry alone.

REFERENCES

[1] Kumar R, Pal P. Turning hazardous waste into value-added products: production and characterization of struvite from ammoniacal waste with new approaches. J Cleaner Prod 2013;43:59–70.
[2] Doyle JD, Parsons SA. Struvite formation, control and recovery. Water Res 2002;36 (16):3925–40.
[3] Myers RH, Montgomery DC. Response surface methodology: process and product optimization using designed experiments. 1st ed. New York: John Wiley & Sons Inc.; 1995.
[4] Bhuiyan MIH, Mavinic DS, Koch FA. Thermal decomposition of struvite and its phase transition. Chemosphere 2008;70:1347–56.
[5] Hameed BH, Tan IAW, Ahmed AL. Optimization of basic dye removal by oil palm fiber based activated carbon using response surface methodology. J Hazard Mater 2008;158:324–32.
[6] Pal P, Bhakta P, Kumar R. Cyanide removal from industrial wastewater by cross-flow nanofiltration: transport modeling and economic evaluation. Water Environ Res 2014;86(8):698–706 (9)
[7] Zinatizadeh AAL, Mohamed AR, Abdullah AZ, Mashitah MD, Hasnain IM, Najafpour GD. Process modeling and analysis of palm oil mill effluent treatment in an up-flow anaerobic sludge mixed film bioreactor using response surface methodology (RSM). Water Res 2006;40:3193–208.
[8] Wang B, Chang X, Ma H. Electrochemical oxidation of refractory organics in the coking wastewater and chemical oxygen demand (COD) removal under extremely mild conditions. Ind Eng Chem Res 2008;47:8478–83.

[9] Frost R, Weier M, Erickson L. Thermal decomposition of struvite: implications for the decomposition of kidney stones. J Therm Anal Calorim 2004;76:1025–33.
[10] Schafer AI, Nghiem DI, Waite TD. Removal of natural hormone estrone from aqueous solutions using nanofiltration and reverse osmosis. Environ Sci Technol 2003;37(1):182–8.
[11] Madaeni SS, Vatanpour V, Monfared HA, Shamsabadi AA, Majdian K, Laki S. Removal of coke particles from oil contaminated marun petrochemical wastewater using PVDF microfiltration membrane. Ind Eng Chem Res 2011;50:11712–19.
[12] Kumar R, Pal P. A membrane-integrated advanced scheme for treatment of industrial wastewater: dynamic modeling toward scale up. Chemosphere 2013;92(8):1375–82.
[13] Ghose MK. Complete physicochemical treatment for coke plant effluent. Water Res 2002;36:1127–34.
[14] Pal P, Kumar R, Srivastava N, Chowdhury J. A visual basic simulation software tool for performance analysis of a membrane-based advanced water treatment plant. Environ Sci Pollut Res 2014;21:1833–49.
[15] Ferry JD. Ultrafilter membranes and ultrafiltration. Chem Rev 1936;18:373–455.
[16] Bowen WR, Welfoot JS. Modeling the performance of membrane nanofiltration-critical assessment and model development. Chem Eng Sci 2002;57:1121–37.
[17] Kumar R, Chakrabortty S, Pal P, Membrane-integrated physicochemical treatment of coke-oven wastewater: Transport modelling and economic evaluation. Environmental Science and Pollution Research, Springer, Environ Sci Pollut Res 2015;22:6010–23.
[18] Pal P, Kumar R. Treatment of coke-wastewater: a critical review for developing sustainable management strategies. Sep Purif Rev 2014;43(2):89–123.
[19] Santos A, Yustos P, Rodriguez S, Simon E, Garcia-Ochoa F. Abatement of phenolic mixtures by catalytic wet oxidation enhanced by Fenton's pre–treatment: effect of H_2O_2 dosages and temperature. J Hazard Mater 2007;146:595–601.
[20] Marsden J, House I. The chemistry of gold Extraction. New York: Ellis Horwood; 1992.
[21] Pal P, Kumar R. Treatment of coke wastewater: a critical review for developing sustainable management strategies. Sep Purif Rev 2014;43(2):89–123.
[22] Kumar R, Pal P. Response surface-optimized Fenton's pretreatment for chemical precipitation of struvite and recycling of water through downstream nanofiltration. Chem Eng J 2012;210:33–44.
[23] Kumar R, Pal P. A novel forward osmosis-nano filtration integrated system for coke-oven wastewater reclamation. Chemical Engineering Research and Design, ELSEVIER, 100 (2015): 542–553
[24] Kumar R, Pal P. Membrane-integrated hybrid system for the effective treatment of ammoniacal wastewater of coke-making plant: a volume reduction approach. Environ Technol 2014;35(16):2018–27.
[25] Braghetta A, DiGiano FA, Ball WP. Nanofiltration of natural organic matter: pH and ionic strength effects. J Environ Eng 1997;123:628–41.
[26] Madaeni SS, Vatanpour V, Monfared HA, Shamsabadi AA, Majdian K, Laki S. Removal of coke particles from oil contaminated marun petrochemical wastewater using PVDF MF membrane. Ind Eng Chem Res 2011;50:11712–19.
[27] Peters MS, Timmerhaus KD. Plant design and economics for chemical engineers. 4th ed. New York: McGraw-Hill; 1991.
[28] Kumar R, Pal P. A membrane-integrated advanced scheme for treatment of industrial wastewater: dynamic modeling toward scale up. Chemosphere 2013;92(10):1375–82.

[29] Bowen WR, Welfoot JS. Modeling the performance of membrane nanofiltration critical assessment and model development. Chem Eng Sci 2002;57:1121–37.
[30] Braghetta A, DiGiano FA, Ball WP. Nanofiltration of natural organic matter: pH and ionic strength effects. J Environ Eng 1997;123:628–41.
[31] Willmott CJ. Some comments on the evaluation of model performance. Bull Am Meteor Soc 1982;63:1309–13.

FURTHER READING

Beg Q, Sahai V, Gupta R. Statistical media optimization and alkaline protease production from *Bacillus mojavensis* in a bioreactor. Process Biochem 2003;39:203–9.
Zhang TLQ, Ding L, et al. Modeling assessment for ammonium nitrogen recovery from wastewater by chemical precipitation. J Environ Sci 2011;23(6):881–90.

SUBCHAPTER 6.3

Treatment of Pharmaceutical Wastewater

6.3.1 TREATMENT OF PHARMACEUTICAL WASTEWATER: CONVENTIONAL TECHNOLOGY

6.3.1.1 Introduction

Pharmaceutical wastewater is a major environmental problem due to its enormous quantity, complexity, and hazardous nature. The major contaminants of pharmaceutical waste include drug residue with high BOD, COD, toxic substances, surfactants, and volatile organic compounds all of which lead to multifarious health hazards. Around the world the medical industry generates huge quantities of pharmaceutical waste, which includes waste from the pharmaceutical industry as well from homes and hospitals that contaminates the environment. Table 6.3.1 describes the various constituents of pharmaceutical wastewater.

Mixing of pharmaceutical waste with other waste in assorted sewage makes the isolation and administration of separate treatment of a particular component difficult. Moreover, even within the pharmaceutical industry, it is extremely undesirable to let the waste from different segments mix together as it leads to creation of complex hybrid wastewater that is difficult and expensive to treat. When individual treatment strategies are ineffective, unknown problems and new types of wastewater whose treatment and disposal measures are largely unmitigated result. Another

Table 6.3.1 Charateristics of pharamceutical wastewater

Parameters	Standard effluent values
BOD	900 – 400 ppm
COD	2000–6000 ppm
pH	1.5–6.0 (variable)
Oil and grease	35–2000 ppm
TDS	1350–7250 ppm
TSS	500–2000 ppm
TKN	800–1000 ppm

problem encountered is the highly variable nature of pharmaceutical wastewater as the composition of the raw materials varies widely for one class of products to another. In many cases, pharmaceutical companies release pharmaceutical waste to the environment without disclosing the details. Often illogical mixing of industrial effluent, domestic sewage, and hospital waste in the public-sewer handling facilities blows the problem out of proportion [1–4].

Despite the existence of the empirical relationships between the concentration of pharmaceutical compounds and influent wastewater characterization parameters (COD, BOD, TSS, TP, and oil) and the relationships between the removal of pharmaceutical compounds with wastewater characterization parameters and operational parameters of flow and HRT, the correlation between the removal of the pharmaceutical compounds and SRT is poor [5], which further complicates the treatment process of the pharmaceuticals. However, there are studies that indicate very high (>80%) or around 100% removal of pharmaceuticals like ibuprofen, ketoprofen, indomethacin, acetaminophen, and mefenamic acid obtained for SRT maintained between 10 and 20 days [6]. The reappearance of traces of pharmaceutical and personal care products in treated water samples and treated output from wastewater-treatment plants points to the failure of these facilities to completely remove the persistent compounds [7]. Fig. 6.3.1 shows the interaction of industrial wastewater with the environment.

6.3.1.2 Composition and Health Hazards of Pharmaceutical Wastewater

Pharmaceutical industry wastewater can be classified roughly as: (a) municipal wastewater, (b) spent liquors from fermentation processes, (c) chemical waste, d) condenser waste from evaporation, and (e) floor and laboratory

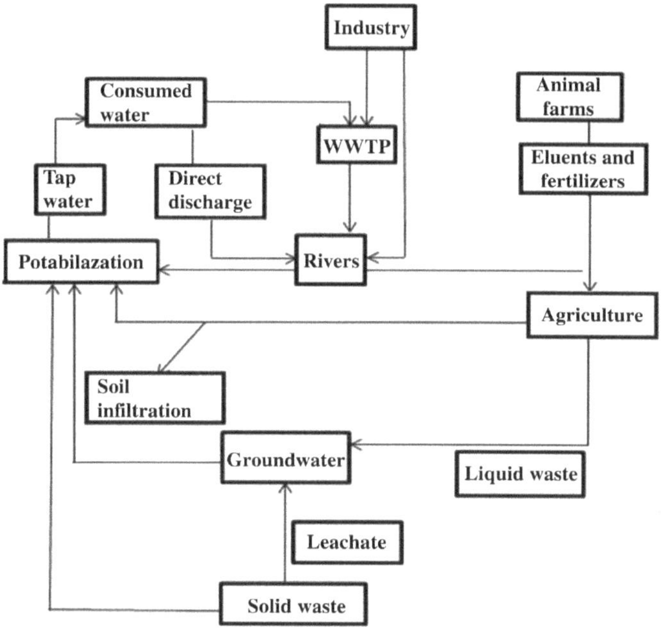

Figure 6.3.1 Schematic showing the interaction of waste in the environment.

washing waste. Biotechnological production of pharmaceuticals leads to discharge of chemicals used as ancillary substances like buffers, chelators, and antibiotics into the aquatic environment. Recent studies indicate that 15%–20% of medicine used in hospitals is potentially bioaccumulative. Added to this is pseudo-persistence, i.e., ecotoxicological impacts of continuous discharge of pharmaceutically active pharmaceutical ingredients (API)], the main noxious constituents. API are complex molecules most of which are polar in nature with molecular weights ranging from 200 to 500/1000 Da and widely used due to specific biological activity which aids in the application of the drugs. Table 6.3.2 presents the composition of pharmaceutical wastewater classified according to different constituents from different sources [8–10].

APIs and the bacteria and fungi that grow on the residue can be highly deleterious. Since 2005 the European Medicines Agency has 2005 that an environmental risk assessment must be included when applying for new marketing authorizations of pharmaceuticals for human use [11]. Hazard mapping and surveys conducted to investigate ecotoxicological levels show a greater risk to workers involved in waste for deadly diseases

Table 6.3.2 Composition of pharmaceutical wastewater

Process liquors	Organic synthesis	Contaminated solvents
Spent fermentation broth	Fermentation processes	Contaminated waters
Spent natural product, raw Materials	Natural product, extraction processes	Leaves, tissues
Spent aqueous solutions	Solvent extraction processes	Contaminated water
Leftover raw material, Containers	Unloading of materials into process equipment	Bags, drums (fiber, plastic, metal), plastic bottles
Volatile organic compounds	Chemical storage tanks	drums Solvents
Spills	Manufacturing and lab operations	Miscellaneous chemicals, mercury, other metals
Wastewater	Equipment cleaning, extraction residue	Contaminated water, phenol-based
Spent solvents Solvent	extraction or wash practices	Contaminated solvents

like lung and skin cancer, bronchitis, tuberculosis, and a host of other such ailments. Evidence suggests that the presence of the ketone group is one of the main ecotoxicity contributors. The contribution of molecular and structural fragments in increasing toxicity in the environment has been established with the help of quantitative interspecies toxicity correlation models for structurally diverse pharmaceuticals (especially for fish). The detrimental effects of these toxic compounds on marine populations and small organisms have been illustrated in several studies [12–17]. Organizations like the WHO have guidelines on the permissible amounts of different compounds in the environment and the health risks of exposure to APIs even when present in dilute concentrations. Researchers have explicitly measured the damage caused to the environment due to sharp decrease in vulture population in the Indian subcontinent due to exposure to diclofenac (an antiinflammatory drug), sensitivity of blue-green algae toward antibiotics, untreated residue of fluoroquinolone contributing to the evolution of fluoroquinolone-resistant *Campylobacter jejuni* (an important human pathogen), and their entry into trophic level of food chains and potential to bioaccumulate. Resistant stocks of APIs have affected the breeding of zebra fish and long-tailed rainbow trout. In some case waste minimization practices have led to the creation of secondary pollution sources. Leachates, fumes, and water-soluble compounds are responsible for widespread degradation of the environment [18–22] impairing soil quality, groundwater, water bodies, and air quality.

6.3.1.3 Challenges in Treatment of Pharmaceutical Wastewater

The adoption of ineffective technology for the treatment of pharmaceutical waste has been reported from several parts of the world [23–25]. Irresponsible and indiscriminate flushing of millions of tons of unused medicine continues all over the world this pharmaceutical waste find its way into municipal wastewater systems adding to the level of hazards through BOD and toxic COD loading. Pharmaceutical residue is potentially dangerous to life and environment since it interferes with biological processes even at extremely low concentrations [26–28]. The total toxicity of pharmaceutical wastewater increases due to chemical reactions and mixing of substances. Even inside the pharmaceutical industry, waste with different physical and chemical characteristics interacts creating complex hybrid wastewater with properties that pose difficulty in analysis, handling, isolation of the components, and overall treatment. Added to this is the illogical method of disposing industrial effluent, domestic sewage, and hospital waste left untreated in public sewer-handling facilities.

In order to access the available pharmaceutical waste-treatment methods and to develop a comprehensive policy to discharge minimal waste in an environmentally friendly manner, pharmaceutical waste can be considered as (1) waste generated by medicinal companies and secondary waste obtained from treatment and recycling plants and (2) medical waste from hospitals and the domestic sector, which substantially contaminates the sewers. The second category of waste includes sources that contribute to active pharmaceutical residue in public sewers and subsequently to the municipal wastewater-treatment facilities, which are commonly ill-equipped to deal with such specialized waste.

6.3.1.4 Treatment Technologies

Autoclaving Technique

Autoclaving is high-temperature treatment of waste to degrade noxious chemicals and microbes that grow on them. The residue of antibiotics with high organic loadings can be treated using autoclaves. However, autoclaving or supercritical carbon dioxide sterilization [29] requires subsequent treatment that is expensive. The challenge thus lies in developing a technoeconomically viable technology that addresses the complexity of this type of waste.

Physicochemical Methods

Physicochemical methods such as ion exchange, adsorption, coagulation–flocculation, frothing, precipitation, chemical reduction, electrochemical processes [30] as well as the combinations of these processes [31] are used to treat pharmaceutical wastewater in different stages from initial production to finally handling the cumulative toxic concentrations in assorted sewage. The efficacy of these conjugate processes lies in their success where direct physical or chemical treatments is not suitable for handling medicinal wastewater due to their poor dissolved COD removal efficiency and introduction of complex chemicals to the system [32]. While processes like precipitation-air floatation show higher COD removal efficiency than coagulation–precipitation the latter costs less (almost 25%). Comparison of the potential of flocculants (e.g., ferric chloride, calcium oxide, chloride, and calcium hydroxide) to remove phosphorus from wastewater shows that the removal rates are dependent on concentration as well as pH [33,34]. The diverse nature of pharmaceuticals makes them treatment-specific; for instance, investigation of the principal removal mechanisms of fluoroquinolone suggested that adsorption of sludge and/or flocks is more effective than biodegradation [35,36].

Technology Based on Advanced Oxidation

Advanced oxidation process (AOP) is especially capable of effective degradation of persistent drugs and compounds like carbamazepine. Conventionally used AOP includes heterogeneous and homogenous photocatalysis based on near UV or solar visible irradiation, electrolysis, ozonation, ultrasonication (US), WAO, and Fenton's treatment. Emerging technologies include ionizing radiation and treatment with microwaves and pulsed plasma. AOP utilizes features such as lytic activity of ozone to simultaneously digest and remove personal care products, stabilization power of Fenton's reagent, and degradation and mineralization kinetics of photocatalysis process. Some technologies are based on ozonation of biorefractory and/or toxic organic pollutants including PhACs treatment with ultraviolet hydrogen peroxide (UV/H_2O_2) to oxidize biodegradation-resistant elusive contaminants in secondary effluent, and Fenton and photoassisted Fenton's oxidation to degrade pharmaceutical residue from laboratory waste [3,4,37–39]. As has been shown, Fenton's treatment in wastewater-treatment plants for stabilization of organic matter including removal of organic contaminants such as PhACs helps in the subsequent removal through oxidation and coagulation while photocatalytic activity of TiO_2 nanofibers not only degrades persistent residue like carbamazepine but also enhances dewatering ability [2,40]. Individual processes such as UV/H_2O_2 treatment can have removal efficiency up to 90% of many pharmaceuticals including carbamazepine at UV dose of 923 mJ/cm^2. The success of AOP is limited to pretreatment only [41].

Adsorption-Based Separation Technology

Adsorption is a useful technique to preferentially remove persisting APIs. Adsorbents can be used to remove selected obnoxious compounds from treated pharmaceutical wastewater, making the wastewater suitable for discharge into water bodies. The simplicity of fabrication and environment friendless of adsorbents derived from plant-based sources have been demonstrated albeit some pertinent doubts that remain regarding the justification in burning tree trunks for the purpose of generating adsorbents for treating large volume of pharmaceutical waste generated in the pharmaceutical industries. A major problem with adsorption-based plants, however, is regeneration of the spent adsorbent material or its disposal. Moreover, most adsorption installations operate in batch mode thus increasing overall operational costs. For successful adsorption, composition of waste needs to be explicitly known because pharmaceuticals and personal care products have widely varying adsorption potentials and are strongly affected by interactions of complex functional groups. Such treatments are also highly pH-dependent [42,43]. Many zeolite-based adsorbents are successful in removal of selected compounds (such as metformine and lincomycine) and micropollutants, which are otherwise difficult to remove by conventional activated carbon-based adsorbents [44]. The crystal structure of zeolite is utilized to remove compounds whose properties (e.g., Stokes diameter) are congruent with that of the silica-alumina lattice of zeolite. Incorporation of magnetic particles and nanotechnology in adsorption has resulted in novel adsorbents for specific adsorption. The use of nanoparticles, nanotubes, and allied components as adsorbents significantly increases the available surface area, tensile strength, and resistance of the material, thus enhancing the efficiency of adsorption. pH sensitivity and high chemical leaching tendency are some of impediments of adsorption. The high surface area of nanoclays (organically modified layered-silicates) and nanoscale particles makes them potent adsorbents of pharmaceuticals [45]. Carbon nanotubes are known to be superior to common adsorbents for removing pharmaceutical compounds like 1,2-dichlorobenzene, trihalomethanes, and microcystines. The magnetic particles are capable of adsorbing compounds from aqueous and gaseous systems and when these are combined with nanocomposite the resulting magnetic nanocomposite materials are characterized by high selectivity and biocompatibility [46–48].

Technology Based on Coagulation and Precipitation

Coagulation and precipitation serve as effective pretreatment steps for the removal of oil and grease, suspended particulate matter, as well as specific compounds from wastewater. Coagulation is a unit operation that removes colloids from water and wastewater. Its principal mode of action is destabilizing colloidal particles by charge neutralization and promoting collisions between neutralized particles, resulting in cohesion. Use of the coagulants decreases COD loading at a relatively low cost and also stabilizes some of the leaching-prone components. Comparative analysis of commonly available coagulants like lime, alum, ferrous sulfate, and ferric chloride has shown that ferric chloride removes BOD_5 and COD to the highest extent. Ferric chloride produces compact sludge with good settlability as reflected in its low sludge volume index (SVI) of 76.3 ± 28.8 mL/g TSS. However, in integrated coagulation–dissolved air flotation (C/DAF) experiments, alum demonstrated higher COD removal ($77.5 \pm 3.2\%$) than ferric chloride ($71.6 \pm 2.9\%$) and ferrous sulfate ($67.7 \pm 3.7\%$). But the removal efficiencies of all the coagulants mentioned above are sensitive to pH changes in the medium and the effectiveness is largely subject to the specific compounds present. For instance, poor COD removal (17%) using polyaluminium chloride at dosage of 0.5 g of concentrated effluent from a traizine manufacturing facility was obtained [49]. The advantages as well as special features of some of the treatment processes are listed in Table 6.3.3.

Electrocoagulation

Electrocoagulation speeds up the conventional coagulation process with electric current and is characterized by the formation of hydroxide ions, which provides a large surface area for adsorption of organic ions and colloidal particles from substrate with subsequent separation of the insoluble flocks by electrofloatation. Processes involving electrocoagulation to treat

Table 6.3.3 Features and limitations of different treatment options [50]

Technology	Features	Limitations
Precipitation	Suitable for large-scale operations	Efficiency depends on the solid–liquid separation step
	Can treat high salt content waste	
Sedimentation	Primary treatment; removes floating and settleable solids.	Large quantity of reagent needed
Coagulation and flocculation	Can be used for building a compact treatment plant or for further purification of treated water	Preliminary chemical
		Coagulation and flocculation are generally not used.
Ion exchange	Good chemical, thermal, and radiation stability; ensuring high selectivity	Blockage problems needs regeneration
Evaporation	Well-established technology	Process limitations include foaming, corrosion
	High volume reduction factor	
	Suitable for a variety of radionuclides	High capital costs
Solvent extraction	Selectivity enables removal and recycling of actinides	Generates aqueous and organic secondary waste

pharmaceutical wastewater have become popular due to improvements in the alternative energy sources used to drive the processes. In electrocoagulation, the anode dissolves due to application of electric potential yielding active coagulant precursors. Deshpande et al. [51] applied electrocoagulation to pharmaceutical wastewater and it significantly decreased the COD loadings by 72% and improved the BOD/COD ratio from 0.18 to 0.3. They illustrated the energy-saving nature, high output in a comparatively small time interval, and overall biodegradability enhancement of the wastewater [52]. Comparison of the COD removal efficiency of electrocoagulation with hybrid associates shows that the removal power increases progressively as follows peroxi-electrocoagulation > peroxi-photoelectrocoagulation > photoelectrocoagulation > electrocoagulation [53]. This also leads to progressive increase in the cost (especially in

Fenton's process where hydrogen peroxide significantly contributes to overall cost) to attain a higher percentage of COD removal. The initial pH of the medium plays a crucial role in process performance because it affects the conversion of cation to higher oxidation states, surface charge of coagulating species, and COD removal efficiency (which peaks only at an optimum pH level depending on whether it follows the adsorption or precipitation mechanism). Thus pH needs careful monitoring. The use of batch systems needs further development and experimentation in continuous mode. One of the challenges is the number of reactions that needs to occur simultaneously and optimally, making the process complex. Improved performance of integrated processes like electrocoagulation and TiO_2 photoassisted treatment [54] or biodegradability enhancement oxidized byproducts on addition of electrocoagulation as a pretreatment [55] may improve the efficiency of this technique. The possibility of flocks and coagulated species being recalcitrant and increasing ecotoxicity or their power to stabilize and prevent the leaching of the hazardous components needs to be explicitly addressed before the high electricity and electrode costs that currently accompany the process can be justified.

Biological Treatments

A significant amount of research has been focused on using microbes to break down pharmaceutical waste and make it to either harmless or useful. Suggested pathways include composting, vermicomposting, and anaerobic and aerobic methods and their combination, some of which yield valuable byproducts. The high COD loadings of pharmaceutical wastewater makes it ideally suited for anaerobic processes. Studies have shown the biodergability potential of antibiotics in an up-flow anaerobic stage reactor, which resulted in a COD reduction of 70–75% for antibiotic-bearing residue. Some studies have shown that when a hybrid process combining an anaerobic sludge blanket and a filter in a hybrid UASB reactor is applied to treat wastewater, it leads to a very high OLR (Organic Loading Rate) of 8 kg $COD/m^3\ d$ together with fairly high COD removal efficiency of 72% [56–61]. Biological treatments, if critically reviewed, would indicate that besides the myriad applications of aerobic and anaerobic methods in reactor form there are two main applications constructed, wastelands and activated sludge and analogous processes, in which different concepts with distinctive characteristics are utilized for effective disposal. Fig. 6.3.2 shows a simple biological treatment scheme.

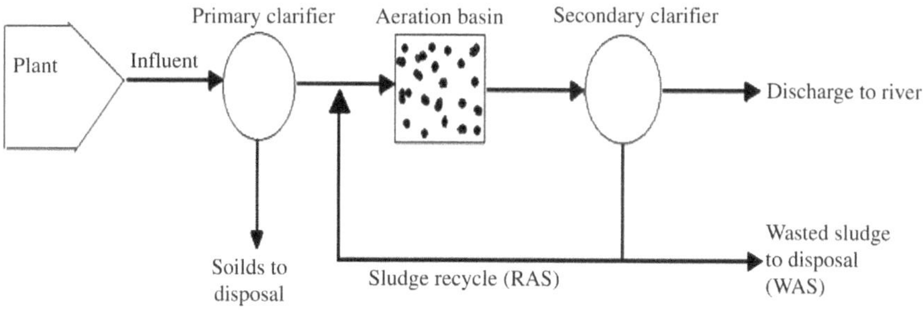

Figure 6.3.2 A typical biological treatment scheme.

Constructed Wastelands

The main challenge in isolating and individually processing pharmaceuticals is that they are mixed thoroughly with other sewage, and then this cumulative waste enters the municipal waste sewers. Biological processes like ASP or constructed wastelands can be used to deal with such composite waste mixtures. Pharmaceuticals present in domestic wastewater have been successfully removed in an activated sludge arrangement followed by individual contaminant processing in a horizontal subsurface flow bed to analyze the removal efficacy of the constructed wasteland of the specific pharmaceutical active compounds.

The constructed wetlands used for the removal of pharmaceuticals from wastewater can be classified into four categories:

1. Surface-Free water Constructed Wetlands (SF-CWs)
2. Horizontal Subsurface Flow Constructed Wetlands (HSSF-CWs)
3. Vertical Subsurface Flow Constructed Wetlands (VSSF-CWs)
4. Hybrid Constructed Wetlands (Hybrid CWs)

The constructed wetland H-SSF (horizontal subsurface flow system) exhibited removal of all therapeutic classes (from 1% for psychiatric drugs to 26% for antihypertensive, on average 16%, with an SD of 8) in the pilot plant setup with *Phragmites australis*. The main problem in implementation is the requirement of vast swathes of land, which might be very difficult to arrange in overpopulated developing countries that limited on space. The use of H-SSF for tertiary treatment increases both cost and complexity. Furthermore, the HRT is about a day at a constant influent flow of 8 m^3/day (assuming a human water daily consumption of 150 L/day implying time constraints that may render it unsuitable when implemented for large volumes of municipal wastewater generated in different areas. These two drawbacks are a major handicap for allied biological processes like lagoons or facultative ponds [62–64].

Activated Sludge and Allied Processes

The use of activated sludge and analogous processes always involves similar structure where the final effluent quality can be improved if subjected to tertiary treatment such as activated carbon adsorption, additional nutrient removal, etc. Overall the activated sludge shows considerable removal efficiency for selected pharmaceuticals; e.g., 68% for tetracycline, 78% for chlortetracycline, and 67% doxycycline. But it has to be effective against at least one particular class of drugs to make it economically viable especially since most of the unused medicine dumped into the garbage enters the sewers unmetabolized thereby posing a direct risk to the environment. Another potential problem associated with ASP is the presence of biotransformation residue in the output, which necessitates further treatment [65–67].

Vermicomposting

One of the most promising techniques for solid-waste management today is vermicomposting, which offers a host of benefits such as degradation of chemical compounds, conversion of organic part of the waste into biofertilizer, and increase in soil aeration. Vermicomposting may be used for the treatment of herbal pharmaceutical residue where the herbal pharmaceutical industrial waste is spiked with cow dung in different amounts using composting earthworm *Eisenia foetida* under simulated conditions in a laboratory environment. Considerable changes in the chemical characteristics of waste mixtures during vermicomposting occur with a reduction in pH of final residual mixtures. After vermicomposting the end product is more stabilized and odor-free with a high range of plant usable forms of soil nutrients with reduction in pH value of all waste [68].

Aerobic and Anaerobic Digestion

Aerobic and anaerobic digestion as well as anaerobic codigestion (AcoD) of pharmaceutical concentrates is an effective disposal technique. AcoD results in a higher degree of degradation of organics compared to using anaerobic digestion on separate substrates and is an economically viable and environmentally benign option. AcoD facilities treating different kinds of biodegradable waste around the globe are found in Denmark, Sweden, Germany, and Switzerland. However, the persistence of active pharmaceutical remains of 23 different drugs including atenolol, hydrochlorothiazide, and diclofenac in the effluent from full-scale integrated treatment of municipal wastewater and organic fraction of municipal solid waste poses questions concerning the success of AcoD alone [69,70].

Technology Using Bioaugmentation

An effective way to combat the effects of pharmaceutical residue is to isolate the particular elements and then to use bacterial treatment to degrade them or convert them into a benign form. Based on this understanding, bioaugmentation technology involves the use of microorganisms to remove contaminants from any waste product. It can also be incorporated to increase functioning of activated sludge systems by promoting flocculation and settling of the activated sludge.

An important aspect often ignored in conventional biological treatment methods such as trickling filter or activated filter is the significantly high concentration of carbon substrate (approximately, higher than 60 mg/L of BOD) needed to maintain a constant concentration of active biomass in the reactor and also during the design of recycling biomass. While bioaugmentation overcomes the problem of continuous supply of inorganic-carbon sources, it is also possible to integrate the benefits of different reactors such as fluidized bed with this scheme to produce valuable biogas as byproduct.

The equipment used for bioaugmentation varies widely but includes a separate reactor called an "enricher reactor" where the suspended microbe culture needed for the process is grown and subsequently transferred to the main reactor in bioaugmentation using real pharmaceutical industry effluent from pharmaceutical industries. This bioaugmentation plant is run in continuous mode to remove antibiotic (cephalexin) from industrial effluent characterized by high COD loading (12,000–15,000 mg/L), organic solvents, and refractory substances hindering conventional mode of treatment. The COD removal efficiency exceeds 88% in continuous mode [71–73].

Technology With Biotransformation Approach

Biotransformation involves chemical modification of a compound often using microorganisms to create chemical reactions that are costly or not feasible otherwise (nonbiologically). One of the advantages of using biotransformation is its power for retroactive treatments of soils contaminated with pharmaceutical residue as well as in dealing with estrogens by converting them into readily degradable compounds. During biotransformation with specialized microorganisms, biofilms are developed with a diverse array of structural and metabolic characteristics that help remove pharmaceuticals, personal care products, heavy metal traces, and toxic minerals from water. Biotransformation can be used to convert APIs such as atenolol, bezafibrate, ketoprofen, metoprolol, ranitidine, and venlafaxine and waste from other biological processes such as from activated sludge.

However, biological methods involve live microbes that are extremely sensitive to changes in pH, temperature, rotational velocity, oxygen levels, substrate concentration, and the toxicity levels of the feed. Maintaining a critical temperature is essential since COD removal efficiency is optimum only for a narrow temperature range. Pharmaceutical wastewater when used as a substrate for an aerobic process shows optimum COD removal efficiency when the reactors are operated at a temperature of 30 degree, with complete shutdown of biological treatment beyond 60 degree. These factors need to be optimized and controlled carefully because the rate of biological reaction is directly dependent on these factors [74–76].

6.3.1.5 Membrane-Based Separation–Purification Technology

Emerging membrane-based schemes are compact, ecofriendly, small, flexible, and easy to install, operate, and maintain. Plants using this technology are known for their high degree of separation due to the high selectivity of the membrane involved, but CP and membrane fouling often create issues. Membrane processes can effectively separate organic loading along with persistent compounds. The modular design, relatively low maintenance costs, high flux, and above all the environmentally friendlessliness make them suitable. After rough separation using MF membrane, UF, NF, or RO can be used as a secondary method to address the residue of drugs present in industrial effluent. The relative characteristics of different membrane processes are listed in Table 6.3.4.

The treatment of waste concentrates from pressure-driven membrane processes in general and RO process in particular has been studied in some detail. Complex pharmaceutical wastewater is best treated in a NF–FO integrated system. Emerging technologies like FO and Membrane Distillation (MD) have been harnessed to decrease the waste volume generated by RO process and to recover water and other components from the RO concentrates. The energy required in these types of proceses is also considerably less. However, the fate of the final concentrated effluent from FO processes is unknown. An NF–RO hybrid setup

Table 6.3.4 Qualitative comparison of different membranes [77]

Process	Membrane	Transfer mode	Pressure (atm)	Flux (L/m^2hour)
Microfiltration	Porous Adsorption	Sieving and Isotropic	0.5–5	500–10,000
Ultrafiltration	Porous Asymmetric	Sieving and preferential	1–10	100–2000
Nanofiltration	Finely porous Asymmetric or Composite	Sieving or Donnan electrostatic hydration or diffusive	7–30	20–200
Reverse Osmosis	Nonporous Asymmetric Or composite	Diffusive	20–100	10–100

for the treatment of pharmaceutical wastewater may send the rejects to a multiple effect evaporator (MEV) for extraction of water and to convert the concentrated solids to a state in which they are easily disposed in the landfills. The reported COD removal efficiencies are around 94% (aerated), 93% (nonaerated) for NF and 99% (aerated), and 98% (nonaerated) for RO, respectively, for aerated and nonaerated feeds. The rejection percentage of TDS is around to 78% and 75% for NF followed by 95% (aerated) and more than 91% (nonaerated) for RO. However, the main challenge of effective disposal of concentrates is mitigated using MEVs, but this approach is extremely energy-intensive and focuses solely of removal of water from reduced effluent. The danger of the ecotoxicity posed by these active pharmaceutical compounds still persists, as does the risk of reaction of these compounds with other agents when they are disposed of in landfills [78–90].

Integrated Process Technology

Membrane Reactor With Conventional Processes

MBR is a powerful agent for treating consolidated forms of pharmaceutical wastewater and municipal sludge. It offers a host of advantages such as higher efficiency, better sludge retention, greater compactness, and capability of withstanding fluctuating pollution loads while maintaining reasonably consistent quality of treated water. The COD removal efficiency of MBR can be greater than 95%, which makes it a particularly lucrative option for treating medicinal waste with high COD loadings. Due to the effectiveness of additives and specific treatments such as electrocoagulation, advanced oxidation, etc., against a particular class of pharmaceuticals, schemes in which MBRs are integrated with conventional techniques have been successful. An entire branch of research is directed toward MBR and possible integration with other existing processes or new schemes such as fungal bioreactor or submerged membrane electrobioreactor. Innovative combinations of integrated MBRs and TiO_2 photocatalysis process for the removal of nondegradable drugs such as carbamazepine from simulated pharmaceutical industrial effluent has been carried out. According to the research if a recycling ratio of 4:1 is adopted a removal rate of 95% for carbamazepine can be achieved. In addition to the reduction of COD effluent and increase in sludge yield, respirometric tests suggest that the oxidation products are mostly biodegradable and do not affect microbial activity adversely. In some studies, conventional physical processes such as adsorption are integrated with bioreactors either by selectively removing some adsorbents and then transferring the bulk to the bioreactors or by segmenting and treating the inorganic constituents in the adsorption chamber and inorganic parts in the reactor chamber. Although this technology has proven to be successful, its effectiveness in large scale remains to be assessed [91–95]. Fig. 6.3.3 presents a typical membrane-bioreactor arrangement.

Technology With Integration of Thermophilic Processes With Membrane Bioreactor

Thermophilic processes (TPPs) are characterized by high removal of organic matter and are ideally suited for treatment of wastewater with hazardous compounds or with high salinity gradient. Thermophilic treatments show high removal kinetics of biodegradable substrates, about 3–10 times higher than those measured in mesophilic conditions. Integrating TPPs with a MR creates system that eliminates the disadvantage of poor sludge sedimentation and the inability to form flocks by the particles in suspension. To deal with high-strength liquid residue (rich in organic matter COD, nutrients, and salinity) from a pharmaceutical industry the thermophilic aerobic membrane reactor appears to be a good candidate, allowing processes to operate with higher concentration of biomass (higher than 50 kg TSS m^{-3}) and resulting in drastic reduction of reactor and aeration tank volumes. Thermophilic MBRs are compact systems but suffer from some disadvantages such as low sludge production and the necessity of biomass acclimatization, which need to be addressed [96–98].

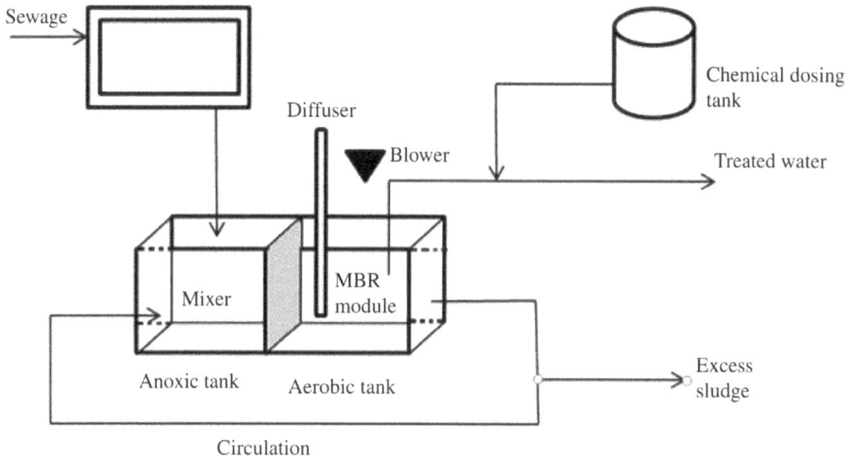

Figure 6.3.3 A compact membrane-bioreactor module.

Technology for Safe Disposal
Stabilization Route

A particular compound or selected group of homologous compounds must be stabilized to render them ineffective to aid in the disposal of real pharmaceutical residue with a host of chemical compounds. Today research has shifted from conventional broad spectrum stabilizers to chelating compounds such octolig (a polyethylene-diamine covalently attached to high-surface area silica gel) to quantitatively remove anions (nitrate, nitrite, phosphate, sulfate) from aqueous solution. However, these chelating agents work only on compounds possessing analogous ions. Thus while octolig completely removes three xanthenylbenzenes (rose bengal, eosin Y, erythrosine) with phenolic and carboxylic acid groups and quantitatively removes zinc phthalocyaninetetra sulfonate and lissamine green B-bearing sulfonate groups, it cannot treat methylene blue and analogous groups with quaternary ammonium compounds. However, the litmus test of several developed model chelating compounds that had a substantial removal in lab-scale studies lies in their effectiveness against industrial wastewater.

One inexpensive and strong stabilizing agent is cement. Portland cement has been used to solidify pharmaceutical waste. It is important to note that pharmaceutical wastewater is primarily characterized by high organic loading whereas binders like cement are primarily suitable against inorganic compounds. Thus successful sludge stabilization must include an organic reduction step. However, due to the diverse chemicals in pharmaceutical waste, it is difficult to perform conventional leaching tests like TCLP to analyze the chances of leakage of the main hazardous components and thereby determine the effectiveness of the stabilization. Thus aerobic sludge stabilization techniques are not successful for disposal of pharmaceutical waste.

There is dissent on the scope of the word "fate" in the context of pharmaceutical residue. In some cases, final concentrations of pharmaceutically active compounds are emphasized while in other cases, the potential dangers posed by chemical compounds gain attention. While primary and secondary treatments have received the most attention, the issues involved in final disposal have been minimally addressed [74–76,78,97,98].

Incineration and Landfilling Approach

The common methods used for disposal are incineration, open burning, and landfilling. The group of processes that use incineration are called warm disposal and high-temperature thermal disposal technologies. Simple incineration often involves open-air burning without the use of any ash disposal practices, or measures to curb any possible spread of disease causing germs. Pharmaceutical waste often requires separate incineration facilities and not combustion with other obnoxious waste such as medical waste. The pitfalls of using ordinary landfills for hazardous waste are well documented. While studies using tools such as the LandSim (landfill performance simulation) modeling program to analyze the environmental consequences of leachate release from a municipal solid waste landfill with household hazardous waste, the needed steps to neutralize the toxic effects of hazardous compounds and the chain reactions in the presence other waste are not fully understood. The endocrine-disrupting nature and power of bioaccumulation make pharmaceutical residue in landfills dangerous. The good removal rates of some wastewater plants do not necessarily result in effective degradation of pharmaceutical compounds. In some cases, sanitary landfills replacing ordinary landfills offer some safeguards against the potential threats but are insufficient to effectively combat the huge volume of medicinal waste worldwide [27,28,79,99–103].

Green Pharmacy and Biopharmaceutical Approaches

The "green pharmacy" approach focuses on the properties of compounds generated with the purpose of making them as innocuous as possible so that rigorous disposal schemes may eventually be turned redundant. This approach uses the principles of "green chemistry" and the entire lifecycle of a compound is taken into account to select areas for risk management and to follow the principle of sustainability. This approaches takes into consideration factors such as easy degradability before the synthesis of compounds.

Green chemistry offers advantages like increased process efficiency, decrease in waste generation, and lower energy consumption. However, the high cost of producing such compounds often stands in the way of adopting green pharmaceuticals route in the conventional medicine manufacturing companies.

Biopharmaceuticals are the drugs produced using biocatalysts, especially isolated enzymes and microorganisms in the production of pharmacologically valuable materials. They not only ensure environmental friendliness but also offer economic advantages over synthetic catalysts. The most direct advantage of this approach is that many of the active pharmaceutical compounds that permeate most waste-treatment plants can be developed into potentially biodegradable forms through biocatalytic processes. In fact, there are already instances of synthesis of small-molecule APIs such as simvastatin, atorvastatin, pregabalin, paroxetine, and levetiracetam using green pharmacy techniques. When compared with conventional processes simvastatin was obtained with high conversion (>99%) and high purity (>98%), while atorvastatin synthesis with biotransformation was characterized by high throughput (200 g/L-d), high yield (90%–95%), and excellent stereo control (98% enantiomeric excess (ee) and 97% diastereomeric excess (de). This demonstrated one of the major benefits of this technique when process steps to atorvastatin was not only truncated but two of the most energy intensive steps were bypassed, which saved energy costs as well as lead to waste reduction (which included material saving of solvents and raw materials) estimated at hundreds of metric tons per year. This technique eliminates the use of hazardous reactions, e.g., in levetiracetam where conventional alkylation is replaced by the benign S_N^2 reaction using pyrrolidinone.

However, there is debate within the scientific community regarding the utility and benefits of biopharmaceuticals. While some believe they are natural products and thus quickly biodegraded or denatured easily, others believe that naturally occurring compounds are not always easily biodegraded and in many cases undergo transformation into forms persistently resistant to biodegradation. Structurally related compounds such as plasmids and their residue are found in the environment. Nevertheless, studies show that it reduces waste production and material consumption in many cases by almost 50%, which can help reduce pharmaceutical waste [101,102,104,105–109].

Value-Addition Approach

Interestingly, pharmaceutical waste is a rich source of chemicals, active intermediates, and residue of well-known compounds. Thus extraction of valuable substances from the concentrates can be valuable. There are successful examples of profitable recovery of valuables from wastewater such as nutrient recovery and metal recovery like nickel. The membrane processes are capable of separating heavy metals, valuable colloids, and organic compounds like phenols and amines. Thus, if feasible, recovery of these components from the concentrates of membrane processes like RO and FO should be attempted. For example, a novel extraction scheme has been developed with phase diagrams comprised of quaternary ammonium compounds and three different salts to treat a solid-state pharmaceutical matrix using absolute to recover paracetamol from pharmaceutical wastewater. Ammonium compounds containing cation cores with longer alkyl chain lengths have higher potential to induce phase splitting. pH of the aqueous phase is also an important factor to determine the phase separation capacity, which decreases with increase in acidity of the substrates. If the pharmaceutical wastewater is rich in phosphates, after it is separated using a membrane process like RO the residual phosphorus can be recovered with the help of a polymeric ligand exchanger (PLE) due to the strong ligand properties of the phosphate group. In a polymeric ligand exchanger, the chelating ligand forms the solid polymer template on which the specific ions of the resins are embedded. These ions form complexes with the target species and are immobilized in the polymer matrix. The phosphate thus engrafted can be precipitated economically in the form of struvite ($MgNH_4PO_4 \cdot 6H_2O$) in addition to attaining almost 80%–85% of the adsorbed phosphate. Phosphorus recovery is also possible by PLE as well as precipitation of phosphorus in the form of struvite. However, the extent of interference from competing ions, the suitability of pharmaceutical wastewater, and the phosphorus content needed to merit separation as well as reactor cost for crystallization of struvite need to be considered. Recovery of phosphorus from fosfomycin (1R-2S-epoxypropyl phosphoric acid) pharmaceutical wastewater can be done using a WAO-phosphate crystallization process for recovery of refractory antibiotic organic phosphorus compounds through their conversion into inorganic phosphate (IP). The results are encouraging as 99% TOP (total organic phosphate) is transformed into IP and COD removal rate of 58% is attained. This approach not only guarantees sustainability but also aids in the wastewater treatment. However, technoeconomic assessment needs to be done to increase scale-up confidence [50,77,110–115].

Moving Toward Advanced Technology

There are no panoptic broad-spectrum treatment methodologies available that can be used to treat and dispose of the diverse pharmaceutical waste from a host of different sources. Options like ozonation, Fenton's oxidation, hydrogen peroxide/UV treatment, and photooxidation can degrade specific compounds but are largely ineffective as a cohesive disposal method. Reducing COD loadings requires an assortment of different techniques and in most cases does not guarantee minimization of waste volume. One of the main challenges is mixing of pharmaceuticals and personal care products. Due to this a large number of research can be found to devise methods to eliminate both. The absence of extensive research on the chemical composition of this type of wastewater and the possible chemical reactions when distinctly different compounds with free radicals and reactive functional groups come into contact means we don't conclusively understand the reaction mechanisms and possible ecotoxicity due to the chain reactions of the free radicals present. Simple sedimentation or coagulation–precipitation using lime, alum, and similar commonly used additives are not helpful in decreasing the ecotoxicity of pharmaceutical wastewater. Among the physicochemical processes adsorption and electrocoagulation can aid in separation and removal of selected compounds but are often costly and energy intensive. Newer innovations in adsorption include the use of nanoparticles and magnetic nanocomposites. However, this is a developing field and successful implementation of magnetic nanocomposite in areas such as magnetic resonance imaging, bioscience, analytical chemistry, and pollutants removal is limited. Biological processes such as vermicomposting and aerobic and anaerobic digestion or their structured applications such as activated sludge and constructed wasteland help in enhancing the overall biodergability, COD removal efficiency, and conversion of hazardous compounds into benign forms. However, the intense chemical-loaded drug residue often prevents their complete biodegradation. For example, persistent drugs like carbamazepine permeate the system and are detected in wastewater treatment plant effluent. Biotransformation can help convert the recalcitrant compounds present in medicinal effluent and bioaugmentation yields almost 88.5% COD removal efficiency. Bioaugmentation also produces biogas (a valuable byproduct) at a rate of 5.62 L/day at 24 hours HRT with a maximum COD concentration of 4000 mg/L. Biological methods provide an alternative to the conventional practices of incineration and burning that are highly polluting. Despite the complexity of biological reactor schemes, this technology is a step toward safe disposal of the waste, although the constraints of time, space, microbe selectivity, and sensitivity are pressing issues that need to be addressed.

Microencapsulation is an alternative to landfills for final disposal of pharmaceutical waste. Stabilization of ionic species and leaching prevention of hazardous compounds are parameters that determine the effectiveness of this process. Chelators such as octolig (a polyethylene diamine covalently attached to high-surface area silica gel) against selected antibiotics as well as in the quantitative removal of anions (nitrate, nitrite, phosphate, and sulfate) can help in further condensation of concentrates. In comparison to the aforesaid processes, membrane technology to separate a host of substances from pharmaceutical wastewater like BOD, COD, heavy metals, colloids, and organic compounds besides promoting ability for dewatering and discoloring; but the problem of concentrate disposal persists. The use of polymeric ligand exchanger aids in the process of recovery of compounds; phosphorus content can be recovered up to 80%–85% along with precipitation in the form of struvite. Recovery of heavy metal ions like lithium, cesium, rubidium, and uranium as well as organic phosphates from RO concentrates proves the necessity of further treatment of membrane rejects before final disposal as well as the scope of recovery of valuable materials. To treat membrane concentrates, processes such as electrochemical oxidation, heterogeneous photocatalysis, Fenton (Fe^{2+}/H_2O_2), and allied processes (Fe^{3+}/H_2O_2) can break down nonbiodegradable compounds, decrease the organic loading of the membrane concentrates, and remove DOC (dissolved organic carbon) by 50% while facilitating residual iron precipitation by increasing pH to 7.5–8.0. New schemes like membrane adsorption using adsorbents such as zeolite (which exhibits almost 97% removal of metal ions like Cu^{2+}, Zn^{2+}) to preferentially bond with compounds followed by separation using a commercial membrane process like MF if used to remove contaminants from concentrates can be a complete treatment option.

Hybrid processes like WAO-phosphate crystallization attains a removal rate of 58% COD, 99% TOP with an increment of BOD/COD ratio to more than 0.5 making it amenable for biological treatment. Pairing of physicochemical techniques like coagulation–precipitation with biological treatments like activated sludge helps increase COD removal and degradation efficiency. MBRs and integrated forms are suitable for achieving high COD removal efficiency since they offer biological treatment of waste and a sustainable membrane-based process for separation. Hybrid processes like AOP and TiO_2-photocatalysis with the MBRs yield a higher degree of degradation and remove persistent compounds like carbamazepine and amoxillin. Recently developed innovative schemes include fungal bioreactor and addition of TPPs to MBRs both of which have shown success in handling medicinal waste.

A modern approach to mitigating the problem of waste disposal involves the use of green technology and biopharmaceuticals to reduce pharmaceutical waste and ensure waste produced has a high percentage of biodegradability. The use of the principles of biopharamacy in the synthesis of APIs such as simvastatin, atorvastatin, pregabalin, paroxetine, and levetiracetam save energy, eliminate hazardous reactions, and reduce waste volume by 50% [116–122].

Conclusion

Industry specific approaches are often the best disposal way to stabilize persistent API and chemicals. Combined treatment of untreated pharmaceutical wastewater with other waste should be avoided. The chemicals used in the synthesis of drugs need to be declared, and the state of the compounds in the effluent and the chemical reactions between colloids and free radicals also needs to be understood. Feasibility surveys for economic recovery or conversion into beneficial products need to be executed on the dominant species. Green pharmacy practices can reduce waste, recalcitrant compounds, hazardous chain reactions as well as result in environmentally friendly catalysts like enzymes. Combining processes like stabilization or microencapsulation as well controlled autoclaving can be effective against selected drugs. However, complete stabilization of waste before discharge, monitoring of leakage, understanding of reactions that might occur, and potential risks need to be considered. In addition, the energy requirement coupled with the paucity of sufficient binding agent for implementation on a large scale also needs to be addressed. Integrating a membrane process with an advanced biological process where bioaugmentation can be administered in a facility like a constructed wasteland is a comprehensive treatment and disposal strategy. Nevertheless such a scheme will lead to single–handed separation of contaminants from effluent, increase dewatering ability and degradability, serve as an alternative to landfills, and release valuable biogas as byproduct while allowing the scope for recovery of valuables from concentrates. Innovative approaches to advanced treatment technology need to be adopted.

6.3.2 ADVANCED TREATMENT TECHNOLOGY: MEMBRANE-INTEGRATED FORWARD OSMOSIS-NANOFILTRATION CLOSED LOOP SYSTEM FOR RECOVERY AND REUSE

6.3.2.1 Introduction

The current decade has seen an increase in research efforts aimed at the recovery and reuse of water from wastewater in light of the scarcity of freshwater worldwide. The goal is to save surface water bodies from the onslaught of hazardous industrial and municipal waste and reduce freshwater consumption through reuse of recycled water thereby creating a sustainable water resource management regime. Unabated pollution from riverside industrial discharge of the thousands of kilometers of river bodies that serve as the vital lifelines of many countries has grown alarming proportions and in many cases this wastewater contains such wide range of toxic and hazardous components, which makes their separation from water extremely difficult. To address this problem, a novel membrane-integrated complete treatment technology was developed by the Environmental and Membrane Technology Laboratory of NIT in Durgapur, India [90]. Polyamide composite membranes have been used in FO and draw solute recovery stages in which DS is 0.5 M NaCl selected through screening of different solutions (of salts of magnesium and sodium) in different concentrations. The scheme achieves more than 97% purification of pharmaceutical wastewater (in terms of COD). Analysis of specific compounds like nebivolol and paracetamol indicates almost 100% removal. This design can significantly reduce CP and RSD resulting in a sustained yield of pure water flux of around 52 L/m^2hour using NaCl as draw solute. A downstream NF module operated at only 12 bar TMP ensures continuous recovery and recycle of 99% of draw solute while producing safe reusable water at the rate of 58–60 $L/(m^2hour)$.

6.3.2.2 Existing Technologies and the Treatment Challenges

The prescribed discharge limit for pharmaceutical waste is 250 ppm expressed in terms of the lumped parameter COD, which includes the large number of contaminants normally present in this type of complex wastewater. It is important to understand the daunting challenge pharmaceutical wastewater (which typically contains COD loading ranging from 2500 to 5000 ppm) poses. Conventional techniques are largely ineffective due to the presence of API and also due to high COD loading. Biological systems can help in decomposing pharmaceutical waste, and MBRs have been used in pharmaceutical waste treatment. However, these systems suffer from disadvantages such as the inability of the microbes to survive in such a highly toxic environment. There are also serious limitations on the volume of wastewater that can be processed. On top of that, time constraints are imposed, which often turn make these systems impractical for large-scale adoption. The myriad of physicochemical processes like advanced oxidation, electrochemical methods, and ozonation have been attempted but the disadvantages of such approaches are high cost, inability to cause complete degradation of waste, and the number of pretreatment steps involved. Disposal practices like incineration and burning are common but create secondary pollution sources. One effective technology is membrane-based processes due to their high degree of separation—purification. Various studies have shown the effectiveness of UF, NF, and RO for removal of pharmaceutical waste. However, significant CP induces flux decline in these systems during long-term operation. Under these circumstances, FO offers an interesting option that not only guarantees high separation efficiency and flux, but also economic viability. Several studies on FO have shown successful desalination and treatment of several different kinds of process streams.

Wastewater discharged by the pharmaceutical industries without effective treatment heavily degrades river and lake water through release of organic solvents, catalysts, reactants, raw materials, drug intermediates, and even complete drugs often expressed as a lumped parameter COD. Transport of these contaminants through the food chain at trophic levels causes adverse effects on the ecological system and on human health. In the last few years, increasing attention has been paid to the widespread occurrence of pharmaceutical pollutants in raw water, wastewater effluent, and even in treated drinking water. High COD compared to low BOD in such wastewater often renders biological treatment ineffective because of the inhibition of microbial activities.

Over the years technologies based on the principles of physicochemical and biological treatments, adsorption, photocatalytic oxidation, and membrane separation have developed aimed at effective treatment of pharmaceutical wastewater. While slow biological treatment schemes often fail to sustain microbial activity in the presence of toxic components other physicochemical schemes fall short of meeting stringent regulations for effluent discharge. Reducing concentrations of toxic and hazardous components in discharged water to permissible levels often requires elaborate and multistep physicochemical treatment plants. Emerging membrane-based schemes on the other hand are compact, ecofriendly, small, and flexible, and easy to install, operate, and maintain. These plants are also known for their high degree of separation due to high selectivity but CP and membrane fouling often prevent sustainable operation at steady flux. High energy consumption in RO is well known. In the backdrop of the problem of ever-increasing pollution in river bodies researchers continue to work on the development of simple, low cost, and easy to implement technologies. Such technologies are needed to close the loop so that the river bodies can be protected from the severe onslaught of industrial waste discharge. In this context, FO in a new fouling-preventive orientation can play a significant role if established through field investigations. This emerging technology has the potential for high degree of separation of almost all types of contaminants with low energy use and the capability of eliminating internal and external CPs and mass transfer resistance. The principle of FO desalination involves transporting water through a semipermeable membrane using a natural osmotic process with a highly concentrated solution (the DS) that draws water from the feed solution. FO has progressed and established itself as a viable technology for water purification in the context of seawater/brackish water desalination, wastewater treatment, and liquid food processing. However, research on effective treatment of pharmaceutical wastewater using this technology is limited. Although the few studies conducted with FO in the context of pharmaceutical wastewater have reported high separation efficiency ranging from 90% to 93%, the achieved flux remains unacceptable. Moreover, studies on schemes involving FO as well as solute recovery resulting in reusable water in a continuous steady-state scheme are rare. The limited studies reported in the literature are on model solution only and the derived data deviates widely from reality in which highly selective nanomembrane and FO membranes are used during treatment of dirty wastewater. Thus issues of internal and external CP (ICP and ECP), membrane fouling, reverse solute flux, and low flux still remain major stumbling blocks in successful use of FO in separation–purification [124–144].

Considering the shortcomings of existing technologies and the emergence of new membranes, a novel technology integrating FO with NF and a new membrane has been developed. This new design largely succeeds in overcoming the major hurdles of FO in the context of removal of contaminants from pharmaceutical wastewater. The scheme consists of an upstream FO loop for separation of the contaminants from wastewater and a downstream NF loop for recovery and recycle of the draw solute while producing safe reusable water at a reasonably high level of flux. The orientations of the membrane modules are different than in widely practiced FO systems. The effectiveness of the new design in achieving the desired performance parameters in the context of pharmaceutical wastewater is well established [123–144].

6.3.2.3 Theory of Mass Transport in FO

The osmotic-pressure differential ($\Delta\pi$), which is the driving force in water transport across the membrane in FO, is often reduced significantly by internal and external CP, mass transfer resistance, and drop in interface concentration due to the largely unperturbed support layer of the membrane. This often limits flux and low flux is the major stumbling block in successful use of FO. Hydraulic pressure differential acts as the driving force in RO and flux depends on this applied pressure. Thus ΔP is equal to zero in the FO process, ΔP is greater than $\Delta\pi$ in the RO process, and in PRO, $\Delta P < \Delta\pi$. The direction of water flow is the same in FO and PRO. In RO, the direction of water flow is reversed by applied ΔP where it is in excess of the osmotic-pressure differential. Water flux in all these processes can be in general described by:

$$J_{\text{exp}} = A(\alpha\Delta\pi - \Delta P) \tag{6.3.1}$$

where A is water permeability constant of the membrane, α is the reflection coefficient, and J_{exp} is the experimental water flux.

The mass transport of water in FO process can be described by [145]:

$$J_{\text{W}} = \left[L_{\text{w}}\pi_{\text{DS}} \times \exp\left(-\frac{J_{\text{W}}}{M_{c,\text{DS}}}\right) - L_{\text{w}}\pi_{\text{FS}} \times \exp\left(\frac{J_{\text{W}}}{M_{\text{c,FS}}}\right)\right] \tag{6.3.2}$$

where J_{W} is the volumetric flux of water, L_{w} is the water permeability, π_{DS} and π_{FS} are the osmotic pressure on the draw and feed side, respectively, and $M_{\text{c,DS}}$ and $M_{c,\text{FS}}$ are denoted as the mass transfer coefficient for external CP in the draw side and feed side, respectively, which can be specified by the following mass transfer equations:

$$M_{\text{c,DS}} = \left(\frac{D_{\text{c}} \times \varepsilon_{\text{p}}}{\tau_{1} \times x_{1}}\right) \tag{6.3.3}$$

The diffusivity coefficient, porosity of the support layer, tortuosity of the support layer, and thickness of the support layer are denoted by D_c, ε_l, τ_l, and x_l, respectively:

$$M_{c,FS} = \left[\frac{1.85 \times D_c \times (\mathrm{Re} \times \mathrm{Sc})^{1/3}}{d_{hd}^{0.67} \times L_{ch}^{0.33}}\right] \tag{6.3.4}$$

Here, d_{hd} and L_{ch} are the hydraulic diameter of the module and channel length. Re and Sc are denoted as the Reynolds number and Schmidt number, respectively.

6.3.2.4 The Treatment Plant

Membrane Materials

Thin-film composite polyamide membranes (NF-1 and NF-2) from Sepromembranes Inc. (USA) were used. Pure water flux of NF-1 at 10 bar is 110 L/m^{2}hour (LMH, liter per square meter per hour) and the same for NF-2 at 195 LMH shows the tighter structure of NF-1 compared to NF-2 membrane. These two membranes carry different negative charges as reflected in their ionic separation behavior. Characterization of the pharmaceutical wastewater used in the investigation before treatment and those of the treated water are listed in Table 6.3.5 [90].

Configuration of the Integrated FO and NF System

The treatment plant consists of two loops—one for FO and the other for NF. The FO loop consists of a flat-sheet cross-flow membrane module with accessories such as a feedwater tank and DS tank made of stainless steel and connected to the FO membrane system via circulating pumps, flow meter,s and pressure gauges as depicted in Fig. 6.3.4 [90].

Peristaltic pumps are used to circulate the feed and DS through the FO module while a diaphragm pump is used to circulate diluted DS through the downstream NF module. The upstream and downstream pressure transducers indicate TMPs. The CFR through the system is monitored and controlled using rotameter and control valves. The module is so designed that the flat-sheet membrane remains in a horizontal plane on a perforated stainless-steel support while feedwater and DS flow tangentially in counter-current directions along the top and bottom surfaces of the membrane, respectively. The bottom chamber of the module is so designed that DS enters the same through one of its inlet at one end and on crossing tangentially along the membrane surface leaves the chamber through an outlet maintained at a much higher level than the level of

Table 6.3.5 Typical characteristics of the pharmaceutical wastewater before and after treatment

Water parameters	Units	Concentration in influent	Treated effluent for DS (DS-0.5 M NaCl)	Treated effluent, DS (DS-0.5 M $MgSO_4$)
pH	–	6.8	6.5	6.5
Salinity	mg/L	8–11	3.3	2.8
Conductivity	ms/cm	11–14	0.16	0.21
Temperature	K	299	299	299
TS	mg/L	960	ND[a]	ND
TSS	mg/L	360	ND	ND
TDS	mg/L	600	12	16
Oil & grease	mg/L	18	ND	ND
Color	TCU[a]	2500	ND	ND
Nebivolol	mg/L	7.8	ND	ND
Paracetamol	mg/L	48	ND	ND
COD	mg/L	3500	105	280
BOD	mg/L	466	ND	ND
Chloride	mg/L	43	1.5	4.2
Sulfate	mg/L	204	4.1	6.5

[a]ND, non detectable; TCU, true color unit

the membrane. This ensures complete immersion of the bottom surface of the membrane in the DS while allowing a continuous sweeping fluid action along its bottom surface. The feedwater that flows tangentially along the top surface of the membrane ensures sweeping fluid action on the top layer. This design minimizes CP, membrane fouling, and back diffusion of the DS thereby ensuring a high unidirectional volumetric flux, i.e., water transport from feed to DS only.

As continuous FO results in dilution of the DS, concentration of the same needs to be done to sustain the FO process. Thus a downstream NF membrane module in flat-sheet, cross-flow mode is operated simultaneously for recovery of clean water from the DS while recycling the draw solute to the system. A diaphragm pump (Milton Roy India Pvt. Ltd.) operated at 10–13 bar maintains circulation of the DS through the flat-sheet, cross-flow NF module. Operating pressure and CFRs are monitored through control valves and rotameter and pressure gauges as illustrated in Fig. 6.3.4 [90].

Plant Operation

The plant is run in a continuous-flow mode using flat-sheet, cross-flow membrane modules with pharmaceutical wastewater. Effective filtration

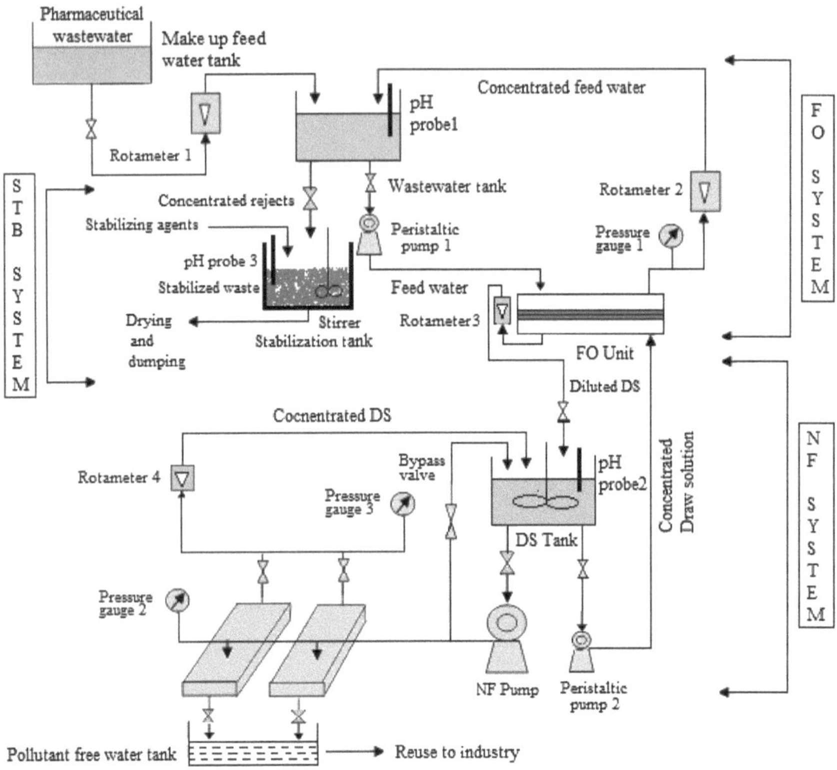

Figure 6.3.4 Schematic of FO–NF hybrid process for removal of COD and toxic compounds from pharmaceutical wastewater [90].

surface of each membrane module is 100 cm^2. Microfiltration is done prior to FO for removing suspended particles in a cross-flow membrane module. Microfiltered pharmaceutical wastewater is used as feed of the FO process where water is extracted in a DS of either NaCl or $MgSO_4$ leaving behind toxic COD and other possible contaminants. The diluted DS is subsequently passed on to another NF membrane module to recover water as the filtrate and recycling the draw solute in the DS loop. Filtration modules can be run long term in continuous mode where feedwater and DS compositions are maintained at the same level through continuous withdrawal of COD rejects and addition of make-up wastewater in the FO loop and recovery of water from the downstream NF module at the same rate of dilution of the DS. Each run of the FO experiment is carried out with the active layer of the membrane facing the feed solution and the support layer facing the DS to afford higher rejection with higher water fluxes due to the reduced internal CP. The feed solutions (pharmaceutical wastewater) and DS flow countercurrently in their respective channels on two sides of the flat membrane. The CFRs for both the draw and feed solution may be varied (e.g., between 21 and 42 cm/second) whereas the pressure of the FO module is in the range of 1–2 bar. The NF module may be run at 10–14 bar operating pressure. Both the feed solution and DS need to be maintained at the same temperature. COD rejection and RSF (Cl, SO_4^-) are continuously monitored through periodic analysis of the samples from the DS tank and feed tank, respectively. NF modules are initially run at different CFRs and at different applied pressures to arrive at the best possible CFR and operating pressure. Thus 12–13 bar operating pressure and around 800 L/hour CFR are used for continuous running of the system at steady state, which ensures high retention of the draw solute while yielding the highest possible pure water flux.

6.3.2.5 Water-Quality Monitoring

Chemical Analysis

COD may be measured using a COD Vario tube test MR (0 to1500 mg/L range) of COD analyzer (e.g., of LoviBond). Detection and quantification of nebivolol and paracetamol may be performed using HPLC. Assay of nebivolol is done with a C18 (4.5 mm × 26 cm) column with mobile phase buffer:acetonitrile:water (45:25:30) at flow rate 1 mL/minute, residence time of 10.9 minutes, and injection volume of 20 μL. Analysis of paracetamol may be done with a μ-Bondapack C8 column with mobile phase 0.01 M KH_2PO_4:methanol:acetonitrile: isopropyl alcohol (84:4:6:6) at flow rate 1 mL/minute, residence time of 4.8 minutes, and injection volume of 10 μL). pH, salinity, conductivity, and TDS measured by InoLab Cond 720, with electrode Tetra Con 325 (WTW, Germany). TSS and total solids (TS) are calculated using standard methods. An ion meter is used to measure the chloride concentrations and the relevant electrodes. BOD, sulfate, color oil, and grease are measured using standard methods.

Analysis of Membrane Morphology

Membrane morphology is analyzed by SEM before and after each filtration operation (over long periods) to determine the degree of membrane fouling. During SEM analysis, membrane pieces are freeze-fractured in liquid nitrogen and then gold-coated in ion sputter at 15 kV.

Determination of Water Flux and RSF

The weight balance is used to determine the water flux by measuring the weight change of the permeate DS tank during each experiment. RSF is determined by periodically analyzing chloride and sulfate concentration in the feed tank.

6.3.2.6 Plant-Performance Analysis Under Different Conditions

Water Flux and COD Removal in FO: Influence of DS Concentration

The osmotic pressure and DS in the FO process are important factors influencing mass transport and overall process performance. A perfect DS has high osmotic pressure, is nontoxic, and is easily recoverable in the reconcentration system. DSs such as those of NaCl and $MgSO_4$ with molar concentrations range varying from 0.1 to 1.0 M may be used. With its high solubility, high osmotic pressure, and low cost, NaCl appears to be the most appropriate DS for the FO process ($MgSO_4$ may also be used as a DS due to its insignificant retarded forward diffusion resulting in negligible reverse solute flux). Fig. 6.3.5 [90] shows how the variations in concentrations of the two different DSs used affect water flux and COD rejection.

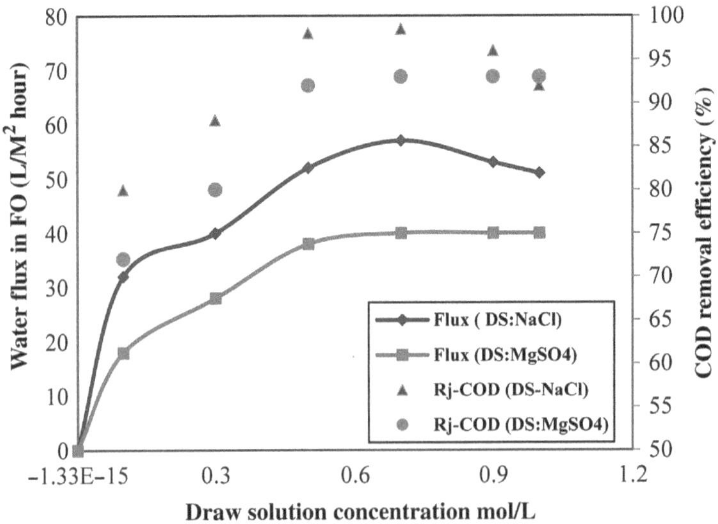

Figure 6.3.5 COD removal and water flux during forward osmosis under varying draw solution (DS) concentration. Experimental conditions: COD concentration of 3500 mg/L, pH 7, DS concentration range 0.1–1.0 mol/L, pressure 1.6 bar, feed side cross flow rate (CFR) 150 L/hour, DS side CFR 10 L/hour, and temperature 308 K [90].

Fig. 6.3.5 shows that DS concentration has a positive correlation with volumetric flux and COD rejection. A concentration 0.5 M of NaCl using NF-2 membrane yields a volumetric water flux of 52 L/m^{2}hour whereas water flux using $MgSO_4$ as draw solute is 40 L/m^{2}hour. Fig. 6.3.5 (primary axis) also shows the effects of DS concentration on COD rejection by NF membrane. COD rejection increases linearly with increasing DS concentration and over 98% COD rejection is observed at a draw solute concentration of 0.5 M NaCl as well as 0.5 M $MgSO_4$. At this DS concentration two major harmful compounds, nebivolol and paracetamol, are completely separated from the feed solution (pharmaceutical wastewater). The DS and the corresponding osmotic pressure it develops are important factors influencing mass transport and overall process performance in FO. A higher concentration of DS produces a higher osmotic pressure or driving force for water transport through the membrane. A higher concentration of DS and hence higher osmotic pressure that forced a larger amount of water as water flux through the membrane eventually reduced the solute flux across the membrane in an uncoupled transport process. This explains the higher COD rejection following higher water flux. The likely reason behind the declination of the curve is the high RSF of NaCl, which increases the dilution of the DS so the osmotic pressure decreases, which decreases the water flux. Of the two types of DS, NaCl appears to be the most useful DS due to its high solubility, low cost, and high osmotic potential. The low diffusivity of $MgSO_4$ compared to that of NaCl is another reason for its significantly reduced RSD.

Effects of TMP: Water Flux and COD Rejection in FO

The variations of water flux and COD rejection with changes in applied TMP are shown in Fig. 6.3.6 [90].

While water transport in an FO system is relies on osmotic-pressure difference, hydraulic TMP plays a significant role in modules like spiral wound or capillary types where it helps overcome the hydraulic resistance of the flow channels. In a flat-sheet, cross-flow membrane module this hydraulic resistance is far less than in other conventional modules. The effect of hydraulic TMP up to a level of 2 bar is seen in Fig. 6.3.6 where the selected NF-2 membrane succeeds in removing around 98% COD at a feed flow rate of 150 L/hour and DS flow rate of 10 L/hour. The magnitude of rejection and flux are greater for NaCl than for $MgSO_4$. The reason for the high flux is the osmotic pressure of the NaCl is higher than that of $MgSO_4$ (due to its lower molecular weight and thus higher effective

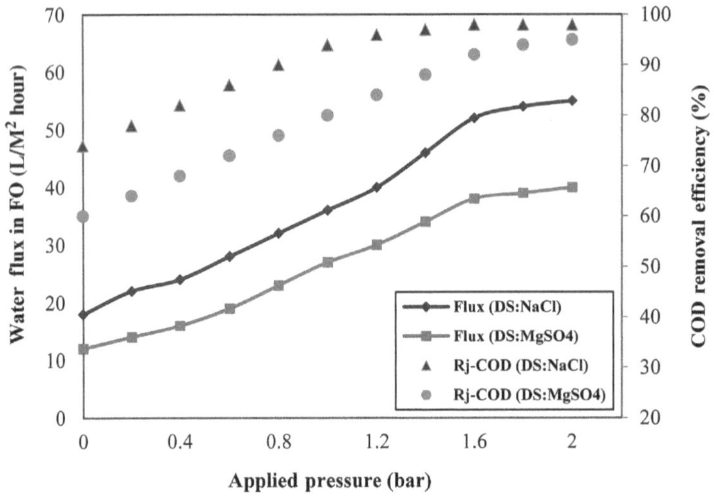

Figure 6.3.6 COD removal and water flux during forward osmosis under varying operational pressure of two different composite membranes. Experimental conditions: COD concentration of 3500 mg/L, pH 7, pressure range 0–2 bar, draw solution (DS) concentration 0.5 mol/L, feed side cross flow rate (CFR) 150 L/hour, DS side CFR 10 L/hour, and temperature 308 K [90].

concentration). Therefore due to the higher water flux the passage of the solutes is comparatively less, which causes the higher rejection of NaCl than for $MgSO_4$ as illustrated by the curves that show greater rejection for NaCl. Since the applied pressure adds to the already high osmotic pressure that forces water through the membrane, transport of other species can't be completely stopped, which explains the less than 100% rejection of COD in permeated water. In the solution-diffusion mechanism of transport through composite polyamide membrane, solute flux and solvent flux are uncoupled, which explains why under enhanced pressure, the solvent flux increases with commensurate decrease of solute flux or rather increase of solute rejection. Fig. 6.3.6 (secondary axis) also shows the effect of TMP on volumetric water flux at a feed flow rate of 150 L/hour and DS flow rate of 10 L/hour in the FO system. The water flux shows a positive correlation with TMP where the highest flux of 52 L/m^2hour is achieved with 0.5 M NaCl DS and 40 L/m^2hour for $MgSO_4$ draw solute.

System Performance Under Varying Feed CFR

The CFR has a significantly positive effect on both water flux and rejection. With an increase in the CFR there is a progressive increase in the sweeping action and thus greater turbulence on the active membrane

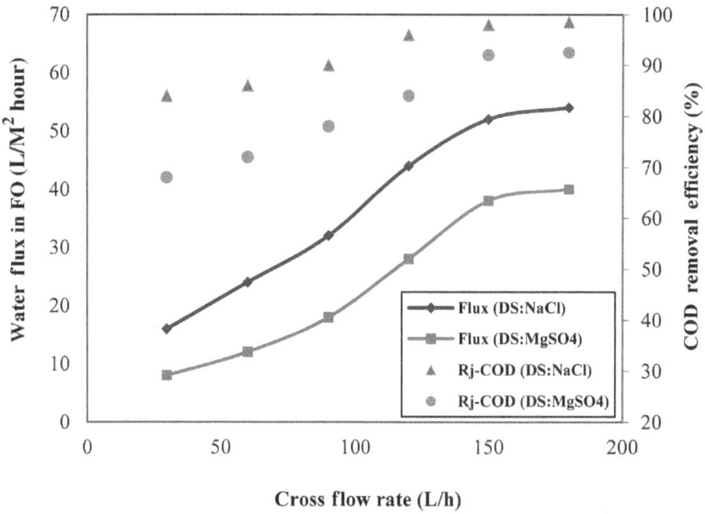

Figure 6.3.7 COD removal and water flux during forward osmosis under different feed side cross flow rate (CFR) of two different draw solution (DS). Experimental conditions: COD concentration of 3500 mg/L, pH 7, feed side CFR range 30–180 L/hour, pressure 1.6 bar, DS concentration 0.5 mol/L, DS side CFR 10 L/hour, and temperature 308 K [90].

surface area, which drives away the accumulated solutes from the membrane surface thereby minimizing the effect of CP and leading to a fouling-free nature that contributes to long operation of the module without any significant drop in flux. Moreover, the increase in CFR decreases the solute build-up, which in turn decreases the CP. This also has a significant effect on increasing the convective force developed by the solvent, which increases the water flux. The performance of the modules under varying CFR can be seen in Fig. 6.3.7 [90].

RSF: Effects of Applied Pressure and DS Concentration

The reverse solute diffusion is an undesirable phenomenon observed during FO. It is caused mainly by the diffusion of the draw solute toward the feed solution due to the concentration difference across the active layer of the membrane. Fig. 6.3.8A [90] shows that with increasing pressure there is very little change in the RSF of $MgSO_4$ in NF-2 whereas a decline of RSF with increasing pressure can be seen for NaCl draw solute. The high RSF for NaCl is attributed to its monovalent ions with high diffusivity. Ions from $MgSO_4$ being bivalent with large hydrated radii have low diffusivity. The decline in RSF with applied pressure may be attributed to physical changes in the membrane active layer as a function of the applied pressure as well as the introduction of a pressure gradient barrier that prevents reverse diffusion.

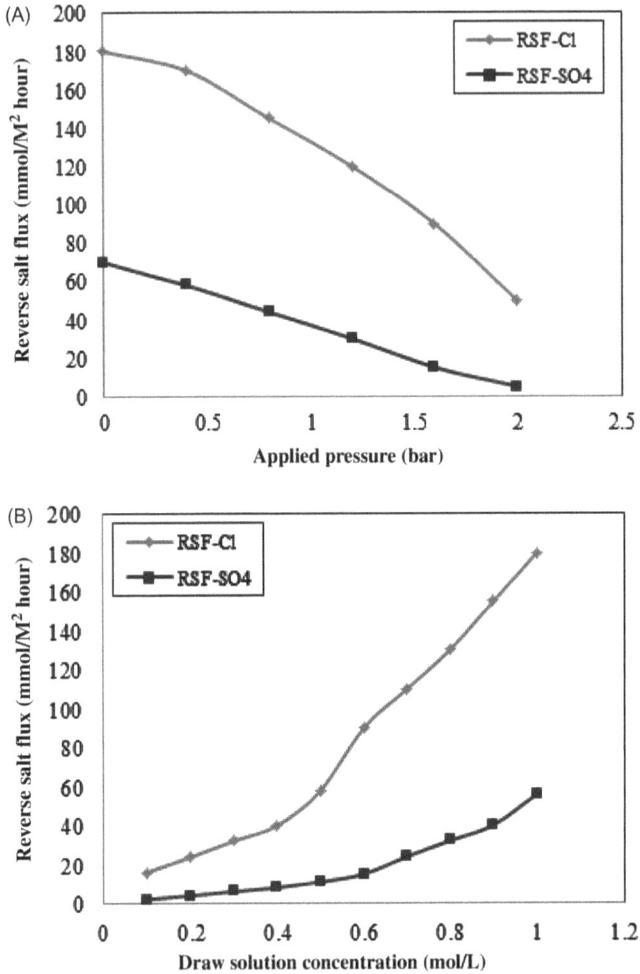

Figure 6.3.8 Reverse salt flux (RSF) during forward osmosis of two different draw solution (DS). (A) Effects of operational pressure on RSF; experimental conditions: COD concentration of 3500 mg/L, pH 7, pressure range 0–2 bar, DS concentration 0.5 mol/L, feed side CFR 150 L/hour, DS side CFR 10 L/hour, and temperature 308K. (B) Effects of DS concentration on RSF; experimental conditions: COD concentration of 3500 mg/L, pH 7, DS concentration range 0.1–0.5 mol/L, feed side CFR 120 L/hour, pressure 1.2 bar, DS side CFR 10 L/hour, and temperature 308K [90].

Increasing pressure at the membrane surface leads to compression of the interface between the thin active layer and support layer of the membranes, thus reducing the possibility of RSF and salt permeability of the membrane, while marginally affecting water flux. In the FO process, salt can diffuse across the membrane from the highly concentrated DS to the feed solution side. Because of this RSD, there is greater salt accumulation near the membrane surface than in the bulk resulting in increased fouling. DS concentration at the interface of the support layer and active rejection layer of the membrane, however, remains at a level much lower than at the bulk DS reducing the available driving force for FO. Fig. 6.3.8B [90] shows that with increase of DS concentration, RSF increases for both NaCl and $MgSO_4$. This positive correlation is well attributed to the fact that the greater the concentration of the draw solute the greater the degree of diffusion. Again, the greater RSF for NaCl is due to its greater diffusivity than that of $MgSO_4$. The relative size of the diffusing species as well as its concentration in the solution largely determines the diffusion.

Effects of Hydraulic TMP on Draw Solute Recovery and Pure Water Flux in the Downstream NF Module

In Fig. 6.3.9A [90] the primary axis indicates the changes in salt-removal efficiency with applied pressure varying from 4 to 13 bar. It is seen that when the pressure is around 13 bar the salt-removal efficiency for both the draw solutes approaches 98%–99%. This is due to the fact that these compounds during the course of the FO process dissociate into positive and negative ions in water. Since NF-1 is a negatively charged membrane, it rejects the negative ions by Donnan exclusion where the negative–negative repulsion force comes into action and helps in separation. The solution–diffusion mechanism of mass transport through NF membrane also plays an important role in solute retention where solute and solvent fluxes are uncoupled and thus an increase in TMP causes an increase in solvent flux, resulting in its higher retention or recovery for recycling.

The differences in the solute fluxes of $MgSO_4$ and NaCl may due to the size of the corresponding ions. Apart from the Donnan exclusion, sieving also plays a significant role in separation of species with large hydrated radii. Again, since the ions from $MgSO_4$ are much larger than their counterparts in NaCl they are retained more resulting in almost 100% recovery and recycle of $MgSO_4$. Recovery of NaCl is also very high by the selected NF membrane, which eventually produces clean water per WHO standards. Fig. 6.3.9A [90] (secondary axis) also shows the effects of applied pressure on volumetric water flux. When the hydraulic TMP is increased from 4 to 13 bar for the NF-1 membrane, the salt-free pure water flux increases from 24 to 58 L/m^2hour.

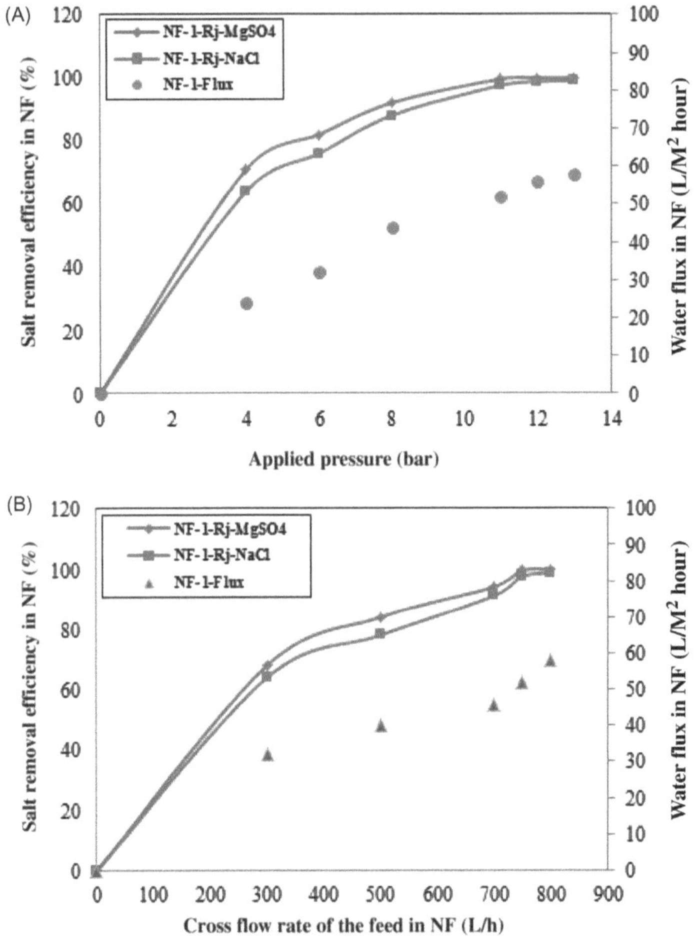

Figure 6.3.9 (A) Salt removal and water flux during nanofiltration (NF) under different transmembrane pressure of two different draw solutions (DSs). Experimental conditions: DS concentration of 0.5 mol/L, pressure range 0–13 bar, pH 7, cross flow rate (CFR) 750 L/hour, and temperature 308K. (B) Salt removal and water flux during NF under different CFR of two different DSs. Experimental conditions: DS concentration of 0.5 mol/L, CFR 0–800 L/hour, pH 7, pressure 13 bar, and temperature 308 K [90].

Effects of CFR on Salt-Removal Efficiency and Permeate Flux in NF System

A pronounced effect of CFR on salt-removal efficiency and permeate flux is observed during NF of DS. In NF, CFR shows a strong positive correlation with salt rejection and pure water flux. The rejection data of the salt in Fig. 6.3.9B [90] show that the removal efficiency of NF-1 increases from 64% to over 99% as CFR increases from 300 to 800 L/hour for the NF membrane with magnesium sulfate as draw solute. A similar trend is observed for NaCl, where the flux increases with increase in CFR from 300 to 800 L/hour at a TMP of 13 bar for all membranes. The NF-1 membrane yields a pure water flux of 58 L/m^2hour under 13 bar TMP and 800 L/hour CFR. With an increase in the CFR, the sweeping action on the active membrane surface area also increases thereby reducing CP. In other flow modes, an increase in solvent flux normally accompanies an increase in CP. However, in the flat-sheet, cross-flow module enhancement of flux following increased cross-flow does not really lead to an increase in CP. Again, reduction of CP results in convective force, which in turn enhances the solvent flux and due to the uncoupling nature of the solute and solvent fluxes also results in higher removal efficiency.

Prolonged Plant Operation: Role of CP on Fouling and Flux Decline

When the filtration system is operated long term build-up of CP that causes decline in water flux is often seen. However, with the use of a cross-flow module, fouling can be largely eliminated by the sweeping action of fluid on the membrane surface, which significantly reduces CP as evident in the overall drop in flux of NF-2 by 11% and that of NF-1 by only 10% over 120 hours of FO and NF, respectively. This indicates significant improvement of the new design over existing designs and remarkable control of CP and membrane fouling ensuring long-term operation at a steady flux. A rare long-term fouling study in a plate-and-frame module was reported in which a flux decline of 20% = in 100 hours of operation against 11% flux decline in the present design over the same duration of filtration operation. The SEM images of the membrane before and after FO and NF show that the membrane does not undergo major morphological changes possibly due to the very flow design of the cross-flow module. Even the insignificant loss in flux of this membrane can be restored after thorough rinsing with 10^{-2} (M) HNO^3 and 0.1(N) NaOH.

Scale-Up and Economic Evaluation

Scale-up and economic evaluation were carried out for a plant capacity of 50,000 L/day wastewater treatment. The closed loop-based FO–NF system yielded 52 L/m^2hour clean water by removing about 97% toxic COD from the wastewater. The clean water produced per day (16 hours of operation) by the plant is $(52 \times 16) = 832$ L/m^2day and $50{,}000/(832 \times 0.3) = 200$ of perplex sheet and SS module for both the system of membrane area 0.3 m^2. The total membrane area needed $(200 \times 0.3) = 60$ m^2 for both the system

Table 6.3.6 Values of the common economic parameters adopted to calculate = scale-up cost [90]

Parameters	**No. of equipment/ specifications**
Plant area	100 m^2 (20 m × 5 m)
Plant life	15 year
Interest rate	8%
Number, type and area of FO module	200, perplex sheet, 0.3 m^2
Type and area of the membrane for FO system	NF-2, 60 m^2
Number, type and area of NF module	200, stainless steel, 0.3 m^2
Type and area of the membrane for NF system	NF-1, 60 m^2
Membrane life	6 months
Name, concentration and amount of draw solutes	NaCl, 0.5 ppm, 50 kg
Number of feed tank and volume	3, 20,000 L
Number of DS tank and volume	1, 5000 L
Number of high-pressure pump	4
Type of high-pressure pump	Diaphragm, Max. pr. 50 bar
Number of low-pressure pump	4
Type of low-pressure pump	Peristaltic pump
Number of rotameter, pressure gauge and pH probe	4, 3, 2
Electricity consumption	2000 KWh/month
Number of labor required	

(FO & NF). The detailed design considerations and specifications are listed in Table 6.3.6.

The sixth-tenth power law was used for scale-up cost estimation and was defined as:

$$\text{Scale-up cost} = \text{Lab scale cost} \times \left(\frac{\text{Scale-up capacity of the system}}{\text{Lab scale capacity of the system}}\right)^{\frac{6}{10}}.$$

The overall cost (investment and operational) assessment is shown in Table 6.3.7.

The cost assessment was based on the annualized investment and operational cost. The annualized investment and operational cost were computed by:

$$\text{Annualized capital cost} = \left[\frac{\text{Total investment (\$)} \times \left(\frac{\mathrm{i}\,(1+\mathrm{i})^{\mathrm{n}}}{(1+i)^{n-1}-1}\right)}{\text{Water recovery in m}^3/\text{year}}\right]$$

Table 6.3.7 Investment and operational cost of a 50,000 L/day capacity wastewater-treatment plant [90]

Investment cost	Price ($)	Operational cost	Price ($/Year)
Civil investment	10,000	Electricity cost	2800
FO module cost	6000	Membrane cost for FO	6000
NF module cost	9600	Membrane cost for NF	6000
Draw solution cost	500	Labor cost	2880
Overall tank cost	5500		
High-pressure pump cost	4800		
Low pressure pump cost	1600		
Others cost	3000		
Total investment cost	41,000	Total operational cost	17,680
		Cost reduction by water reutilization	4380
		Actual operational cost	13,300

$$\text{Annualized operational cost} = \left[\frac{\text{Total operational cost (\$/year)}}{\text{Water recovery in m}^3/\text{year}}\right]$$

where n and i are the plant project life and interest rate, respectively.

Thus the annualized cost for production of 1 m^3 clean water is the sum of the annualized capital cost and annualized operating cost, and is $(0.24 + 0.72) = 0.96\$$.

Treatment Technology: Toward a Sustainable Solution

Keeping thousands of kilometers of river bodies clean is now one of the biggest challenges confronting countries such as India and China. A paradigm shift in treatment policy is absolutely essential for new facilities in industries along the banks of these countries. The treatment scheme must close the loop of treatment plants by protecting the river bodies from the onslaught of hazardous waste discharge. New schemes should treat water to the level of reusable criteria and should integrate FO with NF. The major hurdles of using a low-energy FO system can be overcome through a new design that significantly reduces CP and permits long-term operation without the membrane fouling and drop in flux. For example, more than 97% COD can be removed from pharmaceutical wastewater at a pure water flux greater than 50 LMH, which is a significant improvement over reported studies. In this continuous scheme, 99% draw solute can be recovered and recycled and pure water flux reaches 58–60 LMH. The design of the flat-sheet cross-flow module with horizontal alignment of the separating membrane in the FO loop minimizes the chances of CP and membrane fouling due to the sweeping fluid flow over the membrane surface. This design significantly improves pure water flux in both FO and downstream NF and ensures sustainable operation at a steady flux.

6.3.3 OPTIMIZATION AND CONTROL OF AN NF–FO ADVANCED TREATMENT PLANT USING A VISUAL BASIC PLATFORM

6.3.3.1 Introduction

A Visual Basic Software tool was developed (Eq. 6.3.5) with the purpose of optimization, control, and monitoring of an NF–FO integrated plant (Eq. 6.3.6) for the treatment of pharmaceutical wastewater. As described in the previous section, the NF–FO integrated system promises a high degree of purification at reasonably high water flux. However, optimization and performance analysis of this type of system need to be done. A Visual Basic software tool can be used to monitor and control plant operation, optimize operating variables, and aid in rapid analysis of the performance of process units as well as the integrated system.

Key Features of the Tool (Eq. 6.3.5)

1. Has the ability to scale-up the FO–NF
2. Operating conditions can be optimized
3. Built-in multiwindow system displays performance results, output profiles, and plant scheme
4. Model constants and input model data can be verified in a single window

Steps in Development of the Software

This software was developed for an integrated FO–NF based hybrid system for the treatment of pharmaceutical wastewater, and uses the dynamic mathematical model of the integrated system.

Step 1. In the first step, the mathematical model for the FO system is developed based on the theory of the dilute and concentrate ECP mechanism, and NF model is developed based on the theory of the extended Nernst–Plank equation with a linearized model.

Step 2. All the physicochemical model parameters used in the model are determined either experimentally or by using literature-based standard mathematical corelations.

Step 3. All the model equations are solved using constant data and input and target model data (rejection and water flux) using appropriate numerical solution techniques (e.g., Runge–Kutta fourth-order method, Runge–Kutta–Feldberg method or fixed-point iteration method).

Step 4. The model performance is checked using standard statistical methods and experimental results. The software is validated through comparison of experimental values with model-predicted ones.

Basic Theory of the Transport Model (Eq. 6.3.5)

In pressure-assisted FO, the principal solute flux is due to two transport mechanisms: diffusion and convection. In addition to this, dilute and concentrated external concentration polarization (ECP) are encountered albeit relatively small in magnitude and predominantly reversible in nature. In an NF system, overall flux can be achieved by three different transport mechanisms: diffusion, convection, and electromigration. Based on these mechanisms two different systems (FO and NF) were developed based on the following five major assumptions: (1) the ratio of bulk concentration to the active membrane-surface concentration is assumed equal to the corresponding ratio of their respective osmotic pressures for both the feed and DS sides; (2) the back diffusion of feed solute from the membrane interface to the bulk-feed solution is negligible; (3) two correction factors, diffusivity and tortuosity, are introduced to account for the degree of diffusion of solute and for the anomalies in the alignment of the path, which affect the solute flux; (4) the effective charge density of the solute particle on the membrane surface is assumed to be constant throughout the membrane surface and differs mainly due to the pH and concentration of the feed solution; and (5) the average values of the concentration of the solute and the Donnan potential inside the membrane pores are considered.

6.3.3.2 Model Equations for Software Development

Model Equations of FO System

Incorporating the above assumptions the following model was developed. In the present model the solute transport from the feed side to the permeate side across the membrane is described by:

$$J_{s,\mathrm{mod}} = [(J_{s,\exp} R_{j,\exp} \Delta\Pi)/(1 - \mathrm{R}_{j,\exp})] e^{(-J_{s,\exp}/K_m)} + [(1-\sigma) J_{s,\exp} C_{\mathrm{eff}}] \quad (6.3.5)$$

Here, $J_{s,mod}$ is the solute flux from the feed side to the DS side, and σ and τ are the reflection coefficient and tortuosity, respectively.

Water flux through membrane may be expressed using reflection coefficient as

$$J_w = A(\sigma\Delta\Pi - \Delta P) \tag{6.3.6}$$

where J_w is water flux is, A is water permeability of the membrane, σ is reflection coefficient, Π and ΔP are osmotic pressure and hydraulic press, respectively.

$\Delta\pi$ is the effective difference of the osmotic pressures, which can be calculated by the Vant Hoff's equation:

$$\Delta\Pi = \Pi_{\mathrm{DB}} - \Pi_{\mathrm{FB}} = nRT_{\mathrm{r}}(C_{\mathrm{DB}} - C_{\mathrm{FB}}). \tag{6.3.7}$$

C_{eff} is the effective concentration, which is concentration difference between the active membrane interface on the DS and feed solution sides, respectively, and expressed as:

$$C_{\mathrm{eff}} = C_{\mathrm{DM}} - C_{\mathrm{FM}} \tag{6.3.8}$$

The concentration of the active membrane interface on the DS (C_{DM}) and feed solution (C_{FM}) can be estimated by utilizing the concentrative and dilutive ECP moduli, which is expressed as:

$$C_{\mathrm{DM}} = \left(\Pi_{\mathrm{DB}} e^{(J_{s,\exp}/K_{\mathrm{m}})}\right)/nRT_{\mathrm{r}} \tag{6.3.9}$$

$$C_{\mathrm{FM}} = \left(\Pi_{\mathrm{FB}} e^{(J_{s,\exp}/K_{\mathrm{m}})}\right)/nRT_{\mathrm{r}} \tag{6.3.10}$$

To calculate the rejection performed by the FO membrane, the permeate concentration needs to be evaluated and can be estimated by applying the suitable boundary conditions in the main mass balance equation, which is formulated as:

$$D_{\mathrm{s}} \int_{C_{\mathrm{B}}}^{C_{\mathrm{P,FO}}} \frac{\partial C}{C} = (J_{\mathrm{s,exp}} - J_{\mathrm{s,mod}}) \int_{0}^{\Delta x} \partial Z \tag{6.3.11}$$

Integration and rearrangement of Eq. (6.3.11) results in the following equation from which the permeate concentration can be directly calculated:

$$C_{P,FO} = \left\{ \left[e^{(J_{s,exp}\Delta x/D_s)} \{ (J_{s,exp} C_B) - J_{s,mod} \} \right] + J_{s,mod} \right\} / J_{s,exp} \qquad (6.3.12)$$

The permeate concentration obtained from the above computation is tallied with the iteratively estimated value of the permeate concentration, which establishes the correctness and suitability of the model equation as well as the iterative procedure performed. It also gives an idea of the actual rejection of solute by the membrane.

The model solvent flux can be estimated by:

$$J_{v,mod} = J_{s,mod} / C_{P,FO} \qquad (6.3.13)$$

Model Equations of NF System (Eq. 6.3.5)

The flux for charged particle through the NF membrane can be measured using the ENP Eq. (6.3.9). In this study, draw solutes (Na^+ and Cl^- ions) have been separated and the modified ENP equation has been used for determination of water flux and rejection, which may be expressed as:

$$J_{s,i} = V_{sol} C_{p,i} = (H_{c,i} C_{m,i} V_{sol}) + \left(-D_{c,i} \frac{dC_{m,i}}{dx_m} \right) + \left[\left(-\frac{d\psi_{em}}{dx_m} \right) \frac{F z_i C_{m,i} D_{c,i}}{RT} \right] \qquad (6.3.14)$$

The flux (J_s) of ion i is the sum of the fluxes due to convection, diffusion, and electromigration. $D_{c,i}$ is the diffusion coefficient of i through the membrane pores where $H_{c,i}$ is the challenge factor for convection. V_{sol} is the solvent velocity.

$C_{m,i}$ is the pore inlet concentration of the membrane of ion i and can be computed by the Donnan equilibrium condition:

$$C_{m,i} = C_{F,i} (1 - \lambda_i)^2 e^{\left(\frac{-z_i F \Delta\psi_{pot}}{RT} \right)} \qquad (6.3.15)$$

The concentration gradient of the membrane pore for charged ions ($Cl^{-)}$ may be derived from Eq. (6.3.14) and is defined as:

$$\frac{dC_{m,i}}{dx_m} = \left[\left(\frac{V_{sol} H_{c,i} C_{m,i} - V_{sol} C_{p,i}}{D_{c,i}} \right) \right] - \left[\left(\frac{F z_i C_{m,i}}{RT} \right) \left(\frac{d\psi_{em}}{dx_m} \right) \right] \qquad (6.3.16)$$

The electromigration gradient through the membrane pores may be derived from:

$$\frac{d\psi_{em}}{dx_m}=\frac{V_{sol}RT\left[\left(\frac{z_p(H_{c,p}C_{m,p}-C_{p,p})}{D_{c,H}}\right)+\left(\frac{z_n(H_{c,n}C_{m,n}-C_{p,n})}{D_{c,n}}\right)\right]}{F\,(z_p^2C_{m,p}+z_n^2C_{m,n})} \quad (6.3.17)$$

Considering the electro neutrality conditions, the pore inlet membrane-wall concentration and permeate concentration of the positive ion of the solutions can be described by the following two expressions:

$$C_{m,p}z_p=-\,z_nC_{m,n}-X_{mc} \quad (6.3.18)$$

$$C_{p,p}z_p=-\,z_nC_{p,n} \quad (6.3.19)$$

where X_{mc} is the concentration of electrical group on the membrane surface (mol/m^3).

By rearranging Eq. (6.3.15), the Donnan potential for both the positive and negative ions at the pore inlet is obtained by:

$$\Delta\psi_{dp}(x=0)=-\left[\left(\frac{RT_r}{F}\right)\ln\left(\frac{C_{m,p}}{\varphi_pC_{F,p}}\right)\right]=\left[\left(\frac{RT_r}{F}\right)\ln\left(\frac{C_{m,n}}{\varphi_nC_{F,n}}\right)\right] \quad (6.3.20)$$

Substituting $C_{m,p}$ and $C_{p,p}$ from Eqs. (6.3.18) and (6.3.19) into Eq. (6.3.17) yields Eq. (6.3.20) for finding the concentration gradient of negative ion and is described as:

$$\frac{dC_{m,n}}{dx_m}=\left[\frac{\left\{\left(\frac{V_{sol}H_{c,p}}{D_{c,p}}+\frac{V_{sol}H_{c,n}}{D_{c,n}}\right)\left(C_{m,n}^2-X_{mc}C_{m,n}\right)\right\}-\left(\frac{C_{m,n}C_{p,n}V_{sol}}{D_{c,p}}\right)-\left(\frac{C_{p,n}V_{sol}}{D_{c,n}}\right)(C_{m,n}-X_{mc})}{(2C_{m,n}-X_{mc})}\right] \quad (6.3.21)$$

The previous equation shows that the effects of charge ion concentration at the pore inlet are relatively small as the concentration profile across the pore length is constant due to the higher order of the numerator of the equation. Under these circumstances, the concentration profile can be estimated as follows:

$$\frac{\Delta C_{m,n}}{\Delta x_m}=\left[\frac{\left\{\left(\frac{V_{sol}H_{c,p}}{D_{c,p}}+\frac{V_{sol}H_{c,n}}{D_{c,n}}\right)\left(C_{m,n_{av}}^2-X_{mc}C_{m,n_{av}}\right)\right\}-\left(\frac{C_{m,n_{av}}C_{p,n}V_{sol}}{D_{c,p}}\right)-\left(\frac{C_{p,n}V_{sol}}{D_{c,n}}\right)(C_{m,n_{av}}-X_{mc})}{(2C_{m,n_{av}}-X_{mc})}\right] \quad (6.3.22)$$

$\Delta C_{\mathrm{m,n}}$ and $C_{\mathrm{m,n_{av}}}$ can be computed with the help of Eqs. (6.3.18) and (6.3.20) and are expressed as:

$$\Delta C_{m,n} = C_{m,n}(0) - C_{m,n}(x) = \frac{1}{2}\left[\sqrt{(X_{\mathrm{mc}}^2 + 4\,\varphi_p.\varphi_n.C_{\mathrm{p,n}}^2)} - \sqrt{(X_{\mathrm{mc}}^2 + 4\,\varphi_p.\varphi_n.C_{\mathrm{F,n}}^2)}\right] \tag{6.3.23}$$

$$C_{m,n_{\mathrm{av}}} = \left(\frac{C_{m,n}(0) + C_{m,n}(x)}{2}\right) = \frac{1}{4}\left[2X_{\mathrm{mc}} + \sqrt{(X_{\mathrm{mc}}^2 + 4\,\varphi_p.\varphi_n.C_{\mathrm{F,n}}^2)} + \sqrt{(X_{\mathrm{mc}}^2 + 4\,\varphi_p.\varphi_n.C_{\mathrm{p,n}}^2)}\right] \tag{6.3.24}$$

The following explicit expression for $C_{\mathrm{p,n}}$ can be obtained by rearranging Eq. (6.3.22):

$$C_{\mathrm{p,n}} = \left[\frac{\left\{C_{m,n_{\mathrm{av}}}\left(\frac{H_{\mathrm{c,p}}V_{\mathrm{sol}}\Delta x_m}{D_{\mathrm{c,p}}} + \frac{H_{\mathrm{c,n}}V_{\mathrm{sol}}\Delta x_m}{D_{\mathrm{c,n}}}\right).(X_{\mathrm{mc}} - C_{m,n_{\mathrm{av}}})\right\} + (2C_{m,n_{\mathrm{av}}} - X_{\mathrm{mc}}).\Delta C_{m,n}}{\frac{V_{\mathrm{sol}}\Delta x_m}{D_{\mathrm{c,n}}}(X_{\mathrm{mc}} - C_{m,n_{\mathrm{av}}}) - \left(\frac{V_{\mathrm{sol}}\Delta x_m C_{m,n_{\mathrm{av}}}}{D_{\mathrm{c,p}}}\right)}\right] \tag{6.3.25}$$

Rejection is calculated by:

$$R_{\mathrm{j}} = 1 - (C_{\mathrm{P_n}}/C_{\mathrm{B}}) \tag{6.3.26}$$

The volumetric water flux can be assessed by:

$$J_{\mathrm{v}} = J_{\mathrm{s}}/C_{\mathrm{P_n}} \tag{6.3.27}$$

6.3.3.3 Determination of Unknown Model Parameters (Eq. 6.3.5)

The physicochemical parameters of the model equations were computed using empirical relations as follows:

i. Determination of experimental water flux ($J_{\mathrm{s,exp}}$):

$$J_{\mathrm{s,exp}} = W_{\mathrm{p}} \times \Delta P_{\mathrm{eff}} \tag{6.3.28}$$

where W_{p} is the water permeability and ΔP_{eff} is the effective pressure.

ii. Determination of mass transfer coefficient (K_{m}):

$$K_{\mathrm{m}} = (\mathrm{Sh_w} \times D_{\mathrm{s}})/d_{\mathrm{h}} \tag{6.3.29}$$

where d_{h} is the hydraulic diameter of the feed channel, which is evaluated from the channel geometry and $\mathrm{Sh_w}$ refers to the Sherwood number, which is calculated from the following empirical correlation applicable for turbulent flow regime:

$$Sh_w = 0.04 \times Re^{0.04} \times Sc^{0.75} \tag{6.3.30}$$

iii. Determination of Reynolds (Re) number and Schmidt number (Sc):

$$Re = (d_h \times V \times \rho)/\mu \tag{6.3.31}$$

$$Sc = \mu/(\rho \times D_s) \tag{6.3.32}$$

where V is the CFV of the feed solution, μ is the viscosity of the feed solution, and ρ is density of the feed solution.

iv. Solvent velocity can be computed by the following Hagen–Poiseuille expression:

$$V_{sol} = \frac{r_{mp}^2 \, \Delta P}{8 \, \eta_s \, \Delta x} \tag{6.3.33}$$

v. Computation of hindered diffusivity ($D_{c,i}$):

The hindered diffusivity (D_c) can be computed by multiplication of the bulk diffusivity (D_b) and challenge factor for diffusion (H_d) and expressed by:

$$D_{c,i} = D_{b,i} \times H_{d,i} \tag{6.3.34}$$

The challenge factor for diffusion of ion i can be calculated by:

$$H_{d,i} = (1.0 - 2.3\lambda_i + 1.154\lambda_i^2 + 0.224\lambda_i^3) \tag{6.3.35}$$

where λ_i represents the ratio of solute radius to pore radius and is expressed as:

$$\lambda_i = r_{i,s}/r_{mp} \tag{6.3.36}$$

vi. Computation of challenge factor for convection of ion i ($H_{c,i}$):

It can be estimated by:

$$H_{c,i} = (2 - \varphi_i)(1.0 + 0.054\lambda_i - 0.998\lambda_i^2 + 0.44\lambda_i^3) \tag{6.3.37}$$

where

$$\varphi_i = (1 - \lambda_i)^2 \tag{6.3.38}$$

6.3.3.4 Evaluation of the Model Performance (Eq. 6.3.5)

A standard statistical method was used to estimate model performance in terms of removal efficiency (COD, DS) and water flux data. The RE and Willmott-d-index are the two different error parameters and are computed with the help of model-predicted data and experimental data. The RE was calculated as:

$$\mathrm{RE} = \frac{\text{Root mean square error}}{\text{Mean value of experimental run}} \tag{6.3.39}$$

The RMSE was computed by:

$$\mathrm{RMSE} = \left[\left(\sum_{i=1}^{N} (M_i - E_i)^2\right) / N\right]^{1/2} \tag{6.3.40}$$

where M_i and E_i are the model-predicted value and experimental value, respectively, and N is the number of investigation.

The Willmott-d-index ($d_{\text{will.}}$) was calculated using:

$$d_{\text{will.}} = \left[1 - \left(\sum_{i=1}^{N} (M_i - E_i)^2\right) / \left(\sum_{i=1}^{N} \left[\left|(P_i - \overline{E})\right| + \left|(E_i - \overline{E})\right|\right]^2\right)\right] \tag{6.3.41}$$

where $\overline{E}$ is the mean value of the experimental results.

If the value of RE is $<= 0.10$ and the value of Willmott-d-index is $>= 0.95$ the model is considered to perform well. The performance of the model is presented in a separate window (Table 6.3.8) of the software.

Table 6.3.8 Evaluation of software performance using COD rejection, draw solute rejection, and water flux data (Eq. 6.3.5)

Figure number	System used	Output parameter	Errors value	
			RE	d_{will}
Fig. 6.3.19A	Forward osmosis	Flux	0.095	0.98
Fig. 6.3.19A	Forward osmosis	Rejection	0.096	0.98
Fig. 6.3.19B	Forward osmosis	Flux	0.098	0.97
Fig. 6.3.19B	Forward osmosis	Rejection	0.095	0.98
Fig. 6.3.19C	Forward osmosis	Flux	0.10	0.95
Fig. 6.3.19C	Forward osmosis	Rejection	0.099	0.97
Fig. 6.3.20A	Nanofiltration	Flux	0.095	0.98
Fig. 6.3.20A	Nanofiltration	Rejection	0.095	0.99
Fig. 6.3.19A	Nanofiltration	Flux	0.097	0.97
Fig. 6.3.20B	Nanofiltration	Rejection	0.096	0.98

6.3.3.5 Model Development

The model is used to evaluate the water flux and removal efficiency (COD and DS) of FO and NF. The model solution is found with the following steps:

Step 1. For FO modeling, first the bulk concentrations and osmotic-pressure differences are evaluated using Eqs. (6.3.7) and (6.3.8).

Step 2. The concentration of the active membrane interface on the DS (C_{DM}) and feed solution (C_{FM}) is calculated by the respective CP moduli Eqs. (6.3.9) and (6.3.10).

Step 3. The solute and solvent flux are evaluated using Eqs. (6.3.5) and (6.3.13).

Step 4. The principal mass-balance equation is integrated and suitable boundary conditions are incorporated to calculate the permeate-side concentration. Then an iterative procedure is adopted to find the actual permeate-side concentration until the computed value equals the assumed value.

Step 5. In the NF modeling, first the pore inlet concentration of the draw solute concentration (chloride) at the membrane surface was evaluated using Eq. (6.3.15).

Step 6. Eq. (6.3.23) is used to find $\Delta C_{m,n}$ where $C_{m,nav}$ is calculated using Eq. (6.3.24) with an assumed permeate-concentration value and checking the predicted value of $C_{p,n}$ using Eq. (6.3.25).

Step 7. The removal efficiency of the draw solute (chloride) is then calculated with the help of the permeate-concentration value using Eq. (6.3.26) where the volumetric water flux of the water is calculated using Eq. (6.3.27).

Step 8. Retention of chloride ions and volumetric water flux were evaluated by implementing an iterative method using some predicted value of concentration of electric-charge group on the membrane surface (X_{mc}) until the predicted value converges with the experimental value.

6.3.3.6 Integrated Treatment System, Methods, and Materials (Eq. 6.3.6)

The NF–FO system is shown in the "Plant Diagram" window of the software (Fig. 6.3.11). The upper portion of the flat-sheet, cross-flow membrane module is connected to the feed tank while the DS tank is

Table 6.3.9 Quantity of the characterized parameters of the pharmaceutical wastewater before and after treatment (Eq. 6.3.5)

Characterized parameters	Units	Quantity before treatment	Quantity after treatment	Maximum discharge limit
pH	–	6.4–6.8	6.2–6.3	8.5
Salinity	–	9–11	0.08–0.1	1000
Conductivity	ms/cm	13–16	0.18–0.19	0.005
Temperature	K	299–301	299	313
Total solids	mg/L	945–965	BDL[a]	–
TSS[a]	mg/L	334–362	BDL	100
TDS[a]	mg/L	588–610	38–44	500
Oil & grease	mg/L	16–19	BDL	10
Color	mg/L	2400–2550	NIL	–
COD[a]	mg/L	3450–3580	115–125	250
BOD[a]	mg/L	448–468	BDL	30
Chloride	mg/L	42–48	3.2–4.5	250
Sulfate	mg/L	210–215	18–28	500

[a]BDL, Below detected level; TSS, Total suspended solids; TDS, Total dissolved solids; COD, Chemical oxygen demand; BOD, Biological oxygen demand

connected in the lower portion. Both the system consists of cross-flow and parallel membrane modules number of which is flexible. The optimum conditions are 0.5 M of NaCl DS, 1.6 bar operating pressure, and CFR of 150 L/hour in the FO system, whereas in the downstream stage (NF system) applied pressure and CFR are 12 bars and 780 L/hour, respectively. Table 6.3.9 shows the wastewater characteristics.

Thin-film, flat-sheet polyamide composite-based NF membranes (NF-1, NF-2) from Sepromembranes Inc. (USA) are used. The effective membrane-surface area used in the investigation is 100 cm^2. The temperature, pressure, and pH tolerance ranges of the membrane are 273–323K, 0–83 bar, and 2–11, respectively. The membrane rejects 99.5% sulfate ions and 90% chloride ions experimentally at 10 bar of pressure (data provided by the manufacturer). Removal of COD from wastewater and draw solutes (NaCl) from DS by the FO module and NF module are evaluated using the initial value (C_f) and the residual value (C_p) of COD and DS concentration in the feed stream and treated stream, respectively, and is expressed by:

$$\%\ \text{Removal efficiency} = [1 - (C_p/C_f)]$$

6.3.3.7 Software Interface

Code Description and Data Input

The "PWWT.VB" simulation software has been designed in VB.NET (Microsoft Visual Studio 2008, Version 9.0.21022.8 RTM), which consists of different types of window splash pages. This type of user-friendly, menu-driven software is capable of producing the visual graphics using the output values of the target responses (removal efficiency and water flux). On start-up, you will see a splash page (Fig. 6.3.10) called the "General Window" consisting of four different types of "Buttons," namely the "General Window," "Plant Diagram," "Forward Osmosis Mode," and "Nanofiltration Mode." Tthese "Buttons" are created by using the "Toolbar" and "Properties box" options. "Forward Osmosis Mode" and "Nanofiltration Mode" are used to optimize and analyze the performance of the individual units as well as the overall system. "General Window" also contains eight different "Panels," namely "Mode of Simulation," "Mode of Operation,"

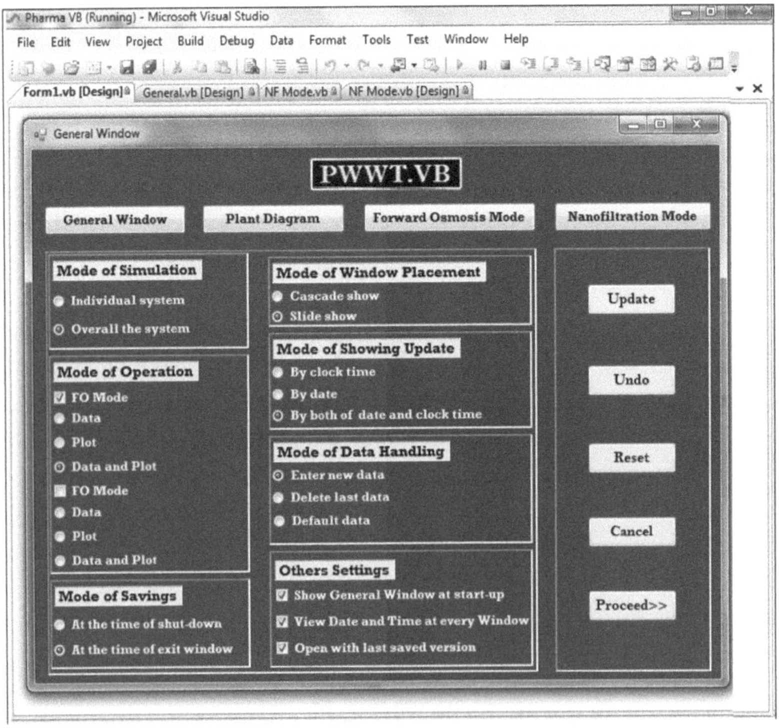

Figure 6.3.10 General window of the "PWWT.VB" software (Eq. 6.3.5).

"Mode of Savings," "Mode of Window Placement," "Mode of Showing Update," "Mode of Data Handling," and "Other Settings."

The "Other Settings" button controls the "PWWT.VB" software. "Mode of Simulation" allows optimization and performance analysis of an FO or NF system and shows the performance results in a data or graphical visualization. Using "Mode of Window Placement" users can choose different window effects such as "Cascade Show" or "Slide Show." "Mode of Savings" allows you to save in updated mode or shut down the software at the time of exit and "Mode of Showing Update" allows you to either update the time to clock time or date or both. The "Mode of Data Handling" option includes the parameters of the input data sheet. Other settings like "Show General Window at start-up," "View Date and Time at every Window," and "Open with last saved version" also help to make the software more user-friendly. Another panel has five different buttons: "Update," "Undo," "Reset," "Cancel, the" and "Proceed." By selecting "Update" tab, a fresh created mode could be kept in an updated version, where by clicking on "Undo" tab, selected modes could be shown in last updated version and by clicking "Reset" button, the mode can be shown in the last saved version and canceling the updated mode by clicking "Cancel" tab. "Proceed" allows further processing of the splash page after any modification and generates the results as data or in graphical mode. The "Plant Diagram" tab allows you to see the schematic (Fig. 6.3.11) of the treatment plant on start-up.

After designing all the operational pages, the program is written in the "Coding Page" with the help of model equations, accessed from the "View Code" option by right clicking on the user interface (Eq. 6.3.5).

In this performance analysis software, dynamic data entry for specific units can be input separately as required. The "FO Mode" window (Fig. 6.3.12) has several panels: "FO constant data," "FO input data," "Choice of DS," "FO output data," and "FO output profile."

The figure shows the effects of different input parameters of the FO system such as operating pressure, operating flow rate, and DS concentrations on COD removal of pharmaceutical wastewater. The output data and profiles are shown in terms of "Rejection" of COD and "Water Flux." Using the input data and constant data of the FO system the output data were evaluated using different model equations, which are written in the "coding page." The input parameters are written in a "Combo box" in which the input data is changed by clicking "Next" and all the data are stacked into the "Combo box." "Flux Profile" and "Rejection Profile" are

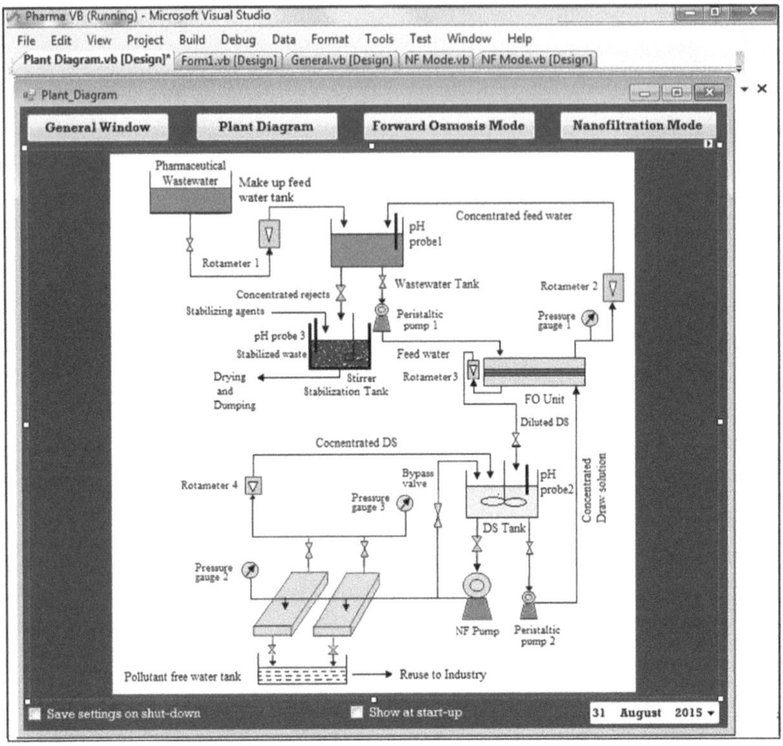

Figure 6.3.11 Schematic of the experimental set-up.

shown the graphical visualization of the respective response that is water flux and removal efficiency of COD. Similarly, operating pressure and CFR were used as input parameters in the NF system (Fig. 6.3.13).

The various input parameters are shown in Figs. 6.3.14A and 6.3.15A. The "Last" option is used to show the older data.

After inputting data and clicking "Next," "Water Flux" and "Rejection" of COD are automatically produced in the "Combo box" of "FO Output data." The various input parameters are shown in Figs. 6.3.17A and 6.3.18A.

By clicking "Next," the "Water Flux" and "Rejection" of the draw solute are calculated in the "Combo Box" and both of the output profiles are found by clicking the "Flux Profile" or "Rejection Profile" options.

For the FO system, an operating flow rate of 0–2 bar and 30–180 L/hour, respectively, and DS concentration from 0.1 to 1.0 mol/L were used. For the NF system, the operating pressure and CFR were varied

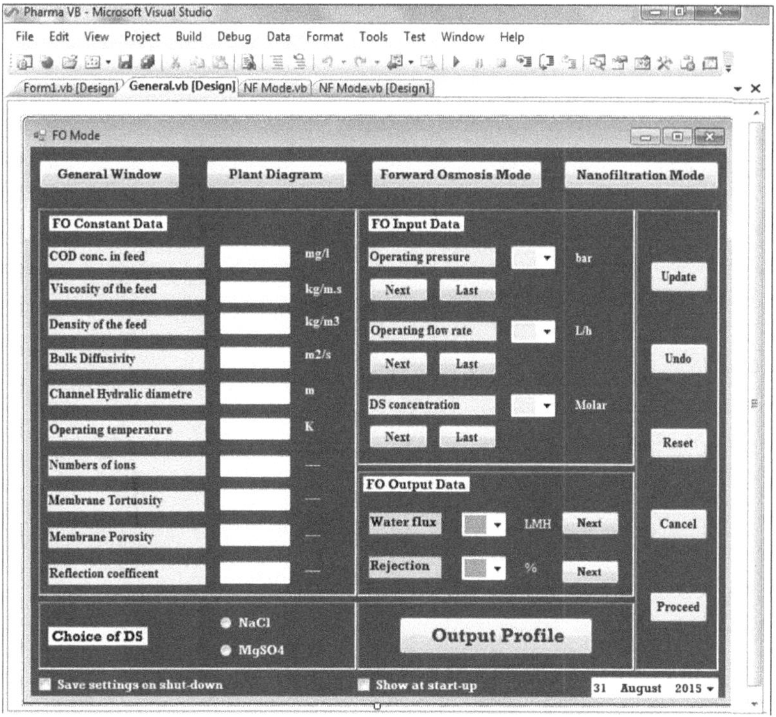

Figure 6.3.12 Input interface of forward osmosis system (Eq. 6.3.5).

from 0 to 14 bar and 0–800 L/hour, respectively. All the necessary input values can be inserted into the respective input box by clicking on the "text" option of the "Properties" window.

6.3.4 SYSTEM PERFORMANCE

Fig. 6.3.14A, B shows the output results of the FO system and output profiles in terms of water flux and removal efficiency of COD, which vary with the operating pressure. The operating pressure to the feed solution increases the solvent permeating through the membrane by increasing the force on it, which enhances the solvent passage and correspondingly, COD retention progressively increases. The water flux becomes constant when it reaches around 1.5 bar pressure, which indicates the optimum operating pressure.

Fig. 6.3.15A, B also shows the effects of draw solute concentration (NaCl) on COD removal and water flux rate. A strong positive correlation

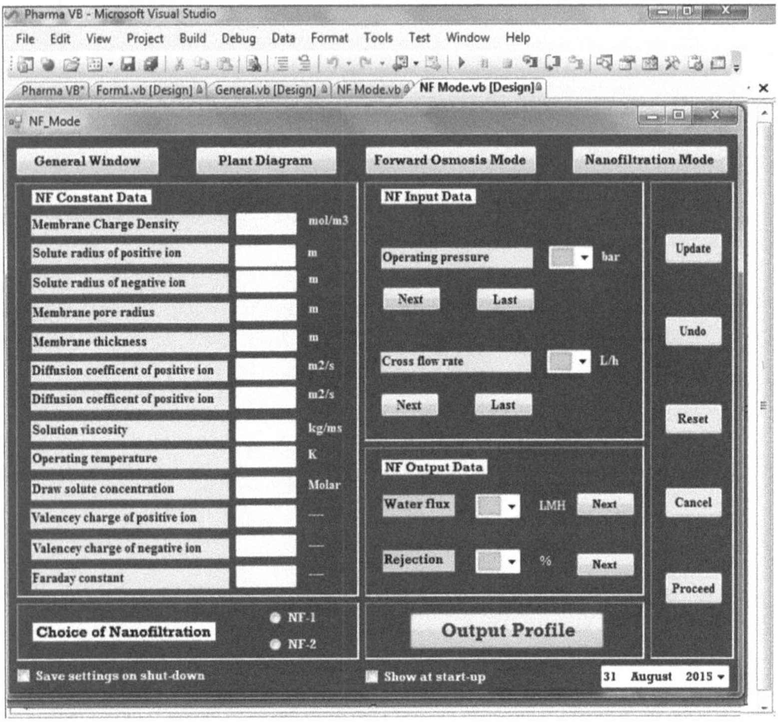

Figure 6.3.13 Input interface of nanofiltration system (Eq. 6.3.5).

between retention of COD and water flux with draw solute concentration is observed. This is due to the fact that the greater the DS concentration, the greater the magnitude of the corresponding osmotic pressure, which means a higher volume of water flux. Subsequently, there is less passage of solute as the solvent occupies most of the membrane pores for water transportation, thus increasing solute removal.

Fig. 6.3.16 shows the effects of operating flow rate and draw solute concentration (NaCl). Water flux and removal efficiency of COD increase with increase in CFR. Due to the increase of CFR, the sweeping action on the active membrane-surface area also increases, which in turn reduces the solute accumulation and decreases the effect of CP. The enhanced CFR not only increases the convective force but also provides the maximum available effective membrane-surface area for separation, both of which contribute to the consistently high water flux and high removal efficiencies.

Similarly for the NF system, Fig. 6.3.17A,B and Fig. 6.3.18A,B show the effects of operating pressure and CFR on water flux rate and removal efficiency of draw solutes, respectively.

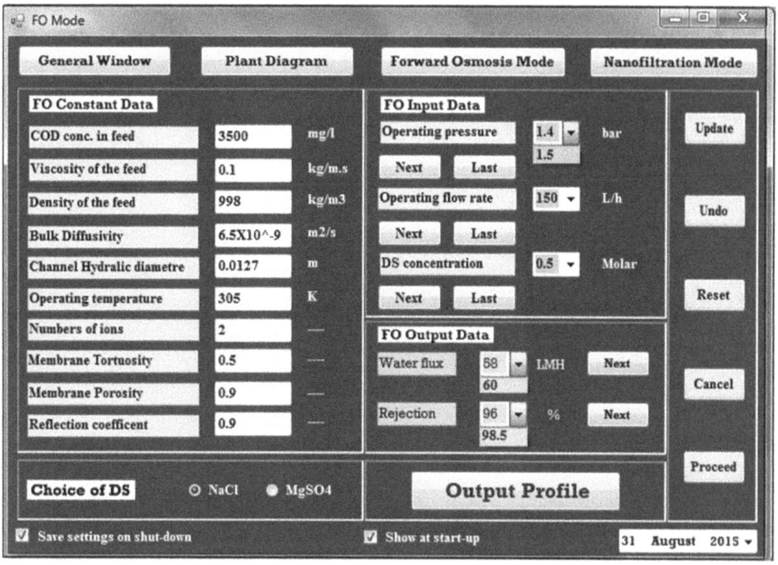

(A)

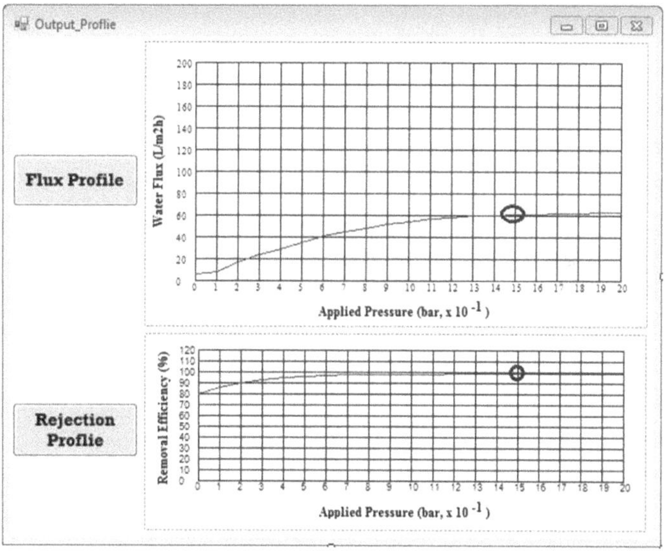

(B)

Figure 6.3.14 The effects of operating pressure on output results of FO system: (A) Effects of operating pressure on removal efficiency and water flux; (B) Output profiles of flux and rejection with applied or operating pressure (Eq. 6.3.5).

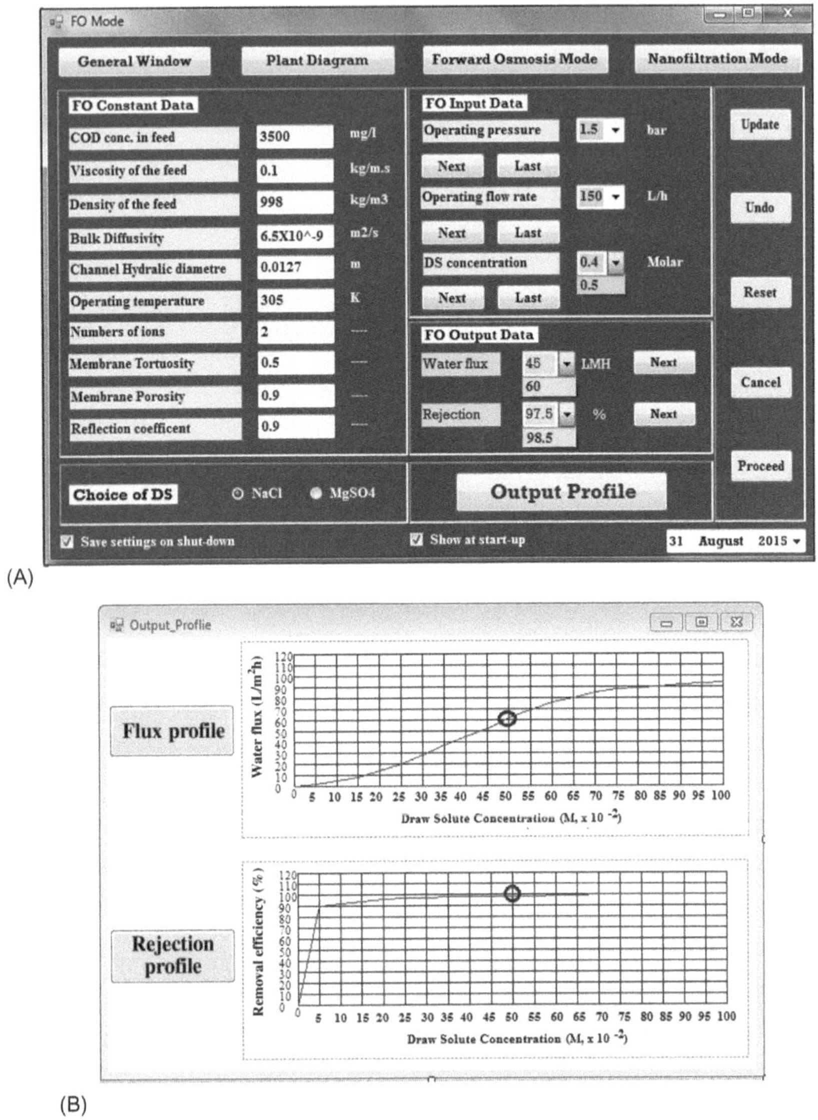

Figure 6.3.15 Effects of draw solute concentration on output results of FO system: (A) Effects of operating pressure on removal efficiency and water flux; (B) Output profiles of flux and rejection with applied or operating pressure (Eq. 6.3.5).

Fig. 6.3.17A,B shows the positive effects on water flux rate and removal efficiency of DS of operating pressure. In the solution–diffusion mechanism of the involved transport process through NF membrane, the solute flux and solvent flux are uncoupled. This implies that with the

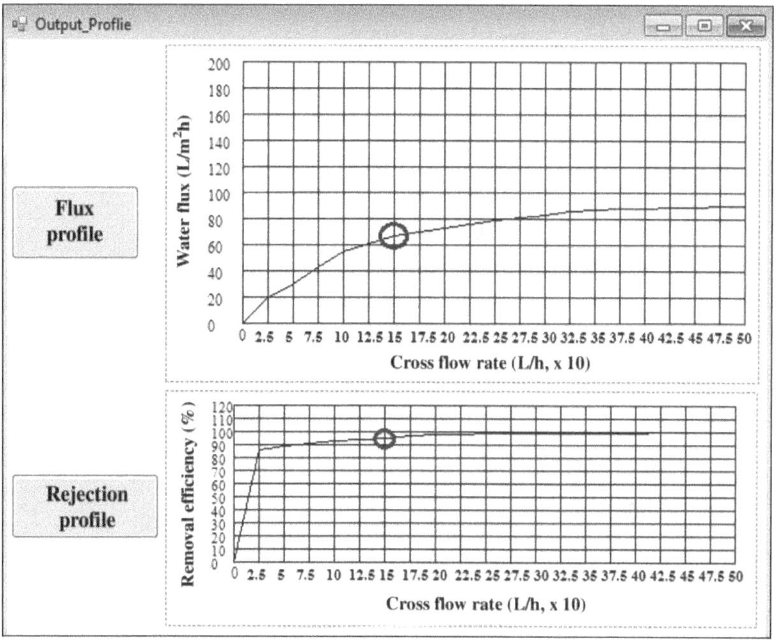

Figure 6.3.16 Graphical visualization of water flux and rejection of forward osmosis system (Eq. 6.3.5).

increase of applied operating pressure, the solvent flux increases without a corresponding increase of solute flux resulting in higher solute rejection (Eqs. 6.3.11–6.3.12). Another reason for this type of output profile is the compressing effect of membrane, in which the steric resistance of the condensed membrane increases with increase in pressure resulting in increase in solute rejection.

A positive correlation is also found between rejection and water flux rate with the CFR (Fig. 6.3.18A,B). The increasing rate of cross-flow of the feed generates a sweeping action over the membrane surface that minimizes the CP effect and keeps the membrane largely fouling-free. Higher removal efficiency can be traced to availability of larger effective membrane-surface area under reduced CP, which increases convective force. All of which contributes to enhanced solvent flux long term.

6.3.4.1 Software Validation Through Experimental Investigation

The software was validated through plant data and model predictions corroborate well with the plant operational data. These comparison results are presented through in Figs. 6.3.19 and 6.3.20.

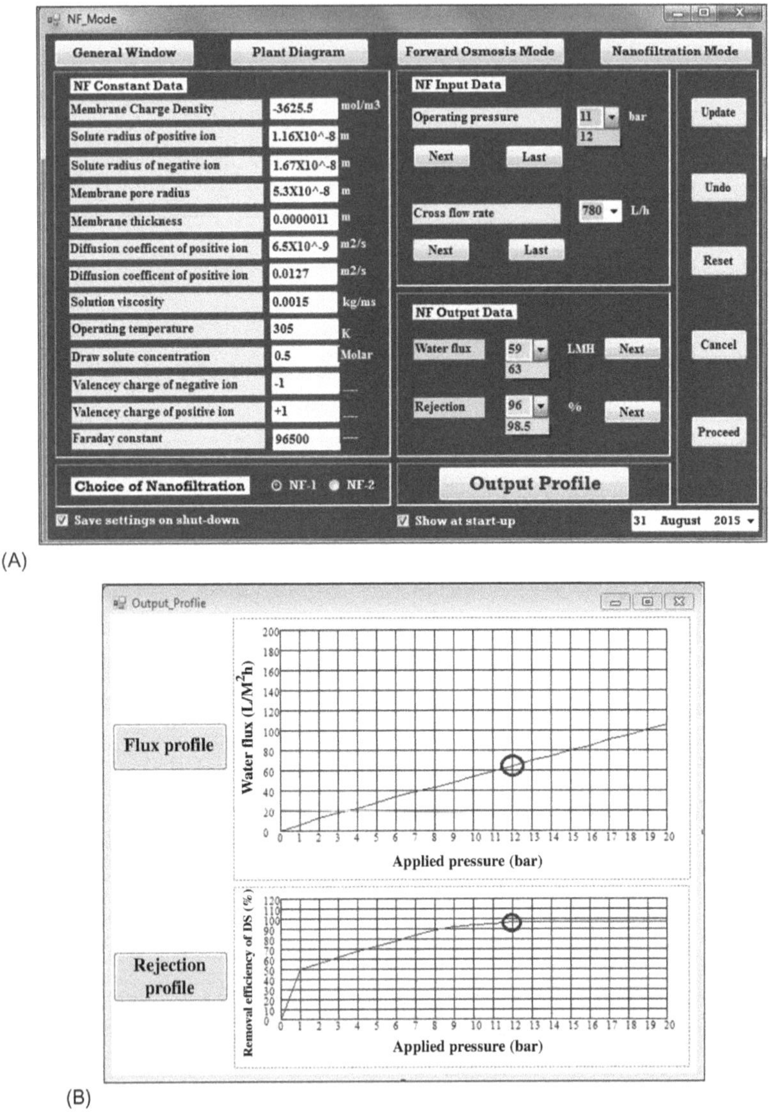

Figure 6.3.17 Effects of operating pressure on output results of NF system: (A) Effects of operating pressure on removal efficiency and water flux; (B) Output profiles of flux and rejection with applied or operating pressure (Eq. 6.3.5).

Fig. 6.3.19A shows the effects of operating pressure on COD removal and water flux. The set of experimental data fits the trend shown by the model-predicted curves, which indicates good corroboration as reflected in the RE (0.09) and Willmott-d-index (0.981).

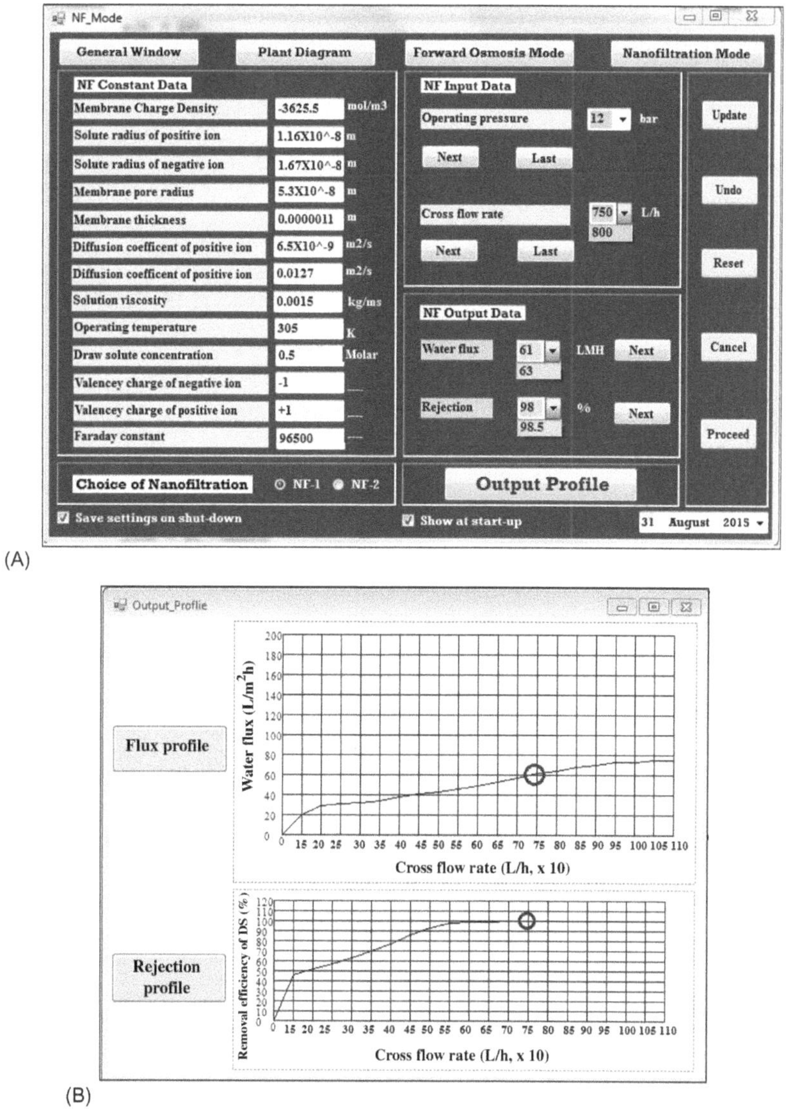

Figure 6.3.18 Effects of cross flow rate (CFR) on output results of NF system: (A) Effects of CFR on removal efficiency and water flux; (B) Output profiles of flux and rejection with CFR (Eq. 6.3.5).

Fig. 6.3.19B compares the model with the experiments and shows the effects of operating CFR on COD removal and water flux as seen by the curves.

A comparison of the experimental data with the model-predicted removal and flux data with draw solute concentration of sodium chloride

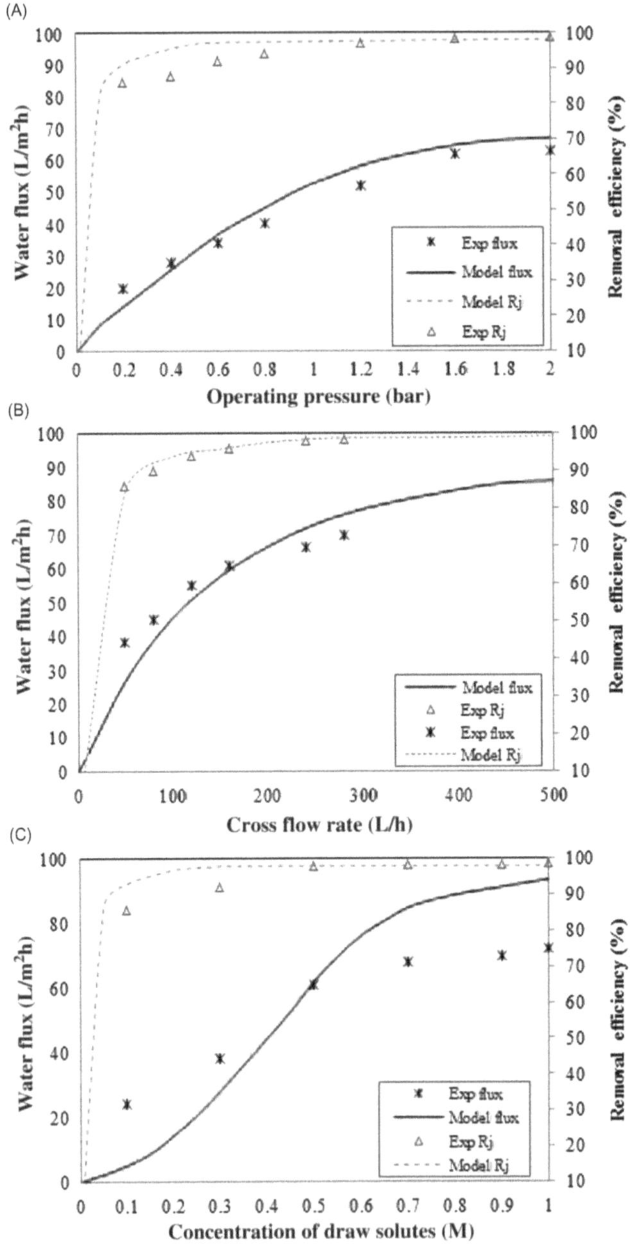

Figure 6.3.19 Comparison of model predicted data and experimental data of water flux and removal efficiency with different operating parameters of FO system: (A) Effects of operating pressure; (B) Effects of cross flow rate; and (C) Effects of draw solute concentration (Eq. 6.3.5).

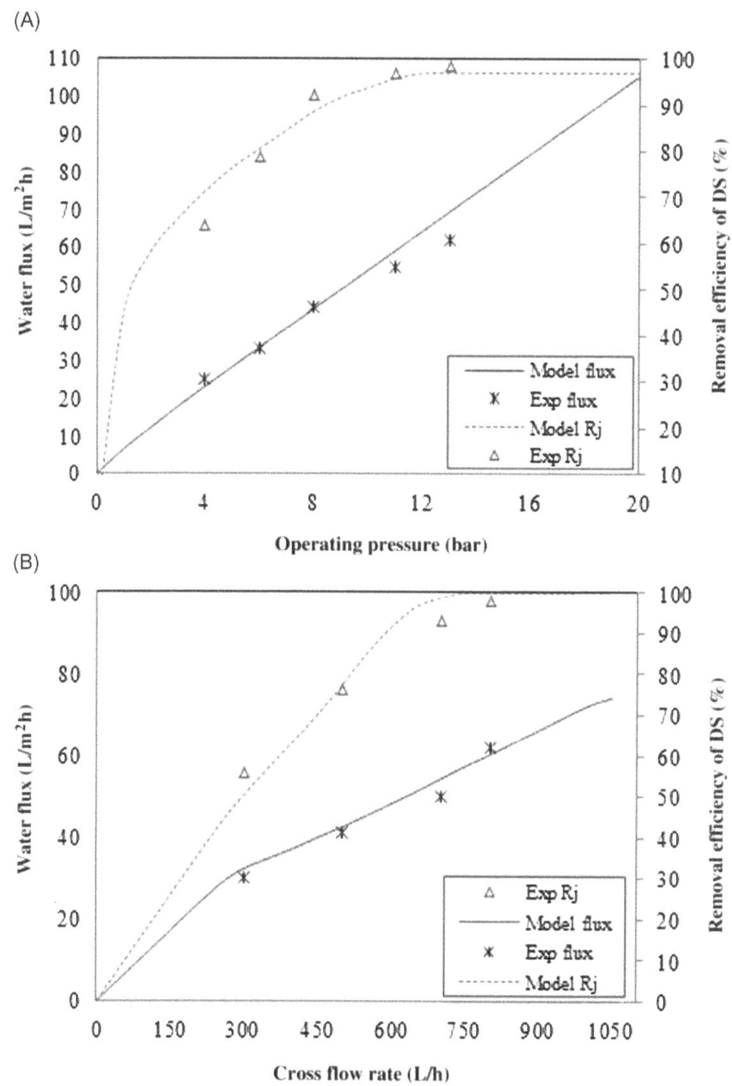

Figure 6.3.20 Comparison of model-predicted data and experimental data of water flux and removal efficiency of DS with different operating parameters of NF system: (A) Effects of operating pressure; (B) Effects of cross flow rate (Eq. 6.3.5).

are shown in Fig. 6.3.19C and indicates the increasing trend of water flux with increase in draw solute concentration. Similar results are shown for the removal efficiency of COD.

Fig. 6.3.20A,B illustrates the retention behavior of NF-1 membrane of draw solute (NaCl) and water flux rate and compares the model data with the experimental data. A strong positive correlation of retention of draw solutes and volumetric water flux with operating pressure and CFR of the feed are observed whereas a negative correlation is found for the feed (DS) concentration. The results of the software for rejection and water flux corroborate well with the plant data as reflected in the RE (0.095) and Willmott-d-index (0.98).

NOMENCLATURE

C_{eff}	Effective concentration difference (mol/m^3)
C_{FB}	Concentration of solute in bulk feed (mol/m^3)
C_{DB}	Concentrations of bulk draw solution (mol/m^3)
$C_{p,FO}$	Permeate concentration with respect to a COD in forward osmosis (mol/m^3)
$C_{p,n}$	Permeate concentration with respect to a particular ion in nanofiltration (mol/m^3)
C_{DM}	Concentration of draw solution at the membrane-draw solution interface (mol/m^3)
C_{FM}	Concentration of draw solution at the membrane-draw solution interface (mol/m^3)
$C_{m,i}$	Concentration of any ion at the membrane wall (mol/m^3)
D_s	Diffusivity of the solute through the membrane (m/s)
d_h	Hydraulic diameter of the feed channel (m)
F	Faraday's constant
$J_{s,mod}$	Solute flux ($L/m^2 \cdot h$)
$J_{s,exp}$	Experimentally obtained solvent flux ($L/m^2 \cdot h$)
$J_{v,\ mod}$	Volumetric solvent flux ($L/m^2 \cdot h$)
J_s	Solute flux for particular ion in nanofiltration ($L/m^2 \cdot h$)
J_v	Total flux for a particular ion in Nanofiltration ($L/m^2 \cdot h$)
K_m	Mass transfer coefficient
n	Number of ions obtained on dissociation
ΔP_{eff}	Effective hydraulic pressure to the feed solution (bar)
ΔP	Transmembrane pressure in Nanofiltration
R	Real gas constant
r_{mp}	Membrane pore radius of the nanofiltration membrane (m)
T_r	Temperature at which the experiment was carried out (K)
Δx_m	Membrane thickness (m)
X_{mc}	Membrane-charge density (mol/m^3)
z	Valency of a particular ion
Ψ_{dp}	Membrane potential
τ	Tortuosity
ε	Porosity
ρ	Density of the feed solution (kg/m^3)
η_s	Viscosity of the solution ($kg/m \cdot s$)
φ	Steric coefficient

REFERENCES

[1] Carballa M, Omil F, Lema JM. Comparison of predicted and measured concentrations of selected pharmaceuticals, fragrances and hormones in Spanish sewage. Chemosphere 2008;72:1118–23.

[2] Santos JL, Aparicio I, Alonso E. Occurrence and risk assessment of pharmaceutically active compounds in wastewater treatment plants. A case study: Seville city (Spain); 2007.
[3] Tong AYC, Peake BM, Braund R. Disposal practices for unused medications around the world. Environ Int 2011;37:292–8.
[4] Vellinga A, Cormican S, Driscoll J, Furey M, O'Sullivan M, Cormican M. Public practice regarding disposal of unused medicines in Ireland. Sci Total Environ 2014;478:98–102.
[5] Santos JL, Aparicio I, Callejón M, Alonso E. Occurrence of pharmaceutically active compounds during 1-year period in wastewaters from four wastewater treatment plants in Seville (Spain). J Hazard Mater 2009;164:1509–16.
[6] Jones OAH, Voulvoulis N, Lester JN. Human pharmaceuticals in wastewater treatment processes. Crit Rev Environ Sci Technol 2005;35:401–27.
[7] Hedgespeth ML, Sapozhnikova Y, Pennington P, Clum A, Fairey A, Wirth E. Pharmaceuticals and personal care products (PPCPs) in treated wastewater discharges into Charleston Harbor, South Carolina. Sci Total Environ 2012;437:1–9.
[8] Künnemeyer J, Terborg L, Meermann B, Brauckmann C, Möller I, Scheffer A, et al. Speciation analysis of gadolinium chelates in hospital effluent and wastewater treatment plant sewage by a novel HILIC/ICP-MS method. Environ Sci Technol 2009;43(8):2884–90.
[9] Jean J, Perrodin Y, Pivot C, Trepo D, Perraud M, Droguet J, et al. Identification and prioritization of bioaccumulable pharmaceutical substances discharged in hospital effluent. J Environ Manage 2012;103:113–21.
[10] Ray MR, Roychoudhury S, Mukherjee G, Roy S, Lahiri T. Respiratory and general health impairments of workers employed in a municipal solid waste disposal at an open landfill site in Delhi. Int J Hyg Environ Health 2005;208(4):255–62.
[11] EMEA, Guideline on the environmental risk assessment of medicinal products for human use. European Medicines Evaluation Agency. Doc.Ref.EMEA/CHMP/SWP/4447/00; 2006.
[12] Verlicchi P, Al Aukidy M, Zambello E. Occurrence of pharmaceutical compounds in urban wastewater: Removal, mass load and environmental risk after a secondary treatment—a review. Sci Total Environ 2012;429:123–55.
[13] Osachoff HL, Mohammadali M, Skirrow RC, Hall ER, Brown LY, van Aggelen GC, et al. Evaluating the treatment of a synthetic wastewater containing a pharmaceutical and personal care product chemical cocktail: compound removal efficiency and effects on juvenile rainbow trout. Water Res 2014;62:271–80.
[14] Galus M, Jeyaranjaan J, Smith E, Li H, Metcalfe C, Wilson JY. Chronic effects of exposure to a pharmaceutical mixture and municipal wastewater in zebrafish. Aquat Toxicol 2013;132–133:212–22.
[15] Sima WJ, Kima HY, Choi SD, Kwond JH, Oh JE. Evaluation of pharmaceuticals and personal care products with emphasis on anthelmintics in human sanitary waste, sewage, hospital wastewater, livestock wastewater and receiving water. J Hazard Mater 2013;248–249:219–27.
[16] Green RE, Newton I, Shult S, Cunningham ZA, Gilbert M, Pain DJ, et al. Diclofenac poisoning as a cause of vulture population declines across the Indian subcontinent. J Appl Ecol 2004;41:793–800.
[17] Luo N, Pereira S, Sahin O, Lin J, Huang S, Michel L, et al. Enhanced in vivo fitness of fluoroquinolone-resistant *Campylobacter jejuni* in the absence of antibiotic selection pressure. Proc Natl Acad Sci U S A 2005;102(3):541–6.
[18] Cooper ER, Siewicki TC, Phillips K. Preliminary risk assessment database and risk ranking of pharmaceuticals in the environment. Sci Total Environ 2008;398:26–33.

[19] Thiele-Bruhn S, Seibicke T, Schulten HR, Leinweber P. Sorption of sulfonamide pharmaceutical antibiotics on whole soils and particle-size fractions. J Environ Qual 2004;3(40):1331−42.
[20] Kar S, Roy K. First report on interspecies quantitative correlation of ecotoxicity of pharmaceuticals. Chemosphere 2010;81:738−47.
[21] Larsson DGJ, De Pedro C, Paxues N. Effluent from drug manufactures contains extremely high levels of pharmaceuticals. J Hazard Mater 2007;148:751−5.
[22] Zuccato E, Chiabrando C, Castiglioni S, Calamari D, Bagnati R, Schiarea S, et al. Cocaine in surface waters: a new evidence-based tool to monitor community drug abuse. Environ Health 2005;14. Available from: http://dx.doi.org/10.1186/1476-069X-4-14.
[23] Castiglioni S, Zuccato E, Crisci E, Chiabrando C, Fanelli R, Bagnati R. Identification and measurement of illicit drugs and their metabolites in urban wastewaters by liquid chromatography tandem mass spectrometry (HPLC-MS-MS). Anal Chem 2006;78:8421−9.
[24] Glassmeyer ST, Hinchey EK, Boehme SE, Daughton CG, Ruhoy IS, Conerly O, et al. Disposal practices for unwanted residential medications in the United States. Environ Int 2009;35:566−72.
[25] Daughton CG, Ternes TA. Pharmaceuticals and personal care products in the environment: agents of subtle change. Environ Health Perspect 1999;107(Suppl. 6): 907−38.
[26] Ankley GT, Brooks BW, Huggett DB, Sumpter JP. Repeating history: pharmaceuticals in the environment. Environ Sci Technol 2007;41(24):8211−17.
[27] Jiang C, Ren Z, Tian Y, Wang K. Application of best available technologies on medical wastes disposal/treatment in China (with case study). Procedia Environ Sci 2012;16:257−65.
[28] Chena Y, Liua L, Fenga Q, Chen G. Key issues study on the operation management of medical waste incineration disposal facilities. Procedia Environ Sci 2012;16: 208−13.
[29] Torres LG, Jaimes J, Mijaylova P, Ramírez E, Jiménez B. Coagulation-flocculation pretreatment of high-load chemical-pharmaceutical industry wastewater: mixing aspects. Water Sci Technol 1997;36(2−3):255−62.
[30] Xing ZP, Sun DZ. Treatment of antibiotic fermentation wastewater by combined polyferric sulfate coagulation, Fenton and sedimentation process. J Hazard Mater 2009;168:1264−8.
[31] El-Gohary F, Tewfik A, Mahmoud U. Comparative study between chemical coagulation/precipitation (C/P) versus coagulation/dissolved air flotation (C/DAF) for pretreatment of personal care products (PCPs) wastewater. Desalination 2010;252:106−12.
[32] Guiying Z, Liwen J. Treatment technology of high-concentration pharmaceutical wastewater containing phosphorus. J East China Jiaotong Uni 2012;3:56−60.
[33] Zorita S, Martensson L, Mathiasson L. Occurrence and removal of pharmaceuticals in a municipal sewage treatment system in the south of Sweden. Sci Total Environ 2009;407:2760−70.
[34] Batt AL, Kim S, Aga DS. Comparison of the occurrence of antibiotics in four full-scale wastewater treatment plants with varying designs and operations. Chemosphere 2007;68:428−35.
[35] Gupta VK, Carrott PJM, Carrott MMLR, Suhas TL. Low-cost adsorbents: growing approach to wastewater treatment—a review. Crit Rev Environ Sci Technol 2009;39:783−842.
[36] Sharma S, Mukhopadhyay M, Murthy ZVP. Treatment of chlorophenols from wastewaters by advanced oxidation processes. Sep Purif Rev 2013;42(4):263−95.

[37] Bernabeu A, Palacios S, Vicente R, Vercher RF, Malato S, Arques A, et al. Solar photoFenton at mild conditions to treat a mixture of six emerging pollutants. Chem Eng J 2012;198–199:65–72.

[38] Laera G, Chong MN, Jin B, Lopez A. An integrated MBR–TiO_2 photocatalysis process for the removal of Carbamazepine from simulated pharmaceutical industrial effluent. Bioresour Technol 2011;102:7012–15.

[39] Mohapatra DP, Brar SK, Tyagi RD, Picard P, Surampalli RY. Analysis and advanced oxidation treatment of a persistent pharmaceutical compound in wastewater and wastewater sludge-carbamazepine. Sci Total Environ 2014;470–471:58–75.

[40] Ternes T, Stuber J, Herrmann N, McDowell D, Ried A, Kapmann M, et al. Ozonation: a tool for removal of pharmaceuticals, contrast media, and musk frgances form wastewater. Water Res 2003;37:1976–82.

[41] Kibbey TCG, Paruchuri R, Sabatini DA, Chen L. Adsorption of beta blockers to environmental surfaces. Environ Sci Technol 2007;41:5349–56.

[42] De Ridder DJ, Verberk JQJC, Heijman SGJ, Amya GL, van Dijk JC. Zeolites for nitrosamine and pharmaceutical removal from demineralised and surface water: mechanisms and efficacy. Sep Purif Technol 2012;89:71–7.

[43] Ghauch A, Tuqan A, Assi HA. Antibiotic removal from water: Elimination of amoxicillin and ampicillin by microscale and nanoscale iron particles. Environ Pollut 2009;157:1626–35.

[44] Peng X, Li Y, Luan Z, Di Z, Wang H, Tian B, et al. Adsorption of 1, 2-dichlorobenzene from water to carbon nanotubes. Chem Phys Lett 2003;376 (1–2):154–8.

[45] Teng X, Yang H. Effects of surfactants and synthetic conditions on the sizes and self-assembly of monodisperse iron oxide nanoparticles. J Mater Chem 2004;14:774–9.

[46] Ponder SM, Darab JG, Mallouk TE. Remediation of Cr (VI) and Pb(II) aqueous solutions using supported, nanoscale zero-valent iron. Environ Sci Technol 2000;34:2564–9.

[47] Cheng H, Xu W, Liu J, Wang H, He Y, Chen G. Pretreatment of wastewater from triazine manufacturing by coagulation, electrolysis, and internal microelectrolysis. J Hazard Mater 2007;146:385–92.

[48] Deshpande AM, Lokesh KS, Bejankiwar RS, Gowda TPH. Electrochemical oxidation of pharmaceutical effluent using cast iron electrode. J Environ Sci Eng 2005;47:21–4.

[49] Deshpande AM, Satyanarayan S, Ramakant S. Electrochemical pretreatment of wastewater from bulk drug manufacturing industry. J Environ Eng 2009;135:716–19.

[50] Zhao D, Sengupta AK. Ultimate removal of phosphate from wastewater using a new class of polymeric ion exchangers. Water Res 1998;32(5):1613–25.

[51] Farhadi S, Aminzadeh B, Torabian A, Khatibikamal V, Fard MA. Comparison of COD removal from pharmaceutical wastewater by electrocoagulation, photoelectrocoagulation, peroxi-electrocoagulation and peroxi-photoelectrocoagulation processes. J Hazard Mater 2012;219(220):35–42.

[52] Boroski M, Rodrigues ACA, Garcia JC, Sampaio LC, Nozaki J, Hioka N. Combined electrocoagulation and TiO_2 photoassisted treatment applied to wastewater effluent from pharmaceutical and cosmetic industries. J Hazard Mater 2009;162: 448–54.

[53] Tekin H, Bilkay O, Ataberk SS, Balta TH, Ceribasi IH, Sanin FD. Use of Fenton oxidation to improve the biodegradability of a pharmaceutical wastewater. J Hazard Mater 2006;136:258–65.

[54] Lillenberg M, Yurchenko S, Kipper K, Herodes K, Pihl V, Lõhmus R, et al. Presence of fluoroquinolones and sulfonamides in urban sewage sludge and their degradation as a result of composting. J Environ Sci Technol 2010;7(2):307–12.

[55] Amin MM, Zilles JL, Greiner J, Charbonneau S, Raskin L, Eberhard M. Influence of the antibiotic erythromycin on anaerobic treatment of a pharmaceutical wastewater. Environ Sci Technol 2006;40(12):3971−7.
[56] Zhou P, Su C, Li B, Qian Y. Treatment of high strength pharmaceutical wastewater and removal of antibiotics in aerobic and anaerobic biological treatment processes. J Environ Eng 2006;132:129−36.
[57] Majumdar D, Patel J, Bhatt N, Desai P. Emission of methane and carbon dioxide and earthworm survival during composting of pharmaceutical sludge and spent mycelia. Bioresour Technol 2006;97:648−58.
[58] Chelliapan S, Wilby T, Sallis PJ. Performance of an up-flow anaerobic stage reactor (UASR) in the treatment of pharmaceutical wastewater containing macrolide antibiotics. Water Res 2006;40:507−16.
[59] Verlicchi P, Galletti A, Petrovic M, Barceló D, Aukidy MAl, Zambello E. Removal of selected pharmaceuticals from domestic wastewater in an activated sludge system followed by a horizontal subsurface flow bed—analysis of their respective contributions. Sci Total Environ 2013;454−455:411−25.
[60] Llorens E, Matamoros V, Domingo V, Bayona JM, García J. Water quality improvement in a full-scale tertiary constructed wetland: effects on conventional and specific organic contaminants. Sci Total Environ 2009;407:2517−24.
[61] Matamoros V, Garcia J, Bayona JM. Organic micropollutant removal in a full scale surface flow constructed wetland fed with secondary effluent. Water Res 2008;42:653−6.
[62] Karthikeyan KG, Meyer MT. Occurrence of antibiotics in wastewater treatment facilities in Wisconsin, USA. Sci Total Environ 2006;361:196−207.
[63] Yang S, Cha J, Carlson K. Simultaneous extraction and analysis of 11 tetracycline and sulfonamide antibiotics in influent and effluent domestic wastewater by solid-phase extraction and liquid chromatography-electrospray ionization tandem mass spectrometry. J Chromatogr A 2005;1097:40−53.
[64] Ruhoya ISR, Daughton CG. Beyond the medicine cabinet: an analysis of where and why medications accumulate. Environ Int 2008;34:1157−69.
[65] Singh D, Suthar S. Vermicomposting of herbal pharmaceutical industry solid wastes. Ecol Eng 2012;39:1−6.
[66] Fountoulakis MS, Stamatelatou K, Lyberatos G. The effect of pharmaceuticals on the kinetics of methanogenesis and acetogenesis. Bioresour Technol 2008;99: 7083−90.
[67] Jelic A, Fatone F, Fabio SD, Petrovic M, Cecchi F, Barcelo D. Tracing pharmaceuticals in a municipal plant for integrated wastewater and organic solid waste treatment. Sci Total Environ 2012;433:352−61.
[68] Li F, Wang J, Nastold P, Jiang B, Sun F, Zenker A, et al. Fate and metabolism of tetrabromobisphenol A in soil slurries without and with the amendment with the alkyl phenol degrading bacterium Sphingomonas sp. strain TTNP3. Environ Pollut 2012;193: 181−8.
[69] Babcock Jr. RW, Chen W, Ro KS, Mah RA, Stenstrom MK. Enrichment and kinetics of biodegradation of 1-Naphthylamine in activated sludge. Appl Microbiol Biotechnol 1993;39:264−9.
[70] Saravanane R, Murthy DVS, Krishnaiah K. Bioaugmentation and treatment of cephalexin drug-based pharmaceutical effluent in an upflow an aerobic fluidized bed system. Bioresour Technol 2001;76:279−81.
[71] Edwards SJ, Kjellerup BV. Applications of biofilms in bioremediation and biotransformation of persistent organic pollutants, pharmaceuticals/personal care products, and heavy metals. Appl Microbiol Biotechnol 2013;97(23):9909−21.

[72] Kern S, Baumgartner R, Helbling DE, Hollender J, Singer H, Loos MJ, et al. A tiered procedure for assessing the formation of biotransformation products of pharmaceuticals and biocides during activated sludge treatment. J Environ Monit 2010;12:2100–11.
[73] Lapara TM, Nakatsu CH, Pantea LM, Alleman JE. Aerobic biological treatment of a pharmaceutical wastewater: effect of temperature on cod removal and bacterial community development. Water Res 2001;35(18):4417–25.
[74] Stull FW, Martin DF. Comparative ease of separation of mixtures of selected nuisance anions (nitrate, nitrite, sulfate, phosphate) using Octolig. J Environ Sci Health A 2009;44:1551–6.
[75] Duan YP, Meng XZ, Wen ZH, Ke RH, Chen L. Multi-phase partitioning, ecological risk and fate of acidic pharmaceuticals in a wastewater receiving river: the role of colloids. Sci Total Environ 2013;447:267–73.
[76] Snip LJP, Alsina XF, Plosz BG, Jeppsson U, Gernaey KV. Modelling the occurrence, transport and fate of pharmaceuticals in wastewater systems. Environ Modell Software 2014;62:112–27.
[77] Petruzzelli D, Dell'Erba A, Liberti L, Norarnicola M, SenGupta AK. A phosphate selective sorbent for the REM NUTs process: field experience at massafra wastewater treatment plant. React Funct Polym 2004;60:195–202.
[78] Sim WJ, Lee JW, Oh JE. Occurrence and fate of pharmaceuticals in wastewater treatment plants and rivers in Korea. Environ Pollut 2010;158(5) 1938–47.
[79] Cunningham VL, Binks SP, Olson MJ. Human health risk assessment from the presence of human pharmaceuticals in the aquatic environment. Regul Toxicol Pharmacol 2009;53(3):9–45.
[80] Pal P, Kumar R. Treatment of coke wastewater: a critical review for evolving a sustainable management strategy. Sep Purif Rev 2014;43:89–123.
[81] Buonomenna MG, Bae J. Organic solvent nanofiltration in pharmaceutical industry. Sep Purif Rev 2015;44(2):157–82.
[82] Kumar R, Pal P. Membrane-integrated hybrid bioremediation of industrial wastewater: a continuous treatment and recycling approach. Desalin Water Reuse 2013;1(3): 26–38.
[83] Kumar R, Pal P. Removal of phenol from coke-oven wastewater by cross-flow nanofiltration membranes. Water Environ Res 2013;85:447–55.
[84] Kumar R, Pal P. Response surface-optimized Fenton's pretreatment for chemical precipitation of struvite and recycling of water through downstream nanofiltration. Chem Eng J 2012;210:33–44.
[85] Chakrabortty S, Roy M, Pal P. Removal of fluoride from contaminated groundwater by cross-flow nanofiltration: transport modelling and economic evaluation. Desalination 2013;313:115–24.
[86] Kumar R, Pal P. Turning hazardous waste into value-added products: production and characterization of struvite from ammoniacal waste with new approaches. J Cleaner Prod 2014;43:59–70.
[87] Kumar R, Bhakta P, Chakrabortty S, Pal P. Separating cyanide from coke wastewater by cross-flow nanofiltration. Sep Sci Technol 2011;46(13):2119–27.
[88] Pal P, Manna AK. Removal of arsenic from contaminated groundwater by solar-driven membrane distillation using three different commercial membranes. Water Res 2010;44(19):5750–60.
[89] Pal M, Chakrabortty S, Pal P, Linnanen L. Purifying fluoridecontaminated water by a novel forward osmosis design with enhanced flux under reduced CP. Environ Sci Pollut Res 2015;22(15):11401–11.

[90] Thakura R, Chakrabortty S, Pal P. Treating complex industrial wastewater in a new membrane-integrated closed loop system for recovery and reuse. Clean Technol Environ Policy 2015;17(8):2299−310.
[91] Radjenovic J, Petrovic M, Barcelo D. Fate and distribution of pharmaceuticals in wastewater and sewage sludge of the conventional activated sludge (CAS) and advanced membrane bioreactor (MBR) treatment. Water Res 2009;43:831−41.
[92] Sipma J, Osuna B, Collado N, Monclús H, Ferrero G, Comas J, et al. Comparison of removal of pharmaceuticals in MBR and activated sludge systems. Desalination 2010;250:653−9.
[93] Chang CY, Chang JS, Vigneswaran S, Kandamasy J. Pharmaceutical wastewater treatment by membrane bioreactor process—a case study in southern Taiwan. Desalination 2008;234(1−3):393−401.
[94] Zhang Y, Geißen SU. Elimination of carbamazepine in a nonsterile fungal bioreactor. Biores Tech 2012;112:221−7.
[95] Hasan SW, Elektorowicz M, Oleszkiewicz JA. Start-up period investigation of pilot-scale submerged membrane zelectrobioreactor (SMEBR) treating raw municipal wastewater. Chemosphere 2014;97:71−7.
[96] Abeynayaka A, Visvanathan C. Mesophilic and thermophilic aerobic batch biodegradation, utilization of carbon and nitrogen sources in high-strength wastewater. Bioresour Technol 2011;102:2358−66.
[97] Rozich AF, Colvin RJ. Design and operational considerations for thermophilic aerobic reactors treating high strength wastes and sludges. In: Alleman JE, editor. Proceedings of the 52nd Industrial Waste Conference, Purdue University. Chelsea: Ann Arbor Press; 1997.
[98] Lapara TM, Alleman JE. Thermophilic aerobic biological wastewater treatment. Water Res 1999;33(4):895−908.
[99] Vilar VJP, Rocha EMR, Mota FS, Fonseca A, Saraiva I, Boaventura RAR. Treatment of a sanitary landfill leachate using combined solar photoFenton and biological immobilized biomass reactor at a pilot scale. Water Res 2011;45(8):2647−58.
[100] Wang F, Smith DW, El-Din MG. Application of advanced oxidation methods for landfill leachate treatment—a review. J Environ Eng Sci 2003;2(6):413−27.
[101] Slack RJ, Gronow JR, Hall DH, Voulvoulis N. Household hazardous waste disposal to landfill: Using LandSim to model leachate migration. Environ Pollut 2007;146:501−9.
[102] Schwarzbauer J, Heim S, Brinker S, Littke R. Occurrence and alteration of organic contaminants in seepage and leakage water from a waste deposit landfill. Water Res 2002;36:2275−87.
[103] Fent K, Weston AA, Caminada D. Ecotoxicology of Human Pharmaceuticals. Aquat Toxicol 2006;76:122−59.
[104] Reinhart DR, McCreanor PT, Townsend T. The bioreactor landfill: Its status and future The bioreactor landfill: Its status and future. Waste Manag Res 2002;20(2):172−86.
[105] Kümmerer K. The presence of pharmaceuticals in the environment due to human use—present knowledge and future challenges. J Environ Manage 2009;90:2354−66.
[106] Lapkin A, Constable D, editors. Green chemistry metrics. Chichester: Wiley-Blackwell Publishers; 2008.
[107] Zaks A, Dodds DR. Application of biocatalysis and biotransformations to the synthesis of pharmaceuticals. Drug Discov Today 1997;2(12):513−31.
[108] Tao J, Xu JH. Biocatalysis in development of green pharmaceutical processes. Curr Opin Chem Biol 2009;13:43−50.

[109] Hu S, Tao J, Xie J. Process for producing atorvastatin, pharmaceutically acceptable salts thereof and intermediates thereof. European Patent, EP 1893767 A1; 2006.
[110] Schlüter A, Szczepanowski R, Pühler A, Top EM. Genomics of IncP-1antibiotic resistance plasmids isolated from wastewater treatment plants provides evidence for a widely accessible drug resistance gene pool. FEMS Microbiol Rev 2007; 31:449–77.
[111] Kümmerer K. Antibiotics in the environment—a review—Part II. Chemosphere 2009;75(4):435–41.
[112] Peng C, Jin R, Li G, Li F, Gu Q. Recovery of nickel and water from wastewater with electrochemical combination process. Sep Purif Technol 2014; 136:42–9.
[113] Petersková M, Valderrama C, Gibert O, Cortina JL. Extraction of valuable metal ions (Cs, Rb, Li, U) from reverse osmosis concentrates using selective sorbents. Desalination 2012;286:316–23.
[114] Francisca SA, Sintra T, Ventura SPM, Coutinho JAP. Recovery of paracetamol from pharmaceutical wastes. Sep Purif Technol 2014;122:315–22.
[115] Kumar M, Badruzzaman M, Adham S, Oppenheimer S. Beneficial phosphate recovery from reverse osmosis (RO) concentrate of an integrated membrane system using polymeric ligand exchanger (PLE). Water Res 2007;41:2211–19.
[116] Qiu G, Song Y, Zeng P, Xiao S, Duan L. Phosphorus recovery from Fosfomycin pharmaceutical wastewater by wet air oxidation and phosphate crystallization. Chemosphere 2011;84:241–6.
[117] Erathodiyil N, Ying JY. Functionalization of inorganic nanoparticles for bioimaging applications. Acc Chem Res 2011;44(10):925–35.
[118] Guo QZ, Mei B, Zhou SX, Shi ZG, Feng YQ, Wu JY. Synthesis, characterization and application of magnetic-zirconia nanocomposites. J Non-Cryst Solids 2009;355 (16–17):922–5.
[119] Roberts PH, Thomas KV. The occurrence of selected pharmaceuticals in wastewater effluent and surface waters of the lower Tyne catchment. Sci Total Environ 2006;356:143–53.
[120] Saravanane R, Murthy DV, Krishnaiah K. Bioaugmentation and anaerobic treatment of pharmaceutical effluent in fluidized bed reactor. J Environ Sci Health A Tox Hazard Subst Environ Eng 2001;36(5):779–91.
[121] Dialynas E, Mantzavinos D, Diamadopoulos E. Advanced treatment of the reverse osmosis concentrate produced during reclamation of municipal wastewater. Water Res 2008;42(18):4603–8.
[122] Westerhoff P, Moon H, Minakata D, Crittenden J. Oxidation of organics in retentates from reverse osmosis wastewater reuse facilities. Water Res 2009; 43:3992–8.
[123] Pal P, Thakura R, Chakrabortty S. Performance Analysis and Optimization of an Advanced Pharmaceutical Wastewater Treatment Plant through a Visual Basic Software Tool (PWWT.VB). Environ Sci Pollut Res 2016. Available from: http://dx.doi.org/10.1007/s11356-016-6238-8.
[124] Pal P, Kumar R. Treatment of coke wastewater: a critical review for evolving a sustainable management strategy. Sep Purif Rev 2014;43:89–123.
[125] Gilron J. Water-energy nexus: matching sources and uses. Clean Technol Environ Policy 2014;16.8:1471–9.
[126] Stackelberg PE, Gibs J, Furlong ET, Meyer MT, Zaugg SD, Lippincott R. Efficiency of conventional drinking-water-treatment processes in removal of pharmaceuticals and other organic compounds. Sci Total Environ 2007;377:255–72.

[127] Pal P, Chakraborty S, Roy M. Arsenic separation by a membrane-integrated hybrid treatment system: modeling, simulation and techno-economic evaluation. Sep Sci Technol 2012;47(8):1091–101.
[128] Boleda MR, Galceran MT, Ventura F. Monitoring opiates, cannabinoids and their metabolites in wastewater, surface water and finished water in Catalonia, Spain. Water Res 2009;43:1126–36.
[129] Fernández C, González-Doncel M, Pro J, Carbonell G, Tarazona JV. Occurrence of pharmaceutically active compounds in surface waters of the Henares-Jarama-Tajo Rive System (Madrid, Spain) and a potential risk characterization. Sci Total Environ 2010;408:543–51.
[130] Cornelissen ER, Harmsen D, Korte KF, Ruiken CJ, Qin JJ, Oo H, et al. Membrane fouling and process performance of forward osmosis membranes on activated sludge. J Membr Sci 2008;319(1–2):158–68.
[131] Kumar R, Pal P. A membrane-integrated advanced scheme for treatment of industrial wastewater: dynamic modelling toward scale up. Chemosphere 2013;92: 1375–82.
[132] Gebhardt W, Schroder HF. Liquid chromatography–(tandem) mass spectrometry for the follow-up of the elimination of persistent pharmaceuticals during wastewater treatment applying biological wastewater treatment and advanced oxidation. J Chromatogr A 2007;1160:34–43.
[133] Kosutic K, Dolar D, Asperger D, Kunst B. Removal of antibiotics from a model; 2007.
[134] Chakrabortty S, Roy M, Pal P. Removal of fluoride from contaminated groundwater by cross-flow nanofiltration: transport modeling and economic evaluation. Desalination 2013;313:115–24 wastewater by RO/NF membranes, Sep Sci Technol(53): 244–249
[135] Liu P, Zhang H, Feng Y, Yang F, Zhang J. Removal of trace antibiotics from wastewater: A systematic study of nanofiltration combined with ozone-based advanced oxidation processes. Chem Eng J 2014;240:211–20.
[136] Kumar R, Pal P. Turning hazardous waste into value-added products: production and characterization of struvite from ammoniacal waste with new approaches. J Clean Prod 2013;43:59–70.
[137] Satyamoorthy S, Andrew C. Assessment of quantitative structural property relationship for prediction of pharmaceutical sorption during biological wastewater treatment. Chemosphere 2013;92:639–46.
[138] Kumar R, Pal P. Removal of phenol from coke-oven wastewater by cross-flow nanofiltration membranes. Water Environ Res 2013;85:447–55.
[139] Shukla KS. Membrane filtration of chlorination and extraction stage bleach plant effluent in Indian paper Industry. Clean Technol Environ Policy 2015;15.2: 235–43.
[140] Kumar R, Pal P. Response surface-optimized Fenton's pretreatment for chemical precipitation of struvite and recycling of water through downstream nanofiltration. Chem Eng J 2012;210:33–44.
[141] Pal P, Chakrabortty S, Linnanen L. A nanofiltration–coagulation integrated system for separation and stabilization of arsenic from groundwater. Sci Total Environ 2014;476–477:601–10.
[142] Pal P, Bhakta P, Kumar R. Cyanide removal from industrial wastewater by cross-flow nanofiltration: transport modeling and economic evaluation. Water Environ Res 2014;86(8):698–709.

[143] Roy M, Chakrabortty S. Developing a sustainable water resource management strategy for a fluorideaffected area: a contingent valuation approach. J Clean Technol Environ Policy 2014;16(2):341–9.
[144] Tang W, Yong Ng H. Concentration of brine by forward osmosis: performance and influence of membrane structure. Desalination 2008;224:143–53.
[145] McCutcheon JR, Elimelech M. Influence of concentrative and dilutive internal concentration polarization on flux behaviour in forward osmosis. J Membr Sci 2006;284:237–47.

FURTHER READING

Jelic A, Gros M, Ginebreda A, Cespedes-Sánchez R, Ventura F, Petrovic M, et al. Occurrence, partition and removal of pharmaceuticals in sewage water and sludge during wastewater treatment. Water Res 2011;45(3):1165–76.
Hossain MS, Santhanam A, Norulaini NAN, Omar AKM. Clinical solid waste management practices and its impact on human health and environment—a review. Waste Manag 2011;31:754–66.
Cheng H, Xu W, Liu J, Wang H, He Y, Chen G. Pretreatment of wastewater from triazine manufacturing by coagulation, electrolysis, and internal microelectrolysis. J Hazard Mater 2007;146:385–92.
Subramani A, Jacangelo JG. Treatment technologies for reverse osmosis concentrate volume minimization: a review. Sep Purif Technol 2014;122:472–89.
Venkata Y, Ravikumar L, Kalyani S, Satyanarayana SV, Sridhar S. Processing of pharmaceutical effluent condensate by nanofiltration and reverse osmosis membrane techniques. J Taiwan Inst Chem E 2014;45:50–6.
Wiesner MR, Aptel P. Mass transport and permeate flux and fouling in pressure-driven processes. In Water treatment membrane handbook. New York: McGraw-Hill Publishers; 1996.
Gur-Reznik S, Katz I, Dosoretz CG. Removal of dissolved organic matter by granular-activated carbon adsorption as a pretreatment to reverse osmosis of membrane bioreactor effluent. Water Res 2008;42:1595–605.
Hammer M, Otterpohl, R. Institute of wastewater management and water protection. Hamburg University of Technology, 21073 Hamburg.
Hill R. WORLD TRADE CENTER San Diego—MEMBRANES JAN, 2011.
Camacho-Muñoz D, Martín J, Santos JL, Aparicio I, Alonso E. Effectiveness of conventional and low-cost wastewater treatments in the removal of pharmaceutically active compounds. Water Air Soil Pollut 2012;223 261-121
Oktem YA, Ince O, Sallis P, Donnelly T, Ince BK. Anaerobic treatment of a chemical synthesis-based pharmaceutical wastewater in a hybrid upflow anaerobic sludge blanket reactor. Bioresour Technol 2007;99:1089–96.
Chakrabortty S, Pal M, Roy M, Pal P. Water treatment in a new flux-enhancing, continuous forward osmosis design: Transport modelling and economic evaluation toward scale up. Desalination 2015;365:329–42.
Jelic A, Gros M, Ginebreda A, Cespedes-Sánchez R, Ventura F, Petrovic M, et al. Occurrence, partition and removal of pharmaceuticals in sewage water and sludge during wastewater treatment. Water Res 2011;45(3):1165–76.

SUBCHAPTER 6.4

Treatment of Pulp and Paper Industry Wastewater

6.4.1 INTRODUCTION

The pulp and paper industry is considered one of the largest industrial sectors along with food, petroleum refineries, refined metals, and chemical and allied process industries in terms of freshwater exploitation [1]. The key material used in paper industries is wood, which is mainly comprised of cellulose, carbohydrates (starch and sugars), and lignin. In the Kraft process, during boiling, the wood chips are treated with white liquor (mixture of sodium hydroxide and sodium sulfide) to enhance delignification while generating wood pulp or slurry that after further processing with bleaching chemicals, dyes, coating materials, and fixation agents turns into paper pulp from which paper is synthesized [2,3].

The black liquor produced from the pulping process contains a large volume of water and the ratio of the liquid waste produced for the production of paper pulp is about 7:1 by weight in tons [4]. On average, about 35 m^3 of freshwater is required for per ton production of any quality of paper while around 36 m^3 of waste Kraft black liquor is produced for the production of the same quantity of paper. An enormous amount of freshwater is needed to maintain different dilutions during the processing of pulp and stock and for washing in the plant set-up as the dirt present in machines may degrade the quality of paper. The bond paper, security papers, and tissue paper mills are the largest users of freshwater among the various types of paper-production mills. These mills together consume as much as 100 m^3 freshwater per ton of paper production to sustain the high quality and texture of these products [5]. According to the annual report of the Central Pollution Control Board (Delhi, India), the whole papermaking sector consumes approximately 905.8×10^6 m^3 of freshwater and discharges about 695.7×10^6 m^3 of wastewater per year. Moreover, due to the use of old technology, the current average water consumption is about 150 m^3/ton of paper and allied products (National Productivity Council, 2006). Although the composition of the Kraft black liquor varies for different mills generally it is a complex solution of 30%–45% ligneous material, 25%–35% saccharinic acids, about 10% formic and acetic acid, 3%–5% extractives (primarily resin acids and fatty acids that get further converted to salts due to the high pH), and many inorganic compounds containing sodium about 17%–20% and sulfur 3%–5% (all % by weight). The composition of paper-mill effluent is listed in Table 6.4.1.

The lignin present in black liquor is the residual amount of it after the Kraft process. The inorganic chemicals are mainly those chemicals used in the process, namely sodium hydroxide (NaOH), sodium sulfide (Na_2S), sodium carbonate (Na_2CO_3), sodium sulfate (Na_2SO_4), sodium thiosulfate ($Na_2S_2O_3$), and sodium chloride (NaCl). The presence of organic matter like the extractives, tannin resins, synthetic dyes, lignin, and degraded residual products formed by chlorination of lignin during the bleaching process increase the values of COD, toxicity, turbidity, TDS, and intense color of the waste. The black liquor is highly alkaline with a pH ranging from 13 to 14. The COD of mill-treated effluent can be as high as 22,000 mg/L, and other variable parameters like color (varies from 2.74 to 1905 PCU), turbidity (ranged from 3.7 to 94.4 ntu), TSS

Table 6.4.1 Composition of paper-mill effluent as reported in the literature

Sl. No.	Parameters	Value	Units
1.	COD	465–2126	mg/L
2.	BOD	142–450	mg/L
3.	TDS	310–5000	mg/L
4.	TSS	40–1300	mg/L
5.	BOD/COD	>10,000	–
6.	pH	8.0–13.0	–
7.	DOC	131–575	mg/L
8.	AOX	8.9–80.2	mg/L
9.	Color	660–3220	mg Pt-Co/L
10.	Conductivity	1200–2000	μS/cm
11.	TOC	234–445	mg/L
12.	Alkalinity	83–86	mg$CaCO_3$/L
13.	Total phenolic compounds	294–440	mg/L
14.	Turbidity	70–108	NTU
15.	Cationic demand (CD)	1065.4	μmol/L
16.	MTBE extract	236.8	mg/L
17.	Lignin	133–265	mg/L
18.	SO_4^{2-}	150–542	mg/L
19.	Cl^-	85–1324	mg/L
20.	Inorganic matter	800–1100	mg/L
21.	Ca^{2+}	17–27	mg/L

(between 6.4 and 113.4 mg/L), and polyphenols (varies from 11.72 to 78.86 mg/L).

The disposal of black liquor in water bodies such as lakes or rivers makes the water toxic to aquatic animals and plants due to its highly alkaline nature. Resin acids and phenolic compounds have fatal effects on algae, fish, plankton, and other biologically active species present in water bodies. Thus such pollutants at even low concentration can have an adverse effect on aquatic life. Compounds containing sulfides, chlorides, and phenolic groups can migrate faster due to their high solubility in water and can contaminate aqueous environments and groundwater. Treatment of this type of wastewater is aimed at the recovery and conservation of renewable resources like water and energy. However, the maintenance of microbial film is tough amidst high chances of complete dying out of the bacterial species due to the toxicity of the produced liquor. Since residual lignin is the main cause of high BOD, COD, and the color of black liquor, research on biological treatments has been widely reported for its degradation and to reduce the load of pulping wastewater. However, reduction of the strong color can not be effectively performed by microbial treatment where the lumped parameters like BOD and COD values could be minimized to a satisfactory level by adapting biological treatment. Moreover, such secondary treatments require high energy consumption with production of a considerable amount of sluddddge and mud inside the biological treatment unit. Membrane-based technologies are emerging with membranes of different regimes such as those belonging to MF and UF [6]. Combined or integrated systems are also developing such as a combination of UF and RO techniques for the reduction of COD level and elimination of the toxic compounds from Kraft black liquor. Through proper management skills, wastewater could be reclaimed and recycled to help meet the ever-increasing demand for freshwater.

6.4.2 TREATMENT TECHNOLOGIES

6.4.2.1 Conventional Treatment Technology

Paper-mill wastewater is conventionally treated in a multistage effluent treatment unit where primary and secondary clarifiers with both aerobic and anaerobic digester units are generally integrated [1]. Using high-pressure pumps, the effluent is first passed through screen differentials that can easily separate and remove large matter and particles carried along by the raw wastewater, which might negatively affect the efficiency of downstream units. Grit chambers, which are simply basins, are subsequently employed for the removal of inorganic particles present in the screened effluent. Thus damage to the pumps and sludge digesters due to accumulation of sludge particles can be prevented. The effluent from grit chambers is sent to the primary clarification units where sedimentation separates the suspended, floating, and settleable solid particles. Inside the primary clarifier, particles with high molecular weights also get sedimented due to gravitational force (gravitational settling). The main purpose of primary treatment is to physically remove as much solid loading from the system as quickly and as cheaply as possible without the use of very high-tech equipment or excessive monitoring and controlling [7]. The overhead liquid effluent primary clarifier unit can be treated with aerobic bacterium in an aerobic chamber in the presence of proper air supply using air diffusers. Microorganisms like bacteria and protozoa use the small particles and dissolved organic matter as food and in turn cause decomposition of the polluting substances of wastewater. After aerobic decomposition by reduction of pollutants the effluent is directed to a secondary clarifier (ASP) where the clean water overflows from the settling unit leaving sludge sediment at the bottom of the clarifier. The clean water obtained from secondary clarifier can be treated in a UV reactor and chlorine disinfection unit for further removal of microbes and pathogens [8]. Effluent from tertiary treatment units can be directly discharged to water bodies. Fig. 6.4.1 shows the total treatment scheme.

Sludge from the primary and secondary clarifiers can be taken to a gravity thickener followed by anaerobic digesters in which sludge is degraded by anaerobic microbes. In the dewatering press unit, the sludge volume is reduced by compression force [7]. The sludge is further dried on cake-drying beds and disposed. A conventional paper-mill wastewater-treatment plant removes up to 80% of solids and bacteria with low cost and low operation and maintenance requirements but such systems require a huge area, robust mechanical unit, and nonflexible processes. Moreover, it demands a special area for sludge disposal because sludge is usually land-spread. Another disadvantage of these systems is the formation of undesired odor, which may affect the health of the operators.

6.4.2.2 Primary Treatment

The primary clarification is carried out in a huge clarifier tank where most of the suspended matter present in the industrial effluent can be

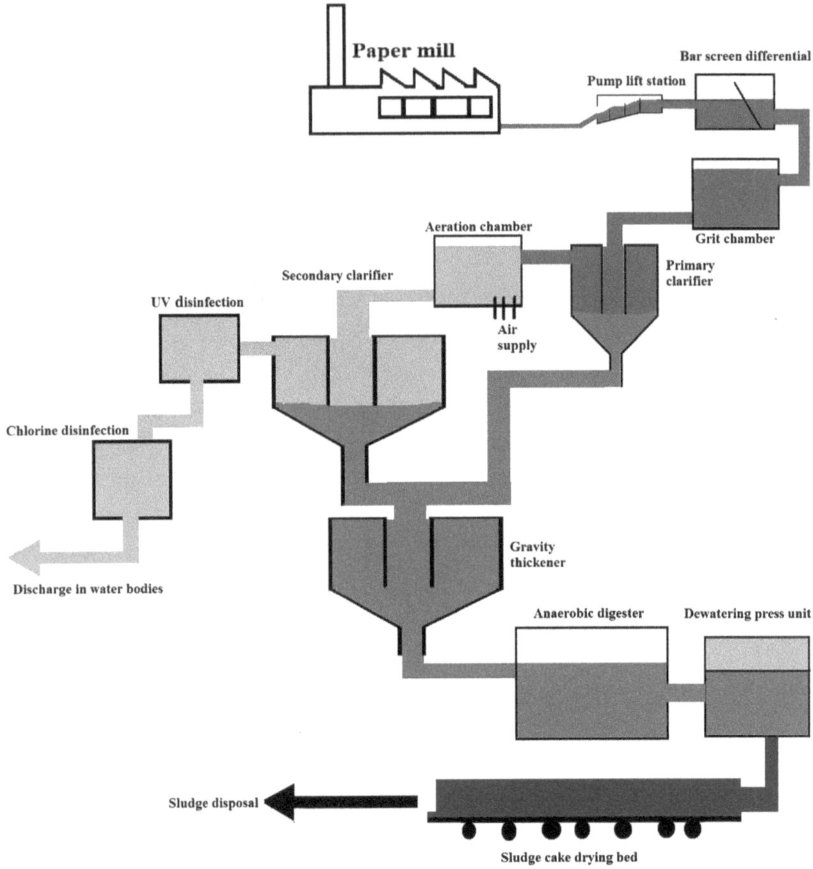

Figure 6.4.1 Flowchart of conventional pulp and paper-mill effluent treatment.

separated out (an average >80%) by either sedimentation or flotation. Sedimentation is most widely practiced. Compounds such as metal salts are added to the effluent in order to destabilize colloidal material while aggregating small particles into larger sized flocks, which can be easily removed. Coagulants of iron and aluminium like ferric sulfate, ferric chloride, aluminium sulfate, and polyaluminium chloride (PAC or liquid alum) are widely used as primary coagulants to promote the formation of aggregates in wastewater treatment and reduce the concentration of particulate matter and dissolved organic compounds. The effectiveness of the process depends on the coagulant dosage, pH, ionic strength, and concentration and nature of the organic compounds present in wastewater in the primary clarifier tank. In flotation process, thickening of waste sludge is done prior to aerobic biological treatment of wastewaters. The depth of the float layer should be at least 150 mm above the level of wastewater to ensure reasonable stability of the float layer during sludge scraping. In general, plants do not allow the float layer to get too thick to ensure stability of the float layer against the destabilizing forces of the sludge scrapers. Dissolved air flotation has been shown to reduce the toxic levels of wastewater in terms of COD (up to 74%) and TS (up to 62%).

6.4.2.3 Secondary Treatment

Secondary treatment techniques involve chemical or microbial unit processes performed on the primary treated wastewater. Chemical unit processes may include advanced chemical oxidation, ion exchange, chemical neutralization, and stabilization. However, pulp and paper mills prefer to treat the primary treated water using anaerobic or aerobic biological unit processes. Biological processes are efficient in removing soluble, colloidal, and particulate matters containing phosphorus, nitrites, and other organics. ASPs and aerated lagoons are the most commonly used aerobic, suspended growth biological techniques where the hydraulic conditions, biomass concentration, and oxygen availability play significant roles in paper-mill effluent treatment.

6.4.2.4 Activated Sludge Treatment

On average, activated sludge treatment of wastewater is able to reduce BOD and COD by 85%–95% and 40%–80%, respectively, from effluent. An activated sludge system could be a plug flow reactor or a complete mixing one or a sequencing batch type reactor dependent on a group of bacteria, which degrade organic matter present in influent stream in the presence of oxygen. An ASP can effectively treat the waste in presence adequate microbial cell density ($\geq 10^6$ cells/L). The specific composition of microfauna inside activated sludge should be based on attached and crawling ciliates with the absence of protozoa. Activated sludge treatment is promising for pulp-mill effluent and can degrade and remove toxic components by more than 94%–98% of the resin acids (like pimaric acid, sandaracopimaric acid, and isopimaric acid) and 41%–67% of the sterols (like campesterol, campestanol, β-sitosterol, and stigmastanol) where about 5% of the resin acids and more than 31% of the sterols are removed in biosludge of the sludge thickener offering good operation stability and flexibility to the process. Effluent from mechanical pulping processes comprise has a quantity of wood fibers than effluent from chemical pulping and with ASP the reduction of wood extractives in the final effluent are found to be more than 90% with a considerable reduction in concentration of fatty acids, resin acids, and sterols (97%–99% for all the three).

Biodegradation of resin acids can be done by gram-negative *Pseudomonas abietaniphila* BKME-9 and *Pseudomonas sp.*A19-6 bacterium in which resin acids like dehydroabietic acid and abietic acid present in effluent can be degraded by more than 90% from an ASP. Treatment of *Pinus radiata*-bleached Kraft mill effluent with activated sludge reduces 90% of BOD5 and 58% of COD when operated under a HRT from 16 to 6 hours. A higher HRT (≥6 hours) increases the control of stable and effective mixed-liquor suspended solids and also increases the degradation of total phenolic compounds, tannin, and lignin, which confirms that compounds with high molecular weight (lignin) show poor degradation in an activated sludge system. Significantly high removal rates of BOD5 (99%) and COD (80%) can be achieved at low organic loading rate (<2 g/L per day) and high HRT (about 24 hours) while the F/M ratio varied from 0.08 to 0.15 and the specific oxygen uptake rate (SOUR) of the biomass usually varies between 0.07 and 0.42 gO_2/g (volatile suspended solid)$^{-1}$. β- Sitosterol is the major sterol in pulp- and paper-mill effluent that can be removed using activated sludge treatment. The other phytosterols include campestrol, stigmasterol, and stigmastanol during phytosterol reduction process using activated sludge system, an 88% increase in stigmasterol is observed, likely due to chemical or biological transformation in the precursor treatment system where it is converted into stigmasterol. However, the growth of a protozoan community is essential to keep activated sludge microfauna in a good physical state.

6.4.2.5 Aerated Lagoons

Suspended-growth lagoons operated on a flow-through basis or solid recycle are shallow earthen basins with mechanical aerators fabricated on some floating or fixed platform to keep the biological solids in suspension. Aerated lagoons are generally operated at considerably higher HRT (4–9 days) compared to that of an activated sludge system (about 8–18 hours). An aerated lagoon exhibits 50%–75% removal efficiency for BOD7, and 10%–50% removal efficiency for chlorinated phenolic compounds. Lagoon systems are even fruitful for the removal of $NH4^+$-N (about 67%) at a pH of 7.3 and within a mesophilic temperature range

(22–35°C). But most Kraft paper-mill wastewater contains insufficient amount of nutrients, which affects microbial growth inside treatment plants resulting in low removal of toxic parameters.

Acclimation of sludge with the most adapted biomass can produce the highest degradation rate. Introduction of support materials facilitating attached growth of microbes increased removal of COD (60%–70%) and phosphorus content (60%–70%) in pilot plants wherein full-scale running lagoons reduced COD and phosphorus only by 30%–40% and 0%–10%, respectively. A lagoon with low HRT (4.5 days) shows significant reduction in concentration of halides (about 65%) but is unable to remove chloroacetic acids, chlorophenols, and chloroguaiacols; in contrast, a lagoon operated at high HRT (45 days) can reduce levels of chloroacetic acids (84%) but is unable to remove chlorocatechols. The use of lime and microbial layer in lagoon-adsorbed organic halides from the pulp and fpaper-mill effluent stream and mass balance of aqueous and solid phase point that more than 99% of the removed halides are mineralized through dehalogenation.

While treating bleached Kraft mill effluent with an aerated lagoon, remarkably good reduction was observed in level of parameter values like mixed function oxygenase, hard-to-degrade organics (chlorinated phenolics, and polychlorinated phenolics) absorbable organic halogens by more than 85%–90% in the treated effluent [7] reported that a series of SBRs (4 in number) were proved to be more robust to shock loads (three times the regular influent concentration) than aerated lagoons as SBRs can recover 50% faster than lagoons, at comparable SRTs (5–10 days) [8]. Sequential reactors with older sludge and larger biomass concentration and high SRTs (10–40 days) do not necessarily provide faster recovery where a shock load eight times the regular influent concentration. Adsorption-based treatments can be efficiently integrated prior to aerated lagoons for the removal of high-molecular-weight dissolved organic matter, COD, and BOD. An exhaustive literature survey indicates that although aerated lagoon systems are being used for the treatment of pulp-mill effluent for considerable reduction in toxic parameters (like BOD, COD, resin acids, phenolic compounds, and TSS) no reasonable reduction on color and recalcitrant pollutants from paper-mill effluent has been shown as the compounds causing the dark brown color (lignin and ligno-cellulosic compounds) are not very biodegradable. Aerated lagoons are easy to operate but the huge operational area needed, elevated land extensions, slowness of the process due to the large HRT, deficiency of microbial nutrients, and odor limit their use.

6.4.2.6 Anaerobic Digestion

Anaerobic digestion involves a series of mechanisms and devices (like anaerobic suspended growth, upflow and downflow attached growth, sludge blanket, and anaerobic lagoons) where the active microbial species degrade the organic matter present in the influent wastewater stream without the presence of oxygen. These anaerobic microbes with production of less biological sludge break down a wide range of organic matter present in influent stream through the catabolic pathways (hydrolysis, acetogenesis, acidogenesis, and methanogenesis) producing 50%–80% of the energy carrier methane, with the remainder being mainly carbon dioxide, turning the process into a potential energy resource. The upward velocity in the UASB reactors is responsible for the development of biomass sludge where partial recirculation of the effluent can also increase flow velocity, tending to improve reactor performance. The upward velocity in UASB reactors should be kept low through provision of high challenge toward mass transfer processes resulting in long hydraulic retention times for the treatment of complex effluent.

In anaerobic treatment of both black liquor and bleach effluent obtained from agro-residue-based pulp and paper mills addition of 1% (w/v) glucose at organic loading rate of 3.6 g COD/(L. d) results in high yield of methane (76%–80%), high level of reduction in BOD (97%), COD (66%–71%), and absorbable organic halides (73%) from the effluent.

6.4.2.7 Advanced Research of Treatment Mechanisms

The primary and secondary treatments are unable to completely degrade organic recalcitrant along with color, which necessitates the use of tertiary treatment of wastewater through processes that include adsorption, precipitation, UV radiation, and oxidative disinfection by ozone or other oxidants (like hydrogen peroxide and halogens like chlorine, chlorine dioxide, bromine, and iodine) with a final step of dehalogination.

6.4.2.8 Adsorption

This process of adhesion leads to formation of a film of the adsorbate on the surface of the adsorbent (e.g., charcoal, silica gel, clay, metal oxides). Several processes (physical, chemical, and biological) occur at the surfaces between two phases, while others commence at the interface where the change in concentration of a given substance due to the accumulation of molecules at the interfacial layer as compared to that of neighboring phases. As a result, the adsorbent (solid or liquid) surface achieves a state of tension due to the unbalanced or residual forces, which leads to the formation of bonds in between the surface molecules (adsorbent) and the molecules in the fluid phase. Adsorption with lignite (cost effective and easily disposable) can be efficiently integrated prior to biological treatment for efficient reduction of levels of BOD, COD, and color as well as for the removal of NaOH, $NaNO_3$, Na_2CO_3, C_6H_5Cl, polyamine, toluenediamine, and diaminodiphenylmethane from wastewater.

The high removal level of color, COD, DOC, and AOX (>90%) is achieved by using activated coke for 4 hours with an adsorbent dosage of 15 g/L. A combination of adsorption (using Bentonite) and coagulation (using polyaluminium silicate chloride) treatment over biological treatment to increase the treatment performance of paper- and pulp-mill effluent has been found to be effective. Adsorption using chemically treated oil palm empty fruit bunch (EFB) fibers can significantly reduce color and organic pollutants (more than 95%) from the paper- and pulp-mill bleaching effluent. EFB fibers are modified chemically by polyethylenimine, which enhances the adsorption capacity of the anionic organic compounds present in the effluent to bind with the cationic amine groups of polyethylenimine. A combined particle-flocculation system comprising bentonite and PAM (polyacrylamide) for the treatment of recycled fiber pulp wastewater focusing mainly on the stickiest substance (mainly due to the presence of methyl tertiary butyl ether, or MTBE) has also been successful.

6.4.2.9 Coagulation Flocculation

Older treatment processes used to treat Kraft black liquor such as chemical pretreatment activated sludge treatment fail to meet regulatory effluent standards. Thus new techniques such as coagulation and flocculation are now being used. Traditionally, the chemical coagulants used in the coagulation and flocculation process are ferric chloride, potassium alum, aluminium chloride, and other polyelectrolytes. Color effluent from paper and pulp mills is highly anionic and is readily neutralized with cationic coagulants. Aqueous particles generally contain some charge (most of the time negative) and repel each other in water. Colloidal particle, being very small in size (approximately 1 nm−1 μm), has a tendency not to settle without help in a reasonable time or not settle at all. Chemicals (coagulants) carrying opposite charge to the colloids are added to reduce the surface charge of the colloid particles and help them destabilize and separate from each other. Rapid mixing ensures the proper dispersion of coagulant particles. Flocculation is the technique of separation of flocks in the effluent treatment in which the destabilized particles aggregate under slow mixing with evolution occurs. Flock quality can be improved by the addition of chemicals (flocculants).

Chloride salt of iron shows high removal of total carbon (>90%), color, and turbidity (90%–98%) among other chloride and sulfate salts of iron and aluminium with a yield of almost clear and colorless effluent. Aqueous solution of polyaluminium chloride dose (8 mL/L) at natural pH (pH 6) results in COD removal by 84% and color removal by 92%. Alum shows the maximum removal of turbidity of 97% followed by Guar gum, Xanthan gum, and locust bean gum with removal efficiency of 95%, 92%, and 93%, respectively. A combination of ferric chloride, PAC, and cationic polymer produces the best results for COD, TSS, and color reduction from paper-mill effluent by 81%, 95%, and 88%, respectively. Loops and tails in PAM get absorbed into particles and create a bridge between different particles and hence increase the size of flock.

6.4.2.10 Advanced Combined Oxidation

Advanced oxidation techniques have proven to be successful in the removal of recalcitrant compounds and when integrated with conventional chemical or biological processes for the treatment of industrial wastewaters the overall treatment efficiency increases. Although the application of such oxidation techniques is an attractive option in pulp and paper-mill wastewater treatment, the greatest disadvantage is the high demand of energy and chemical reagents.

6.4.2.11 Ozonation

Chemical oxygen demand can be effectively reduced using combinations of chemical/biological treatments as described in the previous sections. However, these are not as successful in removing the color where ozone-based treatment with an ozone dosage of 60 mg O_3 L^{-1} is found successful in removing around 90% of the color. Simple ozonation for a contact time of 2–3 hours and pH range of 7.1–8.4 can reduce the organic-carbon level up to 35% and COD by 53%–72%. Treatment of industrial wastewater with ozone as a primary application not only increases wastewater biodegradability but also enhances the chances of reclamation and reuse of water from the effluent. Ozone oxidation (dosage of 0.8 g per hour) of diluted paper-mill (10% diluted) effluent at alkaline condition (pH 9) is more effective in terms of COD removal (about 55%) and the treatment of raw effluent (COD removal about 20%) but a conjugation of ozone and UV treatment at pH 9 can produce considerably good COD reduction of about 98%. A higher pH level (pH 10−12) facilitates high ozone oxidation with 15 L/hour resulting in high removal efficiency of dissolved organic carbon (32.4%), total phenol (84%), and color intensity (67%) in 2 hours. Moreover, higher dosage of ozone increases degradation of COD. Recalcitrant compounds like resin acids (ethylenediaminetetraacetic acid), which are not easily biodegradable, can be easily degraded to 90% and COD by 65% by implementing ozone treatment at a relative dosing of 1:5 of O_3.

Integration of ozone-based technologies as a posttreatment stage with the conventional biological process for pulp- and paper-mill wastewater treatment can ensure high removal of coloring materials (>96%). TOC removal efficiency increases with increasing pH because of self-decomposition of ozone into free hydroxyl radicals, which are able to oxidize the organic compounds more efficiently.

When ozone at a flow rate of 3 g/hour and initial concentration of 150 mg/L are employed, 90% color reduction can be expected. When combined treatment of 3 g/hour ozone and 1.5 g/L carbon is used, 92% color reduction can be obtained whereas 94% color reduction is observed using 3 g/hour ozone and 2.6 mM H_2O_2. Advanced oxidation using ozone is very effective in the reduction of phenol concentration to 70%. Simple ozonation at a rate of 30.9 mg/(L · minute) for 3 hours on bleached Kraft pulp (from softwood) for printing paper effluent at an initial pH of 7.4 shows very high reduction of COD (70%) and TOC (51%).

Ozone-based treatments can be promising as they increase biodegradability of paper-mill effluent while removing TOC, color, and toxic compounds like phenols, resin acids, and COD. However, ozonation cannot be applied as a standalone technology for toxicity removal.

6.4.2.12 Catalytic Oxidation

The reactions that are accelerated due to the presence of light or UV light absorbed by an adsorbing substrate or catalyst are photocatalyzed reaction. Fenton's reagent, hydrogen peroxide, and titanium dioxide (TiO_2) play a major role in semiconductor photocatalysis in such oxidative reactions based on the formation of hydroxide radicals (OH^-), which further reduce/destroy dissolved organic compounds, aromatic compounds, toxic compounds, detergents, pesticides, and many more recalcitrants. Among the various semiconducting materials (oxides or sulfides), TiO_2 is known for its high photocatalytic activity, chemical stability resistance to photo-corrosion, commercial availability, low cost, nontoxicity, and favorable wide band-gap energy (3.2 eV). The removal of COD and impurities by oxidation depends on factors like concentration of H_2O_2 (a strong oxidant and electron scavenger) and TiO_2, solution pH, irradiation time, and integration with a primary biological treatment system.

The addition of H_2O_2 to the UV/ TiO_2 treatment system enhances performance of photocatalytic oxidation of primary clarified and bio-treated paper- and pulp-mill waste.

6.4.2.13 Electrochemical Treatment

There are several electrochemical technologies used in the treatment of wastewater, with the primary focus being electrodeposition (for recovery of heavy metals from the electroplating baths, ethants, and eluates of an ion-exchange unit), electrocoagulation (for the removal of suspended solids as well as oil and grease thereby reducing the turbidity and color from wastewater), electroflotation (for removal of colloids, oil, and grease as well as organic pollutants), and electrooxidation (for oxidative degradation of the refractory pollutants on the surface of a few electrodes). In electroflotation, pollutants float to the surface of a water body by tiny bubbles of hydrogen and oxygen gases generated from water electrolysis. Compared to other floatation technologies such as dissolved air flotation, sedimentation, and impeller flotation it shows better performance. In electrooxidation, titanium-based boron-doped diamond film electrodes (Ti/BDD) show high activity and give reasonable stability. TSS and COD can be removed by more than 90% using electrochemical techniques. The effects of polyelectrolytes as coagulants have been studied using sodium silicate, calcium carbonate, and polyacrylamide at a dose ranging from 0 to 20 mg/L, and the results indicate that polyelectrolyte has no effect on pollutant removal as the extent of color removal both in the presence and absence of polyelectrolyte is similar. Since the supply of current influences the amount of ferrous ions that are produced from the sacrificial electrodes, the effect of current density turns out to be the governing parameter.

Electrochemical degradation of pulp- and paper-mill wastewater can be catalyzed by molybdenum- and phosphate-modified kaolin with graphite as anode and cathode. The results show 95% and 96% COD removal at current density of 30 and 40 mA/cm^2, respectively, at 40 minutes, which shows the occurrence of faster COD removal at higher current density. When Mo–P modified kaolin loaded with Fe^{3+} is used for the electrochemical degradation of mill wastewater at pH of 7, about 76% COD removal is obtained within 40 minutes of the reaction whereas without Fe^{3+} COD removal is found to be 51%. This indicates that Fe^{3+} plays a significant role in electrochemical degradation of wastewater. The initial pH of wastewater also strongly affects the electrochemical degradation efficiency as COD removal reaches 96% within 40 minutes at a pH value of 4. The introduction of NaCl also leads to considerable increase in COD removal from 76 to 93%. The most effective reaction is seen at pH 4 with Mo–P modified kaolin loaded with Fe3+ as catalyst at 30 mA/cm^2 current density, showing 96% COD removal in 40 minutes.

6.4.2.14 Anaerobic Baffled Reactor

In anaerobic baffled reactor (ABR) a series of vertical baffles containing large active microbial mass are arranged to force the wastewater to flow under and over (or through) the baffles covering the full surface of baffles enabling contact between influent wastewater and biomass. These simple, low-cost (construction and operation) reactors are highly stable to shock loads (both hydraulic and organic) providing high void volume with low clogging due to reduced sludge generation because of high solid retention times and low hydraulic retention times. A schematic of the ABR is shown in Fig. 6.4.2.

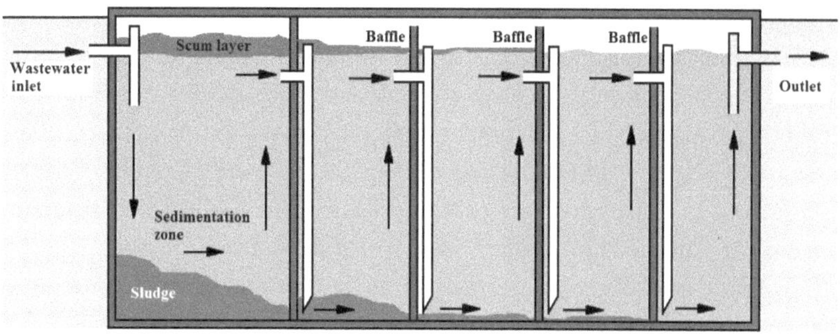

Figure 6.4.2 Treatment using an anaerobic baffled reactor.

At higher organic loading rate, biogas production increases but COD removal decreases due to increasing rate of acidogenesis and nonproportional growth of methanogens. The disadvantages of such reactors are reflected in increased solid loss, long start-up phase, disruption of microbial communities and bioflocks, and requirement of another secondary treatment technique due to low reduction of pathogens.

6.4.2.15 Membrane Technology

MF membranes have the highest pore size (500–50,000 Å) and lowest operating pressure range (1–4 bar) and can easily separate microbial cells where UF membranes of pore size 20–500 Å operated under an operating pressure range of 5–9 bar can separate out cells, proteins, and fats from feed solutions. In both MF and UF, separation is based on size-exclusion principles and sieving effects where the solute particles with larger molecular weight than the MWCO of a MF or UF membrane are almost completely rejected. Thus MF and UF membranes are prone to fouling. NF membranes are designed to have a pore size of less than 2 nm and to be operated under operating pressure range of 10–20 bar whereas RO membranes are usually nonporous and operated under operating pressure over 20 bar. Solute transport through NF and RO membranes generally occurs due to the sieving or steric challenge effect and electrostatic charge repulsion or Donnan effect due to the presence of negatively charged polyelectrolyte (like polyamide) layers on th membrane surface. Solute ions of the same charge get repelled whereas those with opposite charge are attracted and pass through the membrane making these membranes effective for separation of microbial cells, proteins, salts, and metal ions. RO technology can separate the same materials as NF but RO membranes need much higher operating pressure involving higher cost. Fouling on the membrane surface is addressed by proper design of tailor-made flat-sheet membranes and cross-flow type membrane modules. This allows the feed to flow parallel to the active surface of the membrane thus minimizing fouling. This particular design allows the operator to change the feed flow rate enabling high sweeping action over the membrane surface resulting in high permeate flux and long-term fouling-free operation. Different membrane modules are shown in Fig. 6.4.3.

Membrane-based plants are simpler in design and operation because a membrane module, pump, pressure gauge, rotameter, and flow-controlling valves are the only requirements of a membrane-based

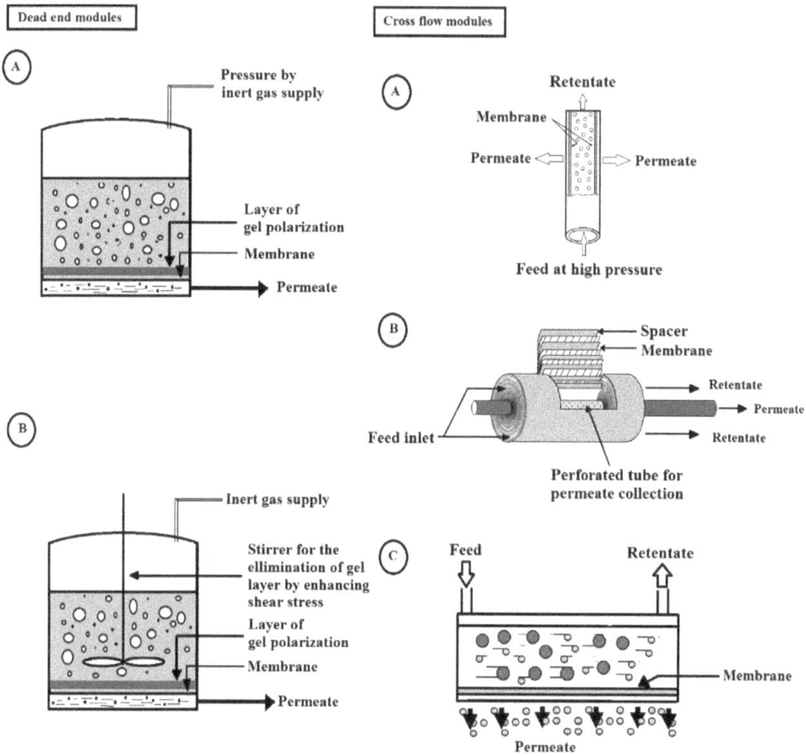

Figure 6.4.3 Schematic of different membrane modules; dead-end modules, (A) simple diagram, (B) shear enhanced dead-end filtration; cross-flow modules, (A) hollow-fiber module, (B) spiral-wound module, (C) flat-sheet, cross-flow membrane module.

separation process. A simple integration using any membrane is shown in Fig. 6.4.4.

Modular design of such processes allows us to increase or decrease the overall output as per demand by flexible addition and subtraction of an effective number of working modules ensuring the process is continuous with the generation of desired permeate with a high level of purity but with less man power or capital investment. Thus implementation of membrane technology could be considered a green technology for clean water reclamation paving the way toward process intensification.

6.4.2.16 Ultrafiltration

Direct UF often results in high fouling although separation of lignin (45–65 g/L) from the Kraft black liquor using UF and subsequent

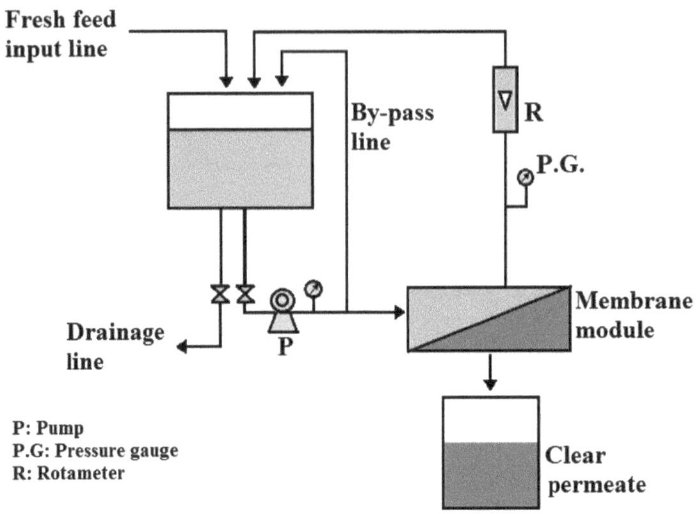

Figure 6.4.4 A simple scheme of membrane integration with a reactor or tank.

diafiltration for concentration of recovered lignin has been attempted. High-purity lignin is extracted by UF and diafiltration with highest content 190 g/L with flux 6 L/m^{2}hour, which could serve as biomass as furnace fuels replacing fossil fuels. COD reduction of the filtrate is higher than BOD with high lignin retention. Metals from chlorine-free Kraft black liquor have been removed using water-soluble polymeric ligands. Polymeric ligands may be used to form chelating compounds with the impurities present in effluent. Higher color removal efficiency has been achieved (97.5%) while using PVA polymer in complexation and further treatment with dead-end stirred cell UF rather than simple UF which removes 88% of the color.

Hemicellulose fraction as well as lignin concentration in permeate is higher when high cut-off membranes are used. The lignin concentration in the product stream is found to be 100 g/L for UF and 165 g/L for hybrid UF and NF systems. UF may reduce COD to a large extent but cannot treat wastewater to the level of reusable criteria.

6.4.2.17 Nanofiltration

NF has been found to be successful in purifying pulp wastewater with around 50 L/m^{2}hour flux with clear water through retention of 90% COD. NF of the ASP discharge water of the pulp and paper industry has been found to yield high flux of 150 L/(m^{2}hour) [9]. The obtained permeate from filtration of discharge water contains only a negligible amount of organic carbon but has relatively high salt concentration whereas filtration of process water contains organic matter of smaller size and are not well retained.

6.2.4.18 Membrane-Integrated Hybrid Process

When a single technology cannot ensure effective treatment of wastewater, an amalgamation of two or more technologies is often adopted. For example, membranes may be integrated with conventional systems. The availability of tailor-made membranes is leading to development of such hybrid technologies where due to their high selectivity the membranes are fractionating the contaminants of pulp wastewater in very compact, simple, flexible plant configurations involving low energy and material cost. NF-integrated electrodialysis [10] and MBR [11] have been found to be effective in pulp- and paper-waste treatment. More than 92% removal of COD has been obtained in such hybrid systems (Fig. 6.4.5).

The treatment of NF concentrate by ozone is found to decrease turbidity, color, and UV-absorbing compounds, and is biodegradability of NF concentrates has also proven to be amplified by 2–4 times from an ozone dosage of 1.2 g/L treated concentrate. Integration of UF as a

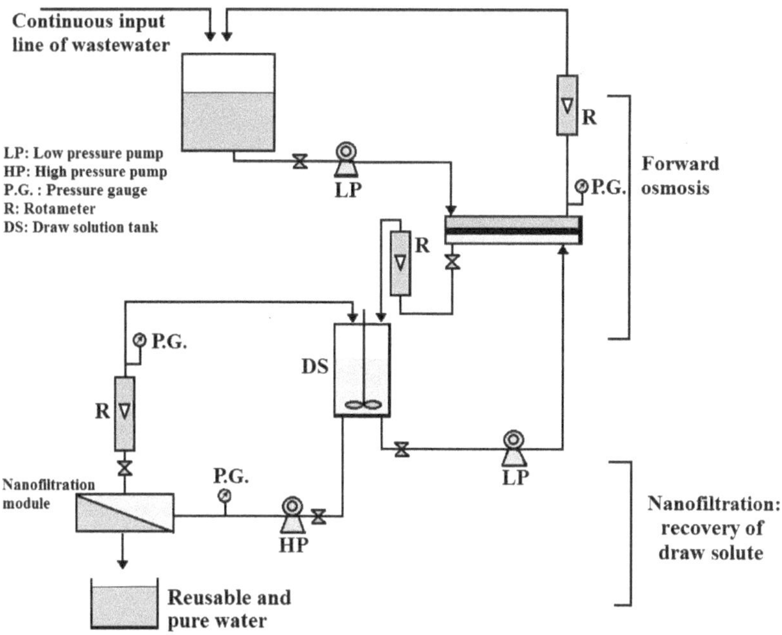

Figure 6.4.5 Schematic of a forward osmosis-nanofiltration integrated system.

pretreatment step of NF can retain about 95% of sulfate ions with substantially high flux. Better separation between hemicelluloses and lignin can be achieved if UF is used as a pretreatment before NF. The flux was also found to increase by almost 100% when permeate from UF is used for NF, which is desirable as higher flux indicates better economy in the process. It is observed that prefiltration using 1 kDa UF membrane before the acid precipitation step results in concentrated lignin fractions by eliminating 75% of the lignin. The purity of the acid increases from 21% to 33% due to retention of acid molecules, which can be explained by Donnan exclusion and electroneutrality principles. Impurities such as sulfate ions and other inorganic compounds are removed by 70%–80%, UV-absorbing compounds like phenolic compounds from lignin degradation are removed by 98%, and TDS is removed by 80% by NF membranes in acidic conditions [9].

6.4.2.19 Membrane-Integrated Advanced Technology

An advanced treatment technology has been developed that intregrates MF, UF, and NF with an ozone-based pretreatment [12]. Ozone treatment on the obtained NF concentrate is carried out to enhance biodegradability. After running the MF or UF, permeate is collected in a storage vessel as shown in Fig. 6.4.6, which is further treated with NF membrane in a plate-and-frame module.

Long-term permeation through NF membrane results in high concentration of impurities in the retentate stream, which can be effectively minimized by ozone oxidation. Subsequent attachment of an ozone reactor increases the biodegradability of concentrated retentate, making it benign and dischargeable into the environment. The overall water flux achieved from the system is about 70 L/(m^2hour) at a pressure of 6.5 bar. The inorganic-carbon retention ranges from 60%–70%. The NF permeate is almost free from color and organic compounds but has more permeated ions such as sulfates and chlorides. The treatment of NF concentrate by ozone decreases turbidity (>80%), color (>50%), and UV-absorbing compounds (>50%). Ozone attacks preferentially double bonds due to which, typically colored compounds and lignin get efficiently oxidized by ozone. It is also observed that biodegradability (ratio of BOD and COD) of NF concentrates is amplified 2–4 times by an ozone dosage of 1.2 g/L on the concentrated retentate. However, ozone oxidation is an expensive process due to the high cost of ozone.

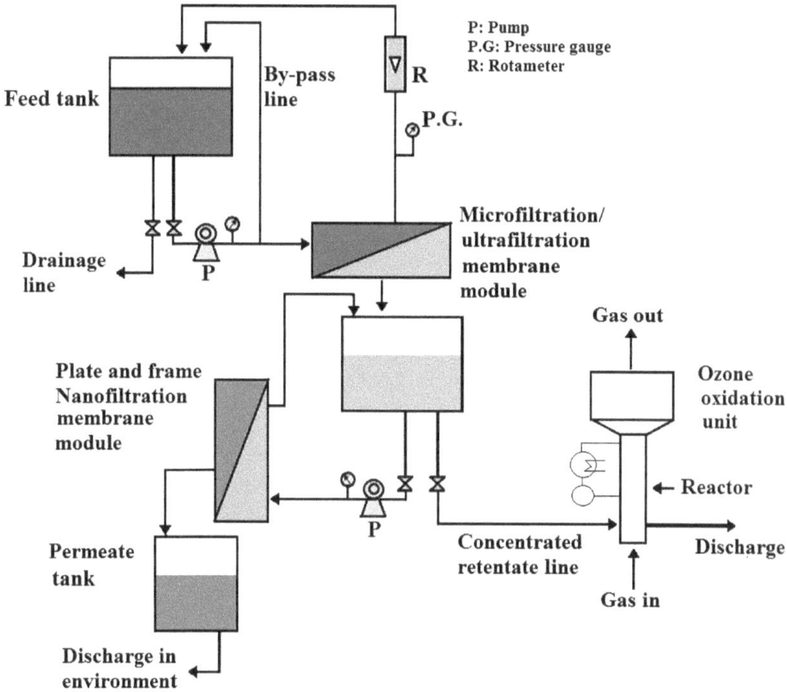

Figure 6.4.6 Advanced membrane-integrated treatment scheme [12].

REFERENCES

[1] Thompson G, Swain J, Kay M, Forster CF. The treatment of pulp and paper-mill effluent: a review. Bioresour Technol 2001;77:275–86.

[2] Holmberg J, Gustavsson L. Biomass use in chemical and mechanical pulping with biomass-based energy supply. Resour Conserv Recycl 2007;52(2):331–50.

[3] Johan G, Fogelholm C-J. Papermaking science and technology: 6A. Chemical pulping. Finland: Tappi Press; 2000. p. 41–2. ISBN 952-5216-06-3

[4] Biermann CJ. Essentials of pulping and papermaking. San Diego: Academic Press, Inc; 1993. ISBN 0-12-097360-X

[5] Olli K, Petri P, Mikko S. Advanced water treatment, recycling lower paper-mill water consumption. <http://www.waterworld.com/articles/iww/print/volume-12/issue-3/feature-editorial/advanced-water-treatment-recycling-lower-paper-mill-water-consumption.html>.

[6] Liu T, Zheng K, Wang D, Li S, Wang H. Bentonite adsorption & coagulation treatment of recycled fiber pulping wastewater. Engineering 2013;5:314–20.

[7] Metcalf & Eddy, Tchobanoglous G, Burton FL. Wastewater engineering: treatment and resource recovery. 5th ed. McGraw-Hill Higher Education; 2013.

[8] Bajpai P. Microbial degradation of pollutants in pulp-mill effluent. Adv Appl Microbiol 2001;48:79–134.

[9] Weeks B, Oleszkiewicz JA. Shock loading lagoons and activated sludge treating bleached kraft mill wastewater. In: Proceedings of the 1993 Environmental Conference. Can. Pulp Paper Assoc., Montreal, Canada; 1993.

[10] Grobicki AMW, Stuckey DC. Performance of the anaerobic baffled reactor under steady state and shock loading conditions. Biotechnol Bioeng 1991;37:344–55.
[11] Manttari M, Nuortila-Jokinen J, Nystrom M. Influence of filtration conditions on the performance of NF membranes in the filtration of paper-mill total effluent. J Membr Sci 1997;137:187–99.
[12] Manttari M, Kuosa M, Kallas J, Nystrom M, Membrane filtration and ozone treatment of biologically treated effluent from the pulp and paper industry. 2008;309 (1–2):112–9.

FURTHER READING

Schnell A, Steel P, Melcer H, Hodson PV, Carey JH. Enhanced biological treatment of bleached kraft mill effluent: I. Removal of chlorinated organic compounds and toxicity. Water Res 2000;34(2):493–500.
Manttari M, Viitikko K, Nystrom M. Nanofiltration of biologically treated effluent from the pulp and paper industry. J Membr Sci 2006;272:152–60.
Zhang Q, Chuang KT. Adsorption of organic pollutants from effluent of a Kraft pulp mill on activated carbon and polymer resin. Adv Environ Res 2001;3:251–8.
Lutchmiah K, Verliefde ARD, Roest K, Rietveld LC, Cornelissen ER. Forward osmosis for application inwastewater treatment: a review. Water Res 2014;58:179–97.

SUBCHAPTER 6.5

Treatment Technology for Leather Industry Wastewater

6.5.1 INTRODUCTION

Leather processing involves several few chemical processing steps through which mechanical strength, durability, surface properties, texture, and color of leather improves significantly. The most widely used chemical ingredient in leather processing is chromium. Chromium salts are used by about 90% of tanneries in the world to produce leather that provides water resistance, flexibility, and avoids putrefaction, properties that are all important for good leather quality. As most chemical treatments involve harsh chemicals and huge quantities of water, the leather processing industry generates an enormous amount of hazardous wastewater that on discharge to open field or water bodies without proper treatment leads to heavy surface-water pollution. The main leather-processing operation is known as tanning and derives its name from the acidic chemical called tannin that is traditionally and widely used in leather processing. The objective of the process is to permanently alter the protein structure of the raw hide on removal of the animal hairs from the hide through liming processes followed by removal of excess fat and meat. The skin or hide is treated with many other chemicals including enzymes to digest the proteins present in them. The treated skin is subjected to two types of tanning operations subclassified as chrome tanning and vegetable tanning. Subsequent to these tanning operations, the hides are subjected to treatment with various dyes and fats to improve their appearance.

The wastewater from the tanning process contains toxic metals like chromium, arsenic, and organic compounds such as organic dyes in high concentrations, which need to be separated out from the wastewater stream before such wastewater is discharged to the environment. This wastewater can be treated to remove and recover the toxic substances present in it. Some of these substances can be used for other purposes, thus improving the economic aspects of the process. The composition varies from source to source, but usually contais a few common pollutants, chromium being one. Chromium escapes into wastewater during the chrome-tanning process and is highly toxic to human beings. Chromium recovered from leather wastewater can be reused in chrome plating and chrome tanning. Apart from chromium, different organic dyes can be recovered and reused in the tanning process. The sustainable treatment approach should ensure not only recovery and reuse of chromium and dye components but also recovery of the treated water in large scale, which through recycle and reuse will save on consumption of freshwater. Understanding the sources of the contaminants at various stages of leather processing is necessary to develop an effective strategy that may be able to replace some chemicals with less harmful substances.

6.5.2 LEATHER INDUSTRY WASTEWATER: CONTAMINANTS AND THEIR ORIGIN

Total processing of leather can be broadly divided into the following four major operations:

1. Skin storage and beam house operation
2. Tanyard operation
3. Posttanning operation
4. Finishing operation

The beaming operation involves pretreatment of raw hides. The hides are first soaked in water, which restores their water content while removing dirt, preservatives, and blood. The dehairing operation includes chemical dissolution of skin into alkaline medium of sulphide and lime in order to dissolve hair and epidermis into the solution. This operation alone contributes most of the chemical contamination expressed in terms of lumped parameter

COD to the leather industry wastewater. Dehairing is preceded by removal of excess meat from the hide. The dehaired alkaline hides are then treated with acid ammonium salts in order to neutralize them and then treated with enzymes to digest off proteins from the hides. This process step is mainly responsible for the large amount of ammonium salts present in the effluent stream. Pickling with acid is also done, which increases the pH value followed by salting (prevents swelling of hide) and degreasing, which again adds to the COD value of the effluent stream. The next important operation is tanning, which may be done through either chrome tanning or vegetable tanning.

6.5.2.1 Chrome Tanning

This process leads to high concentration of chromium ions in the effluent stream when chromium is added to the hide along with a base under varying pH regime.

6.5.2.2 Vegetable Tanning

This process is mainly achieved in tanning vats where vegetable tannins are added to the hide. These tannins are mainly polyphenolic compounds.

6.5.2.3 Posttanning Operations

The posttanning process includes retanning of the tanned hides and then adding dyes and fats to achieve the desired texture and color. The extra water is removed and the hides are sent for splitting and shaving before actual drying is allowed to take place. Organic dyes and fats are added to the effluent streams at this step.

The products of posttanning are further subject to finishing operations for further improvement of quality, color, and texture.

6.5.3 ENVIRONMENTAL IMPACTS OF LEATHER PROCESSING

The industrial processes that produce leather and its goods, fur, etc., from raw hides often involve production of byproducts that heavily pollute natural air and water systems. If this effluent is directly released into water bodies, it will have significant harmful effects on flora and fauna as well as human beings who are either directly or indirectly influenced by it. This effluent often contains hexavalent chromium, which has a mutagenic, carcinogenic, and tetratogenic nature. The other contaminants include salts of trivalent chromium like chromium (III) sulfate, ammonia, sulphides, and different organic compounds like azo dyes. Tanning is usually done with the help of tanning agents that in turn produce highly turbid and color wastewater, which often has a foul smell. The sulfide present in tannery wastewater is one of the major components of effluent and causes major toxicological effects. Tanning wastewater contains a high amount of chromium, which could add up to 4950 mg/L. Hexavalent chromium has a carcinogenic nature. Biodegradation components (expressed in terms of BOD) and other chemical pollutants expressed as a lumped parameter as COD are present in tannery wastewater in high concentrations. The wastewater from the beam-house area contains the highest amount of salt. The COD content of beam-house wastewater may be as high as 27,000–28,000 mg/L. A number of salts are used in the preservation of raw hides and processing of leather from hides. During dehairing, lime and sodium sulfide are used in bulk quantity and in the degreasing step organic solvents are widely used. About 15%–40% of common salt is used to preserve hides, which may be washed away with the effluent stream. Sodium chloride is the major component of wastewater when the salted raw hides are processed further. Chromium (trivalent and hexavalent) spent dyes, sulfonated oils, tannins, and COD are relatively high for retaining and wet finishing streams from plants, while the BOD and TSS levels are relatively low. Generally, tannery effluent streams are rich in organic nitrogen and lean in phosphorus while the TDS values can be as high as 37,000 mg/L. TOC concentration is also an important parameter that needs to be measured and monitored for tannery wastewater since high concentration of chlorides and sulfides can affect the measurement of COD [1–6]. The typical composition of leather industry wastewater is given in Table 6.5.1 [5].

6.5.4 CONVENTIONAL TREATMENT TECHNOLOGY

6.5.4.1 Primary Treatment

The primary treatment eliminates the coarse materials normally present in the raw wastewater that could clog/block pumps, pipes, and possibly sewerlines. Different tannery waste streams should be mixed together to produce homogenized "raw material" of consistent composition that can be treated efficiently in the treatment plant. After proper adjustment of pH, elimination of toxic substances (sulfides), and smoothing out of flow fluctuations, the raw waste stream should be introduced to downstream treatment units, preferably biological treatment units, which are known for their sensitivity to the operational environment. Fig. 6.5.1 [6] shows a simplified flowchart of the full-fledged tannery effluent treatment plant (ETP) the major components of which are the mixing-cum-aeration devices including diffused-air systems, venture ejectors, and fixed or

Table 6.5.1 Typical composition of leather industry wastewater [5]

Wastewater parameters	Unit	Average value
BOD	mg/L	750–800
COD	mg/L	5600–5700
TSS	mg/L	1650–1700
Chromium	mg/L	400–500
Sulfides	mg/L	2850–2900
Chlorides	mg/L	6500–6500
Phosphorus	mg/L	160–170
Ammonium Nitrogen	mg/L	20–280
Fats and Oil	mg/L	50–500
pH		8.5

floating aerators. Elimination of the sulfides is mostly done by catalytic oxidation. To oxidize 1 kg of S^{2-} to thiosulfate, approximately 1 kg of O_2 is needed at an oxygen transfer efficiency of about 1.5 kg O_2/kWh.

The suspended solids from water are removed during primary treatment through coagulation and flocculation. Aluminium sulfate ($AlSO_4$), ferric chloride ($FeCl_3$), ferrous sulfate ($FeSO_4$), and lime (calcium hydroxide) are used as the coagulation/flocculation agents to reduce organic and inorganic pollution loads as well as the suspended solids. To effectively use each coagulant, the pH must be optimized because coagulants are very pH sensitive. The range of pH for each coagulant depends on the characteristics of the wastewater as well as the dosage of the coagulant. By the combined effect of alum and $FeCl_3$, removal of more than 70% of COD is possible. In addition, total chromium concentration can be brought down to below 5 mg/L level by alum and can be removed almost completely using $FeCl_3$ [7,8].

Removal of suspended solids is done in the primary sedimentation or settling unit. The primary settling units or clarifiers are either circular or rectangular in which sludge is removed at the bottom and grease (scum) from the top. The sludge contains only 2%–4% dry-solids indicating large water content. The water content should be drastically reduced by sludge thickeners or through mechanical dewatering in filter presses or through centrifugal actions in sludge decanters (centrifuges). Natural drying is the final step in which dewatering and sludge volume reduction is achieved in sludge-drying beds.

6.5.4.2 Biological (Secondary) Treatment

At this stage, further reduction of organic and partially inorganic mass (expressed as BOD and COD) and other substances still present in the

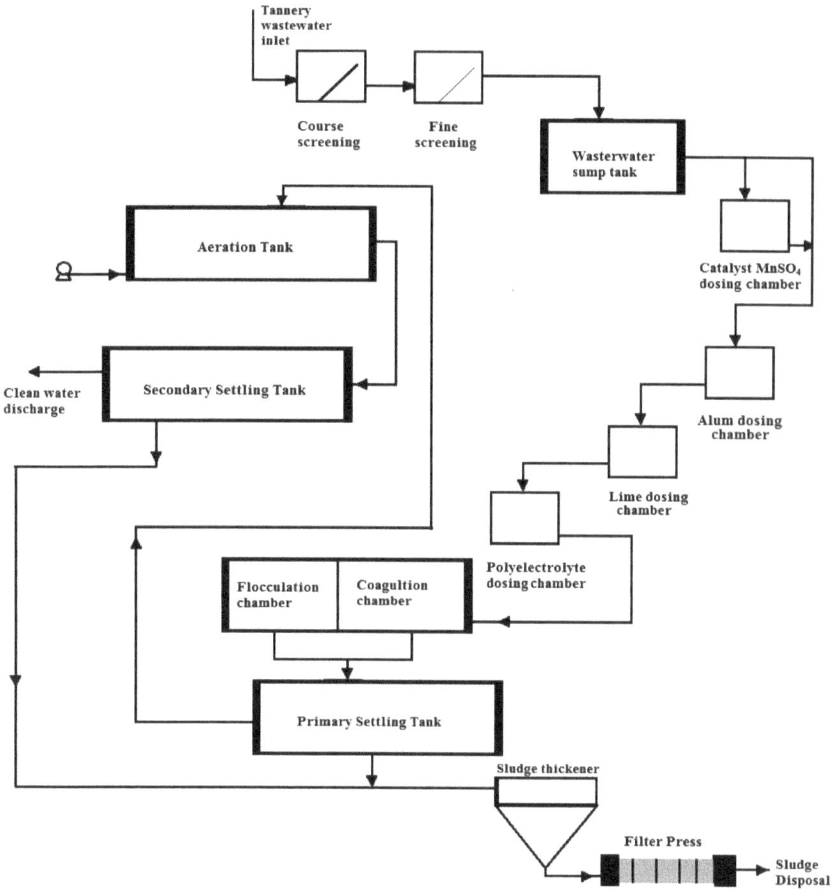

Figure 6.5.1 Simplified flowchart for full-fledged tannery effluent treatment plant [6].

effluent after primary treatment is done to meet effluent discharge standards. The efficiency of such treatment largely depends on the biodegradability of the residual polluting substrates after removal of the suspended and colloidal solids by flocculation and adsorption. Biological treatment may aerobic, facultative, or anaerobic or a combination thereof. The microbial community that does that job comprises various species of bacteria, fungi, and protozoa, and sometimes rotifers and nematodes. The main operational parameters for activated sludge treatment are total influent volume, tank volume, organic loading, MLSSs, loading factor (food/biomass), and hydraulic retention time. A constant inlet flow over the entire period provides optimum conditions for absorbing the effects of toxic substances or shock loads and thus enhancing the efficiency of secondary treatment. Dissolved oxygen (DO) is one of the most important factors determining the efficiency/performance of wastewater biological treatment and should at least be maintained at 2 mg/L level. The optimum pH range for aerobic processes is between 7.0 and 7.5 with an effective process range of 6–9. Adjustment with lime or other alkali is necessary only if the pH drops below 6. As the temperature is increased from 10 to 30°C, the growth rate increases. However, higher temperature negatively affects the water solubility of oxygen and the oxygen transfer rate (solubility decreases with an increase in temperature). For this reason, an increase in the aeration rate is necessary during the hot season.

The nutritional balance of an aerobic system may be maintained by supplying a minimal amount of nitrogen and phosphorus to maintain the cell structure produced by the removal of BOD from waste. A BOD:N:P ratio of 100:5:1 in the waste usually ensures adequate nutrition. Tannery effluent is very rich in nitrogen and sometimes poor in phosphorus.

The SVI provides a good indication of sludge-compacting characteristics and mineralized sludge has SVI <100. The SVI is very helpful for controlling the ASP, especially when determining return sludge pumping requirements to maintain different mixed-liquor concentrations.

6.5.5 ADVANCES IN TREATMENT OF LEATHER INDUSTRY WASTEWATER

6.5.5.1 Aerobic Oxidation Process

Aerobic microbes are used to decompose the organic matter present in the effluent of tannery wastewater. Primarily, the microbes digest the organic carbon present in the effluent stream and convert it to biomass and carbon dioxide. This biomass can be used as a clean fuel for different purposes and the carbon dioxide escapes from the system. In association with physicochemical processes, aerobic oxidation can reduce BOD by 95% and COD by 60%–80%. When combined with chemical ozonation, the efficiency of COD reduction improves substantially to 96%. Similarly it succeeds in removal of total Kjeldahl nitrogen by 92% and TSS by 98%. The ozonation step partially oxidizes the refractory compound thereby increasing its biodegradability. A combination of electrooxidation and aerobic oxidation can also be applied to considerably increase the removal efficiency of the contaminants, particularly the sulfides, COD, and TN [9–11].

6.5.5.2 Anaerobic Oxidation Process

Anaerobic treatment is carried out in the absence of oxygen with or without a supplement like cow dung slurry. The major advantages of this approach include redundant oxygenation arrangement, reduced sludge volume, additional byproduct (biogas in large volume), and substantial reduction of BOD, COD, and chromium. The process releases a large amount of biogas, which is a valuable byproduct, and a little amount of sludge. The BOD, COD, pH, and chromium values are also considerably reduced. However, one difficulty that may be encountered is the presence of sulphide ions, which may reduce the rate of anaerobic oxidation. The process can be run at low cost using low-cost devices like UASB reactor [12,13].

6.5.5.3 Multistage Activated Sludge Treatment

By running biological treatment units such as activated sludge units in multistages with multistage clarifiers in between, very high degree of removal of BOD (>98%), COD (>98%), and chromium (>99%) can be achieved. The byproducts are gases that escape and the biological cell-tissue mass that settles down at the bottom of the tanks. Multistage clarifiers ensure proper clarification of the effluent producing clear solids-free water. The tannery wastewater in this type of multistage treatment scheme may first be fed into the primary clarifier after pH adjustment. After around 45 minutes, the stream may be transferred to the aeration tank where more than 120 hours of residence time is allowed for biodegradation following which the stream is again sent to the secondary clarifier for another 45 minutes [14].

6.5.5.4 Physicochemical Processes

Electrochemical Oxidation

Electrochemical oxidation may be adopted for effective removal of nitrogen compounds and those contaminants from wastewater which are otherwise nonbiodegradable and hard to remove using conventional biological or chemical processes. Direct oxidation of the pollutants occurs on interaction with the anode material when the pollutants from the solution diffuse to the anode surface. In such electrochemical oxidation, Ti/Pt and Ti/Pt/Ir electrodes are used. Ti/Pt/Ir electrodes are found to have electrocatalytical properties for removal of NH^{4+}. However, these electrodes are sensitive to H_2S poisoning. Indirect oxidation occurs when a strong oxidizing agent is electrogenerated at the anode surface, which destroys the pollutants [12,15].

6.5.5.5 Ozonation-Integrated Aerobic Oxidation

Ozone is one of the strongest chemical oxidants available. Ozonation is a chemical treatment of wastewater in which ozone is injected into the water oxidizing the impurities present in the suspension. Integration of ozonation with aerobic disintegration can result in high removal efficiencies for COD, TSS, TKN, surfactants, and color leaving behind residual concentrations far below the discharge limits. The other major advantage of this process is production of extremely low volumes of sludge (0.1 kg dry sludge/m^3 of treated wastewater) [16].

6.5.5.6 Advanced Oxidation Processes

Fenton's Oxidation and Photo-Fenton's Oxidation

Fenton's oxidation is a process of generating hydroxyl radicals in the solution by catalytic decomposition of H_2O_2 by ferrous and ferric cations. The hydroxyl ions readily mineralize a vast range of organic compounds quickly and unselectively. The reaction highly depends on the pH level of the solution. The best results are obtained at a pH range of 2.8–3 and the rate sharply reduce at higher pH values. The photo-Fenton's process is similar to Fenton's process, only difference being that it is carried out in the presence of strong sunlight. The UV-B rays from the incident radiation in the range of 12.2–12.8 are used to oxidize the range of organic impurities present in the effluent stream. The pH is maintained at 2.5 using H_2SO_4. In both these processes 70% removal of COD is achieved [17].

6.5.5.7 Membrane Technology for Tannery Wastewater Treatment

Membrane-based technologies are emerging as fast and efficient separation–purification technologies in the chemical and allied process industries. Membrane processes play a significant role in removing toxic compounds particularly in tannery wastewater treatment. It is observed that toxic compounds such as chromium, sulfides, tannins, COD, and BOD can be efficiently removed by membrane processes like UF, NF, MF, and RO and any combinations of these processes. One of the major problems encountered in all the above processes is the fouling of membranes. Use of different cleaning procedures with chemicals like sodium hypochlorite, sodium dodecyl sulfate, different enzymes, combination processes, and use of the better membrane modules available today can reduce membrane fouling drastically resulting in sustained high flux over long duration. Membrane-based technologies can offer a high degree of separation–purification at reasonably low cost while promising even reusable water from such dirty industrial wastewater. Modular design, availability of tailor-made membranes, new fouling-free modules, and better membrane materials imparting high mechanical strength are making membrane technologies or membrane-integrated hybrid technologies very attractive with the promise of offering sustainable solutions to many pollution problems.

Nanofiltration

Using NANOMAX50 (composite polyamide/polysulfide) NF membranes in spiral-wound membrane module and is a composite membrane of polyamide/polysulfide reduction of 85% of turbidity and reduction of more than 95% COD has been achieved [18]. Using DS5 membranes, 99% removal of chromium at high flux can be achieved [19].

Combined Ceramic MF and RO

Tannery effluent has high concentrations of organic and inorganic materials characterized by about 759 mg/L of BOD_5, 5680 mg/L of COD, etc. A dual-stage treatment containing ceramic MF followed by RO succeeds in removing 91% COD and BOD_5, 62% TOC, and 100% sulfides. The finally treated water is reusable in the tannery industry itself and the system ensures good flux [5].

Hybrid Membrane Bioreactor

On integrating MF with traditional electrocoagulation and biological processes a hybrid membrane bioreactor (HMBR) developed that successfully removes more than 90% COD, 92% color. Incorporation of enhanced electrocoagulation with the ASP and dead-end MF process shows that the hybrid process can treat treatment to within the permissible limits [20].

Membrane-Integrated Advanced Treatment Technology

Conventional chemical and biological processes like coagulation–flocculation and activated sludge fail to remove all dissolved inorganics. Membrane technologies are being increasingly incorporated into conventional treatment schemes to solve the problems of dissolved solids in effluent [21].

Thus treatment schemes incorporating RO membrane have been developed as shown in Figs. 6.5.2 and 6.5.3 [21,22]. However, high concentration of chlorides that causes pickling and to avoid it tanning a RO process becomes necessary to make it reusable [22]. The schemes in Figs. 6.5.2 and 6.5.3 show that exhaustive pretreatment is necessary prior

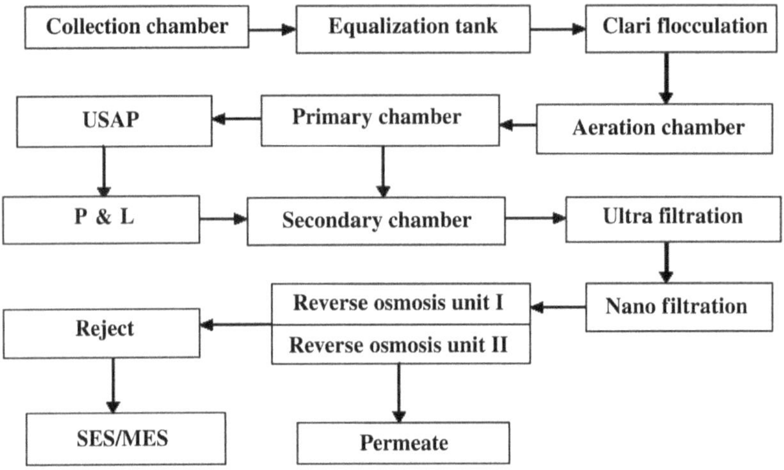

Figure 6.5.2 Process flowchart of emerging technology for tannery wastewater treatment [21].

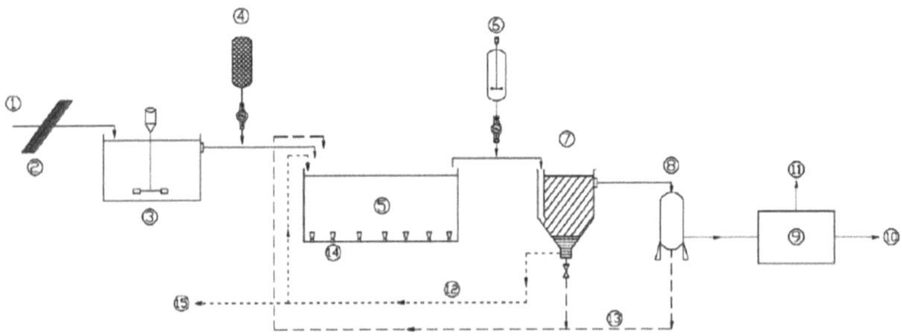

1 Inlet wastewater
2 Bar screens
3 Equalization basin
4 pH Correction
5 Complete mix aeration tank
6 Flocculant
7 Settling tank
8 Sand filter
9 Reverse osmosis
10 Permeate to reuse
11 Concentrate
12 Return activated sludge
13 Filter backwash effluent
14 Diffused air system
15 Effluent sludge

Figure 6.5.3 Flowchart of the treatment system for tannery wastewater [22].

to the RO process. Thus the total scheme includes treatment devices like bar screens, equalization basins, pH correction units (by lime dosing), biological activated sludge (with a hydraulic retention time of 30 hours), secondary sedimentation (assisted by the addition of polyelectrolyte), sand filtration, and an RO process with a plane membrane [22]. In this scheme, the upstream process consists of a chemical-physical process as a pretreatment and a biological ASP with low F/M ratio. The biological treatment is essential for the removal of organic matter, whereas the RO process removes ammonium compounds. Thus this conventional method is very useful to increase the life of the membrane of the RO system, which will result in reduction of cost of effluent treatment. The conventional method efficiently removes the toxic pollutants like chromium and sulfides by oxidation and precipitation. However, TDS removal is very tedious in conventional technology. Thus integration of membrane separation with a conventional system results in a hybrid system that effectively removes all contaminants including solids in all forms. The use of MF/UF and NF/RO membrane during physical and chemical treatment of tannery wastewater not only ensures long life of the membrane but also improves the quality of permeate in a significant measure where BOD in the treated water falls below the detection limit and TDS goes down to the low level of around 230–250 mg/L, which are well within the drinking water standard.

Operating the Integrated Treatment Plant

Wastewater after pH correction is introduced to the aeration tank for aeration, mixing, and bioreactions. The inlet flow rate to the biological reactor may typically be maintained at 25–26 L/hour with HRT of around 30 hours. After biological treatment accumulation and sedimentation are done. The DO and MLVSS concentrations are measured for monitoring at 5 mg/L and 8000 mg/L, respectively, during biological treatment. Temperature is maintained in the range of 24 and 27°C. Downstream purification of biologically treated wastewater is done by RO membrane. During the RO process, the removal percentages of all the parameters are high and thus the permeate stream can be reused in the tannery production cycle. The remaining COD and ammonium-N and nitrate substances are removed by 97%, 96%, and 98%, respectively, whereas salts, chloride, and sulfate removal are 98.8% and 99.8%, respectively.

General Guidelines in Selection of Treatment Technology

In removing suspended solids, sedimentation alone is grossly inadequate and needs to be integrated with advanced membrane separation techniques for producing reusable water. Biological treatment schemes offer treatment at the least cost but often are very slow and highly sensitive to operating conditions. However, when integrated with advanced chemical oxidation and membrane separation it can produce excellent results. Treatment of wastewater by means of coagulation demands precise dose calculations, and requires a complementary process to completely eradicate the harmful species present in the wastewater. Many adsorption processes have been developed in the past from a multitude of materials used as adsorbents. However, the removal efficiency drops significantly over time and replacement of the adsorbent at regular intervals is necessary. A combination of physical, chemical, and biological processes can be developed for maximum removal efficiency. Although the cost analysis to reach an optimized level would be of utmost importance. Vegetable tannins and coloring materials are removed from tannery waste using UF and HMBR (color removal) techniques with better flux recovery rates. NF can be very efficient in the removal of hazardous chemicals like sulfides and chromium and can even result in potable water from wastewater. Integrated membranes of different regimes (ceramic MF and RO) have the potential to reduce water requirements. Sustainable treatment technology demands that treatment loops be closed enabling the plant to reuse the treated water. Leather processing requires novel approaches involving less harmful and environmentally benign chemicals or materials.

REFERENCES

[1] Rahman Z, Singh VP. Cr(VI) reduction by Enterobacter sp. DU17 isolated from the tannery waste dump site and characterization of the bacterium and the Cr(VI) reductase. Int Biodeterior Biodegradation 2014;91:97–103.

[2] Kavouras P, Pantazopoulou E, Varitis S, Vourlias G, Chrissafis K, Dimitrakopulos GP, et al. Incineration of tannery sludge under oxic and anoxic conditions. J Hazard Mater 2015;283:672–9.

[3] Isarain-Chávez E, De la Rosa C, Godínez LA, Brillas E, Peralta-Hernández JM. Comparative study of electrochemical water treatment processes for a tannery wastewater effluent. J Electroanal Chem 2014;713:62–9.

[4] Cooman K, Gajardo M, Nieto J, Bornhardt C, Vidal G. Tannery wastewater characterization and toxicity effects on Daphnia spp. Environ Toxicol 2003;18(1):45–51.

[5] Bhattacharya P, Roy A, Sarkar S, Ghosh S, Majumdar S, Chakraborty S, et al. Combination technology of ceramic microfiltration and reverse osmosis for tannery wastewater recovery. Water Res Ind 2013;3:48–62.
[6] Introduction to treatment of tannery effluent; UNITED NATIONS INDUSTRIAL DEVELOPMENT ORGANIZATION Vienna, 2011; Internet: http://www.unido.org.
[7] Song Z, Williams CJ, Edyvean RGJ. Treatment of tannery wastewater by chemical coagulation. Desalination 2004;164:249–59.
[8] Ates E, Orhon D, Tünay O. Characterization of tannery wastewaters for pretreatment selected case studies. Water Sci Technol 1997;36:217–23.
[9] Sastry CA. Ind J Environ Protect 1986;6:159.
[10] Midha V, Dey A. Biological Treatment of tannery wastewater for sulfide removal. Int J Chem Sci 2008;6(2):472–6.
[11] Szpyrkowiczi L, Kaul SN, Neti RN. Tannery wastewater treatment by electrooxidation coupled with a biological process. J Appl Electrochem 2005;35:381–90.
[12] Tamilchelvan P, Dhinakaran M. Anaerobic digestion treatment of tannery wastewater. Int J Eng Res Appl (IJERA) 2012;2(3):932–6.
[13] Cassano A, Drioli E, Molinari R. Recovery and reuse of chemicals in unhairing, degreasing and chromium tanning processes by membranes. Desalination 1997; 2–3(113):251–61.
[14] Hafez A, Manharwwy MSE. Design and performance of two stage/two pass RO membranes system for removal from tannery wastewater, part 3. Desalination 2004;165:141–51.
[15] Anglada A, Urtiaga A, Ortiz I. Contributions of electrochemical oxidation to wastewater treatment: fundamentals and review of applications. J Chem Technol Biotechnol 2009;84(12):1747–55.
[16] Di Iaconi C, Ramadori R, Lopez A. The effect of ozone on tannery wastewater biological treatment at demonstrative scale. Bioresour Technol 2009;100(23): 6121–4.
[17] Lofrano G, Meriç S, Inglese M, Nikolaou AD, Belgiorno V. Fenton oxidation treatment of tannery wastewater and tanning agents: synthetic tannin and nonylphenol ethoxylate-based degreasing agent. Desalin Water Treat 2010;23:1–3.
[18] Taleb-Ahmed M, Taha S, Chaabane T, BenFare N, Brahimi A, Maachia R, et al. Treatment of sulfides in tannery baths by nanofiltration. Desalination, 185. 2005. p. 269–74.
[19] Chuan MC, Liu CJ. Release behavior of chromium from tannery sludge. Water Res 1996;30(4):932–8.
[20] Keerthi V, Suganthi KV, Mahalakshmi M, Balasubramanian N. Development of hybrid membrane bioreactor for tannery effluent treatment. Desalination 2013;309: 231–6.
[21] Krishnamoorthi S, Saravanan K, Priyenka Devi KS. Integrated effluent treatment in tannery industries - feasibility study. J Ind Pollut Control 2011;27(2):1–5.
[22] De Gisi S, Galasso M, De Feo G. Treatment of tannery wastewater through the combination of a conventional activated sludge process and reverse osmosis with a plane membrane. Desalination 2009;249:337–42.

FURTHER READING

Covington AD. Modern tanning chemistry. Chem Soc Rev 1997;26:111–26.

SUBCHAPTER 6.6

Petroleum Refinery Wastewater Treatment

6.6.1 INTRODUCTION

Petroleum refinery effluent is very harmful to aquatic life due to the presence of high polycyclic aromatic compounds, which are carcinogenic. Moreover, the presence of these compounds reduces the DO content as the microbes whose growth is encouraged by the presence of these organic substances consume DO to decompose the organic waste. Oily wastewater also contains compounds of sulfur, nitrogen, heavy metals, and other chemicals that may be used during petroleum refining. Moreover, being sticky in nature, oil and grease clog drain pipes and sewer lines resulting in nasty smells and corrosion of lines. Wastewater generated from petroleum refinery typically contains oil in the range of 100–1000 mg/L. Industrial oily wastewater is classified as free floating oil, unstable oil/water emulsion, and stable oil–water emulsion. Among these, the first two types can be easily treated by conventional processes. In stable oil–water emulsion where the oil droplets are microsized, conventional processes are unable to separate oil. To overcome this difficulty, methods like thermal demulsification or biological methods have been developed through the last few decades. However, considering the disadvantages of such schemes, development of new technologies is essential to successful treatment of petroleum refinery wastewater [1–6].

6.6.2 PETROLEUM REFINERY WASTEWATER: COMPOSITION AND HEALTH HAZARDS

The major contaminant of petroleum refinery wastewater is the oil itself that escapes from refinery units and processes. This oil mixes with water and eventually has to be discharged from the refinery after treatment in the ETP. The immediate effects of this oil are unpleasant odor and color and the water becomes unfit even for any in the industry itself. Oily water being discharged into water bodies such as rivers/lakes or sea reduces algal growth, destroys water plants, lowers DO level, and adversely affects fish and other aquatic life. Oil adversely affects photosynthesis and absorption of oxygen from atmosphere. The other major contaminants may be phenol, furfural, sulfides, suspended solids, heavy metals from catalyst-handling areas, and total organic and inorganic chemical pollutants, which may be measured in terms of lumped parameters such as BOD and COD. These contaminants eventually make the water unfit for human consumption, industrial or domestic use, and harmful to aquatic life.

6.6.3 EFFLUENT TREATMENT PRINCIPLES

The main aim of an ETP an the petroleum refinery is to recover water in as pure form as practicable for recycle and reuse and discharge the unusable water after treating it to the safe limits prescribed by the concerned pollution control norms of industrial wastewater discharges. Consisting of an emulsion of oil (form various units of plant) and water the effluent is treated via a series of mainly clarification and continuous settling units. Effective chemical and biological treatments are also done to facilitate flocculation and settling and to reduce the BOD and COD of the water. Separation of oil from water is a major challenge particularly when it is present in emulsified form. In the immediate Section 6.5.4, these oil-water separation principles along with individual oil-water separation techniques are described followed by detailed description of an integrated total ETP operation that takes addresses water contaminants from a petroleum refinery in a series of logical unit operations.

6.6.3.1 Conventional Technologies

Oil–Water Separation

API Oil–Water Separators

API oil–water separators are standard in separation of oil from refinery wastewater. An API oil–water separator is a gravity-separation device that functions using Stokes Law to define the increased velocity of oil droplets based on their density and size. Separation is based on the specific gravity difference between the oil and the wastewater where most of the suspended matter settles at the bottom. The light oil overflows, rises, and is separated from the three-phase system.

Plate Separators

Plate separators also to some extent exploit the principle of gravity separation with prior provision of parallel sets of plates where oil drops adhere, coalesce, and become bigger in size and eventually rise to the surface of the main vessel housing the plates. The oily layer is skimmed from the top surface.

Hydrocyclone

A hydrocyclone is a device that is used to separate oil droplets from wastewater based on the ratio of their centrifigal force to fluid resistance. A hydrocyclone has two exit points on the axis—a smaller one on the bottom (underflow or reject) and a larger one at the top (overflow or accept). The underflow consists of the denser or coarser fraction, while the overflow comprises the lighter or finer fraction. The oily water separator based on the vortex principle in which the oily water pumped tangentially into a cone-shaped separator creates a spinning vortex resulting in a separation force. During downward movement of the vortex the oily water is accelerated when the associated centrifugal forces separate the heavier water phase to the outside of the vortex while the lighter oil phase moves to the center. The separated oil is forced out through an orifice positioned in the inlet end and the treated water is collected through an exit point at the other end.

Dissolved Air Flotation

Suspended oil droplets can be recovered from water by gas flotation if the gas bubbles are passed through an oil–water emulsion. The oil droplets attach to the bubbles and are carried to the top of the mixture where they can be removed. Coagulants such as ferric chloride or aluminium sulfate are often added to oily wastewater to facilitate flocculation of the fine oil droplets. A part of the obtained clear water leaving the flotation tank is pumped into a small pressure vessel to which compressed air is passed resulting in air saturation of the effluent water. The air-saturated water stream is recycled to the front of the float tank and flows through a pressure-reduction valve just as it enters the front of the float tank, which results in the air being released in the form of tiny bubbles. The bubbles adhere to the oil droplets, causing them to float to the surface and form a froth layer, which is then removed by a skimmer. The froth-free water leaves the float tank as clear water.

Chemical Coagulation

Suspended oil droplets can be removed from wastewater by adding chemicals that coagulate and flocculate the droplets. These chemicals overcome the electrostatic repulsion charges on the individual droplets, allowing them to coagulate into larger-sized droplets. These larger sized droplets can be more easily removed by gravity separation equipment. Lime, alum, and polyelectrolytes are used as coagulants for oil–water separation.

Electric-Field Separation

In oil–water separation, electric field can very efficiently play its role where under the influence of an electric field, polarization of oil and water molecules quickly takes place. The electric field can be applied by either direct current or alternating current. Oil droplets in the oily wastewater possess negative surface charge, i.e., zeta potential, which can be utilized to remove them. When direct current is applied to the oily emulsion, oil droplets migrate toward the positive electrode. As the migration velocity of the drops is very low, separators containing very narrowly spaced parallel plates can facilitate the process of migration. When an alternating current is applied the droplets flocculate in the presence of metal hydroxide. This phenomenon is known as alternating current electrocoagulation. In this phenomenon, a metal hydroxide is added to the water and an alternating current is used to overcome the electrostatic repulsion charge on the particles. When the electrostatic repulsion charges on the particles are neutralized, they can flocculate and be more easily separated from the water.

6.6.4 TYPICAL REFINERY ETP DESIGN AND OPERATION

All the effluent from various sections of petroleum refinery is subject to a purification process in the ETP as shown in Fig. 6.6.1 [7].

The liquid wastewaters from different units in the refinery are segregated into three basics streams based on the nature of wastewater and the treatment requirement for the removal of the contaminants. The wastewater streams thus segregated are discussed in the following.

6.6.4.1 Oily Sewer

The system is a broad network of the underground pipes (RCC/CS). The network covers the whole refinery. It collects the oil–water mixture from the refinery and offsite areas and delivers it to the influent sump in the ETP. The wastewater is brought here by the pipelines and through tankers. The ETP is in the lower side and all the units are on the upside, thus the oil flows to this network by gravity.

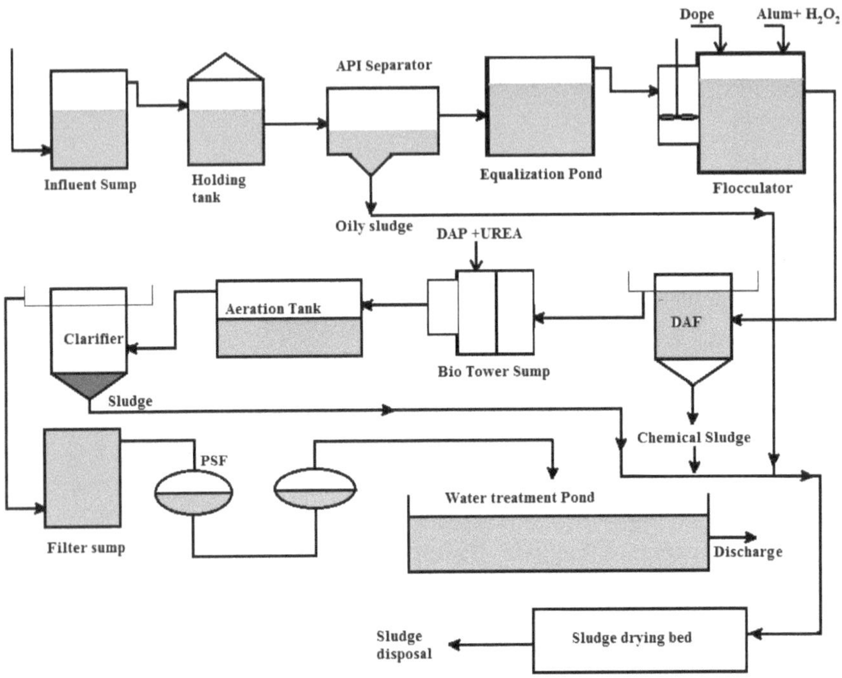

Figure 6.6.1 A typical petroleum refinery effluent treatment scheme [7].

6.6.4.2 Stormwater Sewer

This system is an open-channel network, and covers the whole refinery. Rain and stormwater are collected inside the dyke of the storage tank and drained to the ETP network and sometimes this water gets mixed up with the oil, which is separated in this unit.

6.6.4.3 Domestic Sewage

All the sanitary from toilets and canteens in the refinery (including the administrative building) are connected to this system. This connection is made partly by gravity and partly by pumping.

6.6.5 ESSENTIAL TREATMENTS IN THE CONVENTIONAL SCHEMES

Refinery waste streams collected as common sump for flow equalization and neutralization are subsequently subjected to physical, chemical, and biological treatments in conventional treatment schemes.

6.6.5.1 Physical Treatment

Refinery wastewater contains coarse suspended and floating solids, oils, etc., as well as settleable pollutants. These need to be removed before the wastewater is subjected to chemical and biological treatments. Physical treatment units include rakes and screens, grinder, grid chamber, grease traps, flocculation units, sedimentation basins, dewatering units, and sludge-separation units. In a refinery, bar screens, wire mesh, and API separators are used and in modern installations, tilted plate interceptors are used for physical treatment. Effluent is first made to pass through bar screens and then wire mesh where debris, rags, and large suspended materials are removed and then sent through a grid chamber to remove suspended solids. The purpose of this equipment is to protect the downstream mechanical equipment and avoid deposition in sumps and channels. The wastewater with free oil and sludge is then routed through the API separators followed by primary sedimentation equipment. At this stage, the velocity of the influent is slowed down considerably to a low level facilitating settling of the suspended particles of higher density while floating the low-density oil. In an API separator, around 50%–60% of the suspended solids are removed along with reduction of BOD by 20%–40% at 20°C.

6.6.5.2 Chemical Treatment

Chemical treatment followed by physical treatment reduces colloidal solids, inorganic chemicals, a part of the organic chemicals, and the remaining suspending solids of the effluent. The important unit operations and processes used for this purpose are chemical coagulation, flocculation, and sedimentation. In a refinery, oxidation of both organic and inorganic chemicals (especially sulfides and phenolic compound) by chlorine is followed. In the coagulation process, a chemical coagulant is added to an aqueous system to drive together suspended matter toward formation of aggregates. Flocculation is the second stage in the formation of this aggregate and is achieved by maintaining gentle and prolonged mixing. Coagulation occurs in the preaeration chamber, where, e.g., the coagulant $FeSO_4$ may be added to the solution along with lime solution in the preaeration chamber. Positively charged iron ions neutralize the negative charges of the emulsified oil and hence release the oil from water. These iron ions are hydrolyzed by hydroxides like $Ca(OH)_2$ to form $Fe(OH)_2$ flocks. Dissolved O_2 in the effluent oxidizes the $Fe(OH)_2$ to $Fe(OH)_3$ flocks that settle at a faster rate than $Fe(OH)_2$. For the highest efficiency, a rapid and intimate mixing of $Fe(OH)_3$ and $Fe(OH)_2$ flocks with effluent is necessary before the flocculation process begins, which is done in a flash mixer chamber where a motor-driven stirrer is rotated continuously.

Flocks thus formed are too light to settle under gravity, and therefore from the flash mixer chamber the effluent goes to the clariflocculator where slow stirring is done by two continuously rotating motor driven stirrers enabling flocculation, i.e., agglomeration of small flocks. Entrainment and absorption of suspended particles (such as free oils, FeS) occurs on the large surface area of the agglomerated $Fe(OH)_3/Fe(OH)_2$ flocks that settle at the clarifier zone of the clariflocculator.

6.6.5.3 Biological Treatment

After physical and chemical treatment, biological treatment of wastewater is done under aerobic condition (i.e., under the presence of an excess of free DO (O_2) in a biological system) for further reduction of organic pollutants. The aim is to enhance microbial oxidation or decomposition of the organic pollutants by the naturally occurring bacteria that consume the major part of the organic substances as their food while converting the residuals into new cells and harmless byproducts. At the end of the microbial lifecycle, decay or death of the microbes results in the formation of a substantial amount of solid sludge, which is discharged from the system periodically. The basic biochemical reaction for the stabilization of organic impurities under aerobic condition by microorganisms in wastewater may be represented by:

$$\begin{aligned}&\text{Organic impurities} + \text{microbes} + O_2\\&\quad\rightarrow \text{Microbes} + CO_2 + H_2O + \text{waste energy}\end{aligned}$$

Common systems for biological treatment include trickling filter and activated sludge tank (also called an aeration tank or bioreactor). In a trickling filter (biofilter) system, wastewater is sprayed on the bed of solid supports such as stones or pebbles. The aeration may be mostly by natural draft from the bottom of the stones upward with the temperature difference (of water and ambient air) as the driving force. Bacteria that grows on the stone surface as a film following the attached growth mechanism eats away at the organic impurities while decaying and washing out periodically following their lifecycle and thus paving the way for growth of fresh bacterial biomass on the solid support. In an activated sludge tank, the bacteria are continuously mixed with wastewater and aerated by motor operated aerators. Here also bacteria eat away impurities. The bacteria water (mixed liquor) is then sent to the clarifier from the aeration tank where the bacterial mass is separated from water. The bacterial mass is recycled back to the aeration tank to maintain the required level of microbial cell concentration. The rest of the bacteria may be sent to the biological sludge lagoons periodically for disposal. The water from the clarifier unit goes to the treated water pool, ready for disposal or reuse based on the final quality. Urea is often used as an additional food supplement or nutrient in the bioreactor. Conventional treatment technologies can treat refinery effluent to the permissible discharge limits but fail to close the treatment loop to allow recycle and reuse of the treated water. Thus novel and advanced treatment technologies need to be developed that can not only effectively treat refinery effluent up to the permissible limits but beyond to make the approach sustainable.

6.6.5.4 Advanced Treatment Schemes

Membrane-Integrated Physical and Biological Total Treatment

With the emergence of membrane of versatile materials, the possibility of integration of membrane separation with conventional treatment is imminent. This type of membrane-integrated approach in refinery effluent treatment is culminating in novel treatment plants with a high degree of process intensification resulting in safer, more flexible, more compact, and economically viable green processing plants. Marked improvement in the quality of the finally treated water can be achieved on integration of downstream membrane separation with conventional physicochemical and biological treatment schemes. One such integrated advanced scheme is shown in Fig. 6.6.2 where one MF stage immediately after conventional treatment stages is incorporated followed by a UF/NF membrane module. All the effluent from various sections of the petroleum refinery is subject to a series of purification stages, namely physical and biological followed by membrane separations in two stages.

Microfiltration

Membrane treatment followed by physical treatment and biological treatment reduces TSS, TDS dissolved oil, grease, TOC, and turbidity. Biologically treated wastewater is passed through cross-flow, flat-sheet MF membrane module as pretreatment prior to the UF or NF membrane

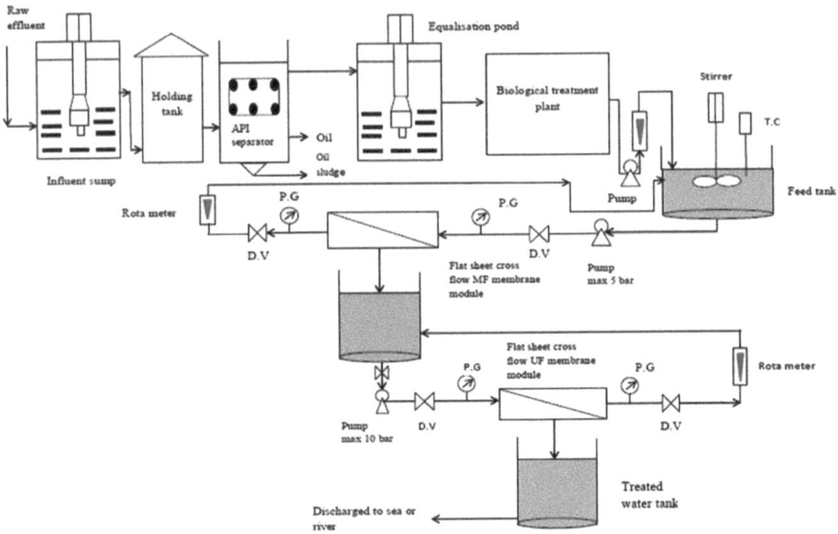

Figure 6.6.2 Membrane-integrated physical and biological treatment plant.

module. Permeate collected from the MF module is stored in a permeate tank. Permeate from MF is then passed through a cross-flow, flat-sheet UF membrane module for further treatment. The whole process runs in a continuous manner. Thus a two-stage membrane-filtration process is replaced by an ecofriendly practice that produces water safe for reuse or discharge to river or sea. A typical MF membrane pore size range is 0.1–10 micrometers (μm) and may be either polymeric or ceramic and capable of producing clean water fit to be used as feed for the UF or NF membrane module. Such MF membranes can effectively remove major pathogens and contaminants such as *Giardia lamblia* cysts, *Cryptosporidium* cysts, and large bacteria. For this application the filter has to be rated for 0.2 μm or less. For mineral and drinking water, the most commonly used format is pleated cartridges usually made from polyethersulfone (PES) media. Microfiltration membrane modules do not need high pressure for operation and can effectively remove turbidity spikes and pathogens without chemical conditioning. With the emergence of cheap, durable membranes, the acceptance and popularity of membrane separation is increasing particularly for more difficult waters that contain more solids and higher levels of dissolved organic compounds. Some of these waters may require chemical pretreatment in conventional schemes including prechlorination often leading to production of harmful byproducts. This further shifts acceptance of purification in favor of membrane-based separation. The filters can be in a submerged configuration or a pressure-vessel configuration. They can be hollow fibers, flat-sheet, tubular, spiral wound, hollow fine fiber, or track etched. These filters are porous and allow water, monovalent species (Na^+, Cl^-), dissolved organic matter, small colloids, and viruses through but do not allow particles, sediment, algae, or large bacteria. MF systems are designed to remove suspended solids down to 0.1 micrometers in size, in a feed solution with up to 2%–3% in concentration, which is suitable for use in place of traditional clarifiers or as a prefilter to a water recycling/recovery RO system.

In CFMF, the suspension is pumped tangentially over the filtration medium. Clear liquid permeates the filtration medium and is recovered as the permeate, while the solids accumulate at the filtration barrier to form a fouling layer, or cake. The cake, constituting an increase in hydraulic resistance, decreases the permeate flux. However, the tangential suspension flow tends to limit the growth of the cake. Thus after an initial rapid increase in cake thickness, cake growth ceases, and the cake thickness becomes limited to some steady-state value. Correspondingly, after an initial rapid decrease, the permeate flux levels off and either attains a steady state, or exhibits a slow, long-term decline with time. This cross-flow process helps minimize the fouling of the surface of the MF membrane. Many different materials have been used for MF but the most common are polysulfone and polyvinylidine fluoride. MF is suited for separate larger sizes, such as suspended solids, particulates, and microorganisms because MF membranes are thought to act as a physical sieve. The membranes are highly porous and have discernible pores even when the surface skins are asymmetric. Therefore the separation is based mainly on size. Membrane material is usually made up of ceramics, teflon, polypropylene, or other plastics.

Ultrafiltration

UF membranes typically have pores in the range 10–1000 Å. The treatment mechanism is based on size exclusion where the TMP required to run the filters is in the range of 3–6 bar. The macromolecules within the range of 10^3–10^6 Da are normally retained by UF membranes that can be operated in a variety of modules such as flat-sheet, cross-flow, spiral-winding, tubular, and hollow-fiber, each with its advantages and disadvantages in terms of parameters such as fouling, CP, durability, mechanical strength, compactness, effective mass transfer area, space requirement, operational flexibility, and maintenance requirement. UF membranes are specified by MWCO.

6.6.5.5 Advances in Oil–Water Separation

Separation of Oil From Wastewater Using Tubular Module

Using a ceramic tubular UF membrane module in stainless-steel housing around 95% removal of oil from water can be achieved under 0.75–1.75 bar operating pressure [8]. The tubular ceramic membrane (α-AL2O3) with a pore size of 0.2 μm although initially successful in yielding more than 250 LMH flux eventually suffered from rapid flux decline following membrane fouling. This membrane fouling can be reduced by chemical cleaning of the membranes as well as backwashing. The cleaning is normally done by rinsing with clean and hot water and then by cleaning with an alkaline cleaner. Backwashing is done to remove the layer of retained material at high pressure of 2 bar. Chemical cleaning is done when the flux decreases by around 50% of its original value. The fouled membrane is first rinsed with water at room temperature and then cleaned with NaOH solution (2%) at 70–80°C to remove organic scales. Again, it is rinsed in water and cleaned with citric acid solution (2%) at 70–80°C to remove inorganic scales. Repeated rinsing is done with clean water. At TMP of 1.25 bar, CFV of 2.25 m/second, and temperature of 32.5°C, MF succeeds in removing 85%, 100%, and 98.6% of oil and grease content, TSS, and turbidity, respectively.

Cascade Spiral-Wound Module MF and UF Module

A pilot-scale membrane cascade system designed using tubular MF of 0.07 micron membrane and UF membrane of 1–5 kDa MWCO in a first stage, where high levels of particulate matter are present, and Sepa flow cells representing spiral wound modules in the second stage succeeded in removing 90% organics with 25–62 L/m^2 per hour flux. Large-channel MF and UF ceramic membranes are used in the first loop while a variety of flat-sheet UF (1–100 kDa MWCO) and NF membranes can be used in the second stage. As has been shown, lower MWCOUF membranes (1–5 kDa) used in the cascade arrangement can retain organics where NF membranes are considered unsuitable due to poor permeate flux performance [9]. In the separation of oil using hollow-fiber, tubular, spiral wound, plate-and-frame (dead-end) membrane modules, the major problem has always been drastic reduction in flux after a few hours of operation, although rejection to the extent of 80%–90% can be achieved.

Separation of Oil From Wastewater Using Glass-Membrane Module

Demulsification of water-in-oil emulsion has been done by using porous glass membranes, which are hydrophilic. Up to 95% rejection of oil with

Table 6.6.1 Properties of different membranes

Parameter	Nylon 0.22	PES-5
Membrane module type	Flat-sheet	Flat-sheet
Membrane-surface area (m^2)	0.012	0.012
Nature of filtration	Microfiltration	Ultrafiltration
Pore size (μm)	0.22	0.001
Membrane thickness (μm)	110–150	165 μm
Porosity	70%	35%
Max process temp (°C)	75	50
pH resistance	2–11	2–11
Molecular weight cut-off (g/mol)	5000–100,000	6000

150 LMH flux can be reached at 90 kPa pressure but decline in flux over time is significant [10].

Separation of Oil–Water Using Flat-Sheet, Cross-Flow Membrane Module With UF Membrane

Membranes

Flat-sheet, CFMF membrane nylon 0.22 (Membrane Solutions, USA) and flat-sheet, cross-flow UF membrane PES-5 (Sepro Membranes, USA) are used in the module. Detailed characteristics of the membranes given in Table 6.6.1.

The porosity of the industrial membranes is calculated manually using Archimedes' principle. At first the membranes are to be dried in an hot-air oven at 50°C for 30 minutes and its dry weight W_1 is measured. Then the membranes are rinsed in water for 24 hours at room temperature. After that the surface water of the membranes is shocked by tissue paper and the wet weight of the membranes W_2 is measured. Then the porosity of the membrane is determined by:

$$\varepsilon = \frac{W_2 - W_1}{W_1} \tag{6.6.1}$$

Membrane Module and Operation

Fig. 6.6.3 represents a MF- and UF-integrated refinery ETP. Oily wastewater from the feed tank first passes through a flat-sheet CFMF module followed by UF membrane modules. The setup is designed and manufactured with high grade stainless steel (SS316). At first MF of the feed is done by nylon 0.22 membrane (Membrane Solutions, USA) and then the permeate was is treated by UF membrane PES-5 (Sepro Membranes, USA) in a continuous manner. Circulation of feed through the

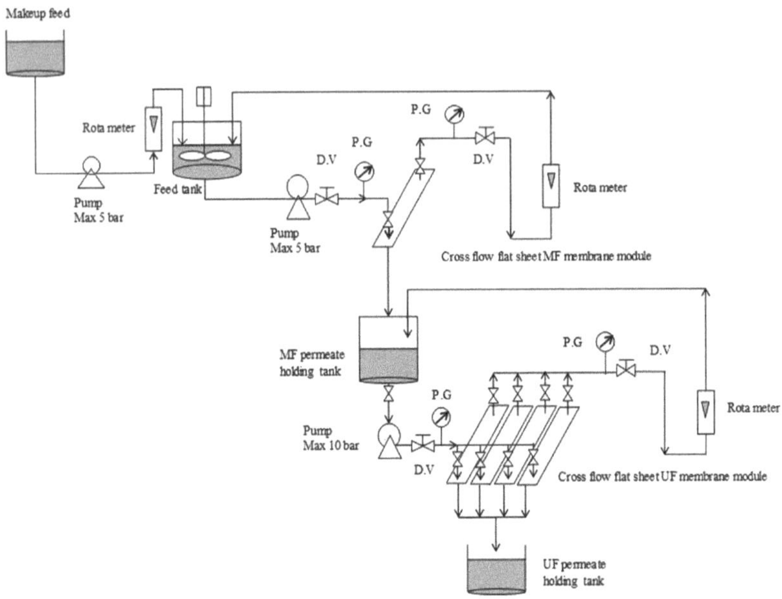

Figure 6.6.3 MF–UF membrane-integrated system of flat-sheet, cross-flow module.

membrane modules is done by diaphragm pumps. The TMP is defined as the average pressure applied across the membrane minus the pressure developed at the permeate side:

$$\Delta P = \frac{P_i + P_o}{2} - P_p \tag{6.6.2}$$

where:

P_i = Inlet pressure

P_o = Outlet pressure

P_p = Pressure at permeate side (negligible)

The membrane fouling that occurs following pore blocking of the membranes by the small oil droplets and due to formation of an oil layer on the membrane surface can be cleaned by conventional methods such as (1) backwashing, (2) rinsing with alkaline solution (0.1 M NaOH), (3) rinsing with acidic solution (2×10^{-2} M HNO_3), (4) sterilizing with 200 ppm NaOCl solution, and (5) rinsing with ultrapure water.

Optimization of Operating Conditions: TMP

Fig. 6.6.3 shows that with increase in TMP, flux increases while rejection decreases. This effect of TMP on permeate flux is explained by Darcy's law.

With increase in pressure oil rejection decreases, which is explained by the sieving mechanism involved in MF and UF where solute flux and solvent flux are coupled and hence increase of solvent flux results in commensurate increase in oil flux reflecting less rejection. Thus there must be an optimum TMP at which the module has to be operated. Through a series of pilot runs, the optimum TMP is normally found.

Figs. 6.6.4 and 6.6.5 show that simultaneous increase of permeate flux and decrease of oil rejection following an increase in TMP result in the same oil concentrations at different CFVs.

Optimization of Cross-Flow Velocity

The effects of CFV on permeate flux and oil rejection are shown in Figs. 6.6.6 and 6.6.7. As can be seen, the permeate flux increases and oil rejection decreases with increase of CFV at different TMPs as well as feed-oil concentrations. A higher CFV leaves a more positive impact on flux as the increased sweeping action following increase in CFV results in long-term operation without significant fouling. Rejection in this coupled flux system will obviously decrease whenever there is increase in solvent flux. However, decline in rejection is not as sharp following a change in CFV unlike the impact of TMP on rejection. This trend again

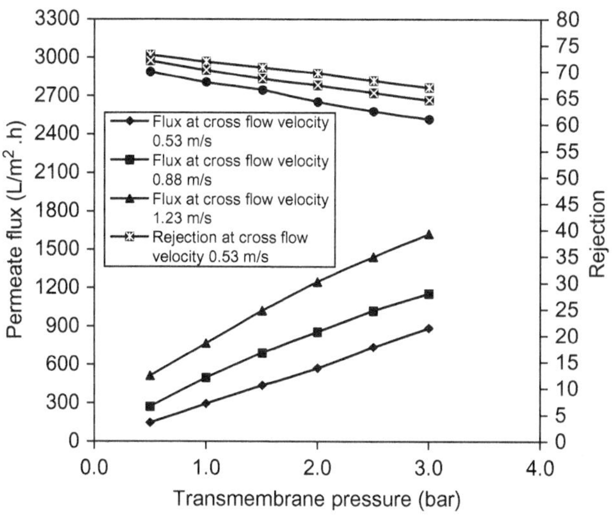

Figure 6.6.4 The effect of TMP on permeate flux and oil rejection for nylon 0.22 membrane at different CFV for 200 mg/L.

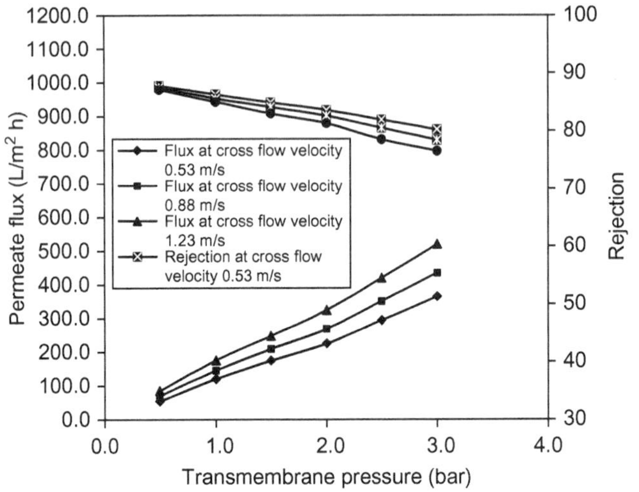

Figure 6.6.5 The effect of TMP on permeate flux and oil rejection for nylon 0.22 membrane at different CFV for 1000 mg/L.

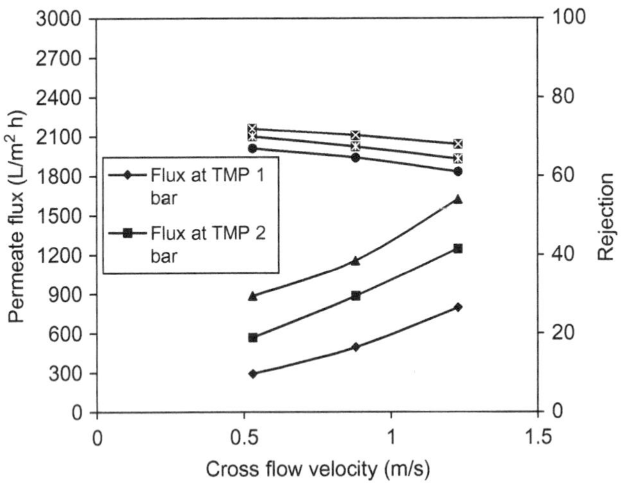

Figure 6.6.6 The effect of CFV on permeate flux and oil rejection for nylon 0.22 for 200 mg/L oil concentration in feed.

indicates existence of optimum CFV that can be arrived at through system run at different CFVs.

Cross-Flow Velocity: Long-Term Flux Behavior

Fig. 6.6.8 shows that at higher CFV long-term flux remains higher.

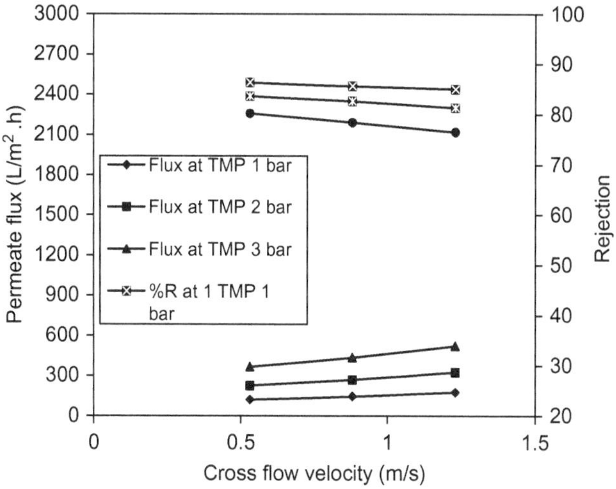

Figure 6.6.7 The effect of CFV on permeate flux and oil rejection for nylon 0.22 for 1000 mg/L oil concentration in feed.

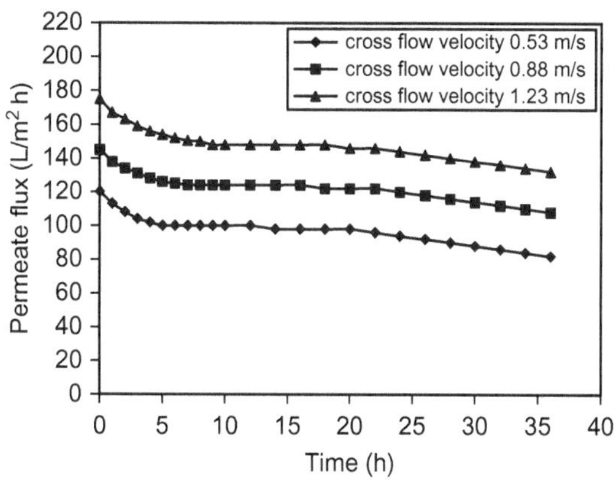

Figure 6.6.8 The effect of CFV on permeate flux at constant TMP 1 bar for nylon 0.22 for 1000 mg/L.

At higher TMP at some point the flux becomes almost stable at some CFV indicating that there must be an optimum CFV at a given pressure at which the steady operation at constant flux is possible.

UF Module: System Optimization

The permeate coming from the MF membrane module is passed through the flat-sheet, cross-flow UF membrane module. The UF membrane used in this second stage is a polyether sulfone type.

Optimization of TMP During UF

Fig. 6.6.9 shows that the flux of pure water increases with increase of TMP. However, the oil rejection decreases with increase in TMP indicating the same trend as with the MF regime.

Optimization of Cross-Flow Velocity in UF

The effect of CFV on permeate flux and oil rejection for PES-5 at 1000 mg/L oil concentration in feed is shown in Fig. 6.6.10. The flux of

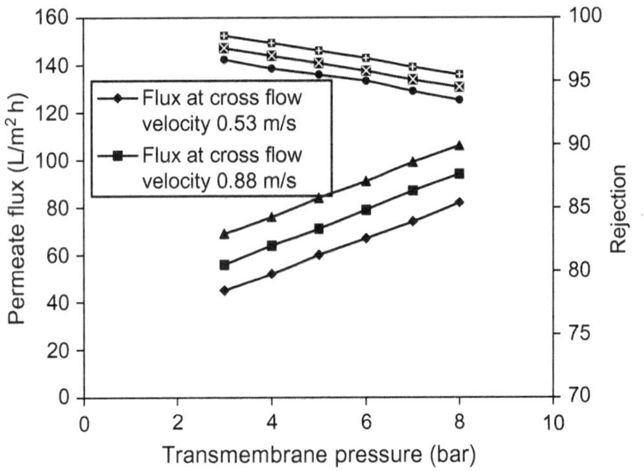

Figure 6.6.9 The effect of TMP on permeate flux and oil rejection for PES-5 membrane at different CFVs for 200 mg/L.

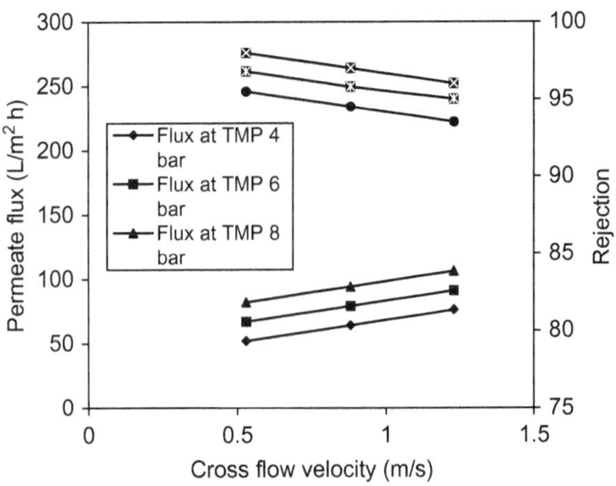

Figure 6.6.10 The effect of CFV on permeate flux and oil rejection for PES-5 for 200 mg/L oil concentration in feed.

pure water shows a positive correlation while oil rejection shows a negative correlation with CFV indicating the existence of an optimum CFV that can be found out through a pilot run. A balance has to be struck between fixing the CFV and TMP as can be seen in Figs. 6.6.9 and 6.6.1. Compared to MF performance, rejection or separation efficiency is much higher in UF membranes. Thus it can be concluded that UF membranes if used in flat-sheet, cross-flow module following a same type MF module, high degree of separation of oil (more than 96%) from water can be achieved and the module can be operated at optimum TMP as well as CFV at set flux mode with reasonably high flux.

REFERENCES

[1] Diya'uddeen BH, Ashri WM, Daud W, Aziz AR. Treatment technologies for petroleum refinery effluent. A review. Process Saf Environ Prot 2011;89:95–105.
[2] Chakrabarty B, Ghoshal AK, Purkait MK. Ultrafiltration of stable oil-in-water emulsion by polysulfone membrane. J Memb Sci 2008;325:427–37.
[3] Um MJ, Yoon SH, Lee CH, Chung KY, Kim KY. Flux enhancement with gas injection in cross-flow ultrafiltration of oily wastewater. Water Res 2001;35:4095–101.
[4] El-Kayar A, Hussein M, Zatout AA, Hosny AY, Amer AA. Removal of oil from stable oil-water emulsion by induced air floatation technique. Sep Technol 1993;3:25–31.
[5] Hosny AY. Separating oil from oil-water emulsions by electroflotation technique. Sep Technol 1996;6:9–17.
[6] Ratlege C. Mini review compilation biodegradation and biotransformations of oils and fats. J Chem Technol Biotechnol 1992;55:397–414.
[7] Refinery Operating Manual, Indian Oil Corporation Limited, Haldia, West Bengal, India.
[8] Abadi SRH, Reza MS, Hemati M, Rekabdar F, Mohammadi T. Ceramic membrane performance in microfiltration of oily wastewater. Desalination 2011;265:222–8.
[9] Peng H, Tremblay AY. The selective removal of oil from wastewaters while minimizing concentrate production using a membrane cascade. Desalination 2008;229:318–30.
[10] Sun D, Duan Xi, Li W, Zhou D. Demulsification of water-in-oil emulsion by using porous glass membrane. J Memb Sci 1998;146.

SUBCHAPTER 6.7

Treatment Technology for Textile Wastewater

6.7.1 INTRODUCTION

Textile industries constitute about 8% of manufacturing goods around the world and consume a huge amount of freshwater, most of which is hazardous wastewater. To cater to the enormous water requirements of these industries, manufacturing plants are often situated on the banks of rivers, which are lifelines in many countries. The generated hazardous waste often escapes into surface water bodies such as rivers or lakes. Not only the liquid and solid waste from the textile plants but also the sludge from the treatment units constitute harmful waste for the biosphere. Textile processing involves dying and finishing steps where the input of a wide range of organic chemicals of complex structure and dyestuffs are necessary. The major pollutants in textile wastewater include very high levels of suspended solids, nitrogen, heavy metals, dyes, COD, color, acidity, and other soluble substances. Textile wastewater is rich in salts and organic compounds such as dyes, pigments, colorants, surfactants, and other substances used to make clothes resistant to physical, chemical, and biological agents. Sludge is generally incinerated or dried by densification and combustion for generation of power. The cotton textile industry residue is generally comprised of the microfibers lost in the industrial processes of spinning and weaving [1–3].

6.7.2 DYE CLASSIFICATION

Dyes are used to impart different colors on textile products and as such impart color to effluent. Chemically, the dyes are organic compunds that adversely affect the environment. Most of these dyes are generally water soluble. However, during the dyeing process in the textile industry, the dyes get hydrolyzed. As a result, about 10%–15% of the initial amount of dye used remains unused and enters the wastewater [4]. Even at low concentration, many dyes are visible in water. In general, textile wastewater contains dye in the range of 10–200 mg/L. In textile finishing, the major dyes used are acid, azoic, basic, direct, disperse, mordant, reactive, solvent, sulfur, and vat. Azo dyes, as the most commonly used dyes, are stable to light, washing, and most chemical oxidants.

Dyes are classified according to color into 25 structural classes and based on parameters known as the "color index" to classify them into those categories, among which the most notable ones are AZO dyes, anthraquinone dyes, and phthalocyanines.

6.7.2.1 AZO Dyes

The largest class of dyes is AZO dyes, which comprise more than 66% of all colorants. The characteristic feature of this group is the presence of one or more AZO groups.

$-N{=}N-$ along with hydroxyl groups and amine and substituted amine groups also known as auxochromes.

The aromatic azo compunds are prepared from aromatic amines through the formation of aromatic diazonium salts. When an aromatic amine is treated with nitrous acid, a diazonium salt is the product of the reaction. Nitrous acid, being unstable, is produced in situ by adding dilute hydrochloric acid to a cool solution of sodium nitrite at a temperature of 278K. In the following example, a solution of benzenediazonium chloride has been prepared from phenylamine (aniline), the simplest aromatic amine:

$$NaNO_2(aq) + HCl(aq) \longrightarrow HNO_2(aq) + NaCl(aq)$$

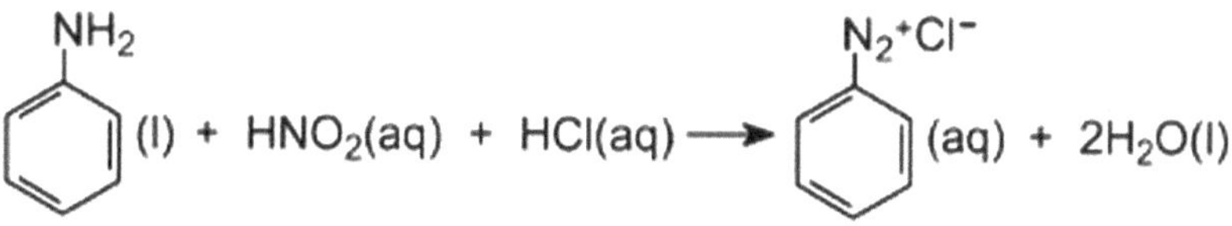

aniline *benzenediazonium chloride*

A solution containing aromatic compounds with hydroxyl or amine group is added to the above solution. The product obtained from this reaction is an AZO compound with a characteristic color. Azobenzene (Fig. 6.7.1) is the chromophore of these AZO dyes.

6.7.2.2 Acid Dyes

Acid dyes are generally used on cotton, wool, or linen as a mordant. The synthetic fiber nylon is frequently dyed with acid dyes when high wash fastness is required. The acid dyes are usually water soluble because of the sulfo or carboxy group present in them. Fig. 6.7.2 shows the molecular structure of an acid dye.

6.7.2.3 Direct (Substantive) Dyes

Direct dyes can be used to dye cellulose fibers without application of mordants. Wool, silk, nylon, cotton, rayon, etc., can be dyed with direct dyes.

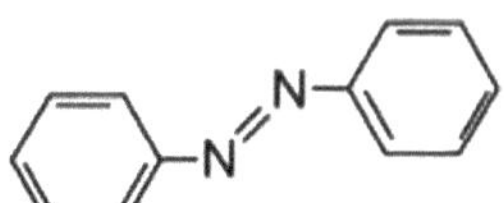

azobenzene

Figure 6.7.1 Molecular structure of azobenzene.

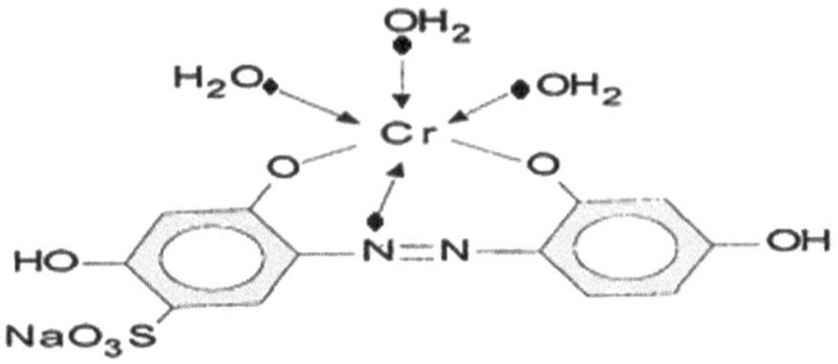

Figure 6.7.2 Molecular structure of a acid dye (disulfonated/dicarboxyiated).

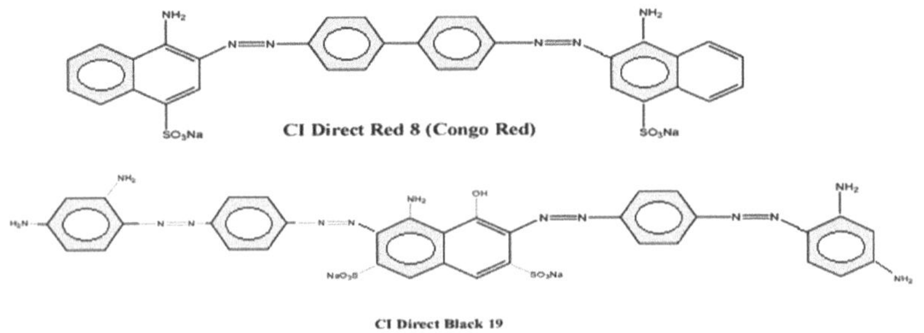

Figure 6.7.3 Molecular structure of two different types of direct dyes.

These dyes do not have bright appearance and have poor fastness to washing, but are fairly fast to light. The dyes in this class are defined as anionic dyes for cellulosic fibers, which are generally applied from an aqueous dye bath containing an electrolyte, either sodium chloride (NaCl) or sodium sulfate (Na_2SO_4), as shown in Fig. 6.7.3.

6.7.2.4 Vat Dyes

Vat dyes cannot be used on fibers directly as they are insoluble in water. In order to make them soluble, they need to be treated with alkaline solution after which they regain affinity toward textile fibers. However, further exposure to air leads to oxidation and thus the dye regains its insolubility. Indigo is an example of a vat dye. Vat dyes are fastest to cotton, linen, and rayon. With other fibers like wool, nylon, and polyester, they are used along with a mordant. Almost any dye can be used alongside a vat dye.

6.7.2.5 Reactive Dyes

Reactive dyes form a new chemical compound when they come into contact with a fiber molecule. Reactive dyes are applied either from a

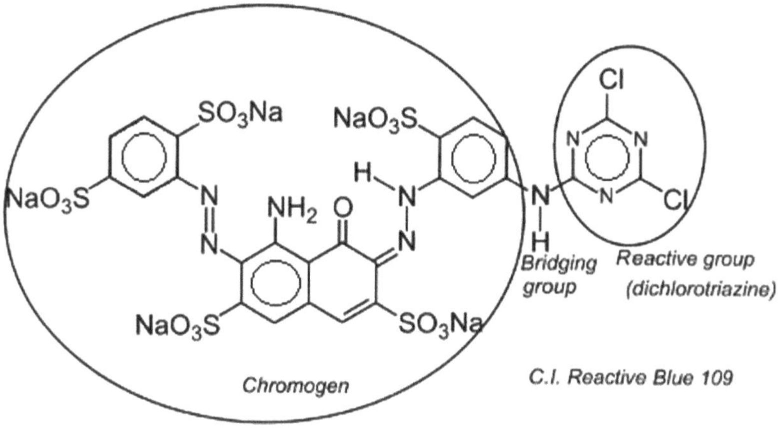

Figure 6.7.4 Molecular structures of a reactive dye.

solution with high pH or from neutral solutions that are later alkalized through a separate process. Sometimes different shades are brought out by applying heat to the dyed textile. Originally, reactive dyes were applied to cellulosic fibers but today other fibers are also dyed using reactive dyes. About 95% of reactive dyes are azo dyes covering an entire range of colors. Blues and greens are also provided by anthraquinone and phthalocyanine structures (Fig. 6.7.4; Table 6.7.1).

6.7.3 TOXICITY DUE TO DYE PRESENT IN GROUNDWATER

One of the most widely used type of dyes are AZO dyes. The complex structures of aminoazo benzene dyes and their various derivatives may lead to mutagenesis, which is a major cause of cancer an intoxication to mammals and alarmingly human beings. Erythrosine, a xanthene dye, is extremely hazardous and is allergic, carcinogenic, DNA damaging, neurotoxic, and xenoestrogenic to a variety of animals including humans. Disperse Orange 1, Disperse Blue 291, and Malachite Green cause genotoxic and mutagenic effects and increase micronuclei in human hepatoma-derived HepG2 cells and significantly decrease in cell viability, total protein content, and colony-forming ability. Reactive dyes are famous for being the most efficient colors but a survey of workers in dye manufacturing industry reveals that these dyes are responsible for increased bladder cancer and "Ardystil syndrome" [6–11].

Table 6.7.1 The average ranges of key chemicals and dyes used in a typical synthetic textile mill [5]

Chemicals used	Average Quantity Range (ton/year)
Acetic acid	20–30
Ammonium sulfate	10–15
P V Acetate	10–15
Wetting Agent	1–2
Caustic Soda	70–80
Organic Solvent	2–4
Organic Resin	60–70
Formic Acid	12–16
Soap	1.5–4.5
Hydrosulfites	75–80
Hydrochloric acid	3–5
Hydrogen peroxide	12–16
Leveling & Dispersing Agent	6–8
Solvent 1425	3–5
Oxalic acid	5–8
Polyethylene Emulsion	14–16
Sulfuric Acid	8–10
Disperse Dyes (polyester)	18–20
Vat Dyes (Viscose)	10–12
Sulfur Dyes	3–5
Reactive Dyes	1–1.5

Modified from Handa BK. Treatment and recycle of wastewater in industry. National Environmental Engineering Research Institute, Nagpur, 1991.

6.7.4 PROCESSING OF RAW FIBERS INTO FINISHED APPAREL AND NONAPPAREL GOODS

The processing of raw fibers in the textile industry is a complex process and is divided into a number of stages including slashing, desizing, caustic kiering, bleaching, scouring, mercierizing, and finishing.

6.7.4.1 Slashing

A sizing agent is used that brings down the chances of yarn breaking during weaving, which is a dry process step and doesn't generate a substantial amount of wastewater. The sizing agent choice varies from one plant to another but generally carboxymethyl cellulose and polyvinyl alcohol are used. The sizing agent is applied on the surface of the cloth. The woven cloth is passed through an arrangement in which a burning flame burns off the minute jagged ends of cotton on the surface of the product and then once dipped into water quickly prevent it from catching fire. The loose cotton dust produced is either burned as fuel or marketed as filling material.

6.7.4.2 Desizing

Mineral acid or enzyme-based compounds are used to remove the sizing agent layer from the surface of the cloth. The cloth is washed with water after desizing is done and is a potential source of wastewater that can be reduced by countercurrent washing. Bleaching, scouring and mercerizing are important fabric processing steps using huge water and chemicals for improvement of whiteness of fibers, removal of wax material and for imparting lustrous appearance respectively.

6.7.4.3 Caustic Kiering

Scouring is done in tall vessels called Kier where the desized products are stacked. The contents of the kier are treated with caustic, i.e., sodium hydroxide, sodium carbonate, sodium silicate, low BOD detergent, and heated to a high-temperature by injecting steam into the Kier. This operation removes the waxes and gums from the surface of the cloth. The product is then washed with water, which is brown in color, has a reasonably high pH, and a frothy appearance. Needless to say, it has a substantial contribution to textile wastewater.

6.7.4.4 Bleaching

The kiered cloth is treated with a bleaching agent like sodium hypochlorite or hydrogen peroxide to remove the pale yellow tinge. After bleaching, the products are washed with water.

6.7.4.5 Scouring

The bleached cloth is acid treated to ensure complete neutralization of any leftover alkali from the surface of the cloth.

6.7.4.6 Mercerizing

This is a finishing step that adds luster to the cloth and has a higher affinity for dye. The cloth is treated with caustic soda solution in a series of steps and is washed with jets of water in the final step from the water from is sent to the caustic recovery section.

6.7.4.7 Finishing

The mercerized cloth is treated with acid to remove any leftover alkali and is then sent for dyeing and printing. There are various finishing processes like softening, cross-linking, waterproofing, and wastewater from all these processes contributes to textile wastewater production.

6.7.5 COMPOSITION OF TEXTILE WASTEWATER

Textile wastewater has a high pH value, high concentration of suspended solids, chlorides, nitrates, metals like manganese, sodium, lead, copper, chromium, iron, and high BOD and COD value (Table 6.7.2). The water also has a dark-brown color. The concentration of different contaminant species varies with source of wastewater. Different stages of operation contribute wastewater of various composition. Table 6.7.3 lists the average composition of wastewater collected from various sectors in a textile manufacturing plant. The difference in composition is due to the variation in processes and to the type of fabric produced and machinery in use. The textile sector requires a high demand of water for its various sectors, and the discharge from these sectors causes environmental damage due to the species of contaminants it carries along with it. The most notable environmental impact is water consumption and wastewater discharge (115–175 kg of COD/ton of finished product, a wide range of organic chemicals, color, salinity, and low biodegradability) (Table 6.7.4).

Table 6.7.2 Composition and quantity of textile wastewater [12]

Wastewater parameters	Unit	Quantity based on literature
Dye concentration	mg/L	700
Chloride	mg/L	15,867
Sulfate	mg/L	1400
Total Nitrogen	mg/L	23
COD	mg/L	1781
BOD	mg/L	363
NH_4	mg/L	17
NO_3	mg/L	2
PO_4	mg/L	17
Ca	mg/L	43
Mg	mg/L	4
Na	mg/L	2900
Fe	mg/L	1.2
pH	–	10

Table 6.7.3 Textile wastewater features for different sectors of a synthetic textile plant [13]

Sector	pH	NO_3^- (mg/L)	TSS (mg/L)	Cl^- (mg/L)	COD (mg/L)	H_2S (mg/L)	NH_4^+ (mg/L)	NO_2^- (mg/L)	BOD (mg/L)	Residue (mg/L)
Burning sector	5.51	2.5	432.8	121.6	3491	1.83	5.67	1.48	800	4353.6
Bleaching sector	11.6	5.54	288.5	516	2689	5.44	8.0	1.45	184	6591
Mercerizi-ng sector	10.8	9.4	105.2	119.5	2788	1.31	8.53	2.77	300	3877
Dyeing sector	8.06	6.06	499.4	213.2	1907	1.62	14.34	0.91	170	2016
Dyeing gauge sector	8.36	4.8	250	150	3606	6.05	18.6	0.2	320	2178.3
Dressing sector	7.15	2.6	273	60	1097	8.6	9.93	1.6	120	358.6

Table 6.7.4 Stepwise removal efficiencies of different pollutants

Steps	BOD (%)	COD (%)	TSS (%)	TDS (%)	NH_3-N (%)
Primary tube settler	34	38	86	58	68
Aeration tank	65	46	21	2.3	44
Secondary clarifier	37	61	59	4.5	43
Total removal	85	87	95		

6.7.6 CONVENTIONAL TECHNOLOGY USED IN A TEXTILE WASTEWATER-TREATMENT PLANT

The ETPs of different textile manufacturing plants vary from one plant to another based on th edesign principles involved, desirability, and affordability of treatment scale, and to some extent the major pollutant in the waste stream. The major and common pollutants are normally represented through lumped parameters BOD, COD, TDS, and TSS, with the most critical pollutant being the dye species. In this small and medium industrial sector, tertiary treatment is often bypassed so the ETPs can have provisions to treat wastewater to a secondary treatment level. However, based on environmental regulations of the area of operation, tertiary treatment plants are also operated often producing reusable water. In some cases, a tertiary treatment scheme is installed for thorough wastewater treatment.

6.7.6.1 Pretreatment

The textile wastewater produced from different steps is routed to and stored in a effluent sump tank that serves as an equalization and neutralization tank forming a composite wastewater pool. Composite wastewater from this tank is then passed onto primary, secondary, and tertiary treatment units in sequential order [14].

6.7.6.2 Primary Treatment Process

The primary treatment process (Fig. 6.7.5) consists of several processes such as coagulation, flash mixing, flocculation, and sedimentation, as discussed in the next section.

Coagulation

The effluent after neutralization and equalization is then sent to a reactor where chemical coagulants (such as such as aluminium sulfate, aluminium chloride, and sodium aluminate; and iron-based, such as ferric sulfate,

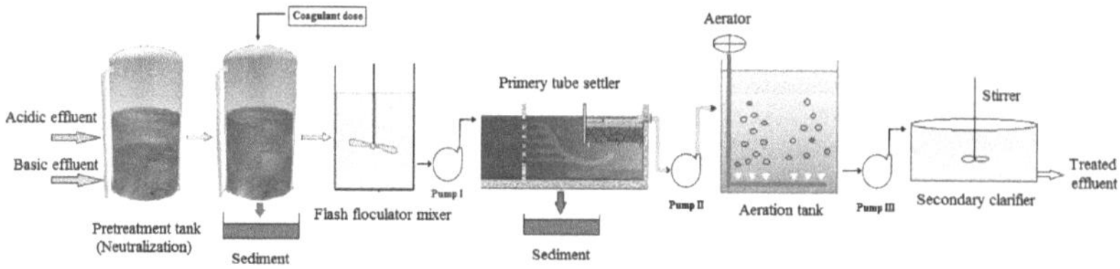

Figure 6.7.5 Schematic of a conventional textile effluent treatment plant [14]. Source: *Modified from Patel S., Rajor A., Jain B.P., Patel P. Performance evaluation of effluent treatment plant of textile wet processing industry: a case study of narol textile cluster, Ahmedabad, Gujarat, Int J Eng Sci Innovative Technol (IJESIT) 2013;2:4.*

ferrous sulfate, ferric chloride, and ferric chloride sulfate) are added to water to destabilize colloidal and finely divided materials and to cause them to begin aggregating. Sometimes other chemicals such as coagulants in the wastewater-treatment process like magnesium carbonate, hydrated lime, etc. The choice of coagulants and dosage depends on the composition of the wastewater. Aluminium and iron coagulants work by forming highly adsorptive multicharged polynuclear complexes. The pH of the system can be manipulated to control the characteristics of the complexes and their effectiveness.

Flash Mixing

The effluent now is sent to a flash flocculator mixer where the coagulants are well mixed with the wastewater with the help of agitators. This step is very important to create the conditions for efficient, effective water treatment. The flash mixing time should be maintained between 30 and 60 seconds. Below 30 seconds the chemicals will not be properly distribute. Again, for a longer period (after 60 seconds), the mixer blades tend to chop or shear the aggregating material back into small particles. To stop the repulsion of like-charged particles and allow the particles to begin bonding and forming larger clumps flash mixing is required after coagulation to neutralize the electrical charge of fine particles. After flash mixing, slow and gentle mixing flocculation begins that brings the fine particles produced during the coagulation step into contact with each other (approximately 30–45 minutes time is needed for the process).

Primary Tube Settler

To improve the clarifiers and sediment basins an existing treatment plant tube settler can be used, and can also be used to reduce the solid loading on downstream filters. Once the mixing is complete, it is sent to a primary tube settler where the essentially large flocks settle. Once the sludge settles, the supernant liquid free from most of the suspended and dissolved solids present in the effluent is sent to the aeration tank.

6.7.6.3 Secondary Treatment Process

The secondary treatment process consists of two different process, i.e., aeration and secondary clarifier, described in the following.

Aeration

For an industrial wastewater treatment, aeration is part of the stage known as the secondary treatment process where aeration provides oxygen to bacteria for treating and stabilizing the wastewater biodegradation process. The supplied oxygen is utilized by bacteria in the wastewater to break down the organic matter containing carbon to form carbon dioxide and water. Without the presence of sufficient oxygen, bacteria is not able to biodegrade the incoming organic matter in a reasonable amount of time. In this study, effluent from the settling tank is brought into the aeration tank where biological treatment of the wastewater commences. Air, containing oxygen as a strong oxidizer, even stronger then chlorine. In an ASP, sending streams of air into the system assists in microbial growth. The microbes eat away the organic waste present in the system forming flocks. The aeration process reduces the BOD and COD of the wastewater and the microbes digest the organic species present in the effluent.

Secondary Clarifier

A secondary clarifier is used to separate the microorganisms so that just clean water is left. Some of the solids collected in the secondary clarifier (return activated sludge) are sent back to the aeration tank to treat more wastewater and the excess (waste activated sludge) is pumped to another location in the plant for further treatment. The clean water that flows out the top of the clarifier is sent along for disinfection. In these studies, the effluent is pumped into a secondary classifier where the flocks settle and form sludge and the supernant liquid is free from most of the pollutant species initially present in the effluent.

Biological Processes

Anaerobic Oxidation

Anaerobic oxidation is carried out with the help of anaerobic activated sludge in the absence of oxygen. With the help of anaerobic sludge treatment, organic components like Kjehdal nitrogen content and TOC can be reduced alongside color, BOD, and COD of the wastewater. Dye reduction has been found to be as high as 88% when coupled with aerobic sequential reactor, in which 75% removal of color was observed, and 60% occurred in the anaerobic step, 46% removal of TOC was observed with 18% removal occurring in the anaerobic step, and 49% of COD reduction was observed where 22% occurred in the anaerobic step. Another study aiming to reduce AZO dyes like Orange II and Reactive Black 3HN from textile wastewater indicated >99% removal of color and COD removal up to 92% in orange dye and 94% in black dye when the reactors where operated for 58 days.

Aerobic Oxidation

Aerobic oxidation is carried out using activated sludge in the presence of air. Many studies have been performed using aerobic oxidation on textile wastewater. Activated sludge cultured from domestic sewage has been used to treat pretreated textile wastewater and reduction of BOD and COD values were observed to be the highest when the retention time was increased. For a 10 day retention time, BOD reduction of 88.2% and COD reduction of 90% was observed. In addition, 79% removal of TSS and 49% removal of TDS was also observed (Iqbal et al., 2006). However, when aerobic and anaerobic oxidation was employed in SBR, the result was more optimized. When sequential anaerobic and aerobic MBRs were used to treat textile wastewater, 90% removal of COD and 99% removal of dye was obtained. Furthermore, the results remained unaffected when increasing salinity of the wastewater.

Physicochemical Processes

Electrochemical Oxidation

Elecctrochemical oxidation involves removal of certain species from wastewater samples by oxidizing them using an electrochemical cell. The chemicals produced possess high oxidizing potential and hence the effect of COD is substantially reduced.

Ozonation

Ozone, as a very strong oxidant, can be used to remove dyes from textile wastewater. Decolorization of the wastewater sample indicates removal of the dye from the sample. The process is very fast, and 98% decolorization can take place within a few minutes of operation. At an ozone flow rate of 9.6 mg/minute at the original pH of the wastewater, 60% reduction in acute toxicity, 90% removal of color, and 50% removal of total COD are observed.

Adsorption

Adsorption uses a substance called adsorbent to introduce active surfaces into the medium that can accommodate certain gases, liquids, or dissolved solids. The adsorption mechanism has been explained using various models such as Langmuir Isotherm, Freundlich Isotherm, BET theory, Kisliuk, and linear. The adsorbent plays a major role in the process, and provides enormous surface area of the active site for the process to take place. Different adsorbents like zeolite, bentonite, activated clay, activated carbon, polymeric isomers, chitosan, montmorrillonite, and fly ash have been used to study the removal of contaminants from textile wastewater. Activated carbon used as an adsorbent is perhaps the most common approach for removal of COD, BOD, and color from textile wastewater. In an investigation where adsorbent behavior of activated charcoal is explained using Langmuir and Freundlich isotherm, the maximum removal of COD, BOD, and color was found to be 87.6%, 81%, and 90%, respectively, for an adsorbent dosage of 11 g/L.

Comparing three different processes including physicochemical, advanced oxidation, and biological techniques (by applying six different techniques) for textile wastewater treatment, only 80% of the influent COD, TSS, and color were removed by a single biological or physicochemical treatment technique where conventional activated sludge treatment followed by effluent polishing with sand filtration and activated carbon adsorption columns (Fig. 6.7.6) was the most promising, removing 81.6%, 88.5%, and 94.5% COD, TSS, and color, respectively.

Membrane Filtration

Disposal of concentrated impurities after membrane filtration continues to be a problem for the environment. As reported MD of NF and RO concentrates with further incineration of MD concentrates is promising. It was reported that by implementing such process, the benefit/cost ratio of 3.58 can be achieved. MF technology is generally employed for removal of suspended particles and colloidal dyes but the unconsumed excess chemicals, dissolved organic pollutants, and other soluble contaminants passing through the membrane to the permeate stream. An asymmetric tubular

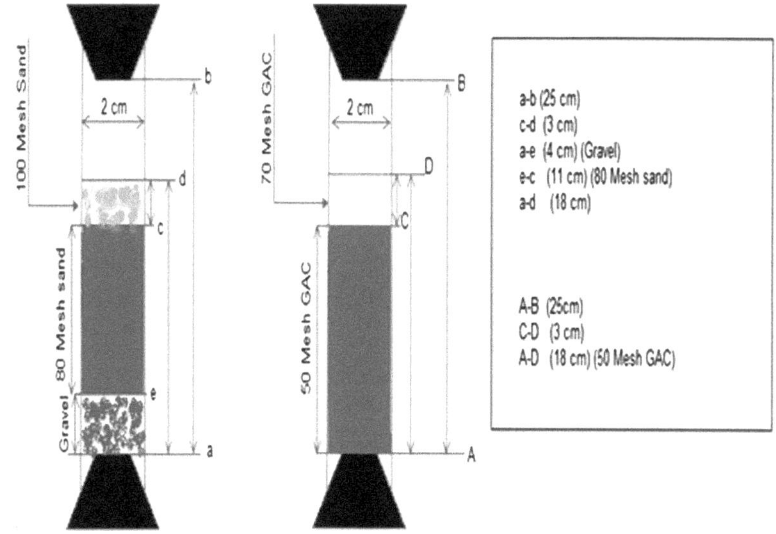

Figure 6.7.6 Combination of sand filtration and activated carbon-based adsorption column.

carbon MF membrane exploiting mineral coal powder and thermosetting resin for the treatment of textile industry wastewater has been developed in which around 50% removal of COD, 30% removal of salinity, and almost complete removal of color and turbidity have been achieved. On amalgamation of two cationic (poly-ethyleneimine, chitosan) and an anionic (poly (acrylic acid)-PAA) polyelectrolyte through layer-by-layer assembly for the development of polyamide MF membranes high rejection was achieved on COD (96%) dyes like methylene blue (80%) and coomassie brilliant blue (87.1%).

Commercial ceramic membranes (molecular weight cut-offs of 30–150 kDa) with high mechanical, chemical, and thermal stability could be viable alternatives to polymeric UF membranes in terms of high flux outcome and rejection to impurities (turbidity by 99% and color by 98%) present in wastewater [15]. A schematic of the treatment plant is shown in Fig. 6.7.7 [15].

A membrane-integrated system consisting of NF and RO membranes in spiral-wound modules is shown in Fig. 6.7.8, in which 99% methyl orange rejection was achieved, although the TDS removal rates, sodium retention, and overall conductivity profiles are mostly identical [9].

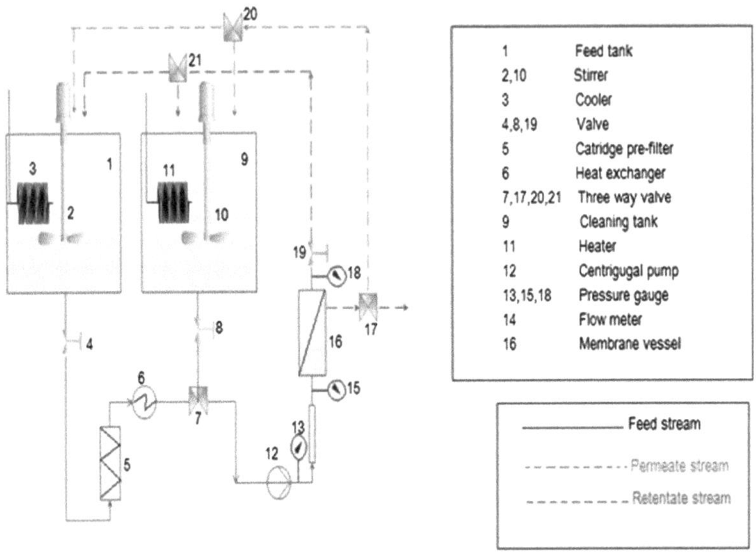

Figure 6.7.7 A UF-based pilot plant scheme for the treatment of textile wastewater [15]. Source:*From Barredo-Damas S, Alcaina-Miranda MI, Iborra-Clar MI, Mendoza-Roca JA. Application of tubular ceramic ultrafiltration membranes for the treatment of integrated textile wastewaters. Chem Eng J 2012;192:211–218.*

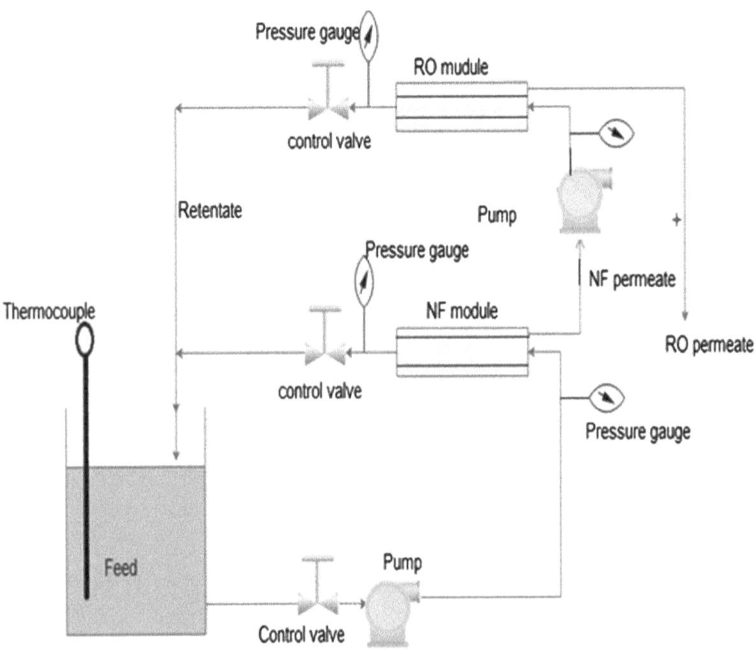

Figure 6.7.8 NF–RO-based textile wastewater-treatment plant.

6.7.7 ADVANCED TECHNOLOGY FOR TEXTILE WASTEWATER TREATMENT

A combined Fenton oxidation and MBR process as advanced treatment of effluent from an integrated dyeing wastewater-treatment plant reduces TOC and color by 40% and 69%, respectively, after 35 minutes of reaction under optimum Fenton oxidation conditions (initial pH 5, H_2O_2 dosage 17 mmol/L, and Fe^{2+} 1.7 mmol/L) [16]. After further purification by the MBR process, COD and TOC removal efficiency reaches 86% at different HRTs. The final effluent of MBR meets the reuse criteria of urban recycling water standards.

The total system consists of three parts: a Fenton oxidation unit where a CSTR tank is used for the oxidation reaction, is a neutralization tank where neutralization of the feed takes place with adjustment of pH, and the MBR. Fig. 6.7.9 depicts the system. Under optimum operating conditions as determined through a "jar test," Fenton oxidation should be carried out. The supernatant is collected into a store tank for further purification by a continuous MBR treatment system. After acclimation, different parameters and running characteristics are evaluated during the long-term operation.

A MBR with incorporation of UF membrane (UP150) from MicrodynNadir for the treatment of textile wastewater can reduce COD by 95% at 50 LMH flux (Fig. 6.7.9). The membrane module is submerged in the activated sludge tank [17]. The detailed describtion of this hybrid process has been shown below [17].

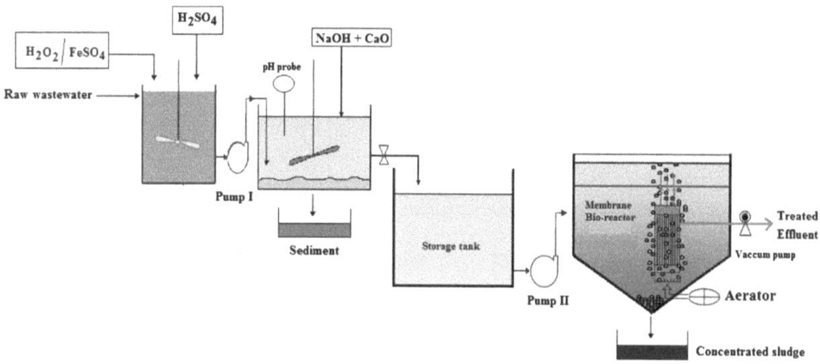

Figure 6.7.9 Fenton's oxidation-based membrane bioreactor system [16].

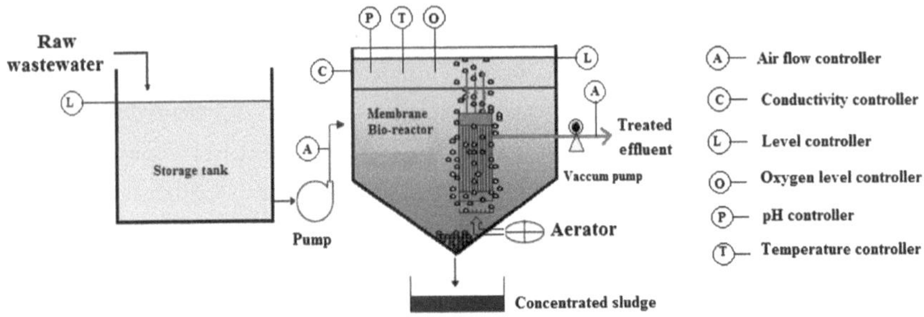

Figure 6.7.10 Schematic of membrane bioreactor [17].

Fig. 6.7.10 shows the MBR. The hydraulic volume is 47 liters after being submerged with the membrane modules. The membrane housing is fitted with two three envelope flat UF membranes. The enveloped membrane stacks are connected to a feed and suction pump and hence can be operated independently. Moreover, the membrane module as well as feed and suction pump of the plant consist of an air compressor and a variety of sensors such as level sensor, differential pressure sensor flow sensor, pH sensor, temperature sensor, conductivity sensor, DO sensor, and air-flow meter.

REFERENCES

[1] Vaidya AK. Globalization: International blocs organizations, Other issues, ABC-CLIO, 2006.
[2] Babu BR, Parande AK, Raghu S, Kumar TP. Cotton textile processing: waste generation and effluent treatment. J Cotton Sci 2000;11:141–53.
[3] US Environmental Protection Agency, Profile of Textile Industry. Washington, USA, 1997.
[4] AI-Degs YS, Khraishen MAM, Allen SJ, Ahmad MN. Effect of carbon surface chemistry on the removal of reactive dyes from textile effluent. Water Res 2000;34:927–35.
[5] Handa BK. Treatment and recycle of wastewater in industry. National Environmental Engineering Re-search Institute, Nagpur, 1991.
[6] Garg A, Bhat KL, Bock CW. Mutagenicity of aminoazobenzene dyes and related structures: a QSAR/QPAR investigation. Dyes Pigm 2002;55:35–52.
[7] Mittal A, Mittal J, Kurup L. Adsorption isotherms, kinetics and column operations for the removal of hazardous dye, Tartrazine from aqueous solutions using waste materials-Bottom.
[8] Tsuboy MS, Anjeli JPF, Mantovani MS, Knasmiiller S, Umbuzeiro GA, Ribeiro LR. Genotoxic, mutagenic and cytotoxic effects of the commercial dye CI Disperse Blue 291 in the human hepatic cell line HepG2. Toxicol In vitro 2007;21:1650–5.
[9] Stammati A, Nebbia C, Angelis ID, Albo AG, Carletti M, Rebecchi C, et al. Effects of malachite green (MG) and its major metabolite, leucomalachite green (LMG), in two human cell lines. Toxicol In Vitro 2005;19:853–8.

[10] Rehn L. Bladder tumours in fuchsin workers. Arch Klin Chir 1895;50:588–600.
[11] Juang Y, Nurhayati E, Huang C, Pan JR, Huang S. A hybrid electrochemical advanced oxidation/microfiltration system using BDD/Ti anode for acid yellow 36 dye.
[12] Jonstrup M, Kumar N, Murto M, Mattiasson B. Sequential anaerobic-aerobic treatment of azo dyes: decolorisation and amine degradability. Desalination 2011;280(1–3): 339–46.
[13] Savin II, Butnaru R. Wastewater characteristics in textile finishing mills. Environ Eng Manage J 2008;7(6):859–64 wastewater treatment. Sep. Purif. Technol., 2013;120:289–295
[14] Patel S, Rajor A, Jain BP, Patel P. Performance evaluation of effluent treatment plant of textile wet processing industry: a case study of narol textile cluster, Ahmedabad, Gujarat. Int J Eng Sci Innovative Technol (IJESIT) 2013;2:4.
[15] Barredo-Damas S, Alcaina-Miranda MI, Iborra-Clar MI, Mendoza-Roca JA. Application of tubular ceramic ultrafiltration membranes for the treatment of integrated textile wastewaters. Chem Eng J 2012;192:211–18.
[16] Feng F, Xu Z, Li X, You W, Zhen Y. Advanced treatment of dyeing wastewater toward reuse by the combined Fenton oxidation and membrane bioreactor process. J Environ Sci 2010;22(11):1657–65.
[17] Luong TV, Schmidt S, Deowan SA, Hoinkis J, Figoli A, Galiano F. Membrane bioreactor and promising application for textile industry in Vietnam. Procedia CIRP 2016;40:419–24.

FURTHER READING

Nataraj SK, Hosamani KM, Aminabhavi TM. Nanofiltration and reverse osmosis thin-film composite membrane module for the removal of dye and salts from the simulated mixtures. Desalination 2009;249:12–17.

CHAPTER 7

Nanotechnology in Water Treatment

7.1 INTRODUCTION

Materials whether ceramic, polymeric, or metallic, acquire unique properties such as electrical, magnetic, or surface plasmon resonance at nanoscale because of the location of the associated atoms and molecules at the surface. Such nanomaterials become extremely reactive due to their coordinative unsaturation. However, handling and arranging such nanoparticles is very challenging because of the high possibility of them escaping to the environment. These nanomaterials and nanoparticles are considered hazardarous to life. Such exposure can be reduced by containing them through glass or polymer embedding. Glass embedding requires high temperature, and the resulting nanostructures are not easily tunable. Polymer embedding, on the other hand, makes handling nanostructures easier. In recent years, hundreds of nanomaterials and nanocomposite materials have been developed for water treatment technologies. This is due, in part, to the world's changing lifestyles and the intensive agricultural and industrial activities now taking place, resulting in pollutants that are often carcinogenic and bioaccumulative. These new-generation nanomaterials are expected to effectively degrade the so-called refractory water pollutants found today. The associated hazards likely to emanate from the release of highly reactive nanomaterials to the environment are also being investigated, but little research has been done on the possible hazards of exposure to nanomaterials. Nonetheless, nanotechnology as a whole has been explored for water treatment for quite some time. This technology may involve the use of nanomembranes (with nanoscale pores) or nanomaterials in different forms like solid adsorbent, ion-exchange material, photocatalytic material, disinfectant material, and nanocomposite membrane [1–3].

Advances in nanotechnology are resulting in the development of next-generation water supply systems that have the potential to overcome many of major difficulties in existing water treatment systems. However,

Industrial Water Treatment Process Technology.
DOI: http://dx.doi.org/10.1016/B978-0-12-810391-3.00007-2

scale-up confidence at this juncture still remains limited in the absence of adequate data on pilot- or industrial-scale operations.

Over the last two decades, several nanomaterials such as nanostructured metal oxides, mixed-metal oxides, nanometals, carbon nanotubes (CNTs), and graphene and graphene-based nanocomposite have been synthesized for application in water purification devices. These nanomaterials have been tested and used in the following nanotechnology-enabled modules:

1. Adsorption
2. Photocatalytic degradation
3. Disinfection
4. Membrane filtration

7.2 NANOMATERIALS AS ADSORBENT IN WATER TREATMENT

The major characteristics of nanomaterials that favor their use as adsorbent are their small size, high surface area-to-volume ratio, high pore volume, high reactivity, large number of active sites, and the possibility of regeneration on exhaustion. Nanoadsorbents act at a much faster rate than their conventional counterparts as often reflected in their fast binding of adsorbate. Nanoadsorbents can broadly be classified as metal- and metal oxide-based material and carbon-based material.

7.2.1 Metal- and Metal Oxide-Based Nanomaterials in Adsorption Module

A diverse class of metal-based nanomaterials comprised of nanoscale metal, metal oxide, mixed-metal oxide, and other metal-doped nanocomposite materials have been found to be effective in the separation of fluoride, arsenic, heavy metals, dyes, and other organic and inorganic pollutants from water in the last two decades. Due to their large surface area, good thermal and mechanical strength and porous structure, enhanced adsorption sites, good dispensability, regeneration ability, nontoxicity, and easy separation mode, these nanomaterials are considered good alternatives to conventional adsorbents. Nanometal oxides (NMOs) in many cases exhibit much faster kinetics and higher selective adsorption capability than conventional macro-sized adsorbents. Among these metal/metal oxide engineered nanomaterials, nanoscale zerovalent iron has been the most widely investigated and applied in water treatment (Table 7.1) [4–5].

Table 7.1 Nanomaterials used in water disinfection and photocatalytic degradation

Nanomaterials used	Water treatment process	Relevant properties of nanomaterials used
Nano TiO_2, Nano WO_3	Photocatalytic degradation	Photocatalytic activity in UV range, stability, low cost, abundance
Nano TiO_2	Solar disinfection (SODIS)	Antibacterial properties
rGO-ZnCdS	Photocatalytic degradation	Photocatalytic activity in UV−visible range
Nano Ag, Nano ZnO, Nano Ce_2O_4, Nano TiO_2	Disinfection of water	Strong antibacterial property, low human toxicity, easy usability

Some of the developed metal and metal oxide-based nanoparticles include alum-impregnated activated alumina, iron oxide, zirconium oxide and hydroxide, hydrous stannic oxide, manganese oxide, and copper oxide. A major hindrance in the large-scale use of NMOs is difficulty in separating them from aqueous medium, excessive pressure drop in the flow systems, and formation of agglomeration leading to loss of surface activity. To overcome these difficulties, NMOs are impregnated in large porous supports like activated carbon or synthetic polymer structure. Synthesis of graphene-based metal nanocomposite (Fe_3O_4/GO) is in this direction. Other important graphene-based nanocomposite adsorbents include G-Fe_3O_4 for Pb^{+2} removal, $MnFe_2O_4$-G for Pb and Cd removal, and $COFe_2O_4$-G and $NiFe_2O_4$-G for Pb and Cd removal. The use of magnetic nanoparticles is another option as such nanoparticles can be separated out easily from water by applying magnetic field. However, modeling and simulation of NMO-based plant operation is difficult as the performance of NMO-based adsorption columns varies significantly with operating conditions (batch or continuous operation), solution chemistry (pH, ion types, temperature), and size of the NMOs. Maintaining a particular size of the nanomaterial is also difficult and depends largely on the involved method of synthesis. However, such nanomaterials in small-scale operation have been found to be efficient in the separation of heavy metals like Pb, Hg, Cr, Cd, Ni, Cu, metalloids like fluoride, arsenic, and radio-nuclides from contaminated water [5−10].

In the removal of various organic pollutants from water through adsorption, the binary and tertiary mixed-metal oxides with modified surface morphology have proven to be effective with organic molecules have also been widely investigated in the search for improved surface morphology and separation behavior. In many cases these mixed-metal oxides have also been effective in detoxification of various inorganic pollutants as well. Such metal and mixed-metal oxides are well illustrated by Fe (III)-Cr (III), Fe-Al materials, Zr-Mn composite material, Mg-doped ferrihydrite, HIACMO, Fe-Zr MMO, Al-Zr impregnated cellulose, CTAB-mediated Mg-doped nano Fe_2O_3, and cobalt-ferrite nanoparticles [11–19]. However, the toxicological effects of NMOs of nanomaterials must be screened thoroughly for possible health hazards prior to field use [20].

7.2.2 Metal-Based Nanoparticles in Multiple Water Treatment Functions

In addition to high adsorption capacity some nanomaterials synthesized from iron, nickel, cobalt oxides, and alloys acquire special ferromagnetic or superparamagnetic properties [5]. Magnetism of such nanomaterials is very much volume-dependent. With a decrease of particle size beyond a critical value of 40 nm, the magnet changes from multiple to single domain with higher magnetic susceptibility. It is observed that with further decrease in size, these magnetic particles turn superparamagnetic and respond to external magnetic field in a way that allows them to be easily separated and recovered by a low-gradient magnetic field. Such magnetic nanoparticles (nanomagnetite; MNPs) can be either used directly as adsorbents or can be used inside the shells made of inorganic components like silica, alumina, or organic molecules such as surfactants and polymers [19,21–26]. The shell provides the desired function, whereas the magnetic core realizes the magnetic field for separation. Multiple functions of magnetic nanomaterials are illustrated in Fig. 7.1.

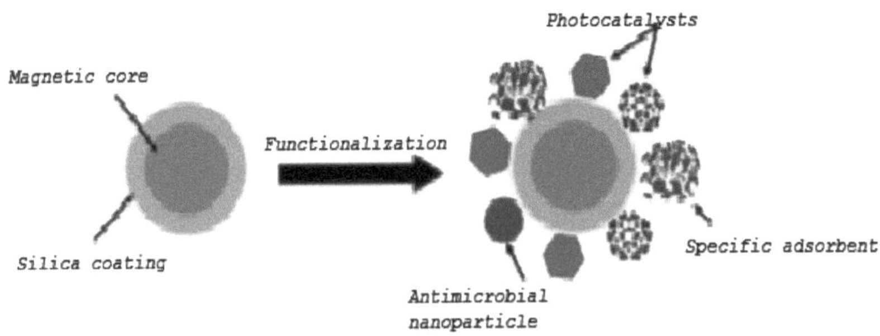

Figure 7.1 Magnetic nanoparticles in multiple water treatment functions [5].

MNPs coated with poly glutamic acid (PGA) are found to be more stable than MNPs without such coating (examined through leaching test) in adsorption of dye. Dye adsorption (78 mg/g) is favored at high pH.

7.2.3 Synthesis Routes of Metal-Based Nanoparticles

Metal oxides and mixed-metal oxide nanoparticles are mainly synthesized by coprecipitation under pH-controlled conditions. Such metal- and metal oxide-based nanoadsorbents are also synthesized through sol gel, citrate gel, ultrasonication as well as catalytic oxidation. Synthesis of magnetic nanoparticles at low cost can be done using low cost iron-ore tailings (IOT) [12,13,27].

7.2.4 Regeneration and Reuse of Nanoadsorbent Materials

Metal-based nanoadsorbent materials can easily be regenerated by changing the pH of the solution. These adsorbents can also be regenerated by treating them with an NaOH solution. After regeneration up to 80% of the original removal efficiency even after 3–5 regenerations can be recovered allowing reuse. However, after five regeneration cycles, adsorption capacity drops significantly [28,29].

7.2.4.1 Adsorption-Based Plant Operation Using NMO

Metal-based nanoadsorbents have been widely explored to remove various organic and inorganic pollutants from water and wastewater. Separation of fluoride and arsenic (III, V) has been done using such adsorbents in packed-bed columns. A moderate-to-high adsorption capacity (64–98 mg g^{-1}) has been achieved using various Zr- and Mg-based nanoadsorbents. Nanosized magnetite and TiO_2 have shown superior arsenic-adsorption performance compared to activated carbon. Metal-based nanoadsorbents cam be introduced into any existing treatment plant without revamping the system.

Though metal-based nanoadsorbents have not been used in large-scale, there are quite a few cases of very successful operation of ArsenX^{np} and ADSORBIA™-based commercial plants. ArsenX^{np} is a hybrid ion-exchange medium consisting of both iron oxide nanoparticles and polymers, whereas ADSORBIA™ is made of nanocrystalline TiO_2 in the form of beads ranging between 0.25 mm and 1.2 mm in diameter [12,14,30,31].

7.2.5 Zeolites as Nanoadsorbent

Natural zeolites are hydrated aluminosilicate frameworks with pores occupied by water, alkali, and alkaline earth-metal cations. Due to their unique 3D porous structure, these materials can acquire exceptional adsorption properties. Zeolite generally acts as a good cation exchanger as a result of its negatively charged surface. By virtue of their cation-exchange ability as well as their molecular sieving properties, natural zeolites can serve as very good adsorbent in separation and purification processes with high selectivity [32].

Several zeolite-modification methods are also available such as acid treatment, ion-exchange, and surfactant functionalization. Modified natural zeolites acquire higher adsorption capacity for organics and anions present in water.

7.2.5.1 Application in Adsorption/Ion-Exchange Module

Natural zeolites, such as clinoptilolite and chabazite, have been used successfully for the removal of heavy metals like Pb, Cd, Cu, Zn, Cr, Ni, Co, and Fe from contaminated water. Synthetic zeolites modified with nanomaterials have also been used in removal of fluoride and arsenic. Al^{3+}-pretreated low-silica synthetic zeolites, H-MFI-24 (H24) and H-MFI-90 (H90), stilbite zeolites modified with $FeCl_3$, are a few such examples with moderate adsorption capability [33–37].

7.2.6 Carbonaceous Nanomaterials for Adsorptive Removal of Water Pollutants

These materials are mainly comprised of carbon atoms. Carbon nanotubes, graphene, and graphene derivatives are some examples of carbon-based nanomaterials with exceptionally high surface area, highly tunable carbon backbone, and hydrophobicity make these materials suitable for adsorptive removal of cyanobacterial toxins, heavy metals, natural organic matter (NOMs), and dyes.

Carbonaceous materials find application as membrane materials due to their high surface area, antifouling property, high selectivity, permeability, high flux, porous structure, and antimicrobacterial property.

7.2.6.1 Carbon nanotubes

Many emerging micro/nanopollutants have creatde a challenge for water treatment plants. In recent years carbon nanotubes (CNTs) have emerged as an efficient water treatment material that can be used both as adsorbent

Figure 7.2 Single- and multiwalled carbon nanotubes [38].

and membrane filter. This is an allotrope of elemental carbon with cylindrical nanostructure belonging to the class of fullerene, and is comprised entirely of sp^{2-}bonded carbon atoms, similar to those of graphite. The huge number of double bonds present in the structure impart unique mechanical strength to this material and are called "nanotubes" because of their long, hollow structure, which is formed by one-atom-thick sheets of carbon called graphene. CNTs may be single-walled carbon nanotubes (SWCNTs) or multiwalled carbon nanotubes (MWCNTs) as shown in Fig. 7.2, which consist of single and multiple layers of graphene sheets, respectively. Due to their unique structure, CNTs have extraordinary thermal, electrical, and mechanical properties. CNT-based nanoadsorbents and membranes have shown excellent performance for water purification and desalination.

Well-defined cylindrical hollow structure, large surface area ($\approx$250 m^2/g), high aspect ratios, hydrophobic wall, and easily modified surfaces make CNTs potentially good adsorbents for removal of many toxic contaminants, micropollutants, and heavy metals like Hg^{+2}, Au^+, Pb^{2+}, Cu^{2+}, and Cd^{2+}. CNTs have also been successful in the removal of cyanobacterial toxins, NOM, dyes, and pathogens. Graphene-modified CNT hybrids are a new addition to the adsorption field for the removal of toxic dyes and heavy metals. Graphene-CNT hybrid aerogels such as graphene/MWCNT and graphene/c-MWCNT using either pristine (MWCNTs) or acid-treated (c-MWCNTs) multiwalled CNTs have shown excellent removal capacity of basic dyes (rhodamine B, methylene blue, and fuchsine) [39–45].

7.2.6.2 Separation Mechanisms in Adsorption by CNTs

The separation mechanism of CNTs is mainly based on electrostatic interaction as well as availability of adsorption sites. The fast adsorption kinetics in the context of adsorption of heavy metals such as Pb^{2+}, Cd^{2+},

Cu^{2+}, and Zn^{2+} on CNTs results in accessible adsorption sites and short intraparticle diffusion distance. Oxidized CNTs can provide better adsorption capacity due to the presence of functional groups like carboxyl, hydroxyl, and phenol groups, which act as extended adsorption sites for metal ions.

The extent of adsorption largely depends on the presence of functional groups on the surface and the nature of the sorbate. Functionalized CNTs with phenolic, carboxylic, and lactonic acid groups adsorb polar compounds efficiently where chemical interaction plays the main role in adsorption. Nonfunctionalzed CNTs, on the other hand, can adsorb nonpolar compounds like polycyclic aromatic compounds where physical interaction plays the dominant role.

The high adsorption of polar organic compounds on CNTs depends uon hydrophobic effect, $\pi-\pi$ interactions, hydrogen bonding, covalent bonding, and electrostatic interactions between contaminants and CNTs. The π-e^-rich CNT surface provides good electrostatic attraction, which facilitates the adsorption of positively charged organic chemicals at suitable pH. Organic compounds containing $-COOH$, $-OH$, and $-NH_2$ groups can also form hydrogen bonding with CNT surface during separation via electronic interaction. The cost of fabricating CNTs is much higher than that of activated carbon powder, although the removal efficiency of CNTs is higher.

7.2.6.3 Regeneration and Reuse of CNTs

Metal-adsorbed CNTs can easily be regenerated by reducing the solution pH without any significant change in metal-adsorption capacity. A metal-recovery rate close to 100% after regeneration has been achieved along with a drop in adsorption capacity of SWCNTs and MWCNTs of less than 25% after 10 regeneration and reuse cycles, which is far better than activated carbon [46,47].

7.2.6.4 Graphene-Based Nanomaterials

As shown in Fig. 7.3, "graphene" is the basic structural element of all carbon allotropes, e.g., graphite, fullerene, CNTs, and charcoal, which have recently started emerging as multifunctional elements in water treatment, and where CNTs have already proven to be most successful.

A monolayer 2D sheet of graphite called "graphene" consists of covalently bonded sp^2 hybridized carbon atoms arranged in a hexagonal

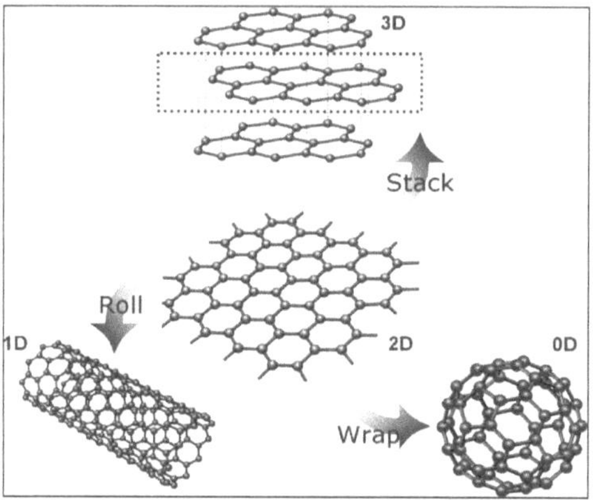

Figure 7.3 Graphene as the mother of all other graphitic forms [48].

honeycomb lattice manner. This single-sheet structure of graphene is also defined by the International Union for Pure and Applied Chemistry (IUPAC) as "a single carbon layer of graphite structure, describing its nature by analogy to a polycyclic aromatic hydrocarbon of quasi infinite size." In 2004, the Novoselov, Geim and group successfully isolated and characterized the mechanically exfoliated graphene monolayer at the University of Manchester. In a hexagonal cell of graphene, each carbon atom is surrounded by three nearest neighbors from different sublattices and bonded together with three σ bonds (in plane) and one π bond (out of plane) making a C–C bond length equal to 0.142 nm as shown in Fig. 7.4. The stability of graphene is actually caused by these tightly packed carbon atoms and sp^2 orbital hybridization. Moreover, delocalized π-electrons over the carbon atoms result in good electrical conductivity [49–53].

7.2.6.5 Advantages of Graphene Over CNTs

Graphene-based nanomaterials are emerging as substitutes for CNTs in environmental remediation and offer the following advantages:

1. Single-layered graphene offers higher adsorption sites than CNTs due to two basal planes available for adsorption and pollutant removal.
2. The inner walls of CNTs remain practically inaccessible to water pollutants due to graphene's huge adsorption sites.
3. Graphene oxide (GO) and reduced GO (rGO) can be easily synthesized through simple chemical exfoliation of low cost graphite material without involvement of complex catalyst, equipment, and system.

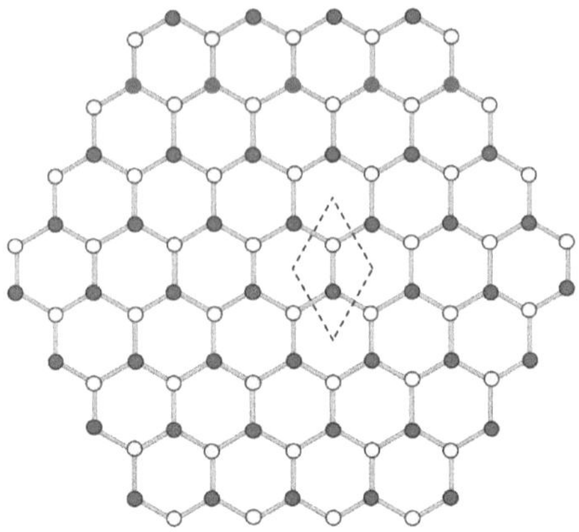

Figure 7.4 A unit cell of graphene (dashed line) containing two atoms from different sublattices [54].

This results in relatively cheap nanomaterial free from toxicological effects in the absence of catalyst.

4. The presence of a large number of oxygen-bearing functional groups in GO makes additional acid treatment for attaching functional groups redundant. This significantly reduces the cost of synthesis and toxicological effects.
5. Composite graphene-based materials such as TiO_2-GO have been successful in the removal of metal ions from water due to the presence of large oxygen-bearing functional groups that interact with metal ions.

7.2.6.6 Synthesis of Graphene Materials

Mechanically exfoliated graphene was first synthesized from graphite, a naturally occurring carbon allotrope, using a simple "Scotch tape" method. Most common methods used to get high-quality, defect-free graphene with excellent physical properties involve direct synthesis from organic precursors (CH_4 and other hydrocarbons), epitaxial growth, and chemical vapor deposition (CVD). Liquid-phase exfoliation, graphite intercalation, and electrochemical exfoliation are other methods used to prepare graphene materials, although these methods can't always ensure high-quality graphene sheets due to the introduction of defects during exfoliation. Preparation of GO by chemical oxidation of graphite, followed by reduction of GO, is another popular method used to synthesize graphene and RGO [55–64].

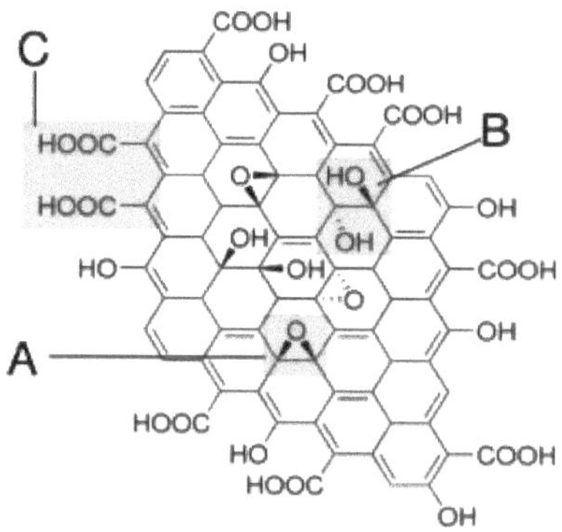

Figure 7.5 GO structure showing multifunctional groups [70].

GO, an important precursor of graphene, can be easily synthesized from graphite by chemical exfoliation. This highly oxidative form of graphene with a variety of oxygen functionalities, such as epoxide, hydroxyl, carbonyl, and carboxylic acids on the basal planes as shown in Fig. 7.5. The π-conjugation of graphene in GO is disrupted by the presence of these functional groups, which in turn makes it electrically insulating. The simultaneous presence of multifunctional groups makes GO widely applicable in different areas of scientific research. Some cost-effective chemical-oxidation methods for GO synthesis include Brodie, Hummers, and Staudenmaier oxidation of graphite with some highly oxidative reagents like potassium chlorate ($KClO_3$), concentrated nitric acid (HNO_3), potassium permanganate ($KMnO_4$), and concentrated sulphuric acid (H_2SO_4). Hummers' method is considered as the most useful for preparing GO, since it involves few steps, is considerably safe, and is less time consuming [65–70].

7.2.6.7 Properties of Graphene Materials

Graphene offers unique properties such as high 3D aspect ratio, large specific BET surface area (2630 $m^2\ g^{-1}$, larger than CNTs), excellent thermal and electrical conductivity, impermeability to small gas molecules, good optical transparency, and outstanding mechanical and chemical prospects, which make them "omni potential materials." The extremely versatile and tunable carbon backbone of the graphene-family coupled with facile functionalization permit the use of these materials in many exciting and revolutionary applications such as structural composites, conducting polymers, nanoelectronics, battery electrodes, supercapacitors, transport barriers, printable inks, and biomedical technologies. The use of graphene-based nanomaterials in water purification both as adsorbent and membrane is relatively new [68–70].

7.2.6.8 Graphene-Family Nanoadsorbents in Batch-Sorption Module

The batch-adsorption behavior of graphene and various graphene-based materials like GO, RGO, and few-layered GO (FGO) for the removal of variety of ecotoxic contaminants from aqueous stream has been well observed [71].

The extremely large surface area, good dispersibility in aqueous medium, and good porosity make graphene materials good adsorbents. GO contains a negatively charged surface that helps it to bind with cationic pollutants. While graphene materials are associated with these advantages there are some difficulties that can limit the use of pure graphene and GO. For example, graphene has a tendency to agglomerate and restack quickly in aqueous medium to form graphite while preceding the adsorption reactions. Moreover, a negatively charged GO surface is to some extent reluctant toward the adsorption of anionic pollutants. To overcome these limitations, functionalization of graphene with nanoparticles or other functional moieties has been studied in a wide range to improve sensitivity, selectivity, and detection limit while treating wastewater. Graphene nanosheets, GO, modified GO, RGO, and nanocomposite graphene materials have been widely used for the separation of toxic heavy metals (Cu, Cr, Cd, As, Pb, Fe, Hg, Co, Ni, and Mn), cationic and anionic dyes (methylene blue, rhodamine B, malachite green, orange G, etc.), fluoride, and other organic pollutants from drinking water. Different materials and methods have emerged to synthesize these adsorbents to create new technologies and improve water quality.

To overcome the difficulties in separation of GO from aqueous medium, several modifications of GO are possible. Polyethersulfone (PES) enwrapped GO porous particles (PES/GO), GO modified with chitosan, Ethylenediamine tetraacetic acid (EDTA), calcium alginate, polypyrrole/GO (PPy/GO) composite nanosheets, sulfonate magnetic GO (SMGO), and GO-fabricated with various metal and mixed-metal oxides are some of these advancements. Moreover, in most of the cases graphene and graphene materials have proved themselves as reusable adsorbents, which can increase their cost-effectiveness from an industrial point of view.

7.3 NANOMATERIALS IN WATER PURIFICATION AS MEMBRANE

7.3.1 CNTs in Membrane Module

CNTs have been developed as membrane due to their hollow structure, large surface area, smooth hydrophobic walls, and antifouling and high fluxing nature. Moreover, the possibility of controlling pore diameter has advanced their application potential as selective semipermeable membrane. The specially aligned structure of CNTs in CNT-based membrane explains the frictionless movement of water molecules through nanochannels [72].

7.3.1.1 Synthesis and Application of CNT Membrane

Both SWCNT and MWCNT membrane can be applied in a wide variety of applications for water desalination. The performance of CNT membrane depends on its processing and fabrication methods. Vertically aligned (VA) and mixed matrix (MM) CNT membrane are two currently available types of nanomembrane materials. VA-CNT membranes can be best synthesized by chemical vapor deposition (CVD) and possess perpendicular CNT arrangements with supportive fillers (epoxy, silicon nitride). On the other hand, cost-effective synthesis of MM-CNT membrane is possible using mixed-polymeric materials. MM-CNT membranes hold great promise to overcome many existing limitations in water separation technology. Success in using CNTs as adsorbent or membrane has been largely possible through functionalization of CNT membrane with various functional groups at the mouth as well as in the interior core (tip functionalized and core functionalized CNT). Such functionalization improves selectivity, mechanical and thermal stability, fouling resistance, pollutant degradation capability, self-cleaning functions, and flux behavior. A range of CNT membrane with Ag, Cu, Au, Pt, Pd, TiO_2 nanoparticles, polymers, and biomolecules with better desalination performance have been developed [73−77].

7.3.2 Graphene-Family Membranes in Different Membrane Modules

In addition to the adsorption behavior, the graphene-family has shown great potential as promising membrane material for both wastewater treatment and water desalination. Recently, various novel graphene-based membranes were developed as better replacements of conventional polymeric membranes, zeolites, ceramics, and even CNT membranes. A class

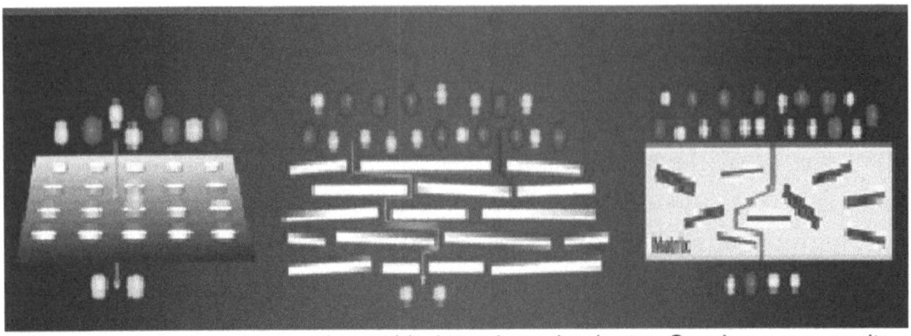

Figure 7.6 Three major graphene-based membranes: (type I) porous graphene layer; (type II) assembled graphene laminates; and (type III) graphene-based composite [83].

of ultra-thin, mechanically robust, high fluxing, and highly selective and antifouling membrane made from graphine is already available. A large variety of graphene membrane with distinct microstructures and transport pathways is being used in different pressure-driven membrane-filtration modules such as UF, NF, and RO [34,78,79].

Initial attempts to make porous graphene membrane for selective passage of water were based on drilling holes in graphene sheets. Graphene, in particular GO nanosheets, can be synthesized with 2D nanochannels for selective permeation of water molecules. Porous graphene layers, assembled graphene laminates, and graphene-based composites as illustrated in Fig. 7.6 are the three main types of graphene-based membranes developed so far [80−82].

Nanoporous graphene or porous graphene membranes with improved permeability facilitate fast water transport under low pressure and variable operational conditions. Molecular dynamic (MD) simulation, in this connection, is one of the newest techniques to create nanopore arrays of consistent size, pores, density, and quality on graphene layers. Water flux through these porous membranes largely depends on pore size, chemistry of the pores as well as pore functionalization. Through MD simulation it was observed that during diffusion of ions such as Li^+, Na^+, K^+, Cl^-, and Br^- through graphene monolayers with functionalized nanopores with negatively charged species the passage of cations is highly favored and positively charged nanopores facilitate the passage of anions, indicating nanoporous graphene monolayers are excellent ion-separation membrane for desalination. During desalination of water using nanoporous single-layer graphene with different pore size and functional groups, it was observed that water permeability of these ion-separation membranes is 2–3 times higher than that of conventional RO membrane with a similar salt rejection of 99% [84–86].

7.3.2.1 Synthesis and Use of Graphene-Based Membranes in Water Purification

In addition to nanoporous graphene, GO nanosheets are also considered suitable as ion-sieving membranes. In spite of the high hydrophobicity of GO nanopores, GO nanosheets themselves are extremely hydrophilic. By differentiating the conditions and creating different spacing between these nanosheets, high-quality GO membrane for water desalination and purification can be fabricated. GO-fabricated polymeric membrane with improved water permeability, antifouling property, and antimicrobial activity has also been developed.

The major methods of fabrication of GO membranes from GO nanosheets include vacuum filtration, layer-by-layer (LbL) deposition, drop casting, and spin coating. Cross-linking GO over polymer matrix rather than simple surface coating of GO on polymeric base support is another recent trend in graphene-based membrane synthesis. These cross-linked composite membranes have shown effective desalination and antibacterial activities. A novel method of synthesis of GO membrane is by LbL deposition of GO nanosheets, cross-linked with polydopamine-coated polysulfone support using 1,3,5-benzenetricarbonyl trichloride as cross-linker. This membrane can deliver 4–10 times higher flux (80–276 LMH/MPa) than most commercial NF membranes, with a low rejection (6–46%) of mono and divalent salts, moderate rejection (46–66%) of methylene blue, and high rejection (93–95%) of rhodamine-WT dye. This GO membrane is emerging as a next-generation, cost-effective, and sustainable alternative to the thin-film composite (TFC) polyamide membrane for water separation applications. Using similar technique, a GO-poly (amide) TFC membrane with high fluxing, antifouling, and superior antimicrobial activity has been developed via an interfacial polymerization (IP) method. Novel antifouling GO-blended Polyvinylidene Difluride (PVDF) ultrafiltration membranes have also been developed with substantially improved permeability over pure PVDF membranes. GO-coated polyamide membrane for RO-based desalination application with inherent improvement in antifouling nature as well as chemical resistance has also emerged. Several other GO-based nanocomposite membranes such as GO/PVDF, GO/PSF, GO/APTS (APTS: 3-aminopropyltriethoxysilane), GO/PPy (PPy: polypyrol), and G/PPy membranes developed in recent years have been found to have excellent desalination, antibacterial, and antifouling activities and high water flux [38–41,87–90].

7.3.2.2 Zeolite-Based Nanomembrane

Thin-film nanocomposite membranes (TFN) with high permeability and selectivity have been developed with nanozeolites as active layer. Such zeolite-based membranes are emerging as good alternatives to polymeric membranes for desalination of seawater and treatment of wastewater. Nanozeolite-doped TFN membranes with increased water permeability (80% higher than in TFC), high rejection (93% salt rejection), and negatively charged surface and thicker polyamide active layer have already proven to be effective in the treatment of water. By doping TFC membrane with 250 nm nanozeolite, more than 99% salt rejection has been achieved in some cases. The enhanced water permeability of these membranes is attributed to the small, hydrophilic pores of the nanozeolites, which create preferential water channels. Several zeolite membranes developed in recent years have been tested in RO membrane module. Metal-doped polymeric membrane has also been invented. Most of these polymeric membranes are operated in microfiltration (MF) and ultrafiltration (UF) modules for effective removal of inorganic contaminants like arsenic from water with higher ease of separation after treatment.

7.3.2.3 Nanomaterial-Incorporated Ceramic Membranes in Water Treatment

Ceramic membranes can generally remove silt, particulates, oil and grease, suspended solids, organic carbon, and metal oxides. Good thermal and mechanical strength, chemical stability, long lifetime, selectivity in separation, high porosity, high fluxing, easy cleaning, minimal maintenance care, and low pretreatment requirements make these membranes popular. However, ceramic membranes mostly purify water with MF and UF regimes. To enhance purification capability beyond the UF range, a range of functional nanomaterials can be incorporated in such ceramic membranes. Ceramic membranes have been modified to incorporate catalytic nanoparticles such as titanium, silica, silver, and similar elements of catalytic activity with simultaneous improvement in permeability, selectivity, and antifouling nature [91].

The structure, surface, and composition of those membranes can also be modified by embedding amino groups to form covalent bonds with silver atoms on silver nanoparticles. Porous alumina ceramic membrane has been fabricated by using acetic acid to stabilize alumina nanoparticles on the ceramic membrane. By incorporating TiO_2 nanoparticles and silver, alumina, iron oxide, and even CNTs as nanoparticles ceramic membranes have been developed in recent years with high flux, selectivity, and antifouling characteristics [85–87].

7.4 NANOMATERIALS IN PHOTOCATALYTIC DEGRADATION OF WATER POLLUTANTS

Photocatalytic oxidation is an advanced degradation process for removal of trace contaminants, pathogens, and organic compounds like dye, which are not easily biodegraded in conventional treatment plants. Photocatalytic degradation may also be adopted to enhance biodegradability of the hazardous and nonbiodegradable pollutants present in contaminated or wastewater. In photocatalytic degradation, oxidation of the target compounds is initiated by the free radicals induced by the electron-hole pair (e^-/h^+) at the photocatalyst surface. For effective photodegradation, three essential conditions must be satisfied:

1. the substance to be degraded should be adsorbed on the photocatalyst surface;
2. the required irradiation and absorption has to take place; and
3. separation and transport of charge carriers (photocarriers) has to occur.

While TiO_2-based nanoparticles have widely been investigated in photodegradation of organic substances like dyes, these methods can work only under UV irradiation because of its associated band-gap energy. Therefore TiO_2-based photocatalyst will not be useful under visible solar radiation. Valence band-gap energy of ZnCdS is 1.0 ev more negative than that of TiO_2, which narrows the bandage energy thereby permitting response to visible energy. The photocatalytic activity in degradation of organic water pollutants can be further enhanced by using rGO as it facilitates fast transport of photocarriers. Thus rGO-ZnCdS nanomaterial has been developed and examined in photodegradation of organic dye. However, the proportion of rGO in the rGO-ZnCdS needs to be controlled as despite initial positive effect on the extent of degradation on increasing rGO may be nullified at higher rGO loading because of hindering visible light absorption by the black color rGO.

Photocatalytic reactors and solar disinfection systems are the two nanotechnology-enabled degradation modules used in this field. Aluminum oxide, TiO_2, and ZnO are the most commonly applied for water treatment through adsorption, chemical degradation, photodegradation, and chemical disinfection. Some nanomaterials widely used as photocatalysts are listed in Table 7.1.

Nano TiO_2 is the most widely used photocatalyst in water and wastewater treatment due to its low toxicity, chemical stability, low cost, and abundance as raw material. Modified nano TiO_2, WO_3, and WO_3 are the other nanomaterials with the potential to be used in photocatalytic water treatment [92–95].

7.5 NANOMATERIALS IN DISINFECTION OF CONTAMINATED WATER

The most widely used water disinfectant is chlorine or chlorine-based compounds because of their effectiveness, residual effects in long-distance transmission, and low cost. However, in the disinfection unit, chlorine may react with NOMs producing toxic disinfection byproducts (DBPs) that are either carcinogens or suspected carcinogens. Ozone-based disinfection though effective at the point of use does not leave any residual effect. The UV-disinfection method requires high dosage for certain viruses such as adenoviruses and is effective at the point of use. To minimize the formation of DBP and to enhance the robustness of this water disinfection technique, several metal-based antimicrobial nanomaterials have been investigated in recent years.

Nano TiO_2-facilitated solar disinfection (SODIS) has also been extensively tested and established as a feasible option to produce safe drinking water in remote areas. The beauty of the SODIS system is its scalability in both small- and medium-scale solar compound parabolic collectors.

Modified UF PES membranes with hyperbranched polyethylenimine (HPEI) GO via phase-inversion method have been developed with effective antibacterial performance against *E. coli.* [96]. Zeolite-based nanomaterials have also emerged with the antimicrobial and antifouling properties of nanozeolites reported in the membrane-separation field.

Nano Ag, nano TiO_2, nano ZnO, and nano Ce_2O_4 are some of the nanomaterials that can show effective antimicrobial properties with minimal DBP formation. Among them, nano Ag is currently the most widely used antimicrobial material due to its strong antimicrobial activity, low human toxicity, and ease of use [44,45].

The antimicrobial property of nano Ag can be attributed to its tendency to release silver ion, which can bind to thiol groups in vital proteins of the microorganisms, resulting in enzyme damage.

Nano Ag is also potentially good for application in Point-of-Use (POU) water treatment. Some commercial devices with this nanomaterial like the MARATHON^R and Aquapure^R systems are already available. A nano Ag-incorporated ceramic microfilter also acts as an antipathogenic system that can be employed in the remote areas.

In using nanomaterials as antibacterial agents in water disinfection the utmost care must be taken to avoid toxicity. While the toxicity of TiO_2 has not been established, other antibacterial nanomaterials such as nanosilver and nanocopper release toxic silver and copper ions leading to serious health hazards.

In water purification, nanomaterials with promising separation capability are emerging rapidly. Of these nanomaterials, graphene-based nanocomposite materials seem to be the most economically viable if integrated with appropriate modules in the form of membranes. Despite the possibility of longer life, these materials, however, offer limited flexibility in matters of application in various modules. Pilot-scale or commercial-scale plant data are limited at this stage of nanotechnology-based water treatment plants, which seriously limits scale-up confidence. Another serious issue arises from the possibility of toxicological effects of released nanoparticles that are extremely reactive and may pose health hazards to aquatic life. However, in matters of removal of specific contaminants from dilute solutions, the efficiency of nanomaterials can be very high and proper embedding holds promise for their safe handling.

REFERENCES

[1] Pal P, Ahammad Z, Pattanayek A, Bhattacharya P. Removal of arsenic from drinking water by chemical precipitation – a modeling and simulation study of the physical chemical processes. Water Environ Res 2007;79(4):357–66.

[2] Rengaraj S, Yeon KH, Kang SY, Lee JU, Kim KW, Moon SH. Removal of chromium from water and wastewater by ion exchange resins. J Hazard Mater 2002;87(1-3):273–87.

[3] Pal P, Chakrabortty S, Linnanen L. A. Nanofiltration-coagulation integrated system for separation and stabilization of arsenic from groundwater. Sci Total Environ 2014;476-477:601–10.

[4] Mishra AK, Ramaprabhu S. Functionalized graphene sheets for arsenic removal and desalination of sea water. Desalination 2011;282:39–45.

[5] Qu X, Alvarez PJJ, Li Q. Applications of nanotechnology in water and wastewater treatment. Water Res 2013;47:3931–46.

[6] Tripathy SS, Raichur AM. Enhanced adsorption capacity of activated alumina by impregnation with alum for removal of As(V) from water. Chem Eng J 2008;138:179−86.
[7] Dou X, Mohan D, Pittman Jr CU, Yang S. Remediating fluoride from water using hydrous zirconium oxide. Chem Eng J 2012;198-199:236−45.
[8] Maliyekkal SM, Sharma AK, Philip L. Manganese-oxide-coated alumina: a promising sorbent for defluoridation of water. Water Res 2006;40:3497.
[9] Goswami A, Raul PK, Purkait MK. Arsenic adsorption using copper (II) oxide nanoparticles. Chem Eng Res Des 2012;9:1387−96.
[10] Basu T, Gupta K, Ghosh UC. Removal of arsenic from aqueous phase by nanoparticle agglomerates of hydrous iron(III)-chromium(III) bimetal mixed oxide: effects of background ions on the As(V) sorption kinetics and equilibrium. Water Qual Res J Can 2010;45(4):437−49.
[11] Giles DE, Mohapatra M, Issa TB, Anand S, Singh P. Iron and aluminium based adsorption strategies for removing arsenic from water. J Environ Manage 2011;92 (12):3011−22.
[12] Mohapatra M, Hariprasad D, Mohapatra L, Anand S, Mishra BK. Mg-doped nano ferrihydrite—a new adsorbent for fluoride removal from aqueous solutions. Appl Surf Sci 2012;258:4228−36.
[13] Biswas K, Gupta K, Goswami A, Ghosh UC. Fluoride removal efficiency from aqueous solution by synthetic iron(III)−aluminum(III)−chromium(III) ternary mixed oxide. Desalination 2010;255:44−51.
[14] Gupta K, Biswas K, Ghosh UC. Nanostructure iron(III) − zirconium(IV) binary mixed oxide: synthesis, characterization, and physicochemical aspects of arsenic(III) sorption from the aqueous solution. Ind Eng Chem Res 2008;47:9903−12.
[15] Barathi M, Kumar KS, Rajesh N. Efficacy of novel Al−Zr impregnated cellulose adsorbent prepared using microwave irradiation for the facile defluoridation of water. J Environ Chem Eng 2013;1:1325−35.
[16] Mohapatra M, Padhi T, Anand S, Mishra BK. CTAB mediated Mg-doped nano Fe_2O_3: synthesis, characterization, and fluoride adsorption behaviour. Desalin Water Treat 2012;50:376−86.
[17] Dey A, Singh R, Purkait MK. Cobalt ferrite nanoparticles aggregated schwertmannite: a novel adsorbent for the efficient removal of arsenic. J Water Process Eng 2014;3:1−9.
[18] Yavuz CT, Mayo JT, Yu WW, Prakash A, Falkner JC, Yean S, et al. Low-field magnetic separation of monodisperse Fe_3O_4 nanocrystals. Science 2006;314(5801):964e967.
[19] Huang Y, Keller AA. Magnetic nanoparticle adsorbents for emerging organic contaminants. ACS Sustain Chem Eng 2013;1:731−6.
[20] Griffitt RJ, Luo J, Gao J, Bonzongo J-C, Barber DS. Effects of particle composition and species on toxicity of metallic nanomaterials in aquatic organisms. Environ Toxicol Chem 2008;27:1972−8.
[21] Yavuz CT, Mayo JT, Yu WW, Prakash A, Falkner JC, Yean S, et al. Low-field magnetic separation of monodisperse Fe_3O_4 nanocrystals. Science 2006;314 (5801):964e967.
[22] Wang P, Shi Q, Shi Y, Clark KK, Stucky GD, Keller AA. Magnetic permanently confined micelle arrays for treating hydrophobic organic compound contamination. J Am Chem Soc 2008;131:182−8.
[23] Karimi MA, Hatefi-Mehrjardi A, Mohammadi SZ, Mohadesi A, Mazloum-Ardakani M, Kabir AA, et al. Solid phase extraction of trace amounts of silver (I) using dithizone-immobilized alumina-coated magnetite nanoparticles prior to determination by flame atomic absorption spectrometry. Int J Environ Anal Chem 2012;92:1325−40.

[24] Faraji M, Yamini Y, Rezaee M. Extraction of trace amounts of mercury with sodium dodecyl sulphate-coated magnetite nanoparticles and its determination by flow injection inductively coupled plasma-optical emission spectrometry. Talanta 2010;81:831–6.
[25] Afkhami A, Moosavi R, Madrakian T. Preconcentration and spectrophotometric determination of low concentrations of malachite green and leuco-malachite green in water samples by high performance solid phase extraction using maghemite nanoparticles. Talanta 2010;82:785–9.
[26] Ashtari P, Wang KM, Yang XH, Huang SS, Yamini Y. Novel separation and preconcentration of trace amounts of copper(II) in water samples based on neocuproine modified magnetic microparticles. Anal Chim Acta 2005;550:18–23.
[27] Tomar V, Prasad S, Kumar D. Microchem J 2013;111:116–24.
[28] Sharma YC, Srivastava V, Singh VK, Kaul SN, Weng CH. Nanoadsorbents for the removal of metallic pollutants from water and wastewater. Environ Technol 2009;30 (6):583e609.
[29] Hu J, Chen GH, Lo IMC. Selective removal of heavy metals from industrial wastewater using maghemite nanoparticle: performance and mechanisms. J Environ Eng ASCE 2006;132(7):709e715.
[30] Deliyanni EA, Bakoyannakis DN, Zouboulis AI, Matis KA. Sorption of As(V) ions by akaganeite-type nanocrystals. Chemosphere 2003;50(1):155e163.
[31] Mayo JT, Yavuz C, Yean S, Cong L, Shipley H, Yu W, et al. The effect of nanocrystalline magnetite size on arsenic removal. Sci Technol Adv Mater 2007;8(1e2):71e75.
[32] Wang S, Peng Y. Natural zeolites as effective adsorbents in water and wastewater treatment. Chem Eng J 2010;156(1):11–24.
[33] Ouki SK, Kavannagh M. Treatment of metals-contaminated wastewaters by use of natural zeolites. Water Sci Technol 1999;39(10-11):115–22.
[34] Wingenfelder U, Hansen C, Furrer G, Schulin R. Removal of heavy metals from mine waters by natural zeolites. Environ Sci Technol 2005;39(12):4606–13.
[35] Onyango SM, Kojima Y, Kumar A, Kuchar D, Kubota M, Matsuda H. Uptake of fluoride by Al^{3+}-pretreated low-silica synthetic zeolites: adsorption equilibrium and rate studies. Sep Sci Technol 2006;41(4):683–704.
[36] Chutia P, Kato S, Kojima T, Satokawa S. Arsenic adsorption from aqueous solution on synthetic zeolites. J Hazard Mater 2009;162:440–7.
[37] Sun Y, Fang Q, Dong J, Cheng X, Xu J. Removal of fluoride form drinking water by natural stilbite zeolite modified with Fe (III). Desalination 2011;277:121–7.
[38] Scarselli M, Castrucci P, De Crescenzi M. Electronic and optoelectronic nanodevices based on carbon nanotubes Published IOP Publishing Ltd J Phys Condens Matter 2012;24(31).
[39] Chowdhury S, Balasubramanian R. Recent advances in the use of graphene-family nanoadsorbents for removal of toxic pollutants from wastewater. Adv Colloid Interface Sci 2014;204:35–56.
[40] Gupta VK, Kumar R, Nayak A, Saleh TA, Barakat MA. Adsorptive removal of dyes from aqueous solution onto carbon nanotubes: a review. Adv Colloid Interface Sci 2013;193-194:24–34.
[41] Li Y-H, Zechao D, Ding J, Wu D, Luan Z, Zhy Y. Adsorption thermodynamic, kinetic and desorption studies of Pb^{2+} on carbon nanotubes. Water Res 2005;39 (4):605–9.
[42] Li Y-H, Ding J, Luan Z, Di Z, Zhu Y, Xu Cailu, et al. Preparation of porous alumina–carbon nanotube composites via direct growth of carbon nanotubes. Carbon 2003;41(14):2787–92.
[43] Upadhyayula VKK, Deng S, Mitchell CM, Smith BG. Sci Total Environ 2009;408 (1):1–13.

[44] Ai L, Jiang J. Chem Eng J 2012;192:156.
[45] Sui Z, Meng Q, Zhang X, Ma R, Cao B. J Mater Chem 2012;22:8767.
[46] Li YH, Di ZC, Ding J, Wu DH, Luan ZK, Zhu YQ. Adsorption thermodynamic, kinetic and desorption studies of Pb21 on carbon nanotubes. Water Res 2005;39 (4):605–9.
[47] Lu CS, Chiu H, Liu CT. Removal of zinc(II) from aqueous solution by purified carbon nanotubes: kinetics and equilibrium studies. Ind Eng Chem Res 2006;45 (8):2850–5.
[48] Bhaskar Garg, Tanuja Bisht, Yong-Chien Ling. Graphene-based nanomaterials as heterogeneous acid catalysts: a comprehensive perspective. Molecules 2014;19 (9):14582–614. Available from: http://dx.doi.org/10.3390/molecules190914582.
[49] Boehm HP, Setton R, Stump E. Nomenclature and terminology of graphite intercalation compounds. Carbon 1986;24:241.
[50] Sing V, Joung D, Zhai L, Das S, Khondaker Sl, Seal S. Graphene based materials: past, present and future. Prog Mater Sci 2011;56:1178–271.
[51] IUPAC Recommended terminology for the description of carbon as a solid (IUPAC Recommendations 1995), 67; 1995 491 [473].
[52] Novoselov K, Geim A, Morozov S, Jiang D, Zhang Y, Dubonos S, et al. Electric field effect in atomically thin carbon films. Science 2004;306(5696):666–9.
[53] Avouris P, Dimitrakopouulos C. Graphene: synthesis and applications. Mater Today 2012;15:86–97.
[54] Zhu Y, Murali S, Cai W, Li X, Suk JW, Potts JR, et al. Graphene and graphene oxide: synthesis, properties, and applications. Adv Mater 2010;22:3906–24.
[55] Rajasekhar Balasubramanian, Shamik Chowdhury. Recent advances and progress in the development of graphene-based adsorbents for CO_2 capture. J Mater Chem A 2015;3:21968–89.
[56] Allen MJ, Tung VC, Kaner RB. Honeycomb carbon: a review of graphene. Chem Rev 2010;110(1):132–45.
[57] Berger C, Song Z, Li X, Wu X, Brown N, Naud C, et al. Electronic confinement and coherence in patterned epitaxial graphene. Science 2006;312:1191–6.
[58] Kim KS, Zhao Y, Jang H, Lee SY, Kim JM, Kim KS, et al. Large-scale pattern growth of graphene films for stretchable transparent electrodes. Nature 2009;457:706–10.
[59] Hernandez Y, Nicolosi V, Lotya M, Blighe FM, Sun Z, De S, et al. High-yield production of graphene by liquid-phase exfoliation of graphite. Nat Nanotechnol 2008;3:563.
[60] Behabtu N, Lomeda JR, Green MJ, Higginbotham AL, Sinitskii A, Kosynkin DV, et al. Spontaneous high-concentration dispersions and liquid crystals of graphene. Nat Nanotechnol 2010;5(6):406–11.
[61] Sun Y, Wu Q, Shi G. Graphene based new energy materials. Energy Environ Sci 2011;4:1113–32.
[62] Iwan A, Chuchmala A. Perspectives of applied graphene: polymer solar cells. Prog Polym Sci 2012;37:1805–28.
[63] Park S, Ruoff RS. Chemical methods for the production of graphenes. Nat Nanotechnol 2009;4:217–24.
[64] Stankovich S, Dikin DA, Piner RD, Kohlhaas KA, Kleinhammes A, Jia Y, et al. Synthesis of graphene-based nanosheets via chemical reduction of exfoliated graphite oxide. Carbon 2007;45:1558–65.
[65] Krishnan D, Kim F, Luo J, Crruz-Silva R, Cote LJ, Jang HD, et al. Energetic graphene oxide: challenges and opportunities. Nano Today 2012;7:137–52.
[66] Staudenmaier L. Verfahren zur Darstellung der Graphitsaure. Ber Dtsch Chem Ges 1898;31:1481–7. Available from: http://dx.doi.org/10.1002/cber.18980310237.

[67] He H, Klinowski J, Forster M, Lerf A. A new structural model for graphite oxide. Chem Phys Lett 1998;287:53 Bibcode: 1998CPL...287...53H. http://dx.doi.org/10.1016/S0009-2614(98)00144-4.
[68] Ruoff R. Graphene: calling all chemists. Nat Nanotechnol 2008;3:10−11.
[69] Stankovich S, Dikin DA, Dommett GHB, Kohlhass KM, Zimney EJ, Stach EA, et al. Graphene-based composite materials. Nature 2006;442:282−6.
[70] Paek SM, Yoo E, Honma I. Enhanced cyclic performance and lithium storage capacity of SnO_2/graphene nanoporous electrodes with three-dimensionally delaminated flexible structure. Nano Lett 2009;9(1):72−5.
[71] Sanchez VC, Jachak A, Hurt RH, Kane AB. Biological interactions of graphene-family nanomaterials: an interdisciplinary review. Chem Res Toxicol 2012;25(1):15−34.
[72] Choi JH, Jegal J, Kim WN. Fabrication and characterization of multi-walled carbon nanotubes/polymer blend membranes. J Membr Sci 2006;284(1):406−15.
[73] Yang HY, Han ZJ, Yu SF, Pey KL, Ostrikov K, Karnik R. Carbon nanotube membranes with ultrahigh specific adsorption capacity for water desalination and purification. Nat Commun 2013;4:3220. Available from: http://dx.doi.org/10.1038/ncomms (Article number: 2220).
[74] Qu X, Alvarez PJ, Li Q. Applications of nanotechnology in water and wastewater treatment. Water Res 2013;47(12):3931−46.
[75] Choi YC, Shin YM, Lee YH, Lee BS, Park GS, Choi WB, et al. Controlling the diameter, growth rate, and density of vertically aligned carbon nanotubes synthesized by microwave plasma-enhanced chemical vapor deposition. Appl Phys Lett 2000;76 (17):2367−9.
[76] Zimmerman CM, Sing A, Kros WJ. Tailoring mixed matrix composite membranes for gas separations. J Membr Sci 1997;137(1):145−54.
[77] De Volder MFL, Tawfick SH, Baughman RH, Hart AJ. Carbon nanotubes: present and future commercial applications. Science 2013;339(6119):535−9.
[78] Moulin C, Bourbigot MM, Taziapain A, Bourdon F. Design and performance of membrane filtration installations: capacity and product quality for drinking water applications. Environ Technol 1991;12:841−58.
[79] Tsuru T. Inorganic porous membranes for liquid phase separation. Sep Purif Methods 2001;30:191−220.
[80] Ciora Jr. RJ, Liu PKT. Ceramic membranes for environmental related applications. Fluid Particle Sep J. 2003;15:51−60.
[81] Hk Oh, Takizawa S, Ohgak S, Katayama H, Oguma K, Yu MJ. Removal of organics and viruses using hybrid ceramic MF system without draining PAC. Desalination 2007;202:191−8.
[82] Kim J, der Bruggen BV. The use of nanoparticles in polymeric and ceramic membrane structures: review of manufacturing procedures and performance improvement for water treatment. Environ Pollut 2010;158:2335−49.
[83] Hofs B, Ogier J, Vries D, Beerendonk EF, Cornelissen ER. Comparison of ceramic and polymeric membrane permeability and fouling using surface water. Sep Purif Technol 2011;79:365−74.
[84] Lv Y, Liu H, Wang Z, Liu S, Hao L, Sang Y, et al. Silver nanoparticle-decorated porous ceramic composite for water treatment. J Membr Sci 2009;331:50−6.
[85] DeFriend KA, Wiesner MR, Barron AR. Alumina and aluminate ultrafiltration membranes derived from alumina nanoparticles. J Membr Sci 2003;224:11−28.
[86] Karnik BS, Baumann MJ, Masten ST, Davies SHR. AFM and SEM characterisation for iron oxide-coated ceramic membranes. J Mater Sci 2006;41:6861−70.
[87] Ding X, Fan Y, Xu N. A new route for the fabrication of $Ti0_2$ ultrafiltration membranes. J Membr Sci 2006;270:179−86.

[88] Das R, Ali ME, Hamid SBA, Ramakrishna S, Chowdhury ZZ. Carbon nanotube membranes for water purification: a bright future in water desalination. Desalination 2014;336:97−109.
[89] Thostenson ET, Ren Z, Chou TW. Advances in the science and technology of carbon nanotubes and their composites: a review. Comp Sci Technol 2001;61(13):1899−912.
[90] Li Y-H, Ding J, Luan Z, Di Z, Zhu Y, Xu Cailu, et al. Competitive adsorption of Pb $^{2+}$, Cu $^{2+}$ and Cd $^{2+}$ ions from aqueous solutions by multiwalled carbon nanotubes. Carbon 2003;41(14):2787−92.
[91] Liu L, Zhao F, Liu J, Yang F. A composite cathode membrane with $CoFe_2O_4$−rGO/PVDF on carbon fiber cloth: synthesis and performance in a photocatalysis-assisted MFC-MBR system. J Membr Sci 2013;437:99−107.
[92] Fujishima A, Zhang XT, Tryk DA. TiO(2) photocatalysis and related surface phenomena. Surf Sci Reports 2008;63(12):515e582.
[93] Ni M, Leung MKH, Leung DYC, Sumathy K. A review and recent developments in photocatalytic water-splitting using TiO_2 for hydrogen production. Renew Sustain Energy Rev 2007;11(3):401e425.
[94] Kominami H, Yabutani K, Yamamoto T, Kara Y, Ohtani B. Synthesis of highly active tungsten(VI) oxide photocatalysts for oxygen evolution by hydrothermal treatment of aqueous tungstic acid solutions. J Mater Chem 2001;11(12):3222e−3227.
[95] Kim J, Lee CW, Choi W. Platinized WO(3) as an environmental photocatalyst that generates OH radicals under visible light. Environ Sci Technol 2010;44(17):6849e6854.
[96] Yu L, Zhang Y, Zhang B, Liu J, Zhang H, Song C. Preparation of a novel PSf membrane containing rGO/PTh and its physical properties and membrane performance. J Membr Sci 2013;447:452−62.

FURTHER READING

Choi W, Choi J, Bang J, Lee J-H, Appl ACS. Mater Interfaces 2013;5:12510−19.
Lee J, Chae H-R, Won YJ, Lee K, Lee C-H, Lee HH, et al. J Membr Sci 2013;448:223−30.
Zhao C, Xu X, Chen J, Yang F. J Environ Chem Eng 2013;1:349−54.
Zhang J, Xu Z, Mai W, Min C, Zhou B, Shan M, et al. J Mater Chem A 2013;1:3101−11.
Xu Z, Zhang J, Shan M, Li Y, Li B, Niu J, et al. J Membr Sci 2014;458:1−13.

CHAPTER 8

Selection of Water-Treatment Technology

8.1 INTRODUCTION

When selecting a water-treatment technology the most important issue to consider is the purpose of the water-treatment. If the purpose of the treatment is to produce safe drinking water, then the desired quality of water is automatically fixed and the quality of water assumes the highest priority in matters of selection of technology as there should be no compromise on quality. However, despite the existence of the guidelines set by the World Health Organization (WHO) on the standards of safe drinking water, there are variations in standards around the world. Some countries have set the standards above the WHO-prescribed limits and other countries have set standards below the WHO standards. Socioeconomic conditions, access to treatment technology, standard of living, reverence for human life, and general perceptions of government play decisive roles in setting water quality standards, which obviously relate to the selection of a treatment technology and vary by country. If the treatment is meant to protect the environment from the onslaught of hazardous wastewater that may be discharged by industries or municipalities, there are a different sets of effluent standards to follow, which are far less stringent than drinking water standards. There are again country-wise variations on such effluent standards, which are largely determined by socioeconomic conditions and reverence for the environment.

The source of water is the second most important issue. Water supply sources broadly fall into two categories: surface water sources and groundwater sources. Lakes, rivers, and reservoirs are the main surface water sources. Underground aquifers constitute the main groundwater sources.

In general, groundwater is usually superior to that of surface water with respect to characteristics such as bacteriological content, turbidity, and total organic concentrations. But the same groundwater may be inferior to surface water in terms of hardness-causing minerals and leached-out metals or metalloids like arsenic, iron, manganese, chromium, and fluoride.

Industrial Water Treatment Process Technology.
DOI: http://dx.doi.org/10.1016/B978-0-12-810391-3.00008-4

Traditionally groundwater sources have been extracted for drinking, agricultural, and industrial purposes in many countries without much need for treatment or for planning. However, with today's ever-increasing population, intensive agricultural and industrial activities drawing huge quantities of water from underground aquifers, rapidly falling groundwater levels along with contamination of the groundwater by leached-out metals, metalloids, and landfill leaching have compounded the problem. Therefore to ensure sustainable and safe water supplies, mapping of all underground aquifers is needed to determine a water-treatment and supply strategy along with periodic assessment of quality of such water. In addition, the possibility of landfill leaching, vegetation, and other industrial activities in the surrounding areas must be considered when developing groundwater sources.

When selecting treatment technology for surface water, wide seasonal variations of both quality and quantity of the water need to be thoroughly assessed. The possibility of eutrophication and algal bloom in surface water should also be taken into consideration. For lake water, the temperature variation in the winter months may cause anoxic situations in the deeper layers compared to surface water, which may solubilize minerals like iron and manganese. The release of anoxic or anaerobic decay products like hydrogen sulfide may significantly deteriorate water quality in terms of color, taste, and odor. In the case of river water, during heavy rains in the monsoon months, there may be substantial deterioration in water quality due to agricultural and industrial runoff. Moreover, there may be accidental spills throughout the year. Thus surface water-treatment must consider additional contaminants when determining the best water-treatment technology.

If the water source is a heavily polluted river or lake, treatment options will be entirely different to address the contaminants in the water. For clean groundwater, a one-step membrane-based treatment may be sufficient. For relatively dirty river water, a series of treatment steps beginning with rough filtration, sand filtration, coagulation—flocculation, membrane filtration, or disinfection may be necessary to produce safe potable water. If treated groundwater is contaminated with a single metal or metalloid like chromium, arsenic, or fluoride, a nanofiltration-based treatment technology may be most suitable. Water required for medicinal purposes such as for production of injection-grade water, a sophisticated treatment technology incorporating membrane is required.

Finally from a technological point of view, selection of a treatment technology should largely be based on characteristics of the raw water and the desired quality of the treated water. Therefore characterization of both raw and treated water following standard methods is essential. The WHO guidelines for safe drinking water as discussed in Chapter 1, Introduction, should be considered for treatment and supply of potable water. For industrial effluents, the country-specific effluent discharge standards are usually considered when deciding on the treatment technology. Temporal and spatial variation in both quality and quantity of water as discussed in Chapter 1, Introduction, may serve as additional guidelines in this context. In the following we discuss how to select a treatment technology based on its capability of improving water quality under various situations.

8.2 AERATION TECHNOLOGY

The simplest water-treatment is aeration alone. When poor quality of water is due to the presence of dissolved gases like H_2S and volatile organic substances (VOCs), a suitable aeration technology as discussed in Chapter 1, Introduction, may be adopted. The presence of these gases and VOCs cause taste and odor problems and make water unfit as potable water. This simple aeration treatment often greatly succeeds in substantially improving the quality of water provided impurities other than these are not present.

8.3 CHEMICAL TREATMENT TECHNOLOGY

Chemical treatment technology as described in Chapter 2, Chemical Treatment Technology, is remarkably fast in treating water where added chemical reagents cause qualitative improvement of water. A chemical treatment technology may incorporate a range of methods for improving water quality such as ion exchange, chemical oxidation, chemical disinfection, chemical precipitation, and neutralization. Ion exchange is particularly suitable for removing hardness-causing elements from water. Chemical oxidation can treat specific chemical contaminants like heavy metals, iron, and manganese. Advanced oxidation addresses highly refractory substances. Chemical neutralization using alkali and acid is often essential for biological treatment and even in chemical precipitation.

Neutralization is a standard provision prior to biological treatment. In some cases, biological treatment cannot succeed in the presence of toxic contaminants like cyanide or phenol in high concentrations. In such cases, degradation of these toxic contaminants by chemical treatment significantly increases the success of a biological treatment scheme. However, chemical treatment invariably leads to generation of a huge amount of sludge and handling and treatment may not be very eco-friendly. Harsh chemicals often need to be added in the treatment process leading to environmental degradation.

8.4 BIOLOGICAL TREATMENT TECHNOLOGY

The idea of biological treatment comes from natural assimilation of many water pollutants or contaminants by natural flora and fauna. Biological treatment is widely suggested for large-scale treatment of large amounts of industrial and municipal wastewater. In the treatment of large amounts of wastewater, biological treatment is often found to be the least costly technology. This treatment technology is highly successful when industrial or municipal wastewater contains high BOD loading. However, when deciding between biological and chemical treatment options, the biodegradability of the wastewater has to be assessed first. Following standard methods, both the BOD and COD of wastewater should be measured.

For successful treatment of wastewater with COD/BOD ratio ≤ 2.5, a pure biological treatment technology based on activated sludge, trickling filter, aerated lagoon, rotating-disk biological contactor, or a submerged bioreactor may be considered suitable. On the other hand, a COD/BOD ratio above 2.5 but below 5.0 may suggest the presence of some molecules refractory to biodegradation. However, biological treatment may still be adopted in this case, with the provision of reasonably high residence time for the wastewater in the biological treatment unit. A COD/BOD that exceeds 5.0 often indicates the presence of toxic substances that are very likely to reduce the metabolic activity of microbes in the biomass. Direct biological treatment in this case should be avoided as the microbes will fail to survive in such an environment. When pretreatment using chemical reagents or adsorbent additive can reduce the COD value substantially, a biological treatment technology that incorporates this type of pretreatment is recommended. The other options may be UV oxidation or advanced oxidation.

8.5 PHYSICOCHEMICAL TREATMENT TECHNOLOGY

In the large-scale treatment of large amounts of surface water for the removal of suspended solids, turbidity and hardness-causing elements, physicochemical treatment such as coagulation–flocculation–settling is effective and economically viable. Physicochemical treatment technology is widely used for water softening. For the removal of heavy metals, treatment technology comprising chemical precipitation coupled with settling and filtration is suggested. For removal of colloidal particulates from water, a treatment technology based on coagulation–flocculation–settling is ideal. The success of the technology depends on the selection of suitable coagulant, maintenance of appropriate doses of coagulants, pH of the medium, required degree of agitation, and settling opportunities. Adsorption-based water-treatment technology has also been traditionally used in water purification for removal of dyes, coloring materials, NOMs, and metals. All adsorption-based technology shows remarkable purification performance initially but over time as the adsorbent bed gets exhausted, the separation efficiency goes down. Without continuous monitoring of water quality, it is difficult to detect the point of system failure. One of the drawbacks of using adsorption-based technology is the need for frequent replacement of adsorbent material and continuous monitoring of the treated water. This treatment technology should be adopted where continuous monitoring of the system performance coupled with regular replacement of adsorbent material is possible. The cost of regeneration/replacement of adsorbent and disposal of pollutant-laden adsorbent need to be considered when selecting an adsorption-based technology.

8.6 MEMBRANE-BASED TREATMENT TECHNOLOGY

Since membranes have been found to be effective in separation–purification beyond nanoscale, the development of tailor-made membranes and largely fouling-free modules in recent years has created unlimited scope of application of membrane-based technology in water purification in cases ranging from groundwater to highly hazardous and dirty wastewater. Membranes belonging to different transport regimes, mechanisms, and sieving regimes are used depending on the separation needs. Membranes being semipermeable barrier during water purification primarily acts as filter though it is not always the sieving mechanism that does the purification job unlike the standard filters. Sand is the best natural filter at the lowest cost, although the degree of separation—purification in this case is only limited to microfiltration range. A full membrane-based technology using membranes of different separation, transport, and sieving mechanisms such as micro, ultra, nano, and reverse osmosis (RO) are now available. In the water-stressed regions of the world, where there are few alternatives to seawater, membrane-based technology such as RO technology is suggested as the best technology for desalination of seawater to produce potable water. Solar-driven RO plant technology has also emerged in recent years as part of water-energy nexus. Membranes can offer almost any degree of water purification at varying costs. Furthermore, the costs of membrane-based solutions are coming down in many countries and higher-quality membranes in terms of flux, rejection, durability, and cost are emerging. Thus membrane-based treatment technologies are viable alternatives to many conventional water-treatment technologies. In general, for low degree of separation, polymer- and ceramic-based membranes belonging to micro to UF regimes are used. For extremely high degree of purification, RO or NF membrane technology is most widely used. Membrane-based technology as discussed in Chapter 5, Water Treatment by Membrane Separation Technology, and Chapter 6, can be used in many water-treatment fields. Integration of membrane from different transport and sieving regimes has created numerous opportunities for the development and deployment of multistage membrane treatment technology in the treatment of high-quality water for use in pharmaceutical, medicinal, and drinking purposes. The largest water-treatment plants today are membrane-based and serve people in the most water-stressed regions of the world. If quality of treated water is the most important concern, a membrane-based technology is likely to be the best choice.

8.7 HYBRID TREATMENT TECHNOLOGY

Sustainable water-treatment technology today is often a hybrid technology. Judicious combination of several treatment schemes has resulted in several hybrid water-treatment technologies. The major strength of these hybrids is capability of handling even wastewater containing the most refractory pollutants. Such hybrid technologies have been successful in closing water-treatment and use cycles in process industries or in highly water-demanding industries. As such hybrid technologies often have nanomembrane or RO membrane as the final polishing step can promise reusable criteria to the treated water permitting their recycle and reuse within the industry. This saves on freshwater consumption and also protects surface water bodies from the contamination of hazardous waste discharge from industries. In Chapter 6, several such hybrid treatment technologies were discussed in the context of industry-specific treatment of wastewater.

8.8 NANOTECHNOLOGY

Nanotechnology in water-treatment has developed over the last two decades, and many nanomaterials developed over this period have been successful at removing specific water contaminants. In particular, nanomaterials for the removal of heavy metals from water have been widely investigated and examined. However, the possibility of release of nanoparticles to the environment has caused concern about the safe use of nanomaterials in water-treatment. For example, the use of nanomaterials in water disinfection has not been successful because of the possible release and toxicological effects of nanoparticles. While the development of nanocomposite membranes has opened up new avenues and possibilities, scale-up confidence is still limited due to lack of plant data on pilot or commercial scale.

8.9 TECHNOLOGY FOR WATER DISINFECTION

The broad technology options for disinfection of water include UV irradiation, ozonation, chlorine-based treatment, and membrane separation. Disinfection is usually required for surface water. However, if the water is transported over a long-distance through pipelines, it is necessary to ensure residual disinfectant is present in the treated water to address any enroute contamination of water. When selecting disinfection technology, the cost of treatment, effectiveness of treatment, presence of residual disinfectant for long-distance transmission, and the possibility of formation of even more harmful disinfection byproducts need to be considered. Ozone-based treatment is very effective for in-situ treatment and use of water. It does not leave any residual effect to take care of enroute contamination. UV radiation is also effective but again offers no residual effect. While UV irradiation and ozonation do not produce any harmful disinfection byproduct, chlorine-based treatment is effective, leaves residual disinfectant, and is also relatively low cost. However, during chlorination, reaction with NOMs may result in formation of disinfection byproducts, many of which are either carcinogens or suspected carcinogens. Water filtration by RO membrane or even NF membrane can ensure pathogen-free water without producing any harmful disinfection byproduct. Thus the choice of the best technology is obvious provided costs can be covered.

8.10 ENVIRONMENTAL REGULATIONS AND COMPLIANCE, PUBLIC AWARENESS

When selecting the best water-treatment technology, environmental regulations, compliance of such regulations, access to better water-treatment technology, public awareness on environmental degradation, and adverse health effects of poor water quality play crucial roles. Under most circumstances, an industrial house will not like to switch over to a novel technology for treatment of its wastewater unless there are compelling regulations. However, the existence of such regulations does not guarantee commissioning of a better treatment plant unless there is governance on strict compliance of environmental regulations. Public ignorance on the hazardous health effects of poor quality water often allows a poor treatment technology to continue despite its failure to treat water to the standards set by the WHO and even the effluent discharge standards of individual countries. The situation may continue despite the availability of new and efficient water-treatment technology.

8.11 COST OF TREATMENT

It is generally known that better-quality water costs more. Therefore selection of a treatment technology will depend on cost. Implementing a novel technology requires capital investment. If cost of treatment can be afforded, the quality of water will not be sacrificed.

8.12 ACCESS AND AWARENESS OF NEW TECHNOLOGY

In many cases, a new water-treatment technology may not involve high costs, but the implementing authority may not be aware of the technology or may not have access to it. The interface problems need to be overcome with proper information, knowledge of recent developments, confidence and courage of venturing into new project. The willingness to embrace the new is essential to implementation of better water-treatment technology.

CHAPTER 9

Design and Construction of Water-Treatment Plants on Novel Technology

9.1 INTRODUCTION

Challenges in water treatment change over time. At the beginning of the twentieth century the challenge was to eliminate pathogens from water, with an emphasis on chlorination. Gradually the technologies improved and were aimed at increased clarification, removing hardness-causing, and discoloring agents from water, in addition to iron and manganese. Compared to the treatment needs of surface water, the needs for treatment of groundwater was very insignificant remaining confined to chlorine-based treatment for maintaining residual chlorine as guard against possible enroute contamination only. The growing population, industrialization, and changes in lifestyles eventually resulted in overexploitation of groundwater and severe pollution of surface-water bodies in several parts of the world. Such developments drastically changed the nature of water contaminants and their concentrations and thus their treatment. Overexploitation of groundwater also led to geological disturbances and leaching out of metallic elements as well as metalloids such as iron, manganese, arsenic, and fluoride from their crystal lattice, contaminating the groundwater make it unpotable. Since the 1980s contamination of groundwater by fluoride and arsenic has affected millions of people across the world, particularly in South-East Asian countries. In view of the carcinogenicity of arsenic, the World Health Organization (WHO) set 10 μg/L as the MCL (maximum permissible contaminant level) for arsenic in potable water. In many arsenic-affected areas, groundwater has been contaminated by leached-out arsenic to levels as high as 3000 μg/L posing a formidable challenge in meeting WHO standards. For fluoride, the MCL is 1.5 ppm. Disinfection byproducts such as total trihalomethanes (TTHM), endocrine-disrupting substances, pharmaceutically active compounds, personal care products, and synthetic organic chemicals, many of which are carcinogenic and suspected carcinogens, are

Industrial Water Treatment Process Technology.
DOI: http://dx.doi.org/10.1016/B978-0-12-810391-3.00009-6

now major concerns in water-treatment domain demanding paradigm shift in treatment strategy through innovative novel technologies.

Today, membrane-based treatment plants using ultra, nano, and reverse osmosis (RO) membranes are becoming more common. In particular, RO technology is now commonly used in water desalination. There has also been a paradigm shift in water-treatment policy that now revolves around closing the loop of water use and recycling, thus protecting surface-water bodies from the onslaught of municipal and industrial wastewater. The water-energy nexus, where water treatment is expected to be energy saving and encompass simultaneous energy-generation through biogas generation in the anaerobic sludge-digestion process, is also a focus today. Such sludge is often a major disposal issue in water-treatment plants.

The design and construction of novel water-treatment plants and innovative technologies have been widely researched and reported on. However, scale-up confidence remains limited. Given the possibility of contamination of both surface and groundwater by new, complex, and hazardous contaminants largely arising from the diverse activities and changing lifestyles of today's ever-growing population, the degree of efficiency of purification of water-treatment plants is critical. The quality of the treated water and the cost effectiveness of the water-treatment technology will be important in designing, commissioning, and sustaining newly designed plants. While new technologies are continually being developed in research laboratories, screening and pilot-scale investigations will have to be brought to the field. The decision-making stages involved in commissioning a new treatment plant based on an emerging technology are shown in Fig. 9.1.

9.2 DESIGN AND CONSTRUCTION OF AN INDUSTRIAL WASTEWATER-TREATMENT PLANT USING NEW TECHNOLOGY

A novel treatment technology based on the principles of FO and nanofiltration (NF) has been developed recently [1] for hazardous industrial wastewater. The treatment scheme, as shown in Fig. 9.2, has been successfully tested in the context of pharmaceutical wastewater. Instead of the usual vertical orientation of the FO membrane, this technology uses horizontal orientation of the FO membrane allowing tangential cross-flow of the wastewater and significantly reducing concentration

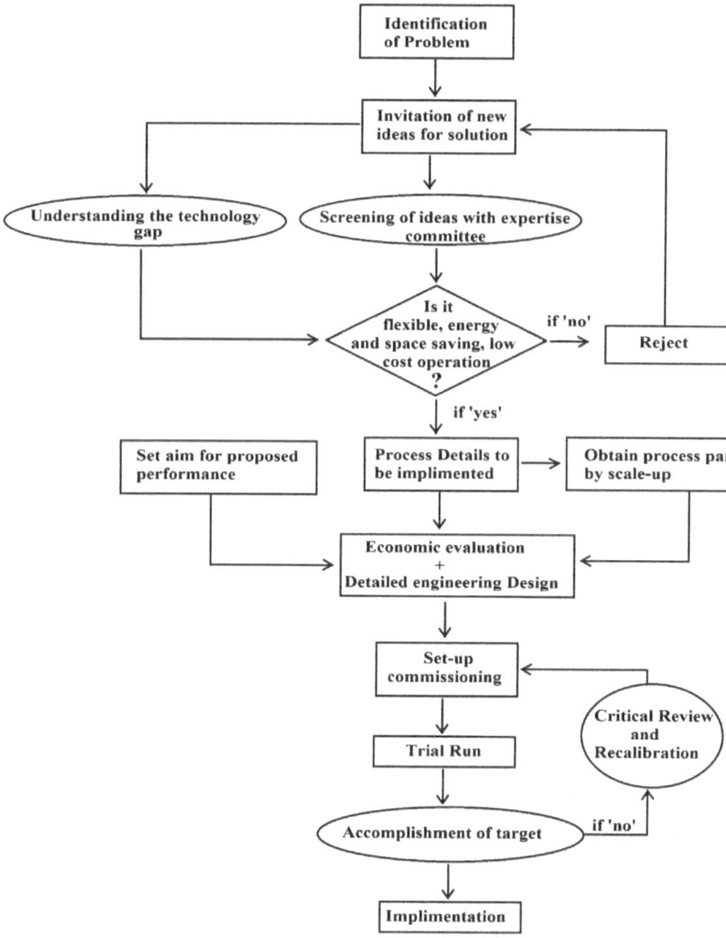

Figure 9.1 Flow chart of the major decision-making stages in the design of a treatment plant.

polarization and hence fouling of membrane, which is considered a major hindrance in membrane-filtration systems.

9.2.1 Design Basis

The design capacity is based on a typical pharmaceutical plant generating 90KL wastewater every day. The target is not only to treat all this wastewater up to the standards of effluent discharge but also to make it reusable.

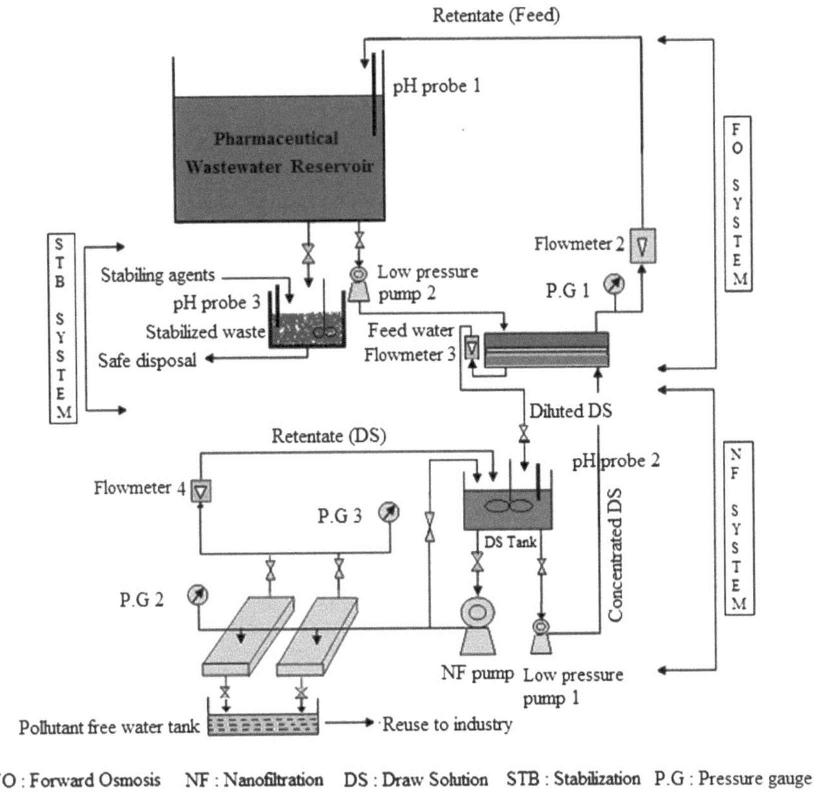

Figure 9.2 FO–NF based wastewater-treatment scheme [1].

9.2.2 System components and Operational parameters

The treatment system comprises a upstream flat-sheet cross-flow FO membrane module and a downstream flat-sheet cross-flow NF membrane module. The draw solution in the FO system is 0.5 M NaCl solution and is used at a cross-flow rate of 10 L/h. Feed wastewater is typically made to flow at a cross-flow rate of 180 L/h under 1.6 bar operating pressure generating a flux of 52 LMH ($L/m^2/h$). For recovering the draw solution and reusable water from the diluted draw solution, an NF membrane as a flat-sheet cross-flow module is adopted for downstream purification where diluted draw solution at a cross-flow rate of 800 L/h is made to flow tangentially over the membrane module under a transmembrane pressure of 12–13 bars generating a reusable pure water flux of 52–54 $L/(m^2h)$. A typical set of commercial membranes may be used. The treatment plant is shown schematically in Fig. 9.2.

Table 9.1 Typically designed parameters for pharmaceutical waste water-treatment plant

Parameters	Material of construction	No. of equipment	Area or volume	Other specifications
FO module	perplex sheet	105	0.3 m^2 each	flat-sheet cross-flow
FO membrane	poly amide	–	125 m^2/year	NF-2
NF module	SS-316	105	0.3 m^2 each	flat-sheet cross-flow
NF membrane	poly amide	–	125 m^2/year	NF-1
Membrane life	–	–	–	3 months
Draw solution	0.5 M sodium chloride + water	–	–	total 50 kg/year
Feed tank	SS-316	5	25,000 L each	Digital pH indicator must be attached
DS tank	SS-316	1	5000 L	–
FO pump	–	8	–	Peristaltic pump (HP)
NF pump	–	4	–	Diaphragm pump, max. pressure 50 bar, 3 HP

9.2.3 Treatment target

At least 98% COD removal and 99% draw solution recovery with reusable water flux of 60 LMH at reasonable price while preventing discharge of 90 KL/day of untreated hazardous industrial wastewater from a typical pharmaceutical plant to the environment.

9.2.4 Scale-Up and Plant Design

Scale-up has been done following a standard procedure to assess the overall investment of a closed loop-based FO–NF integrated pharmaceutical wastewater-treatment plant for treatment where flux of the FO plant is 52 LMH and that of the NF plant is 52–54 LMH. The design parameters are provided in Table 9.1.

9.2.5 Module Design

Treating 90 KL/day requires provision for treating 5625 L/h. On passing 180 L/h feedwater, 52 L/(m^2/h) flux is generated for the pure or treated water.

During 16 h of continuous operation, the volume of treated water produced at the FO module is $52\ \text{L}/(\text{m}^2/\text{h}) \times 16\ \text{h/day} = 832\ \text{L}/(\text{m}^2\text{day})$ of treated water flux, which is equivalent to processing $180(\text{L/h}) \times 16\ \text{h/day}$ wastewater (as feed flow rate is 180 L/h).

Thus 180×16 L/day wastewater is equivalent to a treated water flux of 832 LMH, and 90,000 L/day wastewater is equivalent to a treated water flux of $(52 \times 16 \times 90{,}000)/(180 \times 16) = 26{,}000$ L/day of treated water.

The membrane surface that needs to be put on duty on a daily basis is $26{,}000\ \text{L/d}/832\ \text{L/m}^2/\text{d} = 31.25\ \text{m}^2$. This means membrane modules number of $31.25\ \text{m}^2/0.3\ \text{m}^2 = 105$ (where each module offers $0.3\ \text{m}^2$ effective surface).

Considering an average life of 3 months for the membranes, the total membrane surface required will be $31.25 \times 4 = 125\ \text{m}^2$

The module number will remain the same, i.e., 105.

The overall cost assessment given in Table 9.2 includes equipment installation, instrumentation, and operating cost. The total capital investment cost involved for such a plant calculates to US\$$23.1 \times 10^4$. Direct capital investments accounts for equipment purchase cost, equipment installation cost, piping cost, building cost, electrical instrumentation cost, yard improvements cost, service facilities, and land purchase cost. Total equipment purchase cost (direct) stands at US\$$18.1 \times 10^4$. For the proposed system, costs for pumps and membrane modules contribute about 22% and 13%, respectively, to the total equipment cost. Other associated direct costs (e.g., piping, electrical connections, instrumentation) involve 65% of the total direct cost, which is 78% of the total capital investment. Similarly, indirect costs like engineering supervision, contractor's fees, and contingency contribute about 22% of the total capital investment cost. The cost of equipment may be based on manufacturers' costs or calculated using the following standard equation [2]:

$$\text{Cost of higher capacity equipment} = \text{Cost of lab scale equipment} \times \left[\frac{\text{capacity of high capacity equipment}}{\text{capacity of lab scale equipment}}\right]^n$$

where n represents the scale-up factor and differs for different equipment; the standard values of n are taken from a standard list [2].

Table 9.2 Total investment cost assessment for FO-NF based pharmaceutical wastwater-treatment plant (capacity: 90,000 L/day = 32850 m^3 per annum)

Item	**Cost (US$)**
(A) Capital investment	
1. Direct costs	
1.1. SS 316 membrane modules (nanofiltration)	1.5×10^4
1.2. Perspex modules (forward osmosis)	0.8×10^4
1.3. Pumps (peristaltic and hydra-cell)	4.5×10^4
1.4. Holding tanks for untreated wastewater	2.2×10^4
1.5. Holding tanks for draw solution	1.6×10^4
1.6. Piping	1.6×10^4
1.7. Electrical connections	2.8×10^4
1.8. Instrumentation	3.8×10^4
2. Indirect costs	
2.1. Engineering and supervision	1.6×10^4
2.2. Contractor's fees	1.5×10^4
2.3. Contingency	1.2×10^4
Total capital investment ($/year)	*23.1×10^4*
(B) Operational cost/year	
i. Membrane cost	7.2×10^3
ii. Draw solute (NaCl)	0.5×10^3
iii. Electrical	2.5×10^3
iv. Manpower	16.8×10^3
v. Overhead charges	4×10^3
vi. Maintenance	4.2×10^3
Total annual operating costs ($/year)	*35.2×10^3*

The wastewater-treatment plant is assumed to be continuously working with two operators and two helpers in two shifts daily. The salary for two skilled operators is taken as $400 per month, and for two semiskilled staff, may be taken as $300 per month. Thus the manpower cost per year is US$$16.8 \times 10^3$ per year.

The cost assessment is based on the annualized investment cost and annualized operational cost. The annualized capital cost is computed using:

$$\text{Annualized capital cost} = \left(\frac{\text{Total capital (\$)} \times \text{Cost recovery factor}}{\text{Water flux per year (m}^3\text{/year)}}\right)$$

The cost recovery factor is dependent on the project life of the plant ($n = 25$ years) and interest ($i = 4\%$). This is calculated by:

$$\text{Cost recovery factor} = \left(\frac{i(1+i)^n}{(1+i)^n - 1}\right)$$

$$\text{Cost recovery factor} = \left(\frac{0.04(1+0.04)^{25}}{(1+0.04)^{25} - 1}\right) = 0.06$$

$$\text{Annualized capital cost} = \left(\frac{23.1 \times 10^4(\$) \times 0.06}{32850(\text{m}^3/\text{year})}\right) = 0.42\$/(\text{m}^3/\text{year})$$

The annualized operational cost can be computed by:

$$\text{Annualized operational cost} = \left(\frac{\text{Total operational cost}(\$/\text{year})}{\text{Water flux per year}(\text{m}^3/\text{year})}\right)$$

$$\text{Annualized operational cost} = \left(\frac{35200(\$/\text{year})}{32850(\text{m}^3/\text{year})}\right) = 1.07\$/\text{m}^3$$

Considering all the factors including maintenance, operating supplies, depreciation, and insurance in membrane-integrated plant, the annualized cost for treating 90 KL wastewater per day or 32,850 m^3/year from a typical pharmaceutical plant adds up to US\$ $(0.42 + 1.07) =$ US\$ 1.49 per m^3.

As the treated water from the plant is of reusable criteria, the actual treatment cost will go down further on deduction of the economic value of the treated reusable water from the total annualized cost of treatment of industrial wastewater.

9.2.6 Energy consumption

The energy consumption of membrane-based wastewater-treatment plants is pretty low due to the absence of any phase change or energy-intensive process and due to the simple plant configuration and few treatment steps. Energy involvement in a scaled-up plant can be found using the six-tenths factor rule [2]. The total annual power consumption of the proposed plant (at each stage and at each shed) is shown in Table 9.3, which has been calculated to be approximately 21.8×10^4 kW/h only. Thus the designed system is energy efficient.

Table 9.3 Total estimated energy consumption of the proposed plant (capacity: 90,000 L/Day = 32850 m^3 per annum)

Equipment	Energy (kW h)
FO pumps	3.5×10^4
NF pumps	12.5×10^4
Other (fans, ACs, lights, computers, laboratory instruments, different electrical appliances)	5.8×10^4
Total energy requirement	21.8×10^4
Energy requirement behind per unit of clean water yield	8.8 kWh/m^3 of clean water

9.2.7 Civil Design and Planning of Estate

Generally such a wastewater-treatment plant would be located outside large urban centers where according to environmental policies, easy drainage or emission is permitted. Wastewater-treatment sector belonging to industrial estates should be well planned to prevent haphazard growth of an industry. Membrane systems being modular in design can be arranged in both series and parallel fashion to reduce space problems as shown in the compact design of Fig. 9.3. The working site should be close to a perennial water supply and within reasonable distance from the source of electric power and transport facility (highways and railways). The orientation of buildings surrounding the plant set-up should be closely linked where adequate lighting inside the total estate should be properly arranged.

Typical design criterion of civil construction generally state that: (1) the height of the plinth should be at least 0.03 m above the crown of the road to avoid flooding during heavy rain; (2) height up to tie level should be a minimum of 5 m of roof from the plinth for the working space; and (3) windows and ventilators should not consume more than 15% to 25% of total floor area depending on regional climatic conditions (from humid region to hot dry regions). A minimum of two doors in each shed is essential for safety but should be increased as required. For the main working plant shed area, sliding doors or shutter doors should be 5.5 m × 4.5 m, the main door size should be 2.5 m × 2.5 m, and the size of rear doors should be 2.5 m × 1.8 m. Floors should be robust enough to hold high-capacity tanks with an overall width of 0.06 m concretized by cement. Turbine ventilators could be attached on a pitched roof of galvanized iron sheets or asbestos cement sheets to reduce cost while increasing ventilation rate.

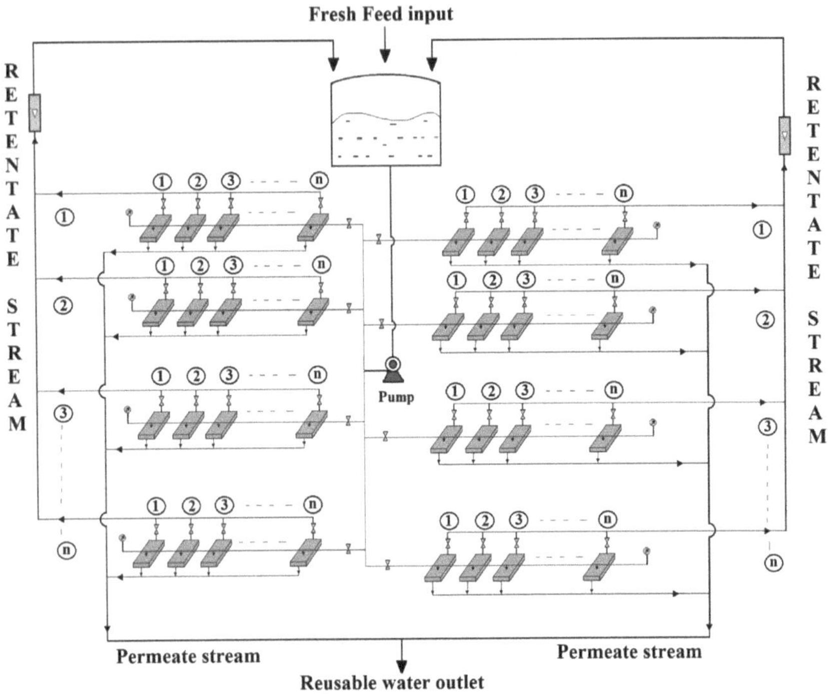

Figure 9.3 Compact series and parallel arrangement of membrane modules.

While sheds of 10 m and 13 m span are mostly adopted in engineering designs, by allowing standard-steel trusses (25–28 kg per m^2 of plinth area), variable length of shed as per requirements may be arranged. As a small but compact wastewater-treatment plant, a minimum area of the covered main work shed of 192 m^2 (16 m × 12 m) should be considered. Using row construction for the work shed, the plant set-up area takes up about 50–60% of the total estate. A control room and a testing laboratory should be built close to the plant set-up area for easy control and analysis of operating parameters and testing of output water samples. A cumulative 8–15% of the total plant area should be dedicated to these departments.

For other departments like maintenance and safety, a resting lounge for laborers, and a warehouse (for quick supply of useful chemicals and consumables), a total of 20–22% of the total estate area should be provided. Approximately 10% of the total estate could be used for management offices. The utilization of the total space for different units is given in Table 9.4, and the complete design layout of the proposed plant is shown in Fig. 9.4.

Table 9.4 Utilization of space in the total estate of such a wastewater-treatment plant

Individual work shed	Required percentage of the total Estate	Actual area (m^2)
Plant set-up	52%	240
Control room and Laboratory	9%	41
Open space	8.3%	38.4
Administrative	5%	19.2
Roads	16%	77
Total covered work sheds	71%	329.2
Total area of the estate		461

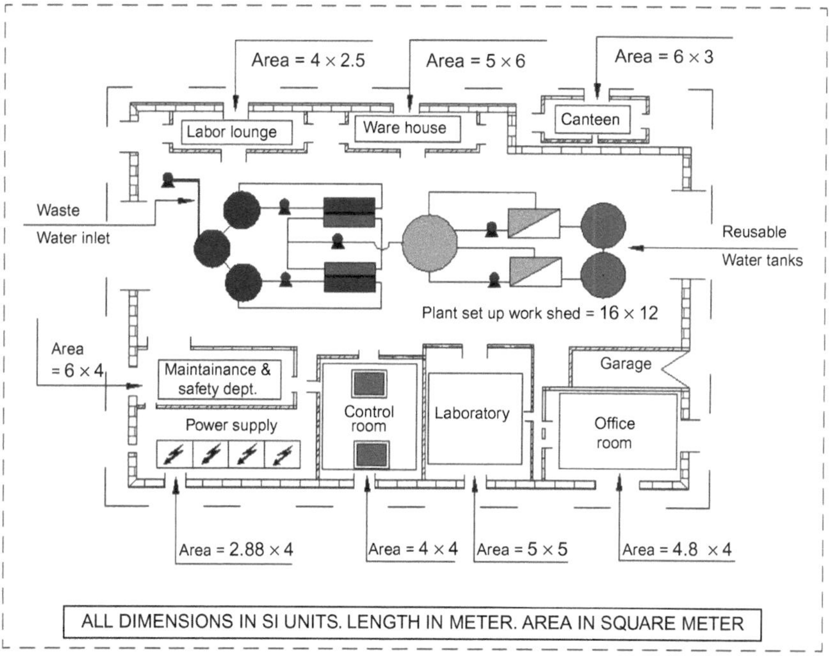

Figure 9.4 Overall plant layout for pharmaceutical wastewater-treatment system.

9.3 CONSTRUCTION OF A GROUNDWATER-TREATMENT PLANT USING NANOFILTRATION-INTEGRATED HYBRID TREATMENT TECHNOLOGY

9.3.1 Introduction

Traditionally groundwater has been used without much treatment as natural adsorbents normally purify water. However, recent reports of widespread contamination of groundwater by leached-out metals and metalloids has caused concern, in part, because reaching permissible MCL (maximum contaminant level) poses a challenge. Adsorption, ion-exchange, coprecipitation, and membrane filtration have been suggested for the treatment of such contaminated water, but in some cases, these treatment schemes fail to meet the MCL and make the water safe. A classic case is that of contamination of groundwater by leached-out arsenic. Arsenic is found both in trivalent (As-III) as well as in pentavalent (As-V) forms. While pentavalent arsenic can be easily separated out by NF membrane, the trivalent type needs to be oxidized to pentavalent form prior to membrane separation. Thus a chemical pretreatment unit for oxidation of trivalent arsenic is needed, resulting in the need for a combined chemical and NF plant. Such a hybrid treatment scheme was recently developed (Indian Patent No. 275244) [3].

Due to limited options for disposal of highly concentrated arsenic particles, many arsenic-removal plants dump the arsenic into the environment, potentially recontaminating underground aquifers through natural percolation. A sustainable treatment technology needs to address the disposal issue of concentrated arsenic while limiting treatment costs. The technology developed by Pal et al. (2014) [3] is a low-cost membrane-based technology for arsenic separation from contaminated groundwater that also provides a solution to the problem of groundwater contamination. This section describes the design and construction of a plant like this for a community of 500 people requiring 5000 L/day of pure drinking water. The NF-based hybrid treatment plant can treat not only arsenic but also similar metallic or metalloid contamination and pathogens. Table 9.5 presents some possible contaminants of water that can be safely treated by such a plant.

9.3.2 Plant Construction and Cost Estimation

Raw Water

Table 9.5 does not present an exhaustive list of contaminants that can be treated by an nanomembrane-integrated hybrid treatment plant. Contaminants like fluoride can also be effectively removed by this design in addition to pathogens if present in the water. Groundwater to be treated can be extracted using a pump based on the plant's capacity. The extracted water may be stored in a water reservoir for subsequent pumping to the membrane module via the oxidation unit at desired pressure.

Table 9.5 Typical contaminants of groundwater that can be removed by the design groundwater used in the water-treatment plant

Water parameters	Units	Quantity before treatment	Quantity after treatment	MCL limit (WHO)
TSS	mg/L	196	BDL	–
TDS	mg/L	205	10	500
Conductivity	μs/cm	598	48	5
Salinity	–	0.45	0.03	1000
pH	–	7.2	6.5	8.5
Chloride	mg/L	190	8.2	250
Manganese	mg/L	1.39	0.06	0.4
Iron	mg/L	4.80	0.15	1
Total arsenic	mg/L	0.18	0.00256	0.01
As-III	mg/L	0.075	BDL[a]	–
As-V	mg/L	0.105	BDL[a]	–

[a]BDL = Below Detection Level

Site Selection

While the plant may be located within the community since it generates little noise, pollution, or health hazard, it is important to do a field survey of the site and soil testing. The following criteria should be used when selecting a site for a groundwater-treatment plant:

- Water levels of the proposed site
- Capacity of the plant
- Location easily accessible by the community
- Location on flood-free high land
- Proper sludge disposal site

Civil Construction

The NF-integrated hybrid treatment plant should be developed within a reasonable distance of the water to be treated. Floor area and height of the building are the main design criteria of the treatment building while height and length are the design criteria of the plant wall. Fig. 9.5 shows a rough layout of the hybrid treatment plant.

The following lists the space requirements as reflected in Fig. 9.5:

1. Labor house: 5 m × 4 m = 20 m^2
2. Concrete storage or feedwater tank: Floor area 20 m^2, height 4.26 m
3. Groundwater reservoir: Concrete feedwater tank with rectangular shape: 6 m^2, Height 2 m

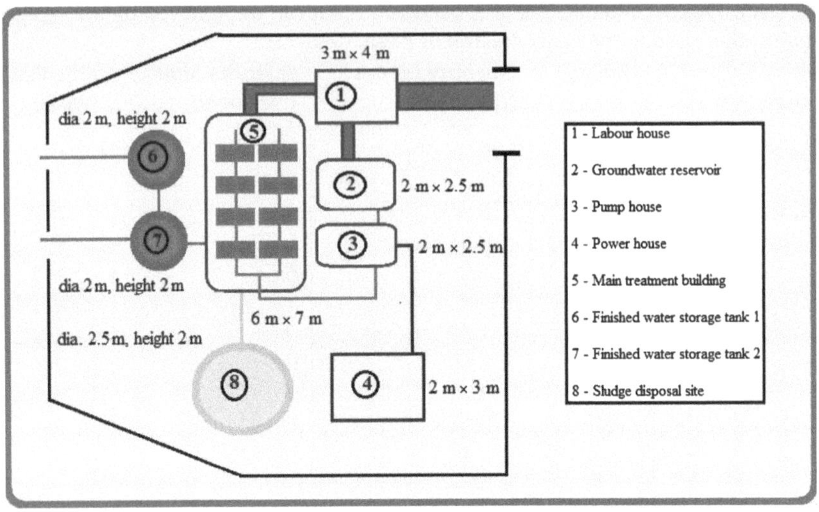

Figure 9.5 Layout of the Hybrid Treatment Plant for arsenic removal.

4. Pump house: Concrete floor and roofing: Floor area (2 m × 2.5 m) 5 m^2, Height 4.26 m
5. Power house: Concrete floor with concrete roofing: floor area 6 m^2, Height 4.26 m
6. Main treatment building: Concrete floor with concrete roofing, floor area (6 m × 7 m) 42 m^2, Height 4.26 m
7. Finished water-storage tank: Two number of concrete tank with cylindrical shape with floor area 3.14 × 2 = 6.28 m^2, height 2 m
8. Sludge disposal site – Floor area 10 m^2

Flooring: Floor may be of 0.04 m to 0.05 m cement concrete or as required.

Roofing: 0.15 m concrete roof.

Height of plinth (base course of a building): 0.3 m above the ground level.

Height up to tie level of roof from base of the building: 4.2 m.

Total space requirement for a community water-treatment plant of 5000 L/day capacity (considering future expansion) is approximately 250 m^2.

Mechanical design of equipment

1. Assessment of volume of treatment

   The developed NF module has a pure water flux of 140 L/(m^2/h) that can yield 2240 L/m^2day considering 16 hours of operation per

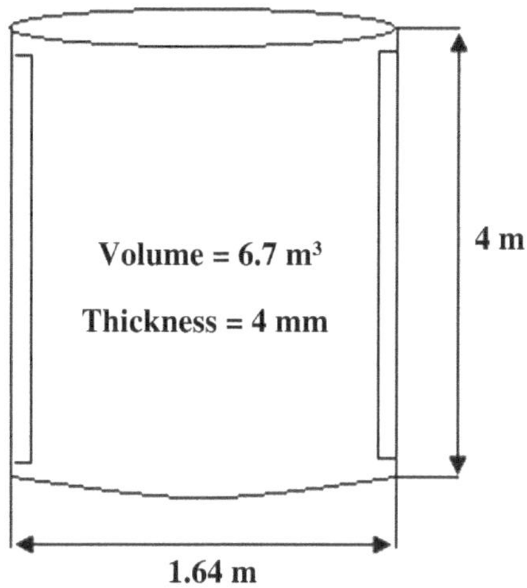

Figure 9.6 Dimensions of the rector.

day. Considering a pure water recovery rate of 80% for a population of 500, the plant has to treat 5000/(80%) or 6250 L/day of raw water to supply 5000 L/day potable water to a community of 500 people (number of people × 10 L/person/day).

2. Volume of the Oxidation reactor

   Considering 75% of the physical volume of the reactor as effective volume, the volume of the reactor to be designed is $V = 6250/0.75$ L−8450 L, which equals a volume of 8.5 m^3.

   The dimensions of the reactor are given in Fig. 9.6.

Membrane Module Design

Considering 16 hours of operation per day, the total volume of water produced by the plant is $140 \times 16 = 2240$ L/m^{2}day as the water flux of the plant is 140 L/m^{2}h.

The effective total membrane area required $= \left(\frac{5000\ \text{L/day}}{2240\ \text{L/m}^2\text{day}}\right) = 2.3\ \text{m}^2$.

As the surface area of each module is 0.1 m^2, the number of membrane modules required to generate 5000 L/day drinking water is $(2.3\ \text{m}^2/0.1\ \text{m}^2) = 23$.

The ratio of membrane module area to effective membrane area is 4.5. Thus the total space required for the membrane module is $(4.5 \times 2.3) = 10.35\ \text{m}^2$. The dimensions of the membrane module are shown in Fig. 9.7.

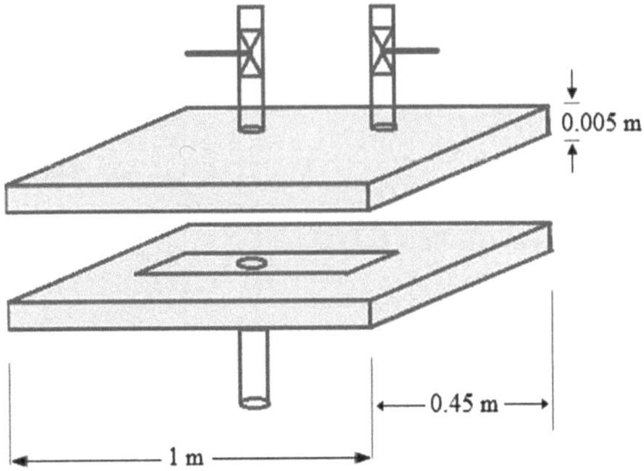

Figure 9.7 Dimentions of the membrane module.

Module Space

For 23 modules, a minimum of two parallel pipelines is required. With 12 membrane modules in each line, the total length (considering the module length and the space between the two modules) is $[(12 \times 0.45)\text{m} + (12 \times 0.15)\text{m}] = 7.2$ m is required. This implies that with some allowance, approximately $7.5\ \text{m} \times 0.076\ \text{m}$ is required for each module in each line. A schematic of the treatment plant is shown in Fig. 9.8.

Materials of Construction

The modules should be made of stainless steel and the piping may be either stainless steel or PVC or equivalent.

Pump

One high-capacity diaphragm pump of discharge capacity 1000 L/h and 30 bar maximum discharge pressure needs to be installed to supply the raw groundwater to the membrane modules. One high-capacity rotameter (maximum flow rate: 2000 L/h) will measure the water flow rate.

Cost Estimation

Capital Costs

Civil investments

In calculating the costs of the labor house, groundwater reservoir, pump house, power house, main treatment building, finished water storage tank, and sludge disposal site, the depreciation period is assumed to be 30 years. In calculating the costs of the membrane modules, reactor,

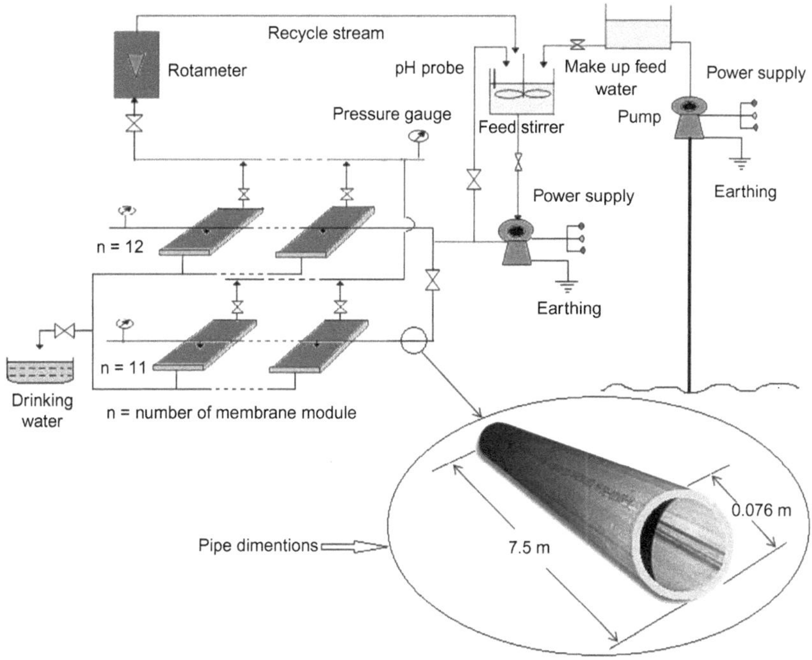

Figure 9.8 Schematic of the treatment plant.

pumps, stirrer, piping, and electrical accessories, the depreciation period is assumed to be 15 years.

In calculating the costs of electrotechnical items such as the costs of energy supply (cabling, transformers, etc.), control engineering, and all electronic components (e.g., pressure gauges, rotameters, and pH probes), the depreciation period is considered to be 10 years.

The sixth-tenths power law has been used for scale-up cost assessment and is defined as:

Operational costs

Energy costs

$$\text{Cost for scale-up} = \text{Cost data for lab scale} \times \left(\frac{\text{Specific design data used in scale-up}}{\text{Specific design data used in lab scale}}\right)^{0.6}$$

- Per day power consumption by the diaphragm pump (3 hp) = $2.3 \times 16 = 36.8$ kwh

- Per day power consumption by the groundwater extraction pump (1/6 hp) $= 0.11 \times 16 = 1.76$ kwh
- Power consumption by others $= 0.5$ kwh

  Total power consumption $= 39.06$ kwh

Cost of Consumables

5 mg/L of potassium permanganate ($KMnO_4$) is required for oxidation of arsenic (from As-III to As-V).

Thus for 5000 L/day arsenic water treatment, the total amount of $KMnO_4$ required $= 5000 \times 5$ mg $= 25$ g/day $= 9.1$ kg/year.

Membrane Costs

Membrane area required $23 \times 0.1 = 2.3\ m^2$

Membrane cost $= 40\ \$/m^2$

Skilled Labor

Labor Costs

The number of skilled staff required is three per day where the plant is to be operated 16 hours per day and the manpower cost is 350\$/month, which is $(350 \times 3 = 1050)$ per month and $(1050 \times 12 = 12{,}600)$ per year.

Additional Costs

Depreciation cost(\$/year) :

$$\left\{\left(\frac{\text{Civil cost}}{\text{Depreciation period 1}}\right)+\left(\frac{\text{Mech. Cost + Electrotech. cost}}{\text{Depreciation period 2}}\right)\right\}$$

Maintenance cost (\$/year): 2% of total investment.

Operation installation cost (\$/year): 2% of total investment.

Investment costs, operational costs, and additional costs with detailed design data of different lab set-ups are summarized in Table 9.6.

The cost assessment is based on the annualized investment cost annualized operational cost. The annualized capital cost was calculated using the following relationship:

$$\text{Annualized capital cost} = \left(\frac{\text{Total capital (\$)} \times \text{Cost recovery factor}}{\text{Water flux per year(m}^3\text{/year)}}\right)$$

Table 9.6 Investment cost, Operating cost, and Additional cost of a community based (5000 L/Day) groundwater-treatment plant

Cost parameters	Specification/materials	Design data	Cost Value
Investment cost			*Cost ($)*
1. Civil infrastructure	Cement construction	Area-84 m^2	5000
2. Membrane module cost	SS, Number of module-23	Module area-0.45 m^2	1700
3. High flow pump	Submersible pump, Number of pump-1	Max. flow rate 5000 L/h	180
4. High-pressure pump cost	Diaphragm pump, Number of pump-1	Max. pr. 50 bar, Max. flow rate 2000 L/h	3000
5. Cost for main feed pipe	Stainless steel	7.5 m long and 0.076 m diameter	300
6. Electrotechnical cost			
i. Rotameter	Number-1	Maximum flow rate-2000 L/h	30
ii. Pressure gauge	Number-1	Number-1/Maximum pressure-50 bar	30
iii. pH probe	Number-1	Number-1/pH range − 2−12	30
iv. Stirring motor	Number-1	Number-1/speed-40−1200 rpm	150
Total Investment Costs			*10420*
Operating costs			*Cost ($/Year)*
1. Electricity Cost		Power consumption-1000 Kwh/month	500
2. Membrane Cost	Commercial NF-1 membrane	Membrane needed-2.3 m^2, Cost-50 $/$m^2$	252
3. Chemical Costs	Potassium permanganate	9 kg/year, Cost of $KMnO_4$ = 14 $/kg	126
	Fe salt	2.5 kg/year, Cost of Fe salt = 15 $/kg	37
	Calcium salt	5 kg/year, Cost of Calcium salt = 12 $/kg	60
4. Labor Costs	Number of skilled staff = 3	Rate-350 $/month	12600
Total Operating Costs			*13575*
Additional Costs			
1. Depreciation cost			527
2. Maintenance cost			208
3. Operation installation cost			208
Total Additional Costs			*943*

The cost recovery factor is dependent on plant project life ($n = 25$ years) and interest ($i = 7\%$) and can be calculated by:

$$\text{Cost recovery factor} = \left(\frac{i(1+i)^n}{(1+i)^n - 1}\right)$$

$$\text{Cost recovery factor} = \left(\frac{0.07(1+0.07)^{25}}{(1+0.07)^{25} - 1}\right) = 0.08$$

$$\text{Annualized capital cost} = \left(\frac{10420(\$) \times 0.08}{1825(\text{m}^3/\text{year})}\right) = 0.45\$/(\text{m}^3/\text{year})$$

Again, the annualized operational cost can be computed by:

$$\text{Annualized operational cost} = \left(\frac{\text{Total operational cost}(\$/\text{year})}{\text{Water flux per year}(\text{m}^3/\text{year})}\right)$$

$$\text{Annualized operational cost} = \left(\frac{13575(\$/\text{year})}{1825(\text{m}^3/\text{year})}\right) = 7.43\$/\text{m}^3$$

Total Annualized Production Costs

Thus the annualized cost of production of 1000 L of drinking water from arsenic-contaminated water is the sum of the annualized capital cost and the annualized operating cost (0.45 + 7.43), which $7.88 (Table 9.6).

REFERENCES

[1] Thakura R, Chakrabortty S, Pal P. Treating complex industrial wastewater in a new membrane-integrated closed loop system for recovery and reuse. Clean Technol Environ Policy 2015;17(8):2299–310, Springer.

[2] Peters Max, Timmerhaus Klaus, West Ronald. Plant Design and Economics for Chemical Engineers. McGraw-Hill Education; 2003, 2003.

[3] Pal P, Chakrabortty S, Linnanen L. A nanofiltration–coagulation integrated system for separation and stabilization of arsenic from groundwater. Sci Total Environ 2014;476–477:601–10.

CHAPTER 10

Sustainable Water-Treatment Technology

10.1 SUSTAINABILITY AND INNOVATION IN WATER-TREATMENT TECHNOLOGY

A sustainable water-treatment technology promises efficient water-treatment at affordable prices while upholding the interests of the people, planet, and organization. In other words, it should ensure long-term environmental, economic, and societal benefits. The technology should aim at improving associated environmental quality itself being ecofriendly. There are two main approaches in water-treatment: the "end-of-pipe" approach and the preventive approach. Obviously the preventive approach is preferred to the "end-of-pipe" approach as the latter only attempts to stabilize the technology in place without moving toward improvement through innovation. The end-of-pipe approach eventually leads to "lock-in." When assessing the sustainability of a water-treatment technology, the most important indices include cost of treatment, quality of treated water, ease of maintenance of the treatment system, reliability of the system, and long-term, trouble-free, ecofriendly, and energy-saving operation with a reasonable profit margin. The objectives of sustainable development require radical departure from existing practices. It is widely believed that science and technology are the root causes of the problems of sustainability but the solutions to the problems at the same time require both science and technology. Sustainable development demands a paradigm shift in regimes, which is possible only through innovation. Moreover, sustainability is not a one-time measurable parameter. What appears sustainable today may not remain so after a decade. Sustainability can be ensured only through continuous innovation.

In the twentieth century, innovation in water-treatment saved millions of human lives by preventing waterborne diseases like cholera, diarrhea, and hepatitis. Innovation in water-treatment technology has a significant impact on the lives of the entire human population regardless of geopolitical boundaries as well as on the earth as a whole. In the following, we discuss some of the innovations in water-treatment technology.

Industrial Water Treatment Process Technology.
DOI: http://dx.doi.org/10.1016/B978-0-12-810391-3.00010-2

10.1.1 Innovation in Industrial Wastewater-Treatment

In industrially developed countries as well as in countries with fast-growing economies and large populations like India and China, keeping thousands of kilometers of river bodies free from the onslaught of hazardous municipal and industrial waste is one of the biggest challenges facing local governments. A water resource-management problem of this magnitude can't be solved by a piece-meal approach. Millions of dollars spent over the years apparently are not reflected in the quality of the prevailing water quality. Thus a paradigm shift in water resource-management policy is absolutely essential. Closing the water use loop via treatment to reusable criteria is one way out. This will not only protect surface-water bodies from being contaminated by industrial discharge of wastewater but will also significantly reduce freshwater consumption. The mindset of using rivers as the sink of all waste materials generated through human activities needs to be transformed. Installation of new treatment facilities based on sustainable technologies along the banks of the rivers of these countries needs to be done. These technologies are capable of treating wastewater to the level of reusable criteria thereby facilitating recycling. One such innovative technology developed recently [1] integrates forward osmosis (FO) with nanofiltration (NF) and can effectively treat industrial wastewater. Some of the hurdles of using a low-energy FO system could be overcome through a membrane module that dramatically reduces concentration polarization thereby ensuring sustainable operation without the problem of any significant membrane fouling and drop in flux. Such a system can remove more than 97% of water pollutants (represented as COD) while sustaining a treated water flux at the level of around 50 LMH (liters per square meter per hour). The draw solution recovering downstream nanofiltration membrane module succeeds in recovering 99% draw solute at treated water flux of around 55–60 LMH. Horizontal alignment of the separating membrane in the FO module coupled with an appropriate hydrodynamic environment comprising sweeping fluid flow over the membrane surface minimizes concentration polarization and membrane fouling. Cost of treatment is reasonably low and the cumulative effect of the plant operation appears to result in continuous improvement in water quality. This innovation is also a sustainable technology.

10.1.2 Innovation in Treating Contaminated Groundwater

Indiscriminate and excessive use of groundwater coupled with other anthropological and geological developments have resulted in large-scale contamination of underground aquifers by leached out minerals like arsenic, fluoride, iron, and chromium. Landfill leaching has only compounded the problems in some areas. The biggest arsenic-affected region in the world is the Bengal-Delta basin spreading over Bangladesh and India and affecting more than 50 million people. Millions of people in Bangladesh, India, China, Taiwan, Thailand, Nepal, Indonesia, and Vietnam are victims of arsenic contamination. Prolonged use of fluoride-contaminated water for drinking cripples human bones, and arsenic contamination leads to almost incurable diseases like hyperpigmentation, keratosis, liver fibrosis, and skin, lung, and liver cancers. Due to its chemical similarity to phosphate arsenic replaces phosphate from ATP, destroys many enzyme activities in the human body, and impedes the Krebs cycle by destroying the functions of pyruvate dehydrogenase (PDH) [2]. The two main valence forms of arsenic in water are arsenite, which is trivalent arsenic (As-III) and arsenate, which is pentavalent arsenic (As-V). As (III) is more toxic than As(V) and is 4–10 times more soluble in water than As(V) it is difficult to separate from water. As the WHO has set 10 ppb as the safe limit of arsenic in water, bringing down contamination from a level often as high as 3000 ppb to such low level is extremely challenging. Socioeconomic conditions of affected areas makes installation of sophisticated treatment systems difficult. Alternate arrangement of supply of surface water from long distance river bodes has in most of the cases been thought as solution where such surface water drawn from river bodies demands proper treatment for pollutants of a wide range other than arsenic. An innovative technology that is efficient, low-cost, and sustainable is needed in these challenging demographic settings to provide safe drinking water for the people. A recently patented innovative technology [3] offers a solution to the long-standing problem of groundwater contamination.

The new technology first converts the tougher arsenite into arsenate form through a very simple preoxidation step and then filters it through a flat-sheet cross-flow nanofiltration membrane module. The arsenic rejects are eventually stabilized in nonleachable solid matrix and can be either used in roadmaking or disposed of safely. The system effectively removes more than 98% of arsenic while offering sustainable flux of around 140–150 LMH. The associated low-cost, cumulative positive effect on the water environment, and trouble-free operation makes this technology sustainable (in 2016, 1000 L of safe arsenic-free water can be obtained at a cost of US$1.4 in a country like India).

10.1.3 Innovation in Recovery of Nutrients from Industrial Wastewater

Eutrophication of surface water from industrial and municipal wastewater often results in the destruction of many natural water bodies around the world. Nutrients like nitrogen and phosphates encourage such eutrophication. Algal bloom makes water bodies unfit for human use. However, eutrophication can largely be overcome by cutting the entry of nutrients to water bodies through municipal and industrial discharge. A recently developed hybrid technology [4] has the potential to solve this problem. This new technology integrates chemical and biological treatments with membrane separation and effectively separates the nutrients from ammoniacal industrial wastewater. The nutrients are recovered as valuable struvite fertilizer for subsequent use in agriculture and planned plantation. This novel approach simultaneously turns waste into a valuable resource, protects surface bodies from hazardous waste contamination, makes the treated water reusable, and adds economy to the wastewater-treatment technology. The associated cost is low and the plant operation promises continuous improvement of the water environment. As this technology is operationally fast, efficient, and economically attractive it is a sustainable technology.

10.1.4 Innovation in Preventing Water Pollution

The best remediation in water pollution is through preventive approaches. This necessitates identifying the municipalities and industries that generate the most wastewater. In such a preventive approach, the process technology that generates the most wastewater would be replaced by a green technology that results in almost zero discharge. A green technology not only reduces water pollution directly but also often reduces water pollution indirectly by demanding significantly less energy than a conventional polluting process technology, which in most cases involves large amounts of water. A green technology that is significantly energy-saving and water-saving is well illustrated by the development [5] of a novel technology in production of lactic acid. The multifaceted environmental benefits are clearly seen. By replacing energy-intensive process units, this membrane-integrated hybrid technology produces more than 98% pure lactic acid through fermentation of a renewable bioresource like sugar cane juice. Water consumption drastically goes down because this technology only uses three water-saving, energy-saving units. By recycling the membrane-separated microbial cells and unconverted sugars to the fermentation unit, high conversion, yield, and productivity are ensured. The final polishing of the nanofiltration unit purifies and concentrates the product to a very high level. Due to its green processing, high productivity, yield, and purity this technology is considered highly sustainable and helps to maintain and improve water quality. The final product of this technology is meant for production of biodegradable polylactide, which again contributes to a better environment by replacing nonbiodegradable polymer, an environmentally damaging substance. In preserving and conserving water, a green technology can play very significant role.

10.2 CASE OF A SUSTAINABLE MANAGEMENT STRATEGY FOR A WATER-TREATMENT SYSTEM

Many water-treatment plants were reportedly set up for treatment of contaminated groundwater in remote areas as well for treatment of municipal sewage and industrial wastewater in many parts of the world. After a period of time many such plants began to fail, which is common in developing countries. While such failure is often attributed to management the technology itself is also sometimes at fault. A water-treatment plant based on a technology can remain sustainable in a sustainable society. And a sustainable society is characterized by integrated decision-making, democratic values, education, knowledge, research, awareness, information, collaboration, equity, justice, shared progress, long-term solutions, compliance of regulations, good leadership, and finally consideration of future generations. In this context, a case study on a sustainable management strategy for a water-treatment plant investigated recently [6] is illustrated here.

To create a sustainable management strategy, a membrane-based water-treatment plant was techno-economically investigated. The treatment plant is proposed for a water-stressed region in Eastern India with groundwater contamination by fluoride. The people of the region draw their drinking water primarily from underground aquifers through tube wells.

The contingent valuation method, a standard economic tool, was applied for cost-benefit analysis to assess the economic viability and hence sustainability of the proposed management strategy. The intention was to involve all stakeholders in the decision-making process and in the management of the new treatment plant. A field survey was conducted among the stakeholders to seek their opinion and to determine their willingness to pay for the new plant and for its sustainable management. The determinants of willingness to pay (WTP) were determined through multivariate regression analysis. The proposed membrane-based treatment technology was considered the best option to provide safe drinking water to the area. While the household economic conditions were found to be directly correlated to the amount stakeholders were willing to pay for improved water quality, most people want to contribute to better water services. Through open-end choice bidding format, the study revealed that the affected people desperately wanted safe drinking water and were ready to pay for it, in spite of previously getting water for free. The most striking revelation of this investigation is that contrary to the belief that an advanced water-treatment technology is expensive the proposed technology offers safe potable water at an affordable price. Despite this, no such water-treatment plant had been previously proposed, reflecting the lack of scale-up confidence, lack of demonstration, lack of awareness, lack of information, and above all lack of initiative. This illustrates the need for a paradigm shift in policy in bringing sustainable technology from laboratory to the field.

10.3 INDISCRIMINATE USE OF GROUNDWATER WITHOUT AQUIFER MAPPING

While two-thirds of the earth is made up of water, freshwater available for sustaining human life on earth is only 2.5%. Safe freshwater preserved in underground aquifers is only a small fraction of the total available water on earth and is thus precious. Where alternate surface water is available, this precious water resource should not be used except for drinking purposes. Water infrastructures in trapping and optimum use of surface water should be developed wherever possible. Groundwater should be used judiciously on being well informed through aquifer mapping. Compulsory and universal aquifer mapping using satellite technology by policy planners should be required in all countries. Data on aquifer mapping should be disseminated throughout society so that informed decisions are made when choosing a water resource for any use.

10.4 ETHICS, COMPLIANCE OF REGULATION

No amount of technology can solve the problems of water pollution unless there are regulations on water use and compliance of said regulations. While compliance of regulations can be enforced to some extent by government, total compliance requires adherence to ethics. Ethics is a source of guidance beyond enforceable law and is not always philosophical. Throughout the ages, ethical decisions have had a profound impact on human society. It was an ethical decision that put an end to 16-hour workdays. It was an ethical decision that puts an end to slavery. It was an ethical decision that put an end to child labor. In matters of water use, we have to also consider ethics. Discharging heavily polluted wastewater into river bodies without proper treatment is an unethical decision on the part of the management. Where windows have to be kept closed to prevent entry of particulate pollutants and hazardous gases from coal-fired thermal power plants, all the involved business houses, policy planners, sanctioning authorities and above all the government are naturally to be blamed for a very unethical decision of running such a plant.

10.5 FREEDOM FOR INNOVATION AND IMPLEMENTATION

Today society is facing unprecedented challenges in matters of environmental pollution and water scarcity. The solutions to these problems do not appear to be found in standard practices, such as the approaches currently being used in many places or through incremental improvements of plants. The solutions will be found in innovation. Many problems of the past have found their solutions through innovations in science and technology. Furthermore, freedom to implement innovations in all practical senses is essential to address the huge problems confronting humanity today.

REFERENCES

[1] Thakura R, Chakrabortty S, Pal P. Treating complex industrial wastewater in a new membrane-integrated closed loop system for recovery and reuse. Clean Technol Environ Policy Springer 2015;17(8):2299–310.
[2] Pal P. Groundwater Arsenic Remediation. Waltham, MA, USA: Elsevier Inc.; 2015.
[3] Membrane-integrated Hybrid Treatment System for Arsenic Removal by Parimal Pal, Indian Patent No. 275244.
[4] Kumar R, Pal P. Turning hazardous waste into value-added products: production and characterization of struvite from ammoniacal waste with new approaches. J Clean Prod 2013;43:59–70.
[5] Pal P, Dey P. Process Intensification in lactic acid production by three stage membrane-integrated hybrid reactor system. Chem Eng Process Process Intensif 2013;64:1–9.
[6] Roy M, Chakrabortty S. Developing a sustainable water resource management strategy for a fluoride-affected area: a contingent valuation approach. J Clean Technol Environ Policy Springer 2013;16(2):341–9 2014

NOMENCLATURE

c_p	Fluid specific heat (J/kg/K)
d	Willmott *d*-index
d_h	Hydraulic diameter of membrane module (m)
D_k	Knudsen-diffusion coefficient of water vapor (m^2/s)
D_{wa}	Diffusion coefficient of water vapor in air (m^2/s)
DCMD	Direct contact membrane distillation
EE	Evaporation efficiency (%)
FVMD	Flash vaporization membrane distillation
g	Acceleration due to gravity (m/s^2)
h	Heat transfer coefficient (W/m^2/K)
H	Enthalpy (J/kg)
$H_L\{T\}$	Liquid enthalpy at temperature T (J/kg)
$H_v\{T\}$	Vapor enthalpy at temperature T (J/kg)
ΔH_{vfb}	Heat of vaporization of water at bulk feed temperature (J/kg)
k	Thermal conductivity (W/m/K)
K_n	Knudsen number
L	Chamber length (m)
ln	Logarithmic mean
M	Molecular weight of water (kg/mol)
MD	Membrane distillation
MDM	Membrane distillation co-efficient
N	Water flux (kg/m^2/s)
Nu	Nusselt number
p	Partial pressure (Pa)
p_{im}^{o}	Interfacial vapor pressure of pure water (Pa)
P_a	Partial pressure of air (Pa)
P_t	Total pressure (Pa)
Pr	Prandtl number
Q	Heat flux (W/m^2)
Q_f^{conden}	Heat transfer to the feed boundary layer from vapor-air film due to condensation (W/m^2)
Q_{mc}	Conduction heat transfer rate through membrane (W/m^2)
r	Pore radius (m)
R	Gas constant (J/mol/K)
R_f	Resistance of feed boundary layer (Pa m^2 h/kg)
R_m	Resistance of the membrane (Pa m^2 h/kg)
R_p	Resistance of permeate boundary layer (Pa m^2 h/kg)
R^t	Total resistance (Pa m^2 h/kg)
Re	Reynolds number
RE	Relative error

RMSE	Root mean square error
T	Temperature (K)
$T_{\text{fm mod}}$	Modified temperature at the feed-membrane surface
T_{m}	Average membrane surface absolute temperature inside the pores
$T_{\text{pm mod}}$	Modified temperature at the permeate-membrane surface
TP	Temperature polarization
TPC	Temperature polarization coefficient
U	Overall heat transfer coefficient (W/m^2/K)
v	Fluid velocity (m/s)
VPC	Vapor pressure polarization coefficient
w	Width of the feed channel (m)

GREEK LETTERS

δ	Membrane thickness (m)
δ_{ft}	Thermal feed boundary layer thickness (m)
δ_{fv}	Vapor-air film thickness (m)
δ_{pt}	Thermal permeate boundary layer thickness (m)
ε	Porosity
μ	Viscosity (Pa s)
ρ	Density (kg/m^3)
τ	Tortuosity

SUBSCRIPTS

f	Feed
fb	At the bulk phase of the feed side
fm	At the membrane surface of the feed side
i	Feed or permeate
L	Liquid phase
m	Membrane
mg	In the membrane pores related to air-vapor
ms	In the membrane solid material
mv	In the membrane pores related to water vapor
p	Permeate
$\mathbf{p_b}$	At the bulk phase of the permeate side
$\mathbf{p_m}$	At the membrane surface of the permeate side
s	Solid phase

SUPERSCRIPTS

M.T.	Heat transfer due to mass transfer
t	Total

INDEX

Note: Page numbers followed by "*f*" and "*t*" refer to figures and tables, respectively.

B

E

G

H

I

K

L

M

O

P

Q

R

S

T

X

Z

FSC
www.fsc.org
MIX
Papier | Fördert gute Waldnutzung
FSC® C083411

Kolibri 360 GmbH
Ferdinand-Jühlke-Straße 7
99095 Erfurt, Deutschland
produktsicherheit@kolibri360.de